Audel™

HVAC Fundamentals
Volume I
Heating Systems, Furnaces, and Boilers

All New 4th Edition

James E. Brumbaugh

Wiley Publishing, Inc.

Vice President and Executive Group Publisher: Richard Swadley
Vice President and Executive Publisher: Robert Ipsen
Vice President and Publisher: Joseph B. Wikert
Executive Editorial Director: Mary Bednarek
Editorial Manager: Kathryn A. Malm
Executive Editor: Carol A. Long
Acquisitions Editors: Katie Feltman
Katie Mohr
Senior Production Manager: Fred Bernardi
Development Editor: Kenyon Brown
Production Editor: Vincent Kunkemueller
Text Design & Composition: TechBooks

For general information on our other products and services please contact our Customer Care Department within the United States at (800) 762-2974, outside the United States at (317) 572-3993 or fax (317) 572-4002.

Wiley also publishes its books in a variety of electronic formats. Some content that appears in print may not be available in electronic books.

Library of Congress Cataloging-in-Publication Data:

ISBN: 0-764-54206-0

Printed in the United States of America

10 9 8 7 6 5 4 3 2

For Laura, my friend, my daughter.

Contents

Introduction

The purpose of this series is to provide the layman with an introduction to the fundamentals of installing, servicing, troubleshooting, and repairing the various types of equipment used in residential and light-commercial heating, ventilating, and air conditioning (HVAC) systems. Consequently, it was written not only for the HVAC technician and others with the required experience and skills to do this type of work but also for the homeowner interested in maintaining an efficient and trouble-free HVAC system. A special effort was made to remain consistent with the terminology, definitions, and practices of the various professional and trade associations involved in the heating, ventilating, and air conditioning fields.

Volume 1 begins with a description of the principles of thermal dynamics and ventilation, and proceeds from there to a general description of the various heating systems used in residences and light-commercial structures. Volume 2 contains descriptions of the working principles of various types of equipment and other components used in these systems. Following a similar format, Volume 3 includes detailed instructions for installing, servicing, and repairing these different types of equipment and components.

The author wishes to acknowledge the cooperation of the many organizations and manufacturers for their assistance in supplying valuable data in the preparation of this series. Every effort was made to give appropriate credit and courtesy lines for materials and illustrations used in each volume.

Special thanks is due to Greg Gyorda and Paul Blanchard (Watts Industries, Inc.), Christi Drum (Lennox Industries, Inc.), Dave Cheswald and Keith Nelson (Yukon/Eagle), Bob Rathke (ITT Bell & Gossett), John Spuller (ITT Hoffman Specialty), Matt Kleszezynski (Hydrotherm), and Stephanie DePugh (Thermo Pride).

Last, but certainly not least, I would like to thank Katie Feltman, Kathryn Malm, Carol Long, Ken Brown, and Vincent Kunkemueller, my editors at John Wiley & Sons, whose constant support and encouragement made this project possible.

James E. Brumbaugh

About the Author

James E. Brumbaugh is a technical writer with many years of experience working in the HVAC and building construction industries. He is the author of the *Welders Guide, The Complete Roofing Guide,* and *The Complete Siding Guide*.

Chapter 1

Introduction

This series is an introduction to the basic principles of heating, ventilating, and air conditioning (HVAC). Each represents a systematic attempt to control the various aspects of the environment within an enclosure, whether it is a room, a group of rooms, or a building.

Among those aspects of the immediate environment that people first sought to control were heat and ventilation. Attempts at controlling heat date from prehistoric times and probably first developed in colder climates, where it was necessary to produce temperatures sufficient for both comfort and health. Over the years the technology of heating advanced from simple attempts to keep the body warm to very sophisticated systems of maintaining stabilized environments in order to reduce heat loss from the body or the structural surfaces of the room.

Ventilation also dates back to very early periods in history. Certainly the use of slaves to wave large fans or fanlike devices over the heads of rulers was a crude early attempt to solve a ventilating problem. Situating a room or a building so that it took advantage of prevailing breezes and winds was a more sophisticated attempt. Nevertheless, it was not until the nineteenth century that any really significant advances were made in ventilating. During that period, particularly in the early stages of the Industrial Revolution, ventilating acquired increased importance. Work efficiency and the health of the workers necessitated the creation of ventilation systems to remove contaminants from the air. Eventually, the interrelationship of heating and ventilating became such that it is now regarded as a single subject.

Air conditioning is a comparatively recent development and encompasses all aspects of environmental control. In addition to the control of temperature, both humidity (i.e., the moisture content of the air) and air cleanliness are also regulated by air conditioning. The earliest attempts at air conditioning involved the placing of wet cloths over air passages (window openings, entrances, etc.) to cool the air. Developments in air conditioning technology did not progress much further than this until the nineteenth century. From about 1840 on, several systems were devised for both cooling and humidifying rooms. These were first developed by textile manufacturers in order to reduce the static electricity in the air. Later, adaptations were made by other industries.

Developments in air conditioning technology increased rapidly in the first four decades of the nineteenth century, but widespread use of air conditioning in buildings is a phenomenon of the post-World War II period (i.e., 1945 to the present). Today, air conditioning is found not only in commercial and industrial buildings but in residential dwellings as well. Unlike early forms of air conditioning, which were designed to cool the air or add moisture to it, modern air conditioning systems can control temperature, air moisture content, air cleanliness, and air movement. That is, modern systems *condition* the air rather than simply cool it.

Heating and Ventilating Systems

Many different methods have been devised for heating buildings. Each has its own characteristics, and most methods have at least one objectionable aspect (e.g., high cost of fuel, expensive equipment, or inefficient heating characteristics). Most of these heating methods can be classified according to one of the following four criteria:

1. The heat-conveying medium
2. The fuel used
3. The nature of the heat
4. The efficiency and desirability of the method

The term *heat-conveying medium* means the substance or combination of substances that carries the heat from its point of origin to the area being heated. There are basically four mediums for conveying heat. These four mediums are:

1. Air
2. Water
3. Steam
4. Electricity

Different types of wood, coal, oil, and gas have been used as fuels for producing heat. You may consider electricity as both a fuel and a heat-conveying medium. Each heating fuel has its own characteristics; the advantage of one type over another depends upon such variables as availability, efficiency of the heating equipment (which, in turn, is dependent upon design, maintenance, and other factors), and cost. A detailed analysis of the use and effectiveness of the various heating fuels is found in Chapter 5 ("Heating Fuels").

Heating methods can also be classified with respect to the nature of the heat applied. For example, the heat may be of the exhaust steam variety or it may consist of exhaust gases from internal combustion engines. The nature of the heat applied is inherent to the heat system and can be determined by reading the various chapters that deal with each type of heating system (Chapters 6 through 9) or with heat-producing equipment (e.g., Chapter 11, "Gas Furnaces").

The various heating methods differ considerably in efficiency and desirability. This is due to a number of different but often interrelated factors, such as energy cost, conveying medium employed, and type of heating unit. The integration of these interrelated components into a single operating unit is referred to as a *heating system*.

Because of the different conditions met within practice, there is a great variety in heating systems, but most of them fall into one of the following broad classifications:

1. Warm-air heating system (Chapter 6)
2. Hydronic heating systems (Chapter 7)
3. Steam heating systems (Chapter 8)
4. Electric heating systems (Chapter 9)

You will note that these classifications of heating systems are based on the heat-conveying method used. This is a convenient method of classification because it includes the vast majority of heating systems used today.

As mentioned, ventilating is so closely related to heating in its various applications that the two are very frequently approached as a single subject. In this series, specific aspects of ventilating are considered in Chapter 6 ("Ventilation Principles") and Chapter 7 ("Ventilation and Exhaust Fans") of Volume 3.

The type and design of ventilating system employed depends on a number of different factors, including:

1. Building use or ventilating purpose
2. Size of building
3. Geographical location
4. Heating system used

A residence will have a different ventilating system from a building used for commercial or industrial purposes. Moreover, the requirements of a ventilating system used to provide fresh air result in fundamental design differences from a ventilating system that must remove noxious gases or other dangerous contaminants from the enclosure.

The size of a building is a factor that also must be considered. For example, a large building presents certain ventilating problems if the internal areas are far from the points where outside air would initially gain access. Giving special attention to the overall design of the ventilating system can usually solve these problems.

Buildings located in the tropics or semitropics present different ventilating problems from those found in temperature zones. The differences are so great that they often result in different architectural forms. At least this was the case *before* the advent of widespread use of air conditioning. The typical southern house of the nineteenth century was constructed with high ceilings (heat tends to rise); large porches that sheltered sections of the house from the hot, direct rays of the sun; and large window areas to admit the maximum amount of air. They were also usually situated so that halls, major doors, and sleeping areas faced the direction of the prevailing winds. Today, with air conditioning so widely used, these considerations are not as important—at least not until the power fails or the equipment breaks down.

Air Conditioning

Although the major emphasis in this series has been placed on the various aspects of heating and ventilating, some attention has also been given to air conditioning. The reason for this, of course, is the increasing use of year-round air conditioning systems that provide heating, ventilating, and cooling. These systems *condition* the air by controlling its temperature (warming or cooling it), cleanliness, moisture content, and movement. This is the true meaning of the term *air conditioning*. Unfortunately, it has become almost synonymous with the idea of cooling, which is becoming less and less representative of the true function of an air conditioning system. Air conditioning, particularly the year-round air conditioning systems, is examined in detail in Chapters 8, 9, and 10 of Volume 3.

Selecting a Suitable Heating, Ventilating, or Air Conditioning System

There are a number of different types of heating, ventilating, and air conditioning equipment and systems available for installation in the home. The problem is choosing the most efficient one in terms of the installation and operating costs. These factors, in turn, are directly related to one's particular heating and cooling requirements. The system must be the correct size for the home. Any reputable building contractor or heating and air conditioning firm should be able to advise you in this matter.

If you are having a heating and ventilating or air conditioning system installed in an older house, be sure to check the construction. Weather stripping is the easiest place to start. All doors and windows should be weather-stripped to prevent heat loss. Adequate weather stripping can cut heating costs by as much as 15 to 20 percent. If the windows provide suitable protection (they should be double- or triple-glazed) from the winter cold, check the caulking around the edge of the glass. If it is cracking or crumbling, replace it with fresh caulking. You may even want to go to the expense of insulating the ceilings and outside walls. This is where a great deal of heat loss and air leakage occurs.

You have several advantages when you are building your own house. For example, you may be able to determine the location of your house on the lot. This should enable you to establish the direction in which the main rooms and largest windows face. If you position your house so that these rooms and windows face south, you will gain maximum sunlight and heat from the sun during the cold winter months. This will reduce the heat requirement and heating costs. The quality of construction depends on how much you wish to spend and the reliability of the contractor. It is advisable to purchase the best insulation you can afford. Your reduced heating costs will eventually pay for the added cost of the insulation. If you suspect that your building contractor cannot be trusted, you can reduce opportunities for cheating and careless work by making frequent and unexpected visits to the construction site.

Career Opportunities

Many career opportunities are available in heating, ventilating, and air conditioning fields, and they extend over several levels of education and training. Accordingly, the career opportunities open to an individual seeking employment in these fields can be divided roughly into four categories, each dependent upon a different type or degree of education and/or training. This relationship is shown in Table 1-1.

Among workers in these fields, engineers receive the highest pay, but they also undergo the longest periods of education and training. Engineers are usually employed by laboratories, universities, and colleges or, frequently, by the manufacturers of materials and

Table 1-1 Relationships between Career Category and Type of Work or Education and/or Training Required

Career Category	*Type of Work*	*Education/Training*
Engineer	Design and development	4 years or more of college
Technician	Practical application	Technical training school and/or college
Skilled worker	Installation, maintenance, and repair	Apprentice program or on-the-job training (OJT)
Apprentice or OJT worker	Training for skilled-worker position	High school degree or equivalency

equipment used in heating, ventilating, air conditioning, and related industries. Their primary responsibility is designing, developing, and testing the equipment and materials used in these fields. In some cases, particularly when large buildings or district heating to several buildings is employed, they also supervise the installation of the entire system. Moreover, industry codes and standards are usually the results of research conducted by engineers.

Technicians obtain their skills through technical training schools, some college, or both. Many assist engineers in the practical application of what the latter have designed. Technicians are particularly necessary during the developmental stages. Other technicians are found in the field working for contractors in the larger companies. Their pay often approximates that of engineers, depending on the size of the company for which they work.

Skilled workers are involved in the installation, maintenance, and repair of heating, ventilating, and air conditioning equipment. Apprentices and OJT (on-the-job training) workers are in training for the skilled positions and are generally expected to complete at least a 2- to 5-year training program. Local firms that install or repair equipment in residential, commercial, and industrial buildings employ most skilled workers and trainees. Some also work on the assembly lines of factories that manufacture such equipment. Their pay varies, depending on the area, their seniority, and the nature of the work. Most employers require that both skilled workers and trainees have at least a high school diploma or its equivalent (e.g., the GED). The requirement for a high school diploma may be waived if the individual has already acquired the necessary skills on a previous job. The pay for skilled workers and trainees is lower than that earned by engineers and technicians but compares favorably to salaries received by skilled workers or equivalent trainees in other occupations.

Pipe fitters, plumbers, steam fitters, and sheet-metal workers may occasionally do some work with heating, ventilating, and air conditioning equipment. Both pipe fitters and plumbers (especially the former) are frequently called upon to assemble and install pipes and pipe systems that carry the heating or cooling conveying medium from the source. Both are also involved in repair work, and some pipe fitters can install heating and air conditioning units.

Steam fitters can assemble and install hot-water or steam heating systems. Many steam fitters can also do the installation of boilers, stokers, oil and gas burners, radiators, radiant heating systems, and air conditioning systems.

Sheet-metal workers can also assemble and install heating, ventilating, and air conditioning systems. Their skills are particularly necessary in assembling sheet-metal ducts and duct systems.

Some special occupations, such as those performed by air conditioning and refrigeration mechanics or stationary engineers, are limited to certain functions in the heating, ventilating, and air conditioning fields. Mechanics are primarily involved with assembling, installing, and maintaining both air conditioning and refrigeration equipment. Stationary engineers maintain and operate heating, ventilating, and air conditioning equipment in large buildings and factories. Workers in both occupations require greater skills and longer training periods than most skilled workers.

It should be readily apparent by now that the heating, ventilating, and air conditioning fields offer a variety of career opportunities. The pay is generally good, and the nature of the work provides considerable job security. Both the type of work an individual does and the level at which it is done depend solely on the amount and type of education and training acquired by the individual.

Professional Organizations

A number of professional organizations have been established for those who work in the heating, ventilating, and air conditioning industries or who handle their products. These organizations (frequently referred to as *associations, societies,* or *institutes*) provide a number of different services to members and nonmembers.

Some professional and trade organizations have established permanent libraries as resource centers for those seeking to improve their skills or wishing to keep abreast of current developments in their fields. In many instances, research programs are conducted in cooperation with laboratories, colleges, and universities.

Many of these organizations address themselves to the problems and interests of specific groups. For example, there are professional organizations for manufacturers, wholesalers, jobbers, distributors, and journeymen. Some organizations represent an entire industry, while others restrict their scope to only a segment of it. Every aspect of heating, ventilating, and air conditioning is covered by one or more of these professional organizations.

Anyone, member or not, can write to these professional organizations for information or assistance. Most seem very willing to comply with any reasonable request. The only difficulty that may be encountered is determining the current name of the particular organization and obtaining its address. Unfortunately, these professional organizations have shown a strong proclivity toward mergers over the years, with resulting changes of names and addresses.

The best and most current guide to the names and addresses of professional organizations is *The Encyclopedia of Associations*, which can be found in the reference departments of most public libraries. It is published in three volumes, but everything you will need can be found in the first one. At the back of this volume is a section called the "Alphabetical & Key Word Index." By looking up the key word (e.g., *heating* or *ventilating*) of the subject that interests you, you can find the page number and full name of the professional organization (or organizations) concerned with the particular area. See Appendix A in this volume for a partial listing of these professional and trade associations.

Some professional organizations of long standing have been merged with others or have been disbanded. For example, the Steel Boiler Institute (formerly the Steel Heating Boiler Institute), which maintained standards in the heating industry with its SBI Rating Code, is now defunct. The Institute of Boiler and Radiator Manufacturers (source of the old IBR Code) merged with the Better Heating-Cooling Council to form the Hydronics Institute. A recent attempt to contact the Steam Heating Equipment Manufacturers Association has resulted in the return of a letter marked "no forwarding address." It seems very likely that it, too, has joined the list of defunct professional organizations.

Appendix A (Professional and Trade Associations) at the end of this book gives a listing of professional organizations. It also contains their present addresses, the names of some of their publications, and a brief synopsis of their backgrounds and whom they represent.

Chapter 2

Heating Fundamentals

There is still considerable disagreement about the exact nature of heat, but most authorities agree that it is a particular form of energy. Specifically, heat is a form of energy not associated with matter and *in transit* between its source and destination point. Furthermore, heat energy exists as such only between these two points. In other words, it exists as heat energy only while flowing between the source and destination.

So far this description of heat energy has been practically identical to that of work energy, the other form of energy in transit not associated with matter. The distinguishing difference between the two is that heat energy is energy in transit as a result of temperature differences between its source and destination point, whereas work energy in transit is due to other, nontemperature factors.

British Thermal Unit

Heat energy is measured by the British thermal unit (Btu). Each thermal unit is regarded as equivalent to one unit of heat (heat energy).

Since 1929, British thermal units have been defined on the basis of 1 Btu being equal to 251.996 IT (International Steam Table) calories, or 778.26 foot-pounds of mechanical energy units (work). Taking into consideration that one IT calorie equals $1/860$ of a watt-hour, 1 Btu is then equivalent to about $1/3$ watt-hour.

Prior to its 1929 redefinition, a Btu was defined as the amount of heat necessary to raise the temperature of one pound of water by one degree Fahrenheit. Because of the difficulty in determining the exact value of a Btu, it was later redefined in terms of the more fundamental physical unit.

Relationship Between Heat and Work

Energy is the ability to do work or move against a resistance. Conversely, work is the overcoming of resistance through a certain distance by the expenditure of energy.

Work is measured by a standard unit called the *foot-pound,* which may be defined as the amount of work done in raising one

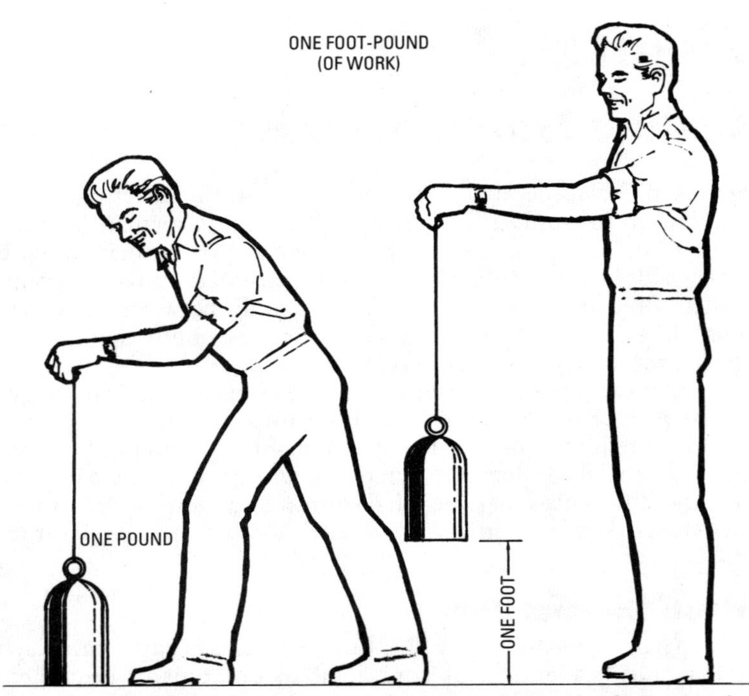

Figure 2-1 Man raising 1 pound 1 foot to illustrate the foot-pound standard unit.

pound the distance of one foot, or in overcoming a pressure of one pound through a distance of one foot (Figure 2-1).

The relationship between work and heat is referred to as the *mechanical equivalent of heat;* one unit of heat is equal to 778.26 ft-lb. This relationship (i.e., the mechanical equivalent of heat) was first established by experiments conducted in the nineteenth century. In 1843 Dr. James Prescott Joule (1818–1889) of Manchester, England, determined by numerous experiments that when 772 ft-lb of energy had been expended on 1 lb of water, the temperature of water had risen 1°F and the relationship between heat and mechanical work was found (Figure 2-2). The value 772 ft-lb is known as *Joule's equivalent.* More recent experiments give higher figures and the value 778 (1 Btu = 778.26 ft-lb). (See the preceding section.)

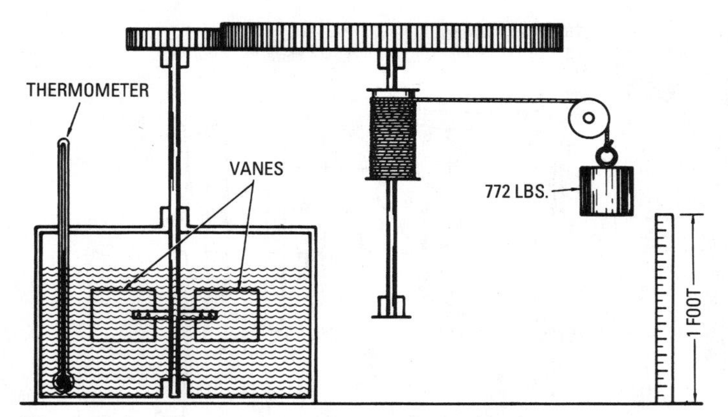

Figure 2-2 The mechanical equivalent of heat.

Heat Transfer

When bodies of unequal temperatures are placed near each other, heat leaves the hotter body and is absorbed by the colder one until the temperatures are equal to each other. The rate by which the heat is absorbed by the colder body is proportional to the difference of temperature between the two bodies—the greater the difference in temperature, the greater the rate of flow of the heat.

Heat is transferred from one body to another at lower temperature by any one of the following means (Figure 2-3):

1. Radiation
2. Conduction
3. Convection

Radiation, insofar as heat loss is concerned, refers to the throwing out of heat in rays. The heat rays proceed in straight lines, and the intensity of the heat radiated from any one source becomes less as the distance from the source increases.

The amount of heat loss from a body within a room or building through radiation depends upon the temperature of the floor, ceiling, and walls. The colder these surfaces are, the faster and greater will be the heat loss from a human body standing within the enclosure. If the wall, ceiling, and floor surfaces are *warmer* than the human body within the enclosure they form, heat will be radiated

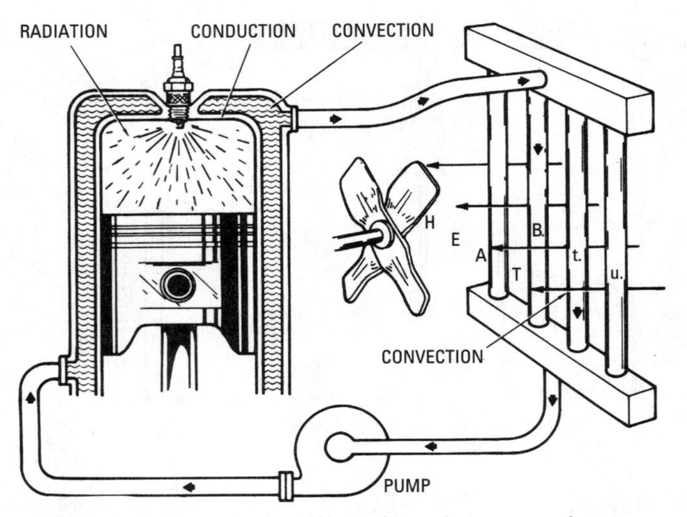

Figure 2-3 The transfer of heat by radiation, conduction, and convection.

from these surfaces to the body. In these situations a person may complain that the room is too hot.

Knowledge of the mean radiant temperature of the surfaces of an enclosure is important when dealing with heat loss by radiation. The *mean radiant temperature* (MRT) is the weighted average temperature of the floor, ceiling, and walls. The significance of the mean radiant temperature is determined when compared with the clothed body of an adult (80°F, or 26.7°C). If the MRT is below 80°F, the human body will lose heat by radiation to the surfaces of the enclosure. If the MRT is higher than 80°F, the opposite effect will occur.

Conduction is the transfer of heat through substances, for instance, from a boiler plate to another substance in contact with it (Figure 2-4). Conductivity may be defined as the relative value of a material, compared with a standard, in affording a passage through itself or over its surface for heat. A poor conductor is usually referred to as a *nonconductor* or *insulator*. Copper is an example of a good conductor. Figure 2-5 illustrates the comparative heat conductivity rates of three frequently used metals. The various materials used to insulate buildings are poor conductors. It should be

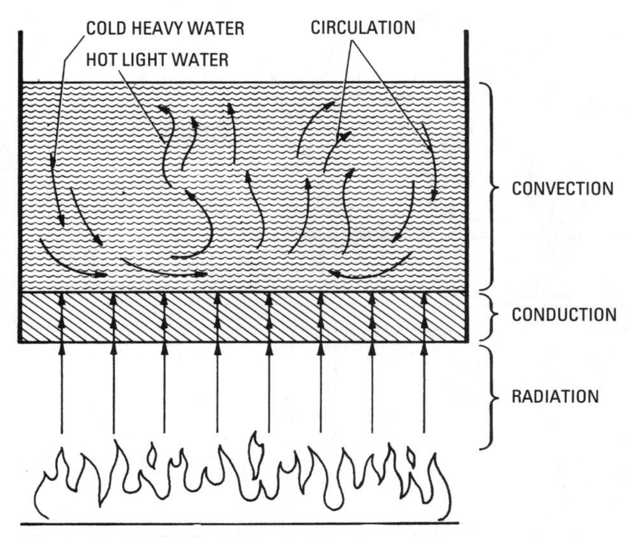

Figure 2-4 Radiation, conduction, and convection in boiler operation.

pointed out that any substance that is a good conductor of electricity is also a good conductor of heat.

Convection is the transfer of heat by the motion of the heated matter itself. Because motion is a required aspect of the definition of convection, it can take place only in liquids and gases.

Figure 2-4 illustrates how radiation, conduction, and convection are often interrelated. Heat from the burning fuel passes to the metal of the heating surface by radiation, passes through the metal by conduction, and is transferred to the water by convection (i.e., circulation). Circulation is caused by a variation in the weight of the water due to temperature differences. That is, the water next to the heating surface receives heat, expands (becomes lighter), and immediately rises as a result of displacement by the colder and heavier water above.

Proper circulation is very important, because its absence will cause a liquid, such as water, to reach the spheroidal state. This, in turn, causes the metal of the boiler to become dangerously overheated. A liquid that has reached the spheroidal state is easy to recognize by its appearance. When liquid is dropped upon the surface of a highly heated metal, it rolls about in spheroidal drops

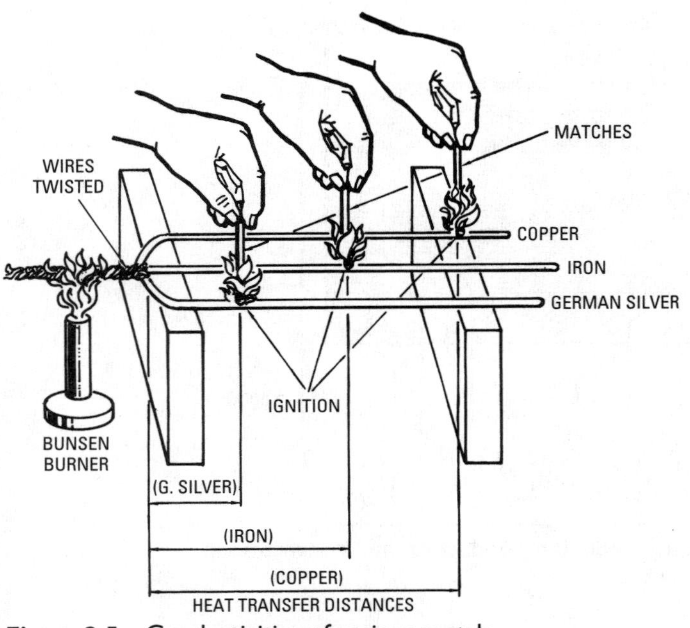

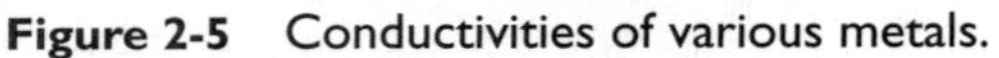

Figure 2-5 Conductivities of various metals.

(Figure 2-6) or masses without actual contact with the heated metal. This phenomenon is caused by the repelling force of heat and the intervention of a cushion of steam.

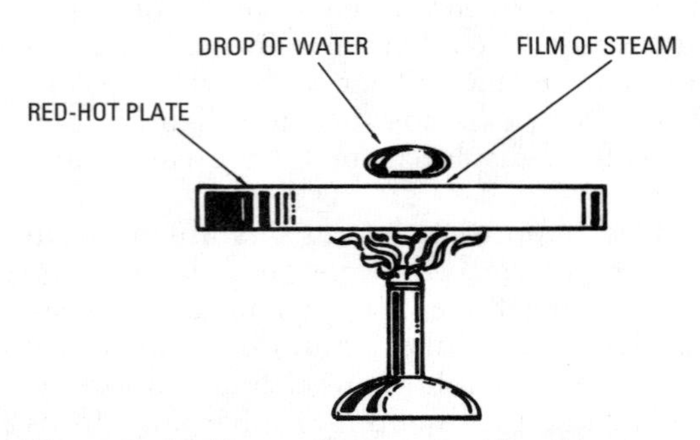

Figure 2-6 Drop of water on a hot plate illustrating the spheroidal state.

Specific, Sensible, and Latent Heat

The *specific heat* of a substance is the ratio of the quantity of heat required to raise its temperature one degree Fahrenheit to the amount required to raise the temperature of the same weight of water one degree Fahrenheit (Figure 2-7). This may be expressed in the following formula:

$$\text{Specific heat} = \frac{\text{Btu to raise temp. of substance } 1^\circ\text{F}}{\text{Btu to raise temp. of same weight water } 1^\circ\text{F}}$$

The standard used in water at approximately 62 to 63°F receives a rating of 1.00 on the specific heat scale. Simply stated, specific heat represents the Btu required to raise the temperature of one pound of a substance one degree Fahrenheit.

Sensible heat is the part of heat that provides temperature change and that can be measured by a thermometer. It is referred to as such because it can be sensed by instruments or touch.

Latent heat is the quantity of heat that disappears or becomes concealed in a body while producing some change in it other than a rise of temperature. Changing a liquid to a gas and a gas to a liquid are both activities involving latent heat. The two types of latent heat are:

1. Internal latent heat
2. External latent heat

These are explained in detail in the next section under *Steam*.

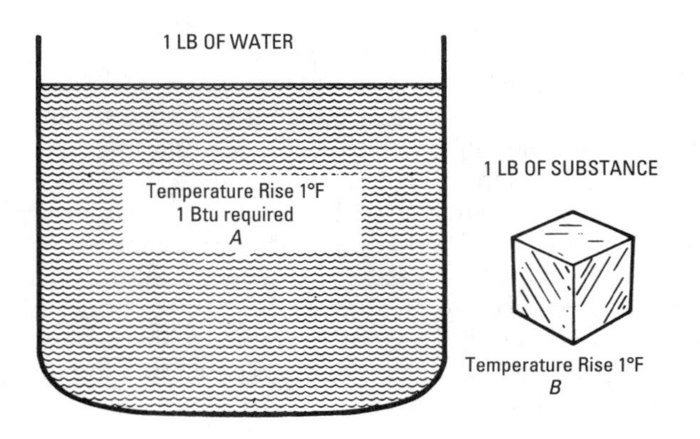

$$\text{SPECIFIC HEAT} = \frac{B\,(\text{Btu})}{A\,(\text{Btu})}$$

Figure 2-7 The principle of specific heat.

Heat-Conveying Mediums

As mentioned in Chapter 1, several methods are used to classify heating systems. One method is based on the medium that conveys the heat from its source to the point being heated. When the majority of heating systems in use today are examined closely, it can be seen that there are only four basic heat-conveying mediums involved:

1. Air
2. Steam
3. Water
4. Electricity

Air

Air is a gas consisting of a mechanical mixture of 23.2% oxygen (by weight), 75.5% nitrogen, and 1.3% argon with small amounts of other gases. It functions as the heat-conveying medium for warm-air heating systems.

Atmospheric pressure may be defined as the force exerted by the weight of the atmosphere in every point with which it is in contact (Figure 2-8), and is measured in inches of mercury or the corresponding pressure in pounds per square-inch (psi).

The pressure of the atmosphere is approximately 14.7 psi at sea level. The standard atmosphere is 29.921 inches of mercury (in Hg) at 14.696 psi. "Inches of mercury" refers to the height to which the column of mercury in a barometer will remain suspended to balance the pressure caused by the weight of the atmosphere.

Atmospheric pressure varies due to elevation by decreasing approximately ½ lb for every 1000 ft ascent above sea level. Atmospheric pressure in pounds per square-inch is obtained from a barometer reading by multiplying the barometer reading in inches by 0.49116. Examples are given in Table 2-1.

Gauge pressure is pressure whose scale starts at atmospheric pressure. *Absolute pressure,* on the other hand, is pressure measured from true zero or the point of no pressure. When the hand of a steam gauge is at zero, the absolute pressure existing in the boiler is approximately 14.7 psi. Thus, by way of example, 5 lb pressure measured by a steam gauge (i.e., gauge pressure) is equal to 5 lb plus 14.7 lb, or 19.7 psi of absolute pressure.

When air is compressed, both its pressure and temperature are changed in accordance with Boyle's and Charles' laws. According to Robert Boyle (1627–1691), the English philosopher and founder of

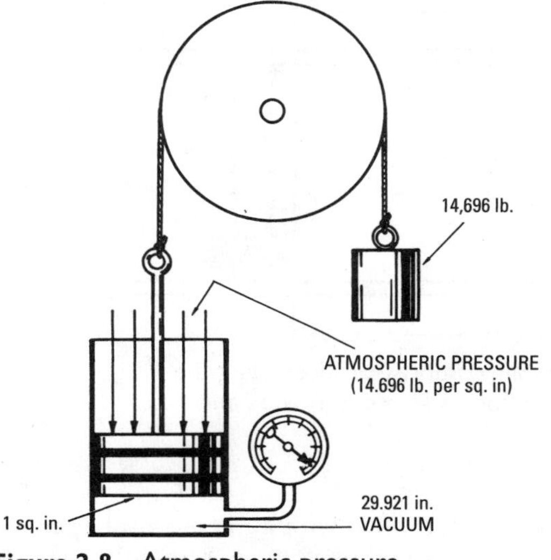

Figure 2-8 Atmospheric pressure.

modern chemistry, the absolute pressure of a gas at constant temperature varies inversely as its volume. Jacques Charles (1746–1823) established that the volume of a gas is proportional to its absolute temperature when the volume is kept at constant pressure.

Table 2-1 Atmospheric Pressure per Square-Inch for Various Barometer Readings

Barometer, in Hg	*Pressure, psi*	*Barometer, in Hg*	*Pressure, psi*
28.00	13.75	29.921	14.696
28.25	13.88	30.00	17.74
28.50	14.00	30.25	14.86
28.75	14.12	30.50	17.98
29.00	14.24	30.75	15.10
29.25	14.37	31.00	15.23
29.50	14.49		
29.75	14.61		

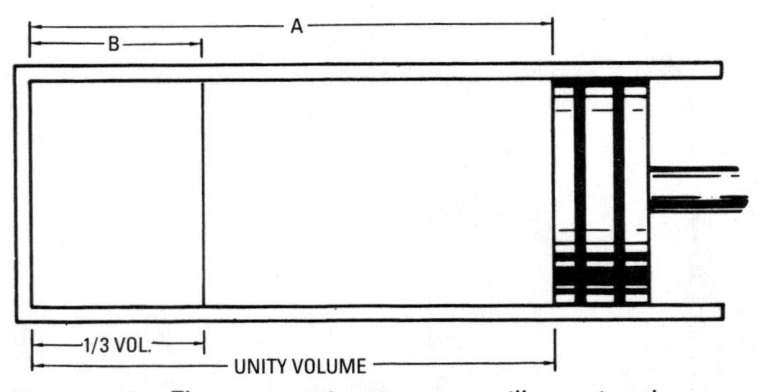

Figure 2-9 Elementary air compressor illustrating the phenomenon of compression as stated in Boyle's and Charles' laws.

If the cylinder in Figure 2-9 is filled with air at atmospheric pressure (14.7 psi absolute), represented by volume A, and the piston B moved to reduce the volume to, say, ⅓ A, as represented by B, then according to Boyle's law, the pressure will be tripled ($14.7 \times 3 = 44.1$ lb absolute, or $44.1 - 14.7 = 29.4$ gauge pressure). According to Charles' law, a pressure gauge on the cylinder would at this point indicate a *higher* pressure than 29.4 gauge pressure because of the increase in temperature produced by compressing the air. This is called *adiabatic compression* if no heat is lost or is received externally.

Steam

Those who design, install, or have charge of steam heating plants certainly should have some knowledge of steam and its formation and behavior under various conditions.

Steam is a colorless, expansive, and invisible gas resulting from the vaporization of water. The white cloud associated with steam is a fog of minute liquid particles formed by condensation, that is to say, finely divided condensation. This white cloud is caused by the exposure of the steam to a temperature lower than that corresponding to its pressure.

If the inside of a steam heating main were visible, it would be filled partway with a white cloud; in traversing the main, the little particles combine, forming drops of condensation too heavy to remain in suspension, which accordingly drop to the bottom of the main and drain off as condensation. This condensation flows into

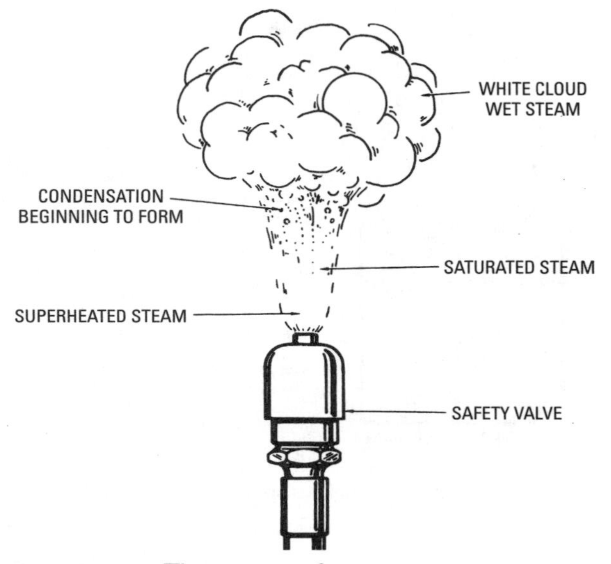

Figure 2-10 Three types of steam.

a drop leg of the system and finally back into the boiler, together with additional condensation draining from the radiators.

Although the word "steam" should be applied only to saturated gas, the five following classes of steam are recognized:

1. Saturated steam
2. Dry steam
3. Wet steam
4. Superheated steam
5. Highly superheated or gaseous steam

Three of these classes of steam (wet, saturated, and superheated) are shown in the illustration of a safety valve blowing in Figure 2-10. It should be pointed out that neither saturated steam nor superheated steam can be seen by the naked eye.

Saturated steam may be defined as steam of a temperature due to its pressure. Steam containing intermingled moisture, mist, or spray is referred to as *wet steam*. *Dry steam* is steam containing no

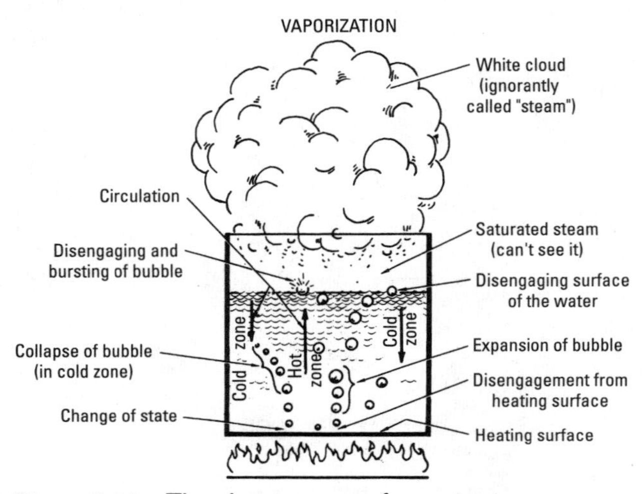

Figure 2-11 The phenomenon of vaporization.

moisture. It may be either saturated or superheated. Finally, *superheated steam* is steam having a temperature higher than that corresponding to its pressure.

The various changes that take place in the making of steam are known as *vaporization* and are shown in Figure 2-11. For the sake of illustration, only one bubble is shown in each receptacle. In actuality there is a continuous procession upward of a great multiplicity of bubbles.

The amount of heat necessary to cause the generation of steam is the sum of the sensible heat, the internal latent heat, and the external latent heat. As mentioned elsewhere in this chapter, sensible heat is the part of the heat that produces a rise in temperature as indicated by the thermometer. The *internal latent heat* is the amount of heat that water will absorb at the boiling point without a change in temperature—that is, before vaporization begins. *External latent heat* is the amount of heat required when vaporization begins to push back the atmosphere and make room for the steam.

Another important factor to consider when dealing with steam is the boiling point of liquids. By definition, the *boiling point* is the

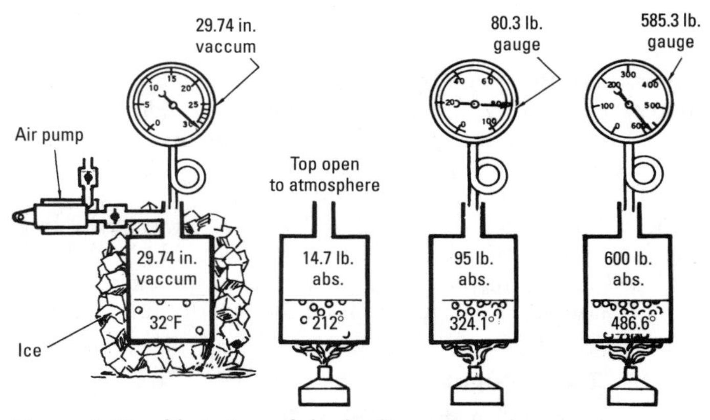

Figure 2.12 Variation of the boiling point when pressure changes.

temperature at which a liquid begins to boil (Figure 2-12), and it depends upon both the pressure and the nature of the liquid. For instance, water boils at 212°F, ether at 9°F, under atmospheric pressure of 14.7 psi.

The relationship between boiling point and pressure is such that there is a definite temperature or boiling point corresponding to each value of pressure. When vaporization occurs in a closed vessel and there is a temperature rise, the pressure will rise until the equilibrium between temperature and pressure is reestablished.

One's knowledge of the fundamentals of steam heating should also include an understanding of the role that condensation plays. By definition, *condensation* is the change of a substance from the gaseous to the liquid (or condensate) form. This change is caused by a reduction in temperature of the steam below that corresponding to its pressure.

The condensation of steam can cause certain problems for steam heating systems unless they are designed to allow for it. The water from which the steam was originally formed contained, mechanically mixed with it, 1⁄20, or 5 percent, of air by volume (at atmospheric pressure). This air is liberated during vaporization and does not recombine with the condensation. As a result, trouble is experienced in heating systems when one attempts to get the air out and keep it out. Suitable air valves are necessary to correct the problem.

Water

Water is a chemical compound of two gases, oxygen and hydrogen, in the proportion of two parts by weight of hydrogen to 16 parts by weight of oxygen, having mixed with it about 5 percent of air by volume at 14.7 lb absolute pressure. It may exist as ice, water, or steam due to changes in temperature (water freezes at 32°F and boils at 212°F when the barometer reads 29.921 in).

One cubic foot of water weighs 62.41 lb at 32°F and 59.82 lb at 212°F. One U.S. gallon of water (231 in^3) weighs 8.33111 lb (ordinarily expressed as $8\frac{1}{3}$ lb) at a temperature of 62°F. At any other temperature, of course, the weight will be different (Table 2-2).

Table 2-2 Weight of Water per Cubic Foot at Different Temperatures

Temp., °F	*lb per ft³*	*Temp., °F*	*lb per ft³*	*Temp., °F*	*lb per ft³*	*Temp., °F*	*lb per ft³*
32	62.41	55	62.38	78	62.23	101	61.98
33	62.41	56	62.38	79	62.22	102	61.96
34	62.42	57	62.38	80	62.21	103	61.95
35	62.42	58	62.37	81	62.20	104	61.94
36	62.42	59	62.37	82	62.19	105	61.93
37	62.42	60	32.36	83	62.18	106	61.91
38	62.42	61	62.35	84	62.17	107	61.90
39	62.42	62	62.35	85	62.16	108	61.89
40	62.42	63	62.34	86	62.15	109	61.87
41	62.42	64	62.34	87	62.14	110	61.86
42	62.42	65	62.33	88	62.13	111	61.84
43	62.42	66	62.32	89	62.12	112	61.83
44	62.42	67	62.32	90	62.11	113	61.81
45	62.42	68	62.31	91	62.10	114	61.80
46	62.41	69	62.30	92	62.08	115	61.78
47	62.41	70	62.30	93	62.07	116	61.77
48	62.41	71	62.29	94	62.06	117	61.75
49	62.41	72	62.28	95	62.05	118	61.74
50	62.40	73	62.27	96	62.04	119	61.72
51	62.40	74	62.26	97	62.02	120	61.71
52	62.40	75	62.25	98	62.01	121	61.69
53	62.39	76	62.25	99	62.00	122	61.68
54	62.39	77	62.24	100	61.99	123	61.66

(continued)

Table 2-2 (continued)

Temp., °F	*lb per ft³*	*Temp., °F*	*lb per ft³*	*Temp., °F*	*lb per ft³*	*Temp., °F*	*lb per ft³*
124	61.64	160	60.99	196	60.21	380	54.47
125	61.63	161	60.97	197	60.19	390	54.05
126	61.61	162	60.95	198	60.16	400	53.62
127	61.60	163	60.93	199	60.14	410	53.19
128	61.58	164	60.91	200	60.11	420	52.74
129	61.56	165	60.89	201	60.09	430	52.33
130	61.55	166	60.87	202	60.07	440	51.87
131	61.53	167	60.85	203	60.04	450	51.28
132	61.51	168	60.83	204	60.02	460	51.02
133	61.50	169	60.81	205	59.99	470	50.51
134	61.48	170	60.79	206	59.97	480	50.00
135	61.46	171	60.77	207	59.95	490	49.50
136	61.44	172	60.75	208	59.92	500	48.78
137	61.43	173	60.73	209	59.90	510	48.31
138	61.41	174	60.71	210	59.87	520	47.62
139	61.39	175	60.68	211	59.85	530	46.95
140	61.37	176	60.66	212	59.82	540	46.30
141	61.36	177	60.64	214	59.81	550	45.66
142	61.34	178	60.62	216	59.77	560	44.84
143	61.32	179	60.60	218	59.70	570	44.05
144	61.30	180	60.57	220	59.67	580	43.29
145	61.28	181	60.55	230	59.42	590	42.37
146	61.26	182	60.53	240	59.17	600	41.49
147	61.25	183	60.51	250	58.89	610	40.49
148	61.23	184	60.49	260	58.62	620	39.37
149	61.21	185	60.46	270	58.34	630	38.31
150	61.19	186	60.44	280	58.04	640	37.17
151	61.17	187	60.42	290	57.74	650	35.97
152	61.15	188	60.40	300	57.41	660	34.48
153	61.13	189	60.37	310	57.08	670	32.89
154	61.11	190	60.35	320	56.75	680	31.06
155	61.09	191	60.33	330	56.40	690	28.82
156	61.07	192	60.30	340	56.02	700	25.38
157	61.05	193	60.28	350	55.65		
158	61.03	194	60.26	360	55.25		
159	61.01	195	60.23	370	54.85		

Water changes in weight with changes of temperature. That is, the higher the temperature of the water, the less it weighs. It is this property of water that causes circulation in boilers and in hot-water heating systems. The change in weight is due to expansion and a reduction in water volume. As the temperature rises, the water expands, resulting in a unit volume of water containing less water at higher temperature than lower temperature.

Fill a vessel with cold water and heat it to the boiling point. Note that boiling causes it to overflow due to expansion. Now let the water cool. You will note that when the water is cold, the vessel will not be as full because the water will have contracted.

The point of maximum density of water is 39.1°F. The most remarkable characteristic of water is its expansion below *and* above its point of maximum density. Imagine 1 lb of water at 39.1°F placed in a cylinder having a cross-sectional area of 1 in^2 (Figure 2-13). The water having a volume of 27.68 in^3 will fill the cylinder to a height of 27.68 in. If the water is cooled, it will expand, and at, say, 32°F (the freezing point) will rise in the tube to a height of 27.7 in before freezing. If the water is heated, it will also expand and rise in the tube; and at the boiling point (for atmospheric pressure 212°F) it will occupy the tube to a height of 28.88 in.

The elementary hot-water heating system in Figure 2-14 illustrates the principle of thermal circulation. The weight of the hot

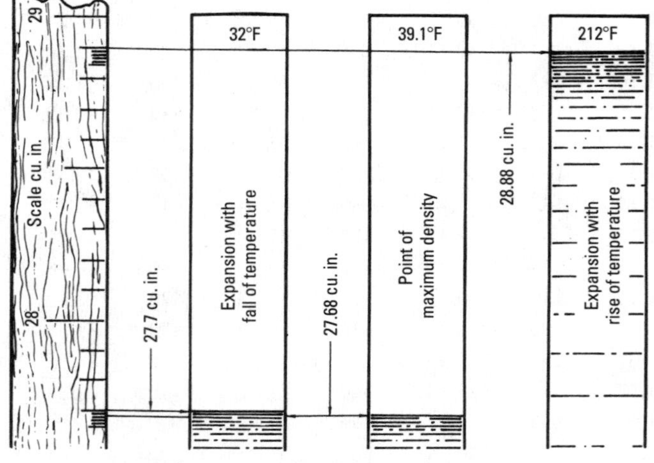

Figure 2-13 The point of maximum density.

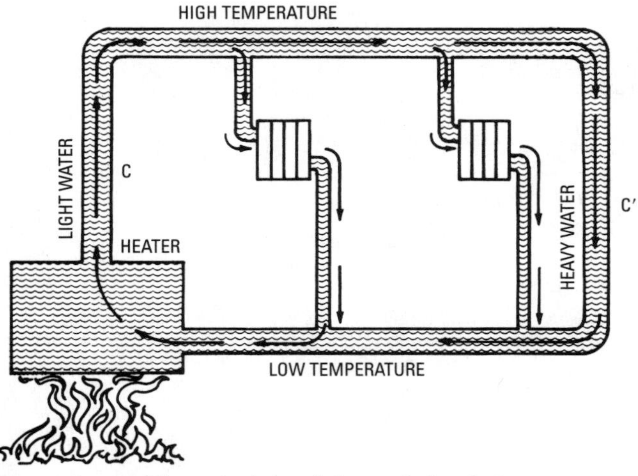

Figure 2-14 The principle of thermal circulation.

and expanded water in the upflow column C, being less than that of the cold and contracted water in the downflow column C′, upsets the equilibrium of the system and results in a continuous circulation of water as indicated by the arrows. In other words, the heavy, low-temperature water sinks to the lowest point in the boiler (or system) and displaces the light, high-temperature water, thus causing continuous circulation as long as there is a temperature difference in different parts of the boiler (or system). This is referred to as *thermal circulation.*

Electricity

Electricity differs from air, steam, or water in that it does not actually convey heat from one point to another; therefore, including it in a list of heat-conveying mediums can be misleading at first glance.

Electricity can best be defined as a quantity of electrons either in motion or in a state of rest. When these electrons are at rest, they are referred to as being *static* (hence the term *static electricity*). Electrons in *motion* move from one atom to another, creating an electrical current and thereby a medium for conveying *energy* from one point to another. Many different devices have been created to change the energy conveyed by an electric current into heat, light, and other forms of energy. Electric-fired furnaces and boilers are examples of devices used to produce heat.

Chapter 3

Insulating and Ventilating Structures

This chapter is not concerned with the theoretical aspects of heat loss or gain, or with heating and cooling calculations per se. This information is covered in Chapter 2 ("Heating Fundamentals") and Chapter 4 ("Sizing Residential Heating and Air Conditioning Systems"). The purpose of this chapter is to describe the various *types* of insulation used in construction, the methods of applying insulation, and the thermal properties of building materials.

Insulating Structures

An uninsulated structure or one that is poorly insulated will experience a heat gain in the summer (or a heat loss in the winter) as high as 50 percent. The exact percentage will depend upon the materials used in its construction, because all construction material will have some insulating effect. In any event, it has been observed that approximately 30 percent of heat gain (or loss) is experienced through ceilings and about 70 percent occurs through walls, glass, and window and door cracks or as a result of air ventilation.

Any attempt to heat or cool an *uninsulated* structure will be terribly inefficient *and expensive*. Obviously, the possibility of anyone being so shortsighted as to attempt to install a heating or cooling system in such a structure is fairly remote, but inadequate insulation in existing structures or the failure to provide for proper insulation in proposed construction is a common occurrence. The provision for some form of barrier to reduce the rate of heat flow to an acceptable level is therefore of prime importance. Insulation materials serve this purpose with varying degrees of effectiveness. As much as 80 to 90 percent of heat loss (or gain) through ceilings and 60 percent of heat loss (or gain) through walls can be prevented by properly insulating these areas.

The inefficiency of ordinary building materials in resisting the passage of heat brought about the need for the development of materials specifically designed for insulation. A *good* insulating material should be lightweight, contain numerous air pockets, and exhibit a high degree of resistance to heat transmission. The material should also be specifically treated to resist fire and the attacks of rodents and insects. Finally, a good insulating material should react well to excess moisture by drying out and retaining its resistance to heat flow, instead of disintegrating.

Many inexperienced workers seem to feel that the more insulation one uses, the better the insulating effect. Unfortunately, this does *not* hold true on a one-to-one basis. For example, twice as much insulation does *not* insulate twice as well. Apparently a law of diminishing returns operates here. The first layer of insulation is the most effective, with successive layers decreasing in the effectiveness to impede the rate of heat flow.

The capacity of a heating or cooling system is determined by the amount of heat that must be supplied (heating applications) or removed (cooling applications) to maintain the desired temperature within the structure or space. In heating applications, the amount of heat supplied to a space at constant temperature should roughly equal the amount of heat lost. In cooling applications, the heat removed from a space should be roughly equal to the amount of heat gained. Both heat loss and heat gain will be controlled by the way in which the structure is insulated. Good insulation will reduce the rate of heat flow through construction materials to an acceptable level, and this will mean that a heating or cooling system will be able to operate much less expensively and with greater efficiency. It will also mean that the capacity of the heating or cooling unit you wish to install will be less than the one required for a poorly insulated structure. This, in turn, will reduce the initial equipment and installation costs. As you can see, this type and quality of insulation in a structure is an important factor to be considered when designing a heating or a cooling system.

The roof or attic floor of any structure can be insulated, as can the walls of any frame building, whether of stucco, clapboard, shingles, brick, or stone veneer. The form in which the insulation is applied depends on whether the structure is already built or is being constructed.

Principles of Heat Transmission

Heat flows through a barrier of building materials (e.g., walls, ceilings, and floors) by means of conduction and radiation. When it has passed to the other side of this barrier, it continues its movement away from the source by means of conduction, radiation, and convection. The *direction* of heat flow depends on the relationship of outdoor and indoor temperatures because warmed air will always move toward a colder space. Thus, in the summer months, the warmer outdoor air will move toward the cooler indoor spaces. In the winter months, on the other hand, the reverse is true: The warmer indoor air moves to the cooler outdoors. The purpose of insulation is to reduce the rate of heat flow to an acceptable level. This, in turn, will reduce the amount of energy required to heat or cool the structure.

Heat Transfer Values

The American Society of Heating, Refrigerating, and Air-Conditioning Engineers (ASHRAE) has conducted considerable research into the thermal properties of building materials, and these data are used in the calculation of heat loss and heat gain. Tables of computed values for many construction materials and combinations of construction materials are made available through ASHRAE publications. One table gives the design values (i.e., conductivity values, conductance values, and resistance values) of various building and insulating materials. These values are used to calculate the coefficients of heat transmission (*U*-values) for different types of construction (e.g., masonry walls and frame partitions). Examples of these two tables have been taken from the *1960 ASHRAE Guide* (see Tables 3-1 and 3-2).

The four heat transfer values contained in these tables are: (1) the *k*-value (thermal conductivity); (2) the *C*-value (thermal conductance); (3) the *R*-value (thermal resistance); and (4) the *U*-value (overall coefficient of heat transmission). Each of these heat transfer values will be examined before their use in heating and cooling calculations is explained.

Thermal Conductivity

The *k-value* (thermal conductivity) represents the amount of heat that flows through 1 square foot of a homogeneous material 1 inch thick in 1 hour for each degree of temperature difference between the inside and the outside temperatures.

To illustrate, we shall use an example from the ASHRAE data shown in Tables 3-1 and 3-2. You will note that the *k*-value for face brick listed in Table 3-1 is 9.0. In other words, 9.0 Btu of heat can pass through 1 ft^2 of 1-in-thick face brick each hour. The thermal resistance (*R*-value) of this same 1-in-thick face brick is the reciprocal of its *k*-value:

$$R = \frac{1}{k} = \frac{1}{9} = 0.11$$

Table 3-1 Conductivities (*k*), Conductances (*C*), and Resistances (*R*) of Various Building and Insulating Materials

Description					*Conductivity or Conductance*		*Resistance (R)*	
Material				*Density (lb per ft³)*	*(k)*	*(c)*	*Per Inch Thickness 1/k*	*For Thickness Listed 1/C*
Air Spaces	Position	Heat Flow	Thickness					
	Horizontal	Up (winter)	¾–4 in	—	—	1.18	—	0.85
	Horizontal	Up (summer)	¾–4 in	—	—	1.28	—	0.78
	Horizontal	Down (winter)	¾ in	—	—	0.98	—	1.02
	Horizontal	Down (winter)	1½ in	—	—	0.87	—	1.15
	Horizontal	Down (winter)	4 in	—	—	0.81	—	1.23
	Horizontal	Down (winter)	8 in	—	—	0.80	—	1.25
	Horizontal	Down (summer)	¾ in	—	—	1.18	—	0.85
	Horizontal	Down (summer)	1½ in	—	—	1.07	—	0.93
	Horizontal	Down (summer)	4 in	—	—	1.01	—	0.99
	Sloping, 45°	Up (winter)	¾–4 in	—	—	1.11	—	0.90
	Sloping, 45°	Down (summer)	¾–4 in	—	—	1.12	—	0.89
	Vertical	Horizontal (winter)	¾–4 in	—	—	1.03	—	0.97
	Vertical	Horizontal (summer)	¾–4 in	—	—	1.16	—	0.86

Air Surfaces	Position	Heat Flow					
Still air	Horizontal	Up	—	—	1.63	—	0.61
	Up sloping (45°)	Up	—	—	1.60	—	0.62
	Vertical	Horizontal	—	—	1.46	—	0.68
	Sloping (45°)	Down	—	—	1.32	—	0.76
	Horizontal	Down	—	—	1.08	—	0.92
15 mph wind	Any position—any direction (for winter)		—	—	6.00	—	0.17
7½ mph wind	Any position—any direction (for summer)		—	—	4.00	—	0.25
Building Board							
Boards, panels, sheathing, etc.							
	Gypsum or plaster board	3/8 in	50	—	3.10	—	0.32
	Gypsum or plaster board	1/2 in	50	—	2.25	—	0.45
	Plywood		34	0.80		1.25	—
	Plywood	1/4 in	34	—	3.20	—	0.31
	Plywood	3/8 in	34	—	2.12	—	0.47
	Plywood	1/2 in	34	—	1.60	—	0.63
	Plywood or wood panels	3/4 in	—	—	1.07	—	0.94

(continued)

Table 3-1 (continued)

Material	Description		Density (lb per ft³)	Conductivity or Conductance (k)	(c)	Resistance (R) Per Inch Thickness l/k	For Thickness Listed l/C
	Wood-fiber board,		26	0.42	—	2.38	—
	laminated or homogenous	31	0.50	—	2.00	—	
	Wood fiber—hardboard type		65	1.40	—	0.72	—
	Wood fiber—hardboard type	1/4 in	65	—	5.60	—	0.18
	Wood—fir or pine sheathing	25/32 in	—	—	1.02	—	0.98
	Wood—fir or pine	1 5/8 in	—	—	0.49	—	2.03
Building Paper	Vapor—permeable felt		—	—	16.70	—	0.06
	Vapor—seal, 2 layers of mopped 15-lb felt		—	—	8.35	—	0.12
	Vapor—seal, plastic film		—	—	—	—	Negl
Flooring Materials	Asphalt tile	1/8 in	120	—	24.80	—	0.04

	Carpet and fibrous pad		—	—	0.48	—	2.08
	Carpet and rubber pad		—	—	0.81	—	1.23
	Ceramic tile	1 in	—	—	12.50	—	0.08
	Cork tile		25	0.45		2.22	—
	Cork tile	1/8 in	—	—	3.60	—	0.28
	Felt, flooring		—	—	16.70	—	0.06
	Floor tile or linoleum—av. Value	1/8 in	—	20.00	0.05		
	Linoleum	1/8 in	80	—	12.00	—	0.08
	Plywood subfloor	5/8 in	—	—	1.28	—	0.78
	Rubber or plastic tile	1/8 in	110	—	42.40	—	0.02
	Terrazzo	1 in	—	—	12.50	—	0.98
	Wood subfloor	25/32 in	—	—	1.02	—	0.98
	Wood, hardwood finish	3/4 in	—	—	1.47	—	0.68
Insulating Materials	Cotton fiber		0.8–2.0	0.26	—	3.85	—
Blankets and batts	Mineral wool, fibrous form, processed from rock, slag, or glass		1.5–4.0	0.27	—	3.70	—
	Wood fiber		3.2–3.6	0.25	—	4.00	—
	Wood fiber, multilayer, stitched expanding		1.5–2.0	0.27	—	3.70	—

(continued)

Table 3-1 (continued)

Material	Description		Density (lb per ft³)	Conductivity or Conductance (k)	(c)	Resistance (R) Per Inch Thickness 1/k	For Thickness Listed 1/C
Board	Glass fiber		—	9.5	0.25	—	4.00
	Wood or cane fiber						
	Acoustical tile ½ in		—	0.84	—	1.19	—
	Acoustical tile	¾ in			0.56	—	1.78
	Interior finish (plank, tile, lath)		15.0	0.35	—	2.86	—
	Interior finish (plank, tile, lath)	½ in	15.0	—	0.70	—	1.43
	Roof deck slab						
	Approx	1½ in	—	0.24	—	4.17	—
	Approx.	2 in	—	—	0.18	—	5.56
	Approx.	3 in	—	—	0.12	—	8.33
	Sheathing (impreg. or coated)		20.0	0.38	—	2.63	—
	Sheathing (impreg. or coated)	½ in	20.0	—	0.76	—	1.32
	Sheathing (impreg. or coated)	25/32 in	20.0	—	0.49	—	2.06

Board and slabs	Cellular glass		9.0	0.40	—	2.50	—
	Corkboard (without added binder)		6.5–8.0	0.27	—	3.70	—
	Hog hair (with asphalt binder)		8.5	0.33	—	3.00	—
	Plastic (foamed)		1.62	0.29	—	3.45	—
	Wood shredded (cemented in preformed slabs)		22.0	0.55	—	1.82	—
Loose fill	Macerated paper or pulp products		2.5–3.5	0.28	—	3.57	—
	Mineral wood (glass, slag, or rock)		2.0–5.0	0.30	—	3.33	—
	Sawdust or shavings		8.0–15.0	0.45	—	2.22	—
	Vermiculite (expanded)		7.0	0.48	—	2.08	—
	Wood fiber: redwood, hemlock, or fir		2.0–3.5	0.30	—	3.33	—
Roof Insulation	All types						
	Preformed, for use above deck						
	Approx.	½ in			0.72	—	1.39
	Approx.	1 in			0.36	—	2.78
	Approx.	1½ in			0.24	—	4.17

(continued)

Table 3-1 (continued)

Material	Description		Density (lb per ft³)	Conductivity or Conductance (k)	(c)	Resistance (R) Per Inch Thickness 1/k	For Thickness Listed 1/C
	Approx.	2 in			0.19	—	5.26
	Approx.	2½ in			0.15	—	6.67
	Approx.	3 in			0.12	—	8.33
Masonry Materials	Cement mortar		116	5.0	—	0.20	—
Concretes							
	Gypsum-fiber concrete 87½% gypsum, 12½% wood chips		51	1.66	—	0.60	—
	Lightweight aggregates, including expanded		120	5.2	—	0.19	—
	shale, clay or slate; expanded slags;		100	3.6	—	0.28	—
	cinders; pumice; perlite; vermiculite;		80	2.5	—	0.4	—
	also cellular concretes		60	1.7	—	0.59	
			40	1.15	—	0.86	
			30	0.90	—	1.11	

			20	0.70	—	1.43	
	Sand and gravel or stone aggregate (oven dried)		140	9.0	—	0.11	
	Sand and gravel or stone aggregate (not dried)		140	12.0	—	0.08	
	Stucco		116	5.0	—	0.20	
Masonry Units	Brick, common		120	5.0		0.20	
	Brick, face		130	9.0		0.11	
	Clay tile, hollow:						
	1 cell deep	3 in	—		1.25		0.80
	1 cell deep	4 in	—		0.90		1.11
	2 cells deep	6 in	—		0.66		1.52
	2 cells deep	8 in	—		0.54		1.85
	2 cells deep	10 in	—		0.45		2.22
	3 cells deep	12 in	—		0.40		2.50
	Concrete blocks, three oval core: Sand and gravel aggregate						
		4 in	—	—	1.40	—	0.71
		8 in	—	—	0.90	—	1.11
		12 in	—	—	0.78	—	1.28
	Cinder aggregate	3 in	—	—	1.16	—	0.86
		4 in	—	—	0.90	—	1.11
		8 in	—	—	0.58	—	1.72

(continued)

Table 3-1 (continued)

Material	Description		Density (lb per ft^3)	Conductivity or Conductance (k)	(c)	Resistance (R) Per Inch Thickness 1/k	For Thickness Listed 1/C
	Gypsum partition tile:						
	3 × 12 × 30-in solid						
			—	—	0.79	—	1.26
	3 × 12 × 30-in 4-cell		—	—	0.74	—	1.35
	3 × 12 × 30-in 3-cell						
			—	—	0.60	—	1.67
		3 in	—	—	0.79	—	1.27
	Lightweight aggregate (expanded shale, clay, slate, or slag; pumice)	4 in	—	—	0.67	—	1.50
		8 in	—	—	0.50	—	2.00
		12 in	—	—	0.44	—	2.27
	Stone, lime, or sand		—	12.50	—	0.08	—
Plastering Materials	Cement plaster, sand aggregate		116	5.0	—	0.20	—
	Sand aggregate	½ in	—	—	10.00	—	0.10

	Sand aggregate	¾ in	—	—	6.66	—	0.15
	Gypsum plaster						
	Lightweight aggregate	½ in	45	—	3.12	—	0.32
	Lightweight aggregate	⅝ in	45	—	2.67	—	0.39
	Lightweight aggregate on metal lath	¾ in	—	2.13	—	0.47	
	Perlite aggregate		45	1.5	—	0.67	—
	Sand aggregate		105	5.6	—	0.18	—
	Sand aggregate	½ in	105	—	11.10	—	0.09
	Sand aggregate	⅝ in	105	—	9.10	—	0.11
	Sand aggregate on metal lath	¾ in	—	—	7.70	—	0.13
	Sand aggregate on wood lath		—	—	2.50	—	0.40
	Vermiculate aggregate		45	1.7	—	0.59	—
Roofing	Asphalt roll roofing		70	—	6.50	—	0.15
	Asphalt shingles		70	—	2.27	—	0.44
	Built-up roofing	⅜ in	70	—	3.00	—	0.33
	Slate	½ in	—	—	20.00	—	0.05
	Sheet metal		—	400+	—	Negl	—
	Wood shingles		—	—	1.06	—	0.94

(continued)

Table 3-1 (continued)

Material	Description		Density (lb per ft³)	Conductivity or Conductance (k)	(c)	Resistance (R) Per Inch Thickness 1/k	For Thickness Listed 1/C
Siding Materials (On flat surface)	Shingles						
	Wood, 16-in 7 ½-in exposure		—	—	1.15	—	0.87
	Wood, double, 16-in, 12-in exposure		—	—	0.84	—	1.19
	Wood, plus insulation backer board	5/16 in	—	—	0.71	—	1.40
	Siding						
	Asphalt roll siding		—	—	6.50	—	0.15
	Asphalt insulating siding (½-in bd.)		—	—	0.69	—	1.45
	Wood, drop, 1 × 8 in		—	—	1.27	—	0.79
	Wood, bevel, ½ × 8 in, lapped		—	—	1.23	—	0.81

	Wood, bevel, ¾ × 10 in, lapped	—	—	0.95	—	1.05
	Wood, plywood, ⅜ in, lapped	—	—	1.59	—	0.59
	Structural glass	—	—	10.00	—	0.10
Woods	Maple, oak, and similar hardwoods	45	1.10	—	0.91	—
	Fir, pine, and similar softwoods	32	0.80	—	1.25	—

Courtesy ASHRAE 1960 Guide

Table 3-2 Coefficients of Heat Transmission for Solid Masonry Walls

Resistances Used		
Construction	***Resistance***	***(R)***
1. Outside surface (15 mph wind)		0.17
2. Face brick (4-in)		0.44
3. Common brick (4-in)		0.80
4. Air space		0.97
5. Gypsum lath (3/8-in)		0.32
6. Plaster (sand aggregate) (1/2-in)		0.09
7. Inside surface (still air)		0.68
Total resistance:		3.47
$U = 1/\mathbf{R} = 1/3.47 =$		0.29
See value 0.29 in boldface type in table below.		
Assume plain wall—no furring of or plaster.		
Total resistance:		3.47
Deduct 4. Air space	0.97	
5. Gypsum lath (3/8-in)	0.32	
6. Plaster (sand aggregate) (1/2-in)	0.09	1.38
Total resistance:		2.09
$U = 1/\mathbf{R} = 1/2.09 =$		0.48

Exterior Construction		Interior Finish											No.
		None	Plaster 5/8-in on Plaster on Furring		Metal Lath and 3/4-in Plaster on Furring		Gypsum Lath (3/8-in) and 1/2-in Plaster on Furring			Insulation Board Lath (1/2-in) and 1/2-in Plaster on Furring		Wood Lath and 1/2-in Plaster (Sand agg.) 0.40	
Resistance			(Sand agg.) 0.11	(Lt. Wt. agg.) 0.39	(Sand agg.) 0.13	(Lt. Wt. agg.) 0.47	No plas. 0.32	(Sand agg.) 0.41	Lt. wt. agg.) 0.64	No plaster 1.43	(Sand agg.) 1.52		
Material	R	U	U	U	U	U	U	U	U	U	U	U	
		A	B	C	D	E	F	G	H	I	J	K	
Brick (face and common)													
(6 in)	0.61	0.68	0.64	0.54	0.39	0.34	0.36	0.35	0.33	0.26	0.25	0.35	1
(8 in)	1.24	0.48	0.45	0.41	0.31	0.28	0.30	0.29	0.27	0.22	0.22	0.29	2
(12 in)	2.04	0.35	0.33	0.30	0.25	0.23	0.24	0.23	0.22	0.19	0.19	0.23	3
(16 in)	2.84	0.27	0.26	0.25	0.21	0.19	0.20	0.20	0.19	0.16	0.16	0.20	4
Brick (common only)													
(8 in)	1.60	0.41	0.39	0.35	0.28	0.26	0.27	0.26	0.25	0.21	0.20	0.26	5
(12 in)	2.40	0.31	0.30	0.27	0.23	0.21	0.22	0.22	0.21	0.18	0.17	0.22	6
(16 in)	3.20	0.25	0.24	0.23	0.19	0.18	0.19	0.18	0.18	0.16	0.15	0.18	7

(continued)

Table 3-2 (continued)

Exterior Construction		Interior Finish											No.
		None	Plaster 5/8-in on Plaster on Furring		Metal Lath and 3/4-in Plaster on Furring		Gypsum Lath (3/8-in) and 1/2-in Plaster on Furring			Insulation Board Lath (1/2-in) and 1/2-in Plaster on Furring		Wood Lath and 1/2-in Plaster (Sand agg.) 0.40	
Resistance			(Sand agg.) 0.11	(Lt. Wt. agg.) 0.39	(Sand agg.) 0.13	(Lt. Wt. agg.) 0.47	No plas. 0.32	(Sand agg.) 0.41	Lt. wt. agg.) 0.64	No plaster 1.43	(Sand agg.) 1.52		
Material	R	U	U	U	U	U	U	U	U	U	U	U	
		A	B	C	D	E	F	G	H	I	J	K	
Stone (lime and sand)													
(8 in)	0.64	0.67	0.63	0.53	0.39	0.34	0.36	0.35	0.32	0.26	0.25	0.35	8
(12 in)	0.96	0.55	0.52	0.45	0.34	0.31	0.32	0.31	0.29	0.24	0.23	0.31	9
(16 in)	1.28	0.47	0.45	0.40	0.31	0.28	0.29	0.28	0.27	0.22	0.22	0.29	10
(24 in)	1.92	0.36	0.35	0.32	0.26	0.24	0.25	0.24	0.23	0.19	0.19	0.24	11
Hollow clay tile													
(8 in)	1.85	0.36	0.36	0.32	0.26	0.24	0.25	0.25	0.23	0.20	0.19	0.25	12
(10 in)	2.22	0.33	0.31	0.29	0.24	0.22	0.23	0.22	0.21	0.18	0.18	0.23	13
(12 in)	2.50	0.30	0.29	0.27	0.22	0.21	0.22	0.21	0.20	0.17	0.17	0.21	14

Poured concrete												
30 lb/ft^3												
(4 in)	4.44	0.19	0.19	0.18	0.16	0.15	0.15	0.15	0.14	0.13	0.13	0.15	15
(6 in)	6.66	0.13	0.13	0.13	0.12	0.11	0.11	0.11	0.11	0.10	0.10	0.11	16
(8 in)	8.88	0.10	0.10	0.10	0.09	0.09	0.09	0.09	0.09	0.08	0.08	0.09	17
(10 in)	11.10	0.08	0.08	0.08	0.08	0.07	0.08	0.08	0.07	0.07	0.07	0.08	18
80 lb/ft^3													
(6 in)	2.40	0.31	0.30	0.27	0.23	0.21	0.22	0.22	0.21	0.18	0.17	0.22	19
(8 in)	3.20	0.25	0.24	0.23	0.19	0.18	0.19	0.18	0.18	0.16	0.15	0.18	20
(10 in)	4.00	0.21	0.20	0.19	0.17	0.16	0.16	0.16	0.15	0.14	0.14	0.16	21
(12 in)	4.80	0.18	0.17	0.17	0.15	0.14	0.14	0.14	0.14	0.12	0.12	0.14	22
140 lb/ft^3													
(6 in)	0.48	0.75	0.69	0.58	0.41	0.36	0.38	0.37	0.34	0.27	0.26	0.37	23
(8 in)	0.64	0.67	0.68	0.53	0.39	0.34	0.36	0.35	0.32	0.26	0.25	0.35	24
(10 in)	0.80	0.61	0.57	0.49	0.36	0.32	0.34	0.33	0.31	0.25	0.24	0.33	25
(12 in)	0.96	0.55	0.52	0.45	0.34	0.31	0.32	0.31	0.29	0.24	0.23	0.31	26

Courtesy ASHRAE 1960 Guide

Remember, this is the thermal resistance for a 1-in-thick face brick. Because the masonry wall in Table 3-2 contains a 4-in thickness of face brick, you will have to multiply the *R*-value (0.11) of the 1-in-thick face brick by 4 (4 in) to obtain the *R*-value (0.44) for the larger (4-in) section.

Thermal Conductance

The *C-value* (thermal conductance) represents the amount of the heat flow through a square foot of material per hour per degree of temperature difference between the inside and outside temperatures. The *C*-value is based on the stated construction or thickness of the material. This distinguishes it from the *k*-value, which is based on a 1-in thickness of a homogeneous material having no fixed thickness. For example, under "loose fill" in Table 3-1, you will see vermiculite listed and given a *k*-value of 0.48. Why a *k*-value? It is given a *k*-value (rather than a *C*-value) because vermiculite is a loose-fill insulating material that can be poured into a space to form whatever thickness is desired. It is convenient, then, to give numerical values (*k*-values) based on 1-in thicknesses of these materials. If, say, vermiculite were used to insulate the top-floor ceiling (that is to say, the attic floor) and poured to a depth of 4 in, the *R*-value (thermal resistance) would be calculated as follows:

$$k\text{-value} = 0.48$$

$$R\text{-value} = \frac{1}{0.48} = 2.08 \text{ (1 in)}$$

$$R\text{-value} = 2.08 \times 4 = 8.32 \text{ (4 in)}$$

Under "masonry units" in Table 3-1, you will find hollow clay tile listed with six different *stated* thicknesses (e.g., "1 cell deep, 3 in; 1 cell deep, 4 in"). Each is given a *C*-value rather than a *k*-value because the thickness of the material listed in the table is the same thickness that will be used in the construction of a structural section.

The masonry wall in Table 3-2 also contains gypsum plaster (sand aggregate) ½ in thick. The *C*-value is given as 11.10, and its *R*-value as 0.09 (Table 3-1). In other words, the *R*-value (thermal resistance) is calculated as follows:

$$R = \frac{1}{C} = \frac{1}{11.10} = 0.09$$

Examine the data in Table 3-1 once more. You will probably note that each building material has either a *k*-value or a *C*-value, but not both. Each material is offered on the basis of either a stated thickness or an arbitrary thickness, as was just explained. Note also that each type of building material has an *R*-value determined by either of the following two equations:

$$R = \frac{1}{k}$$

$$R = \frac{1}{C}$$

The *R*-value is simply the reciprocal of either the *k*-value or the *C*-value.

Thermal Resistance

The *R-value* (thermal resistance) of a building material is its *resistance* to the flow of heat. As explained in the preceding subsection, the *R*-value is the reciprocal of the *k*-value or the *C*-value and represents heat flow resistance expressed as a numerical value. The higher the *R*-value of a material (or a combination of materials in a wall, ceiling, floor, or roof), the greater its resistance to the flow of heat. The minimum recommended *R*-values are *R*-30 for an attic, *R*-11 for walls, *R*-19 for ceilings and raised floors, and *R*-42 for ductwork (Figure 3-1). Energy savings and lower fuel costs will result if insulation with an *R*-value higher than the recommended minimum is installed. This holds true for both new construction and existing structures.

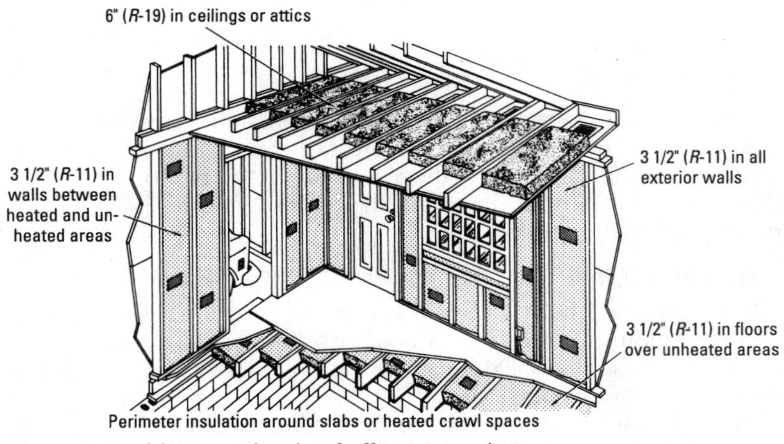

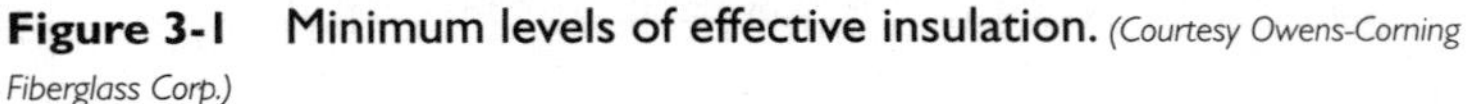

Figure 3-1 **Minimum levels of effective insulation.** *(Courtesy Owens-Corning Fiberglass Corp.)*

Returning to the masonry wall illustrated in Table 3-2, you will note that all seven R-values have been added to produce the total resistance of a solid masonry wall (3.47). In other words, the *total* resistance to heat flow through a solid masonry wall of the type of construction illustrated in Table 3-2 is equal to the sum of the resistances of the wall components:

$$R_t = R_1 + R_2 + R_3 + \cdots + R_n$$

The total resistance (R_t) of a particular type of construction is used to determine its U-value, which represents the total *insulating* effect of a structural section. The U-value is the reciprocal of the total resistance (R_t). Using the data supplied for the masonry wall illustrated in Table 3-2, its U-value is determined as follows:

$$U = \frac{1}{R} = \frac{1}{3.47} = 0.29$$

Overall Coefficient of Heat Transmission

The *U-value,* or *U-factor* (overall coefficient of heat transmission), represents the amount of heat (in Btu) that will pass through 1 square foot of a structural section per hour per degree Fahrenheit. This is the difference between the air on the inside and the air on the outside of the section. The U-value is the reciprocal of the sum of the thermal resistance values (R-values) of each element of the structural section. The total resistance (R_t) for the solid masonry wall illustrated in Table 3-2 is 3.47. Since the U-value is the reciprocal of R_t), the following is obtained:

$$U = \frac{1}{R_t} = \frac{1}{3.47} = 0.29$$

The overall coefficient of heat transmission (U-value or U-factor) is used in the formula for calculating heat loss or heat gain. The use of this formula is described in Chapter 4 of this book ("Heating Calculations") and in Chapter 9 of Volume 3 ("Air Conditioning Calculations").

Condensation

Air will always contain a certain amount of moisture (water vapor), and the amount of moisture it contains will depend upon its temperature. Generally speaking, warm air is capable of containing more moisture than cold air.

Air will lose moisture in the form of condensation when its temperature falls below its dew point. The dew point is the temperature at which moisture begins to condense, in the form of tiny droplets or dew. Any reduction in temperature below the dew point will result in condensation of some of the water vapor present in the air.

The attic is one place where condensation often occurs. In the winter, the warm air from the occupied spaces leaks past the insulation on the top-floor ceiling and comes into contact with the cooler air in the attic. This causes the water vapor to condense. The moisture resulting from condensation works its way down through the ceiling and walls, eventually causing damage to the insulation, wood, and other building materials.

The solution to condensation in the attic is twofold. Adequate ventilation should be provided in the attic to remove the humid air. This can be accomplished by installing louvers at each end of the attic or in roof overhangs to provide cross ventilation (Figure 3-2). The louvers should have a total area of 1 ft^2 for each 300 ft^2 of attic floor space. For example, if the attic has an area of 900 ft^2, the total louver area should be at least 3 ft^2. In addition to suitable attic ventilation, you should install a vapor retarder between the rafters. This is particularly true if the attic is occupied or used extensively.

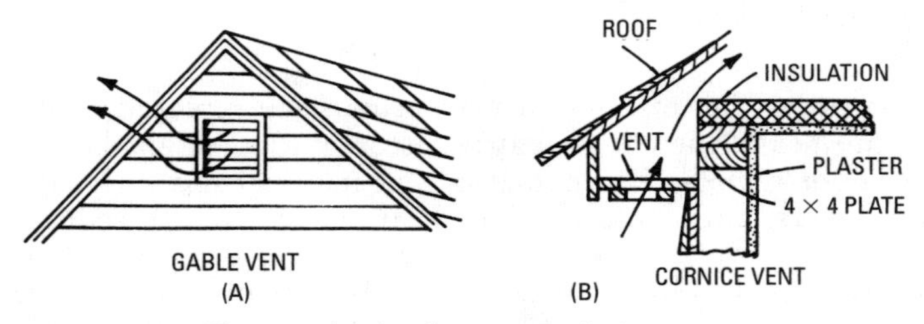

Figure 3-2 Two methods of attic ventilation.

Foundation crawl spaces are also subject to condensation. A vapor retarder between the floor joists prevents excess moisture from accumulating and causing dampness and eventually damage to the structure. Ventilating the crawl space so that the humid air can be removed is also very effective. Finally, placing an insulating paper (e.g., roofing paper) on the ground will reduce the amount of moisture released by the soil (Figure 3-3).

Vapor Retarders

Insulation must be protected against moisture, or it will lose its ability to reduce thermal transmission. An unprotected insulation material will absorb moisture and lose its insulating value. This can

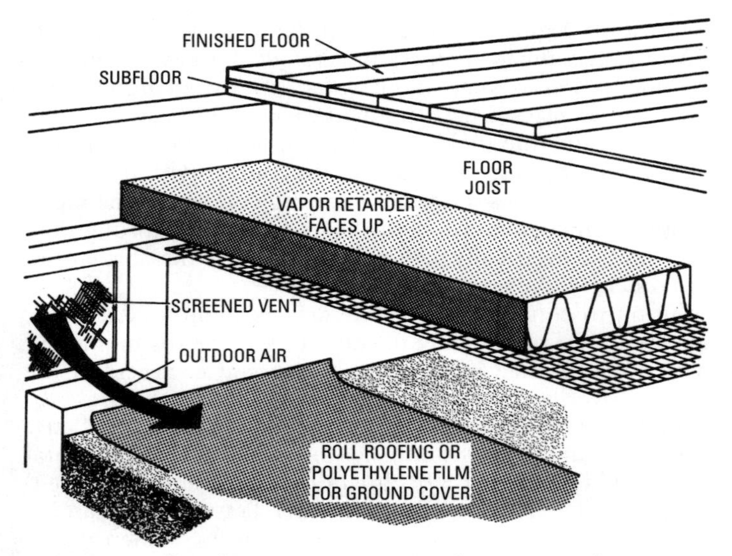

Figure 3-3 Crawl-space construction showing vent location and vapor retarder installation.

be very expensive, because the insulation must be replaced completely. In most cases there is no way to restore it to its original condition. A *vapor retarder* is a nonabsorbent material (e.g., asphalt laminated papers, aluminum foil, or plastic film), used to reduce the rate at which water vapor can move through a construction material (Figure 3-4). Note that it *reduces* the rate of water vapor

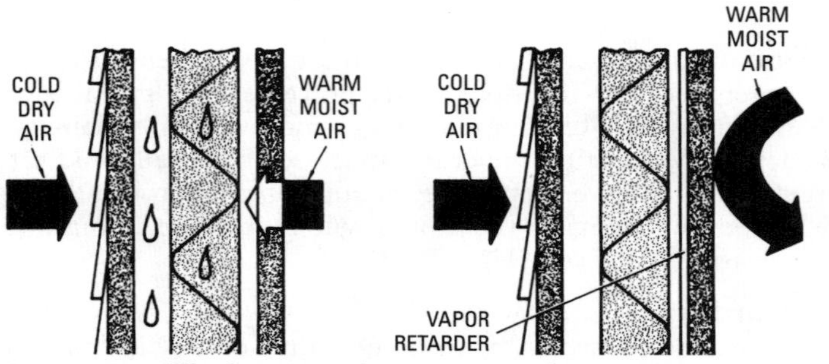

Figure 3-4 The location of a vapor retarder in respect to insulation. Note that the vapor retarder faces the warm, moist air.

movement. No material can completely stop moisture transfer. For that reason, the term *vapor retarder* has now replaced the older term *vapor barrier,* although the latter is still found in many publications and is widely used in the field by technicians. A vapor retarder protects the insulation by either completely enclosing it or by being installed between the insulation and the interior of the structure; that is to say, it is installed so that it faces inward toward the heated rooms and spaces.

Note

> A vapor retarder is more accurately referred to as a vapor diffusion retarder (VDR) because the moisture it cannot stop is diffused through the material.

In addition to protecting insulation from absorbing moisture, a vapor retarder is an effective means of retaining moisture in the rooms and spaces in the structure or preventing it from entering these areas. As a result, it functions as an important factor in maintaining interior humidity conditions.

If you add insulation over an existing vapor retarder, you should make a number of cuts in the surface of the vapor retarder to avoid moisture entrapment. Remember, the "name of the game" is *dry insulation.* Take whatever measure necessary to protect it from moisture.

Air Barriers and Air/Vapor Retarders

Air barriers are used to block random air movement through structural cavities while at the same time allowing the moisture in the air to diffuse back out. Some common materials used as air barriers are fibrous spun polyolefin (commonly called *house wrap*), gypsum drywall board, plywood, and foam board. Air barriers are installed to the outside of the structure, with all joints sealed with tape.

Caution

> House wrap materials react poorly with the wood lignin in certain types of exterior wood sidings, such as cedar, redwood, and manufactured hardboard siding. Substitute 30-lb impregnated paper for the house wrap in these applications. It will not react to the wood lignin.

Air/vapor retarders are materials designed to control simultaneously both air movement and water vapor transmission. Air/vapor retarders are very common in regions with hot, humid weather, such as that common in the southern United States. An air/vapor retarder can be a single material, but it is more commonly a combination of polyurethane plastic sheets, builder's aluminum foamboard insulation, and other exterior sheathings. Like the house wrap, it is applied to the exterior of the structure, with all joints carefully sealed.

Insulating Materials

Insulating materials are specifically designed to reduce the rate of heat transmission through ordinary construction materials to an acceptable level. A dry material of low density is considered a good insulator; however, in addition to this characteristic, it must also have a conductivity value of less than 0.5.

As described earlier, the conductivity value of a material is a purely arbitrary one determined by the amount of heat that flows in 1 hour through a 1-inch thickness of a material 1 square foot in area with the temperature exactly 1 degree Fahrenheit higher on one side of the material than on the other.

Air spaces, or air spaces bounded by either ordinary building materials or aluminum foil, also provide some insulation, but not to the degree formerly thought possible. Dead air spaces in building walls were once considered capable of preventing heat transmission in a manner similar to the space between the walls of a Thermos bottle. Later research proved this to be a somewhat false analogy, because the air in such spaces often circulates and transmits heat by convection.

Air circulation can be checked by filling the hollow space with an insulating material that contains a great number of *small, confined* air spaces per unit volume. This stoppage of air circulation is what produces the insulating effect, and not merely the existence of the air space. Under these circumstances, it is obvious that the most practical method of insulation is to fill the area in the walls with a material containing these minute air spaces.

Several manufacturers produce insulating materials in a variety of shapes and forms for installation in houses and other buildings. Frequently, instructions for the installation of the products will also be provided by the manufacturer. Local building supply outlets and lumber yards will often be very helpful, too, and will usually recommend the best way to install the insulation material. Some of these materials and their applications are described in the following paragraphs.

Rigid Insulation Board

Rigid insulation board (Figure 3-5) is a rigid or semirigid synthetic product commonly referred to as structural insulating boards (SIP)

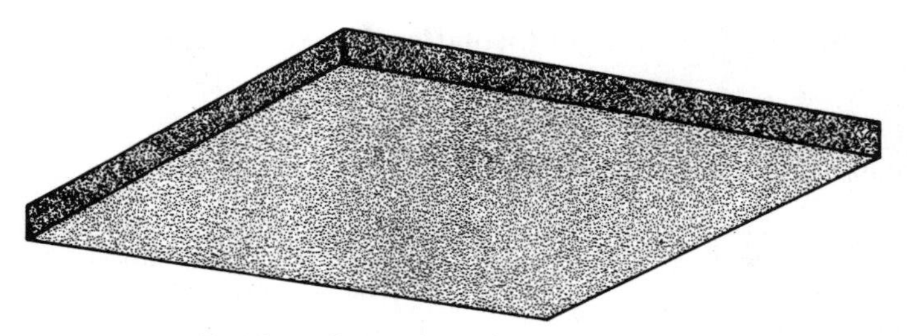

Figure 3-5 Rigid insulation board.

or panels, rigid boards, foam insulation boards, rigid foam insulations, slab insulation, or sheathing boards. They are produced under a wide variety of trade names. Rigid insulation boards or panels are commonly manufactured in 2′ × 4′, 4′ × 4′, and 4′ × 8′ sheet sizes with thicknesses of ½″, ¾″, and 1″ nonlaminated and laminates of ½″, 1″, and 2″. All edges are butted. Rigid insulation boards are available faced or unfaced. Reflective facings function as vapor retarders.

Caution

> Rigid insulation board is flammable and must be covered with a fire-resistant material, such as gypsum wall board, when it is installed on interior walls.

Some rigid insulation board is produced specifically for use on interior walls. This type is commonly made from wood or cane fibers with one side prefinished. Rigid insulation board designed for use as structural insulation is produced from compressed fibrous materials other than wood and does not have a prefinished surface. Rigid board panels made from polyurethane, polystyrene, polyisocyanurate, and phenolic are commonly used for exterior wall sheathing in new construction and to insulate basement and crawl space walls. Some *R*-values for common type of rigid insulation board are listed in Table 3-3.

Note

> Rigid board insulation has a high *R*-value and can be used in areas where space is limited, but it is also has a higher cost per inch than other insulating materials.

Caution

> Rigid insulation board must be installed with tightly butted edges. Fill gaps of 1⁄16 inch or greater with pieces of insulation

Table 3-3 Some Rigid Insulation Board *R*-Values

Types of Rigid Insulation Board	*R-Value per Inch of Board Thickness*	*Common Uses*
Molded or expanded polystyrene board (sometimes referred to as bead board)	*R*-3.6 to *R*-8	Exterior sheathing, interior basement walls, siding backer board, or suspended ceiling panels
Extruded polystyrene board (sometimes referred to as blue or pink board)	*R*-4.5 to *R*-5	Exterior foundation walls or surfaces, exterior wall sheathing, or interior applications (above grade)
Unfaced polyurethane and polyisocyanurate board	*R*-5.6 to *R*-6.3	Impregnated with asphalt for hot-mopped flat roof applications; vinyl-faced for beam ceilings
Faced urethane and polyisocyanurate board	*R*-7.1	Foil-faced exterior wall sheathing
Semi-rigid fiberglass panels	*R*-4 to *R*-5	Below-grade basement or crawl space exterior wall surfaces

board cut to fit. Do not fill a gap between boards with other noninsulatng materials or any type of adhesive.

Reflective Insulation

Reflective insulation (Figure 3-6) is designed to reflect heat rather than resist its rate of flow. This characteristic distinguishes it from other insulating materials. Reflective insulation is available in the form of foil-faced paper, foil-faced cardboard, foil-faced polyethylene bubbles, and foil-faced plastic film. Aluminum foil or a special

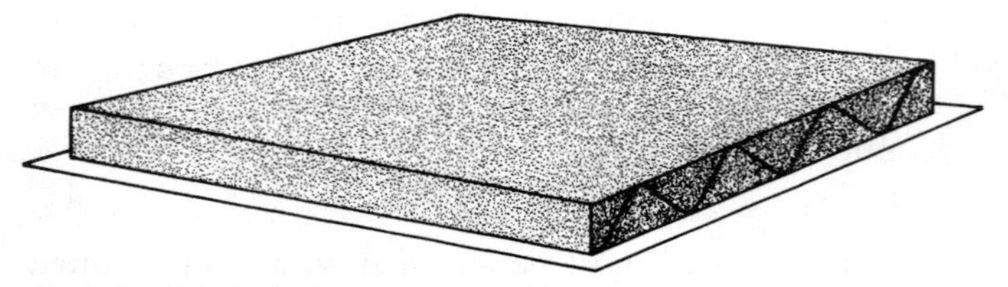

Figure 3-6 Reflective insulation board.

coating is commonly used as the facing (reflective surface). It is commonly used to insulate unfinished ceilings, walls, and floors in existing structures.

Air spaces must be provided on either side of a layer of reflective insulation. This type of insulating material has proven to be a good moisture barrier and exhibits a high degree of resistance to heat transmission.

Note

> Reflective insulation is designed to reflect heat and not interfere with its flow. For that reason, its effectiveness is highest during the cooling season.

Blanket or Batt Insulation

Blanket or batt insulation is commonly made from rock wool or fiberglass. Standard rock wool blankets or batts have an *R*-value of 3.2 per inch thickness; fiberglass blankets or batts, an *R*-value of 3.1. Some manufacturers now produce medium- and high-density blankets and batts with higher *R*-values, ranging from *R*-11 to as high as *R*-38, depending on the thickness.

Note

> The term *mineral wool,* which is often used in reference to blankets and batts, refers to the following three basically similar types of insulation: (1) *fiberglass* or *glass wool,* produced from recycled glass; (2) *rock wool,* produced from basalt; and (3) *slag wool,* produced from steel-mill slag.

Blanket insulation (Figure 3-7) is produced in the form of a continuous flexible strip and is generally available in rolls or packs ranging in thickness from 1 to 3 in. Widths are commonly 15 in, but wider blankets or batts with widths up to 33 in can be obtained on special order. Blanket insulation lengths range from 36 to 48 ft. Batts (Figure 3-8) are essentially smaller examples of blanket insulation, ranging in thicknesses from 1 to 6 in and available in 4- and 8-ft lengths.

Blankets and batts commonly are manufactured with a vapor seal paper attached to one face, covering the entire surface and extending 1½ in outside the blanket or batt. These 1½-in laps, neatly folded against the membrane backing in manufacture, are turned out and tacked or stapled to the studs, rafters, or joists in application. The paper backing helps to prevent the passage of vapor and resists the penetration of moisture from excess water in fresh plaster or other sources.

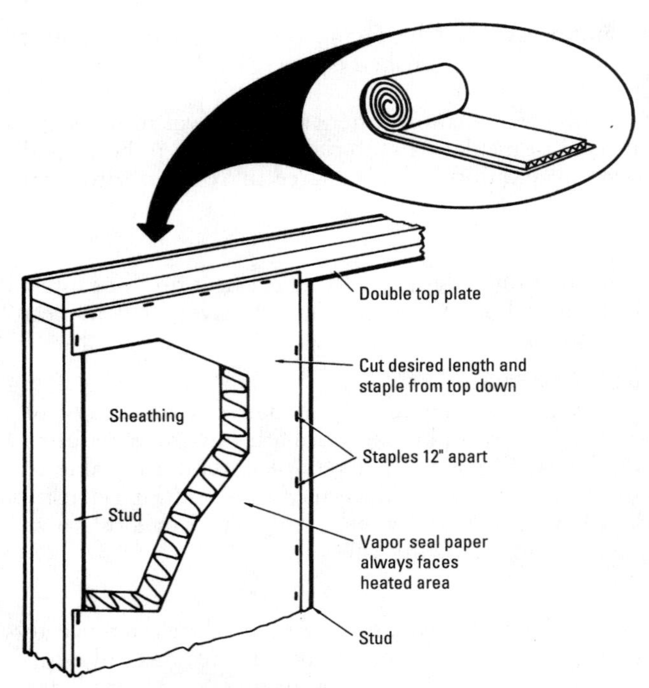

Figure 3-7 Blanket insulation.

Loose-Fill Insulation

Loose-fill insulation (Figure 3-9) is available in granular, fibrous, or powdered form and is sold in bulk, usually in bags (Table 3-4). It is commonly produced from vermiculite, perlite, glass fiber, cellulose, or rock wool. Because loose-fill insulation is so easy to install, it is often used to insulate the cavities between joists in unfinished attics.

Note

A major disadvantage of loose-fill insulation is that it eventually tends to settle and lose some of its effectiveness as an insulating material.

Note

Vermiculite and perlite loose-fill insulation are seldom used today as an insulating material in residential construction.

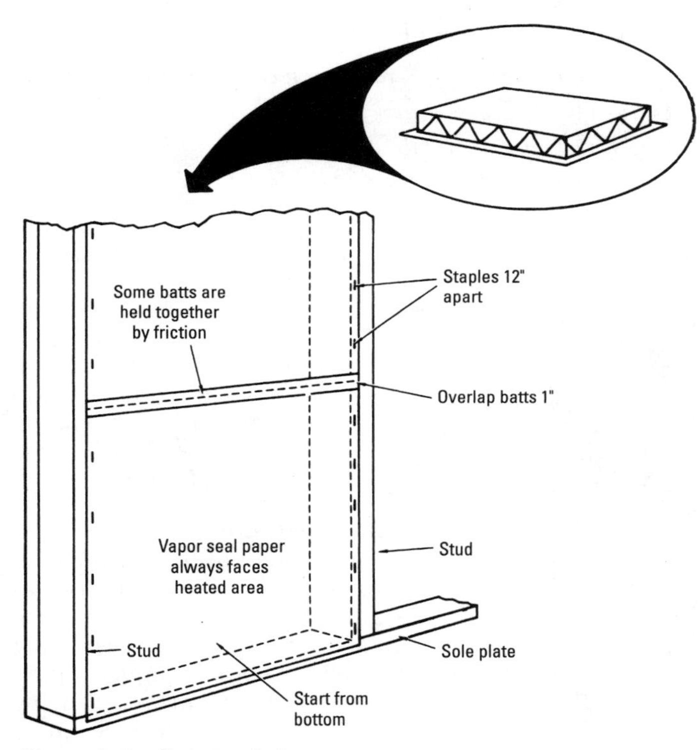

Figure 3-8 Batt insulation.

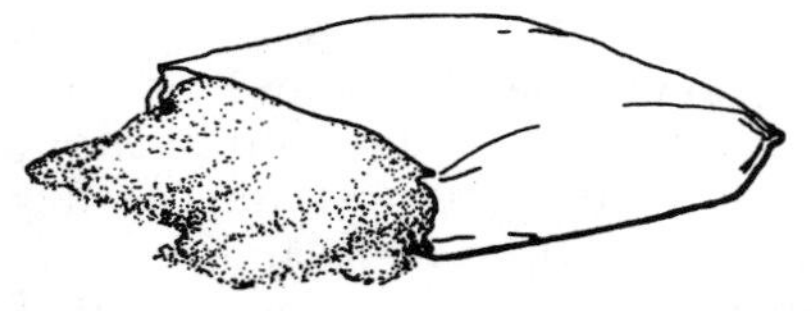

Figure 3-9 Loose-fill insulation.

Blown-In Insulation

Blown-in insulation (Figure 3-10) is a loose-fill insulating material, typically fiberglass, rock wool, or cellulose (see *Loose-Fill Insulation*). The insulation material is blown under pneumatic pressure into the spaces between joists and wall studs in an existing structure. Water vapor retarders are applied first and are available in the form of a moisture-resistant, spray-on paint. Blown-in insulation is most commonly used to insulate interior walls, especially in existing structures. Blown-in insulation also can be used in new construction, but the wall cavities first must be covered with temporary containment sheeting to hold insulation in place. The insulation is then blown into the wall cavities, and the studs are covered with the appropriate finishing material.

Table 3-4 Loose-Fill Insulation R-Values

Type of Loose-Fill Insulation	*R-Value per Inch Thickness of Insulation*
Fiberglass loose fill insulation	*R*-2.3 to *R*-2.7
Rock wool loose-fill insulation	*R*-2. to *R*-3.0
Cellulose loose-fill insulation	*R*-3.4 to *R*-3.7
Vermiculite	*R*-2.4
Perlite	*R*-2.8

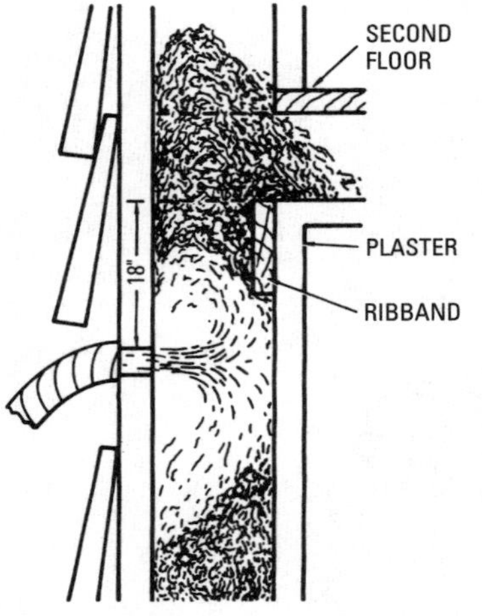

Figure 3-10 Blown-in insulation.

Note

> Because special equipment must be used for its application, blown-in insulation should be applied by individuals trained in its use.

Foam Insulation

Foam insulation is available in spray-in-place systems as well as rigid boards. Spray-in-place foams have very high insulating *R*-values. For example, spray polyurethane foam has an *R*-value range of between *R*-5.6 to *R*-6.3 per inch of thickness. The spray equipment requires skilled workers to operate. Common spray-in-place foams include polyurethane foam, urea-formaldehyde (UF) foam, and phenolic foam. These foam insulations expand and set almost immediately after their application. They will eventually shrink and settle in the wall cavities, but their shrinkage should be not more than 5 percent if they are properly applied.

Caution

> Foam insulation is flammable and requires a fireproof covering material if it is installed on the interior of a structure.

Building Construction and Location

A new building (particularly a residence) should be located so that the large windows in the main rooms face south to receive the

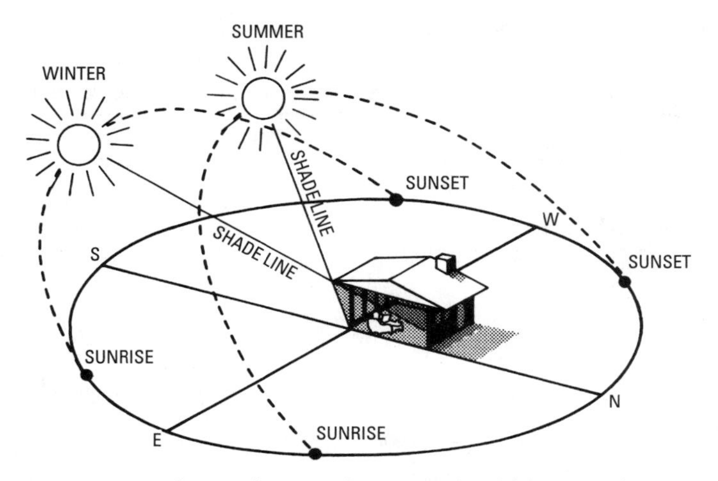

Figure 3-11 Large glass window area should face south to take advantage of the winter sun. *(Courtesy U.S. Department of Agriculture.)*

maximum sunlight during the winter months (Figure 3-11). If possible, the building should be built in a location that offers some natural protection from the prevailing winter winds. Tight, well-insulated construction should be incorporated in the design of the building from the beginning. Although the initial costs will be somewhat higher, they will be effectively offset by the reduction in heating and cooling costs.

If a new heating or cooling system is planned for an *older* structure, the existing insulation should be checked and, if necessary, repaired or replaced before the new system is installed.

Recommended Insulation Practices

The recommendations found in the following paragraphs will contain references to *U*-values and *R*-values. As was already stated in this chapter, the *U*-value is the overall coefficient of heat transfer and is expressed in Btu per hour per square foot of surface per degree Fahrenheit difference between air on the inside and air on the outside of a structural section. The *R*-value is the term used to express thermal resistance of the insulation. The *U*-value is the reciprocal of the sum of the thermal resistance values (*R*-values) of each element of the structural section. Table 3-5 lists the insulation recommendations for structures both in terms of *R*-values and overall coefficients of heat transfer in terms of *U*-values.

Table 3-5 Insulation Recommendations (*R*-Values) and Overall Coefficients of Heat Transfer (*U*-Values) for Major Areas of Heat Loss and Heat Gain in Residential Structures

Type of Construction	*Opaque Sections Adjacent to Unheated Spaces*			*Opaque Sections Adjacent to Separately Heated Dwelling Units*		
	Walls	*Floors*	*Ceilings*	*Walls*	*Floors*	*Ceilings*
Frame	*R*-11 *U*:0.07	*R*-11 *U*:0.07	*R*-19 *U*:0.05	*R*-11 *U*:0.07	*R*-11 *U*:0.07	*R*-11 *U*:0.07
Masonry	*R*-7 *U*:0.11	*R*-7 *U*:0.11	*R*-11 *U*:0.07	*R*-7 *U*:0.11	*R*-7 *U*:0.11	*R*-11 *U*:0.07
Metal section	*R*-11 *U*:0.07	Not applicable	Not applicable	*R*-11 *U*:0.07	Not applicable	Not applicable
Sandwich	*R*-[a] *U*:0.07	*R*-[a] *U*:0.07	*R*-[a] *U*:0.05	*R*-[a] *U*:0.07	*R*-[a] *U*:07	*R*-[a] *U*:07
Heated basement or unvented crawl space	*R*-7 *U*:0.11	Insulation not required	Insulation not required			
Unheated basement or vented crawl space	Insulation not required	See "Floors"	Insulation not required			

NOTE: Insulation R-values refer to the resistance of the insulation only.

Courtesy Electric Energy Association.

[a]*Since the thermal resistance of sandwich construction depends upon its composition and thickness, the amount of additional insulation required to obtain the maximum U-factor must be calculated in each case.*

Frame Walls

Frame walls (Figure 3-12), with either wood or metal studs, are commonly insulated with rolls and batts of mineral, glass, or wood-fiber insulation material. Sprayed-on or foamed-in-place rigid insulation is also being used more and more. Whatever the insulation decided upon, exterior opaque walls and walls between separately heated units should have a maximum *U*-factor of 0.007, which will require insulation rated at *R*-11 or greater (see Table 3-5). A thermal resistance of *R*-11 can be provided by 3½-in fiber batts or other equally suitable materials.

Masonry

A masonry wall is a particularly poor thermal barrier. A typical masonry cavity wall (4-in face brick on the exterior, an air space,

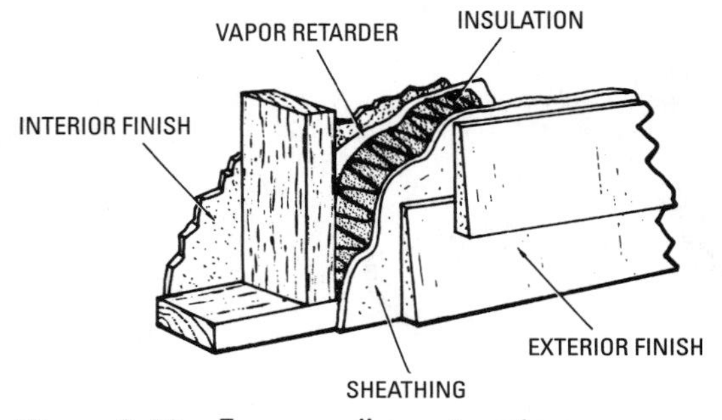

Figure 3-12 Frame wall construction.

4-in concrete block, an air space, and a layer of gypsum wall board) has a thermal resistance equal to approximately 1 in of rigid insulation. Exterior opaque masonry walls should have maximum *U*-value of 0.11, requiring insulation rated at *R*-7 or greater. This can be accomplished in the following ways:

1. Place a layer of preformed rigid insulation between the exterior face brick (or other finishing material) and the concrete block during construction (Figure 3-13). Another method is to leave the space empty and then fill it with a foam insulation before capping the wall.

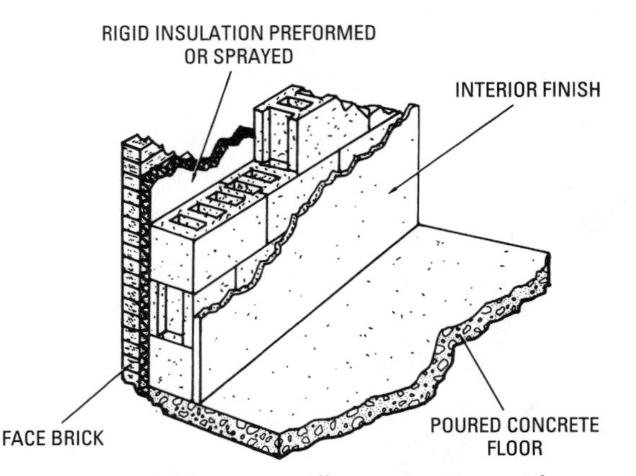

Figure 3-13 Masonry wall construction with insulation installed on the outside portion.

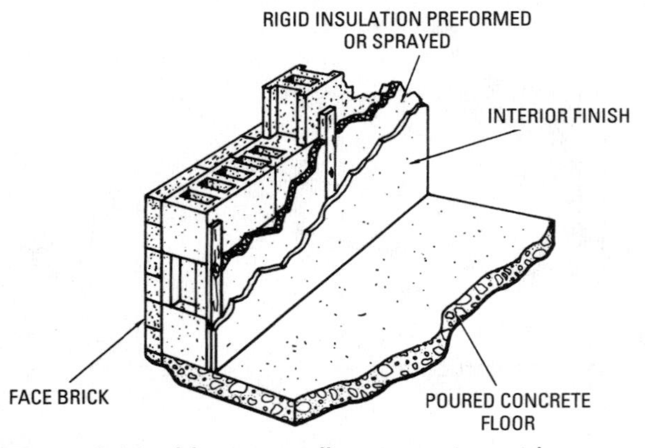

Figure 3-14 Masonry wall construction with insulation installed on the inside portion.

2. If no air space exists between the exterior face brick or other exterior finish and the concrete block, it will be necessary to add a layer of suitable insulating material (rated *R*-7 or better) on the inside surface of the block (Figure 3-14). The insulating material can be rigid insulation attached to furring strips or preformed rigid insulation secured in place with adhesives. An interior wall finish is then applied over the insulating material.

If the wall is constructed entirely of concrete, apply preformed rigid insulation to either side by adhesive bonding and cover the layer of insulation with a wall-finish material (Figure 3-15). The insulation must be rated *R*-7 or better.

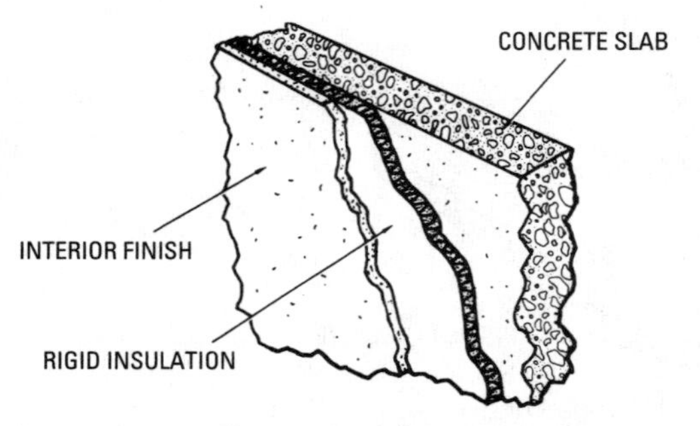

Figure 3-15 Concrete slab construction.

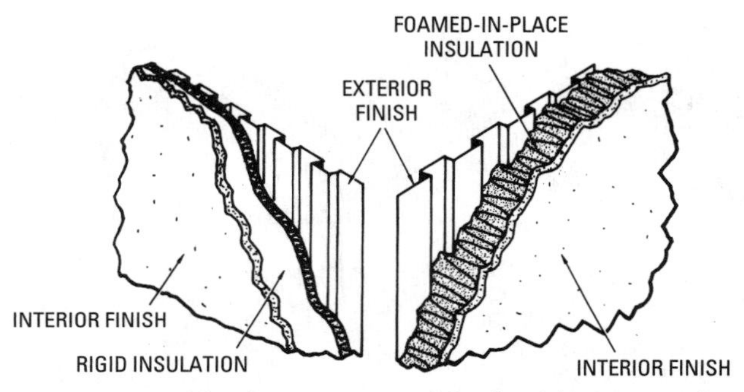

Figure 3-16 Metal sections used for load-bearing walls and certain walls in residential construction.

Metal

Exterior opaque metal walls (Figure 3-16) should have a maximum *U*-value of 0.07. This will require insulation having a thermal resistance rated at *R*-11 or better, which can be provided by a layer of rigid insulation either preformed or foamed in place.

Sandwich Construction

Sandwich or layered construction will vary in composition and thickness but will typically consist of three layers: two facings and a core. Exterior opaque walls of a sandwich construction should have a maximum *U*-value of 0.007. The thermal resistance (*R*-value) must be calculated on the basis of its composition and thickness.

Basement Walls

There is no need to insulate the exterior walls of unheated basements, because there is no need for reducing heat loss. The exterior walls of *heated* basements (or interior walls separating heated from unheated areas) are a different matter, because heat loss is a concern here. An insulating layer having a thermal resistance rated at *R*-7 or greater should be applied. Figure 3-17 illustrates two methods for insulating these walls.

Crawl Space Exterior Walls

The exterior walls of *unvented* crawl spaces should be insulated with a material providing a thermal resistance of at least *R*-7 or better. A method for doing this is illustrated in Figure 3-18. No insulation is required for the exterior walls of *vented* crawl spaces.

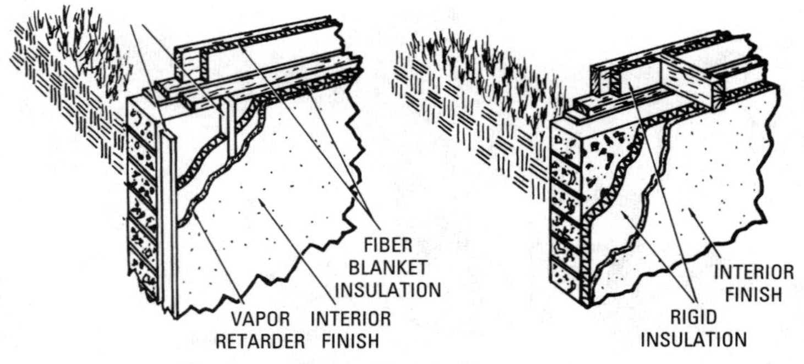

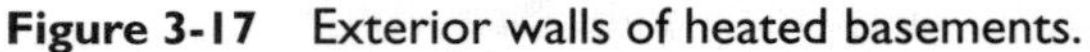

Figure 3-17 Exterior walls of heated basements.

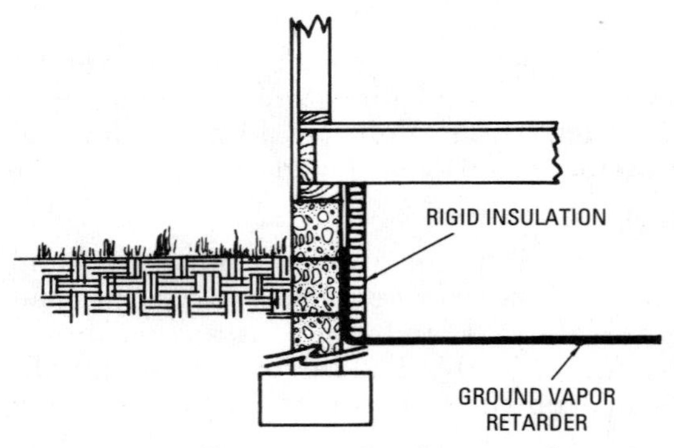

Figure 3-18 Exterior walls of unvented crawl space.

Walls Between Separately Heated Dwelling Units

Where walls separating adjacent but independently heated dwelling units occur, use insulation with a thermal resistance of at least *R*-11. The insulation may be placed in either wall or divided proportionally between them (Figure 3-19).

Wood or Metal Joist Frame Floors

Wood or metal joist floors over unheated basements or vented crawl spaces should have a maximum *U*-value of 0.07 (requiring insulation rated at *R*-11 or greater). Three methods for providing insulation are illustrated in Figures 3-20 through 3-22.

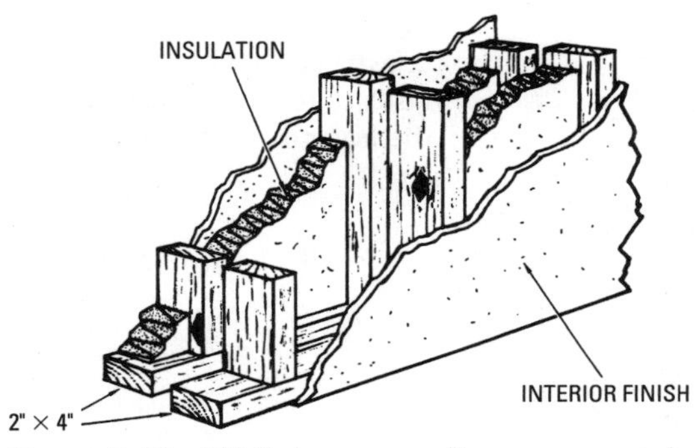

Figure 3-19 Walls between adjacent separately heated dwelling units.

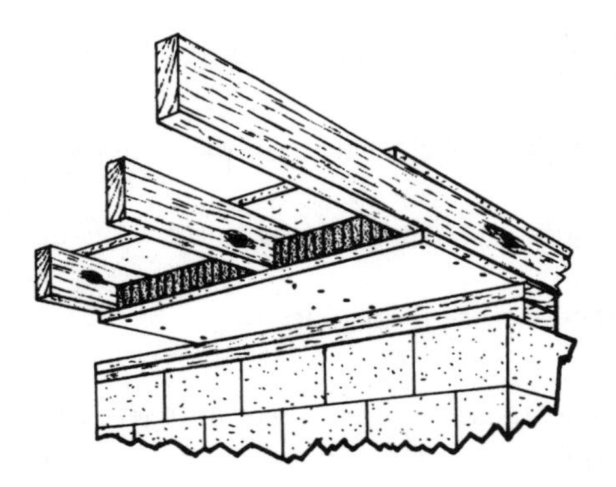

Figure 3-20 Frame floor construction with insulation between the joists.

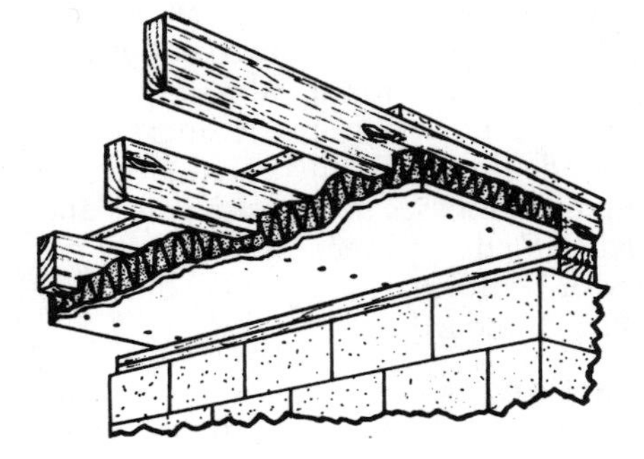

Figure 3-21 Frame floor construction with insulation covering the joists.

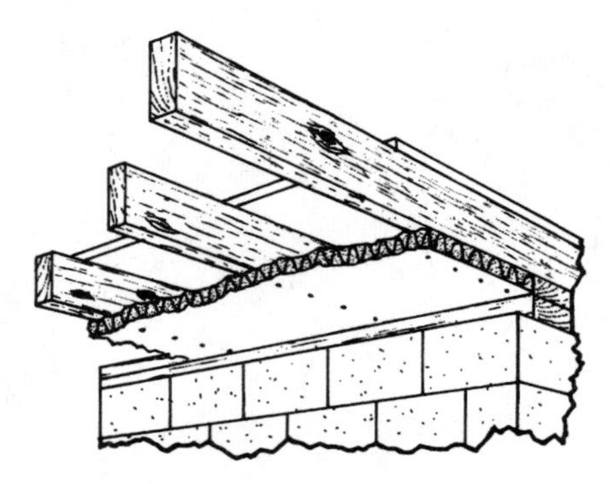

Figure 3-22 Frame floor construction with rigid insulation panels attached to the bottom of the joists.

Concrete Floors

Concrete floors over unheated spaces should have a maximum *U*-value of 0.11, requiring insulation rated at *R*-7 or greater. The following three methods are recommended for applying a layer of insulation:

1. Spray rigid insulation in a suitable thickness on the *underside* of the floor.
2. Bond preformed rigid insulation to the top or bottom of the surface.
3. Form concrete around a core of rigid insulation.

Slab-On-Grade Floors

Structures constructed on concrete slabs are in direct contact with the ground. Unless they are properly insulated, the slab can become extremely cold and damp. Some form of insulation should be placed between the slab and the ground before the concrete is poured. Insulation can be added to an existing structure by excavating around the edges of the foundation to a level below the frost line and adding 2 in of waterproof insulating material to the sides of the slab.

Insulation rated *R*-5 is required for all slab-on-grade floors. Two methods for insulating slab-on-grade floors are illustrated in Figure 3-23. Table 3-6 lists the maximum heat losses in terms of watts and Btu/h per linear foot of exposed slab edge, as recommended by the Electric Energy Association. More information can be obtained from the FHA Minimum Property Standards.

Floors of Sandwich Construction

Floors of sandwich construction should be insulated in the same manner as sandwich walls.

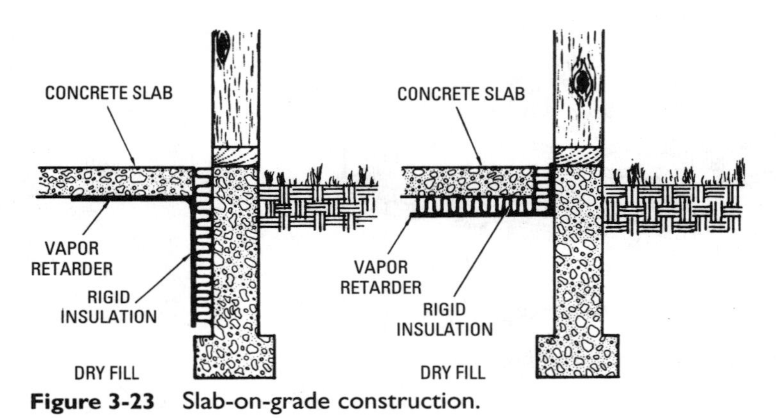

Figure 3-23 Slab-on-grade construction.

Frame Ceilings and Roofs

Any ceiling below an unheated space should be insulated with materials having a rating of R-19 or better. The ceilings should have

Table 3-6 Slab Edge Heat Loss

Outdoor Design Temperature, °F	*Total Width of Insulation, in Rated R-5*	*Heat Loss per Foot of Exposed Slab Edge*	
		Watts	*Btu/h*
−30 and colder	24	10.0	34
−25 to −29	24	9.4	32
−20 to −24	24	8.8	30
−15 to −19	24	8.2	28
−19 to −14	24	7.9	27
−15 to −9	24	7.3	25
0 to −4	24	7.0	24
+5 to +1	24	6.4	22
+10 to +6	18	6.2	21
+15 to +11	12	6.2	21
+20 to +16	Edge only	6.2	21
+21 and warmer	None	—	—

Courtesy Electric Energy Association

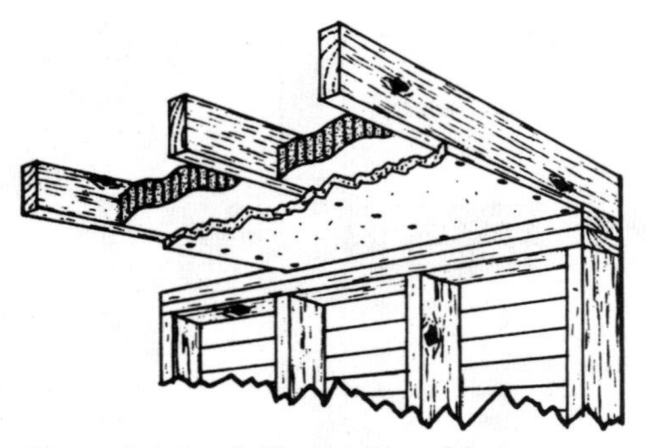

Figure 3-24 Ceiling and roof frame construction.

a maximum *U*-value of 0.05. Some methods for providing such insulation are the following:

1. Staple blankets or batts to the joists before the finished ceiling is applied (Figure 3-24).
2. Blow or manually apply loose insulation between ceiling joists from above.
3. Spray on insulation between the joists from above.

Concrete Ceilings

Concrete ceilings below unheated spaces should have a maximum *U*-value of 0.07, requiring insulation rated *R*-11 or greater. The following methods are suggested for insulating concrete ceilings:

1. Attach preformed rigid insulation to either surface with adhesives.
2. Apply sprayed-on rigid insulation to either surface.
3. Make a sandwich construction with preformed rigid insulation formed around a concrete core.
4. Apply rigid insulation to the *top surface* where the ceiling also serves as the roof of the structure.

Sandwich Ceilings

Ceilings of sandwich construction below unheated spaces should have a maximum *U*-value of 0.05, requiring insulation rated at *R*-19 or better. Sandwich ceilings that also serve as the roof of a structure meet the same requirements.

Windows and Doors

Windows and doors can account for at least 30 percent of the total heat loss (or gain) in a structure. Much of this results from air infiltration through cracks around the windows and doors. This problem can be corrected by installing weather stripping, applying caulking, or installing double-glazed windows, solid wood doors, or doors with built-in insulation.

An exterior door constructed with a core of rigid insulation and faced with metal or acrylic film will have a very low *U*-value (as low as 0.074). Infiltration heat loss around the edges of the door can be effectively reduced with suitable weather stripping.

A typical 1-in solid wood door will have a *U*-value of approximately 0.64 (compare this to the *U*-value for the door described in the preceding paragraph). Doubling the thickness of the door will reduce the *U*-value to about 0.43. The addition of metal of glass storm doors will further reduce the *U*-value (0.39 and 0.29, respectively).

Windows with double glazing will have a *U*-value of 0.58, as compared to 1.13 for single glazing. Double glazing will also reduce noise transmission.

Insulating Attics, Attic Crawl Spaces, and Flat Roofs

Most houses and many other structures are constructed with an attic or attic crawl space between the roof and the ceiling of the top floor. Unless this space is properly insulated and ventilated, it will produce the following two problems:

1. Condensed moisture formation
2. Excessive heat loss or gain

During the winter months, the heat from the occupied lower spaces moves upward (because warm air is lighter) into the cooler attic or attic crawl space. As it moves, the heat follows a number of different pathways, including:

1. Through poorly insulated ceilings
2. Around cracks in attic stairway doors or attic pulldown stairways
3. Through ceiling fixture holes
4. Along plumbing vents and pipes
5. Along air ducts
6. Through air spaces within interior partitions

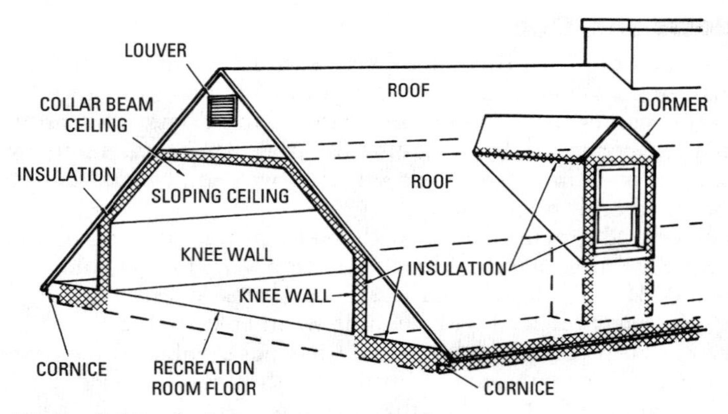

Figure 3-25 Attic insulation methods.

Each of these pathways for heat transmission can be blocked by insulation. Typical attic insulation methods are illustrated in Figure 3-25. Attic door cracks can be blocked with weather stripping. Pipe, duct, vent, and fixture holes can be sealed off by stuffing them with loose insulation torn from batts or blankets.

The ceiling of the top floor should be covered with a minimum of 6 in of suitable insulation material to be properly effective in reducing the heat loss from the occupied spaces during the winter months or the heat gain during the summer. This can be installed during the initial construction stages of a new house or building, or it can be added to the attic floor surface of an existing structure. Sealing or filling the air spaces within interior partitions is relatively easy to do in new construction. An existing structure presents a more difficult problem that generally requires considerable experience in installing insulation materials.

Sometimes a half-wall (a knee wall) is installed between the attic floor and the roof. This is most commonly insulated with blanket insulation, as illustrated in Figure 3-26.

The condensation of moisture in the attic or attic crawl space is caused by the moisture in the warm air rising from the lower occupied spaces. This moisture comes in contact with the colder air of the attic, condenses, and causes the insulation and other building materials to become damp. In time, this dampness will have a damaging effect. Ventilation is an effective method of reducing or eliminating condensation in attics and attic crawl spaces. The air enters and leaves through vents placed in the gables and cornices. The size of the vents will be determined by the area being ventilated.

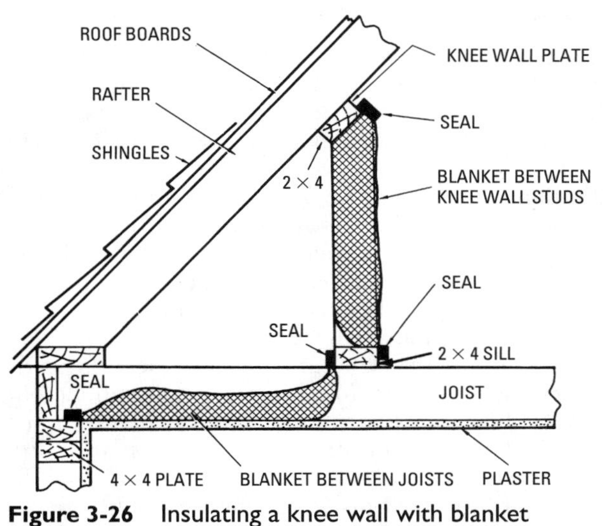

Figure 3-26 Insulating a knee wall with blanket insulation.

A flat roof can be effectively insulated by installing 2-in-thick rigid insulation board before the roofing material is applied. Roofing felt should be placed between the bare wood boards of the roof deck and the insulation. Additional insulation is placed *under* the deck boards (i.e., between the deck boards and the ceiling of the room below). (See Figure 3-27.)

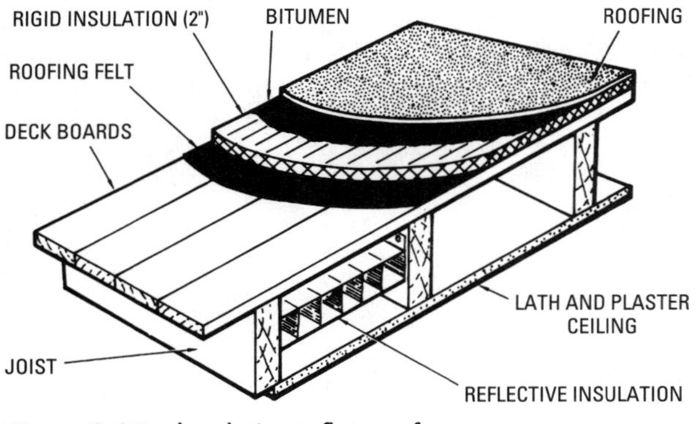

Figure 3-27 Insulating a flat roof.

Ventilating Structures

A properly ventilated structure increases the comfort level for its occupants and eliminates indoor pollutants by exchanging stale indoor air with fresh air from the outdoors. This exchange of air (air circulation) can be accomplished by either natural or mechanical ventilation.

Natural ventilation is the *unaided* movement of air into and out of a house or building. Unaided air movement from the outdoors into the interior of the structure through cracks in the walls and around windows or doors is called *infiltration*. The unaided movement of air from the interior to the outdoors is called *exfiltration*. The temperature differences between the indoor and outdoor air, the wind, the orientation of the structure, and the location of the windows and doors, must all be factored in when determining natural ventilation rates. Because so many variables are involved, it is very difficult to calculate an appropriate natural ventilation rate accurately for a structure.

Before the advent of well-insulated houses and central heating/cooling systems, natural ventilation was an acceptable means of exchanging outdoor and indoor air. Structures were naturally drafty because of the minimal use of insulation. The air could move relatively freely through the wall cavities between the outdoors and the interior of the structure. As fuel prices increased over the years, however, homeowners also increased the levels of insulation in order to reduce energy costs. The higher levels of insulation, tighter construction, and the growing reliance on central space conditioning required the use of centralized mechanical ventilation systems to circulate the air between the outdoors and the indoors. If a central heating/cooling system is properly sized, it can provide the optimal number of air change rates required to maintain fresh, healthy interior air. Ventilation principles and detailed descriptions of centralized mechanical ventilation systems are covered in Volume 3.

Chapter 4

Sizing Residential Heating and Air Conditioning Systems

The size of the heating system is directly related to the amount of heat lost from the house or building. *All* structures lose heat to the outdoors or to adjacent unheated or partially heated spaces when the temperatures of the outdoor air or adjacent spaces are colder than those inside the structure. The heat within the building is normally lost by transmission through the building materials and by infiltration around doors and windows.

The loss of heat from a structure must be replaced at the same rate that it is lost. Consequently, determining the correct size of the heat system and the rated capacity of the heating plant required by the steam are very important. Unfortunately, many heating and/or cooling systems are either undersized or oversized, with the latter being the most common mistake. *Undersizing* means that the heating and/or cooling equipment does not have the capacity (output) to meet the heating and cooling requirements of the structure. *Oversizing* means that the heating and/or cooling equipment has more capacity than required. Both undersizing and oversizing are caused by using *guestimates* or rule-of-thumb sizing calculation methods.

The Results of Oversizing and Undersizing

Oversized heating equipment is more expensive to install, operates inefficiently, uses more energy resulting in higher fuel bills, creates uncomfortable indoor temperatures by providing more heat than the structure requires, and produces wide temperature swings. Oversized heating equipment also breaks down more often. Oversized units require larger air flow, resulting in noisier operation.

Undersized heating equipment lacks the capacity to provide sufficient heat, especially during extreme cold spells.

Oversized air conditioners and heat pumps create higher than normal humidity levels indoors because they do not run often enough to dehumidfy the air. This is known as *short-cycling*. The dampness in the air can also result in unhealthy mold growth indoors.

This chapter describes several methods for calculating heat loss, ranging from rule-of-thumb methods to the more precise method of using overall coefficients of heat transmission (*U*-values) computed for the various construction materials and combinations of construction materials through which heat is commonly transmitted.

Rule-of-Thumb Methods

There is nothing wrong with using a rule-of-thumb method when making an initial rough estimate of the heating/cooling load requirements for a structure. But it should be used as nothing more than a benchmark from which the true and precise calculations are made. It provides the equipment installer/contractor and the homeowner with some idea of how much the heating and/or cooling system will cost. But correct sizing will involve the consideration of many different factors, including but not limited to the following:

- Size, shape, and orientation of the structure
- Local climate
- Type, location, and number of windows
- Type and amount of insulation
- Number and ages of occupants
- Structure design
- Construction materials
- Planned use of the heating/cooling system
- Condition of the distribution system (ducts or pipes)
- Air infiltration rates

There are three common rule-of-thumb methods used by contractors/installers to size heating/cooling equipment.

Upgrade Method. A common rule-of-thumb method is to install a furnace or boiler the same size or larger than the original one in an existing house or building. The problem with this approach is that the original equipment may have been incorrectly sized. Furthermore, many changes have probably been made to an existing structure over the years, and these changes will have changed the load requirements for new equipment. The house has most likely had its insulation levels increased, because adding insulation in the attic, caulking around windows and doors, or installing double-glazed windows are relatively inexpensive upgrades. If the HVAC contractor looks at the metal furnace or boiler tag (nameplate) specifying its output, Btus per hour, and so forth, and advises you to purchase and install one of the same or higher output, get another estimate. Unfortunately, this is an all too common method used by many to size equipment. Correct sizing involves the consideration of many different factors.

Sizing by Square Footage Method. Another rule-of-thumb method is to size by the square foot area of the house or building. This is called the "sizing by square footage" method. It is one of the most commonly used of the inaccurate sizing methods. It involves taking the square footage area of the structure and multiplying by a specific value. For example, a typical value assigned to air conditioning equipment is 1 ton (12,000 Btu/h) per 500 square feet of space (46 m2). It fails to take into consideration any of the variables listed above.

Chart Method. The so-called chart method of sizing heating/cooling equipment involves filling in the blanks on a prepared chart. The chart lists the following:

- Floor area of each of the heated rooms and spaces
- Insulation levels in the floors, walls, and ceilings
- House category (closest description of the type of house)

When all the required information is entered on the prepared form, multiply both the upper and lower values for heat loss in Btu per hour per square foot (from the data table used in conjunction with the chart) by the floor area of the house to estimate the required heating range. This estimating method does not take into consideration house location, design, or many of the other factors listed above.

Tip

Always insist on a correct sizing calculation before signing a contract with an HVAC contractor or installer. See the section *Manual J and Related Materials Used for Sizing Heating/Cooling Systems* later in this chapter.

Sizing Systems Using Coefficients of Heat Transmission

Calculating heat loss from a structure requires a thorough knowledge of the thermal properties of many materials and combinations of materials used in its construction. The term *thermal property* is used here to mean the overall coefficient of heat transmission (i.e., the rate of heat flow through a material). Each construction material (or combination of materials) will have its own coefficient of heat transmission. Before continuing any further, it would be advisable to review the sections of Chapter 3, "Insulating and Ventilating Structures," that specifically pertain to the problem of heat loss, particularly *Principles of Heat Transmission*.

ASHRAE and other authorities suggest the following basic steps (in the sequence given) for calculating heat loss:

1. Decide upon the desired inside air temperature for the structure.
2. Obtain the winter outside design temperature for the location of the structure from published lists or a local weather bureau.
3. Determine the design temperature difference (the difference between the temperatures found in Steps 1 and 2).
4. Identify on the heat loss worksheet each room or space in the structure.
5. List every structural section in each identified room or space that has an outer surface exposed to the outdoors or to an unheated or partially heated space.
6. Determine the coefficient of heat transmission (*U*-value) for each structural section (e.g., walls, glass, ceiling).
7. Calculate the infiltration heat loss for each identified room or space.
8. Calculate the *total* area for each exposed surface (i.e., total outside wall area, total floor area).
9. Calculate the area for each door and window in the walls and add these figures together to obtain the total area for these openings.
10. Subtract the total area for doors and windows from the *gross* outside wall area (obtained in Step 8) to determine the total *net* outside wall area. Enter the amount on the heat loss worksheet.
11. Multiply the total net wall area determined in Step 10 by the *U*-value (Step 2) by the design temperature difference (Step 3) to obtain the total heat loss for walls (expressed in Btu/h).
12. Make the same calculations for the other surface areas (floors, ceilings, etc.) determined in Step 8.
13. Add the heat loss figures calculated for each surface category and the infiltration heat loss to obtain the total heat loss for the identified room or space.

14. Repeat the procedure outlined in Steps 1 to 11 for each identified room or space in the structure.
15. Add the various totals to obtain the total heat loss from the structure (expressed in Btu/h).

These 15 basic steps for calculating heat loss are provided as a useful means of reference for the more detailed description of the procedure contained in the paragraphs that follow.

Outside Design Temperature

In heating calculations, the *outside design temperature* is the coldest outside temperature expected for a *normal* heating season. It is *not* the coldest temperature on record, but the lowest one recorded for a particular locale over a 3- to 5-year period.

Lists of outside design temperatures are published for selected localities throughout the United States (Table 4-1). If a locality is not included on the list of winter outside design temperatures, check with a local weather bureau. Using an outside design temperature from the nearest locality on the list can be misleading, because even closely located cities can differ widely in weather conditions as a result of different altitudes, the effects of large bodies of water, and other variables.

Caution

> Make sure the HVAC contractor/installer is using the correct local outside design temperature and humidity for the area. Air conditioners are often oversized by using a higher than required summer design temperature. Undersizing the air conditioner is caused by underestimating the latent load (i.e., the energy used by the unit to remove moisture from the air).

Inside Design Temperature

The desired inside design temperature will depend upon the intended use of the space. In a large structure, such as a hotel or a hospital, there will be more than one inside design temperature because there is more than one type of space usage. For example, hospital kitchens will generally have an inside design temperature of about 66°F. The wards, on the other hand, will range from 72 to 74°F.

Residences will generally have a single inside design temperature for the entire structure, with 70 or 71°F the most commonly used temperatures.

Table 4-1 Winter Outside Design Temperatures for Major Cities in the United States

State	*City*	*Outside Design Temperature Commonly Used*	*State*	*City*	*Outside Design Temperature Commonly Used*
Alabama	Birmingham	10	New Hampshire	Concord	−15
Arizona	Tucson	25	New Jersey	Atlantic City	5
Arkansas	Little Rock	5		Trenton	0
California	San Francisco	35	New Mexico	Albuquerque	0
	Los Angeles	35	New York	Albany	−10
Colorado	Denver	−10		Buffalo	5
Connecticut	New Haven	0		New York City	0
District of Columbia	Washington	0	North Carolina	Asheville	0
Florida	Jacksonville	25		Charlotte	10
	Key West	45	North Dakota	Bismarck	−30
Georgia	Atlanta	10	Ohio	Akron	0
	Savannah	20		Cincinnati	0
Idaho	Boise	−10		Columbus	−10
Illinois	Cairo	0	Oklahoma	Oklahoma City	0
	Chicago	−10	Oregon	Portland	10
Indiana	Indianapolis	−10	Pennsylvania	Erie	−5
Iowa	Des Moines	−15		Harrisburg	0
	Sioux City	−20		Philadelphia	0

Kansas	Topeka	−10	Rhode Island	Providence	0
Kentucky	Louisville	0	South Carolina	Charleston	15
Louisiana	New Orleans	20	South Dakota	Huron	−20
Maine	Portland	−5	Tennessee	Knoxville	0
Maryland	Baltimore	0	Texas	Abilene	15
Massachusetts	Boston	0		Austin	20
Michigan	Detroit	−10		Brownsville	30
	Escanaba	−15		Corpus Christi	20
	Sault Ste. Marie	−20		Dallas	0
Minnesota	Duluth	−20		Houston	20
	Minneapolis	−20	Utah	Salt Lake City	−10
Mississippi	Vicksburg	10	Vermont	Burlington	−10
Missouri	Kansas City	−10	Virginia	Lynchburg	5
	St. Louis	0		Norfolk	15
Montana	Billings	−25	Washington	Seattle	15
	Helena	−20		Spokane	−15
	Miles city	−35	West Virginia	Parkersburg	−10
Nebraska	Lincoln	−10	Wisconsin	Green Bay	−20
	Valentine	−25		Madison	−15
Nevada	Reno	−5	Wyoming	Cheyenne	−15

Design Temperature Difference

The *design temperature difference* is the variation in degrees Fahrenheit between the outside and inside design temperatures. It is used in the heat transmission loss formula (see below) and is a crucial factor in heating calculations.

Be careful that you obtain the degree *difference* between the two temperatures; do not simply subtract the smaller figure from the larger one. For example, if the outside and inside design temperatures are −20 and 70°F, respectively, the design temperature difference will be 90° (20° below zero plus 70° above zero).

Determining Coefficients of Heat Transmission

The *overall coefficient of heat transmission,* or *U*-value, as it is commonly designated, is a specific value used for determining the amount of heat lost from various types of construction. It represents the time rate of heat flow and is expressed in Btu per hour per square foot of surface per degree Fahrenheit temperature difference between air on the inside and air on the outside of a structural section. Furthermore, the *U*-value is the reciprocal of the total thermal resistance value (*R*-value) of each element of the structural section and may be expressed as

$$U = \frac{1}{R_r}$$

Professional societies such as the American Society of Heating, Refrigerating, and Air-Conditioning Engineers (ASHRAE) have already determined the *U*-values for a wide variety of floor, ceiling, wall, window, and door construction. Tables of these *U*-values are made available through ASHRAE publications (e.g., the latest edition of the *ASHRAE Handbook of Fundamentals*) found in many libraries. Many manufacturers of heating equipment also provide tables of *U*-values in their literature. When calculating heat loss, you have the option of selecting *U*-values from the tables provided by manufacturers and certain professional associations or computing them yourself. The latter method, if done correctly, is the more precise one.

Calculating Net Area

Having determined the design temperature difference and computed (or selected) an overall heat transmission coefficient for each construction material or combinations of materials, you are now ready to calculate the net area of each surface exposed to the outside or adjacent to an unheated or partially heated space.

The best procedure for calculating surface area is to work from a building plan. If one is not available, you will have to make your own measurements. Measurements for calculating net area are taken from *inside* surfaces (i.e., inside room measurements). You will not be concerned with structural surfaces (e.g., walls, ceilings, floors) between rooms and spaces heated at the same temperature, because no heat transmission occurs where temperatures are constant.

Calculating the total area for each surface should be done as follows:

1. Multiply room length by room width to determine floor and ceiling area.
2. Multiply room length (or width) by room height to determine the outside wall area for each room.
3. Multiply door width by door height to determine the surface area for each door.
4. Multiply window width by window height to determine the surface area for each window.

Whether you use room length or room width in calculating the outside wall area will depend upon which wall surface is exposed to the outside. In some cases (e.g., corner rooms), both are used and require at least two separate calculations (i.e., room length × room height, and room width × room height).

Add the calculated surface area of each outside wall (see Step 2 in *Sizing Systems Using Coefficients of Heat Transmission*) to obtain the *gross wall area* for the structure. Subtract the sum of all door and window surface areas from the gross wall area. The result will be the *net wall area* for the structure. Multiply the net wall area by the heat loss in Btu per hour per square foot to calculate the heat loss through the walls.

Heat Transmission Loss Formula

The heat loss (expressed in Btu) of a given space is determined by multiplying the coefficient of heat transmission by the area in square feet by the design temperature difference (i.e., the difference between the indoor and the outdoor design temperatures). Because a heating system must supply an amount of heat equal to the amount of heat lost in order to maintain a constant indoor design temperature, heat loss is approximately equal to heat required. This may be expressed by the following formula:

$$H_t = AU(t_i - t_o)$$

where

H_t = heat loss transmitted through a structural section (e.g., roof, floor, ceiling) expressed in Btu per hour, representing both the heat lost and the heat required.

A = area of structural components in square feet.

U = overall coefficient of heat transmission.

T_o = outdoor design temperature.

T_i = indoor design temperature.

Computing Total Heat Loss

The commonly accepted procedure for calculating the total heat loss from a structure is to calculate the heat loss for each room or space *separately* and then add the totals.

The heat loss calculation worksheet you use should contain a column in which each room or space can be identified separately. Under the identification are listed those structural sections through which heat transmission losses occur. When applicable, these will include all or most of the following:

1. Walls
2. Glass
3. Ceiling
4. Floor
5. Door(s)

TYPE OF ROOM OR SPACE	TYPE OF STRUCTURAL SECTION	
BEDROOM NO. 1	WALLS	
	GLASS AREA	
	CEILING	
	FLOOR	
	DOOR AREA	
	INFILTRATION	

Figure 4-1 Tabulation worksheet.

In addition to the five structural sections listed above, each identified room or space should also include a line for heat loss by air infiltration. Figure 4-1 illustrates how your worksheet should appear at this point.

Now that you have identified the room or space, listed the structural sections through

TYPE OF ROOM OR SPACE	TYPE OF STRUCTURAL SECTION	NET AREA/ AIR VOLUME	COEFFICIENT	TEMP. DIFF.	HEAT LOSS (BTU/h)
BEDROOM NO. 1	WALLS	235 SQ.FT.	0.26	90	5599
	GLASS AREA	45 SQ.FT.	0.45	90	1823
	CEILING	300 SQ.FT.	0.17	40	2040
	FLOOR	—	—		
	DOOR AREA	—	—		
	INFILTRATION	1800 CFH	0.018	90	2916

Figure 4-2 Tabulation worksheet.

which heat loss occurs, and calculated the rate of air infiltration, you must determine the net surface area and the *U*-value for each structural section. The design temperature difference must also be determined and entered for each structural section. Except where surfaces are exposed to partially heated spaces (e.g., a garage, attic, or basement), each structural section will have the same design temperature difference. An example of how your worksheet should look at this point is illustrated in Figure 4-2.

Each identified room or space should also include the calculated air infiltration heat loss (see *Infiltration Heat Loss* following). The amount of heat loss due to air infiltration will depend upon the size, type, and number of windows and doors and other variables.

The bedroom given as an example in Figure 4-1 has two exposed walls (8 ft × 15 ft and 8 ft × 20 ft). The number of air changes suggested for a room with two exposed walls is 1½ changes per hour.

The volume of air infiltration for the bedroom can be calculated as follows:

$$\begin{aligned}\text{Vol. of air infiltration} &= \text{No. of air changes} \times \frac{\text{ceiling area} \times \text{ceiling height}}{2} \\ &= 1.5 \times \frac{300\ \text{ft}^2 \times 8\ \text{ft}}{2}\end{aligned}$$

Knowing that the volume of air infiltration is 1800 ft^3/h, the heat loss in Btu/h can be calculated as follows:

$$\begin{aligned}\text{Heat loss} &= 0.018 \times Q(t_i - t_o) \\ &= 0.018 \times 1800 \times 90 \\ &= 2916\ \text{Btu}\end{aligned}$$

The *total* heat loss for bedroom No. 1 (as it would be designated on the worksheet) is 12,378 Btu/h. If there is a door in one of the outside walls, its heat loss also has to be calculated and entered on the worksheet. Furthermore, the door area (in square feet) would have to be subtracted from the surface area for the walls, because the latter is always a *net* figure.

After calculating the heat loss for each room or space in the structure, the results are added to obtain the total heat loss (in Btu/h) for the structure.

Loss in Doors and Windows

It is not easy to compute the coefficient of heat transmission (*U*-value) for a door because door construction varies considerably in size, thickness, and material. A fairly accurate rule-of-thumb method is to use the *U*-value for a comparable size *single-pane* glass window. If you wish to compute the coefficient of heat transmission for a door, the necessary data can be found in ASHRAE publications. The same source can be used for computing the coefficients of heat transmission of windows.

Loss in Basements

The amount of heat lost from a basement will depend upon a number of different variables. An important consideration, for example, is ground temperature, which will function as the outdoor design temperature in the heat loss transmission formula. The ground temperature will vary with geographical location (Table 4-2).

The usual method for calculating heat loss transmission through basement walls is to view them as being divided into two sections (Figure 4-3). The upper section extends from the frost line to the basement ceiling and includes those portions of the wall exposed to outdoor air temperatures. The heat loss for this section of the basement wall is calculated in the same way that other surfaces exposed to the outside air are calculated. The section of the wall extending

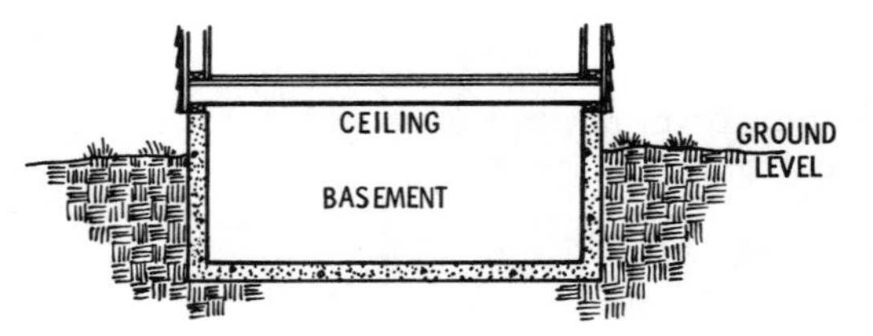

Figure 4-3 Basement cross-section.

Table 4-2 Ground Temperatures Below the Frost Line

State	*City*	*Ground Temperature Commonly Used*
Alabama	Birmingham	66
Arizona	Tucson	60
Arkansas	Little Rock	65
California	San Francisco	62
	Los Angeles	67
Colorado	Denver	48
Connecticut	New Haven	52
District of Columbia	Washington	57
Florida	Jacksonville	70
	Key West	78
Georgia	Atlanta	65
Idaho	Boise	52
Illinois	Cairo	60
	Chicago	52
	Peoria	55
Indiana	Indianapolis	55
Iowa	Des Moines	52
Kentucky	Louisville	57
Louisiana	New Orleans	72
Maine	Portland	45
Maryland	Baltimore	57
Massachusetts	Boston	48
Michigan	Detroit	48
Minnesota	Duluth	41
	Minneapolis	44
Mississippi	Vicksburg	67
Missouri	Kansas City	57
Montana	Billings	42
Nebraska	Lincoln	52
Nevada	Reno	52
New Hampshire	Concord	47
New Jersey	Atlantic City	57
New Mexico	Albuquerque	57
New York	Albany	48
	New York City	52

(continued)

Table 4-2 (continued)

State	*City*	*Ground Temperature Commonly Used*
North Carolina	Greensboro	62
North Dakota	Bismarck	42
Ohio	Cleveland	52
	Cincinnati	57
Oklahoma	Oklahoma City	62
Oregon	Portland	52
Pennsylvania	Pittsburgh	52
	Philadelphia	52
Rhode Island	Providence	52
South Carolina	Greenville	67
South Dakota	Huron	47
Tennessee	Knoxville	61
Texas	Abilene	62
	Dallas	67
	Corpus Christi	72
Utah	Salt Lake City	52
Vermont	Burlington	46
Virginia	Richmond	57
Washington	Seattle	52
West Virginia	Parkersburg	52
Wisconsin	Green Bay	44
	Madison	47
Wyoming	Cheyenne	42

from the frost line to the basement floor and the floor itself can be calculated on the basis of the ground temperature. The ground temperature is substituted for the outside design temperature when calculating heat loss.

Loss in Slab Construction

The heat loss for houses and small buildings constructed on a concrete slab at or near grade level is computed on the basis of heat loss per foot of exposed edge. For example, a concrete slab measuring 20 ft × 25 ft would have an exposed edge of 90 ft. This represents the measurement completely around the perimeter of the exposed edge of the slab. The heat loss will depend upon the thickness of insulation along the exposed edge and the outside design temperature *range*. This type of information is available from ASHRAE publications. To obtain the heat loss in Btu/h, simply multiply the total length of exposed edge by the heat loss in Btu/h per lineal foot (lin ft). For example, a 90-ft exposed edge with 2-in-edge insulation at an outdoor design temperature of 35° would be calculated as follows:

90 lin ft × 45 Btu/h/lin ft = 4050 Btu/h

Infiltration Heat Loss

During the heating season, a portion of heat loss is due to the infiltration of cooler outside air into the interior of the structure through cracks around doors and windows and other openings that are not a part of the ventilating system. The *amount* of air entering the structure by infiltration is important in estimating the requirements of the heating system, but the *composition* of this air is equally important.

A pound of air is composed of both dry air and moisture particles, which are combined (*not* mixed) so that each retains its individual characteristics. The distinction between these two basic components of air is important, because each is involved with a different type of heat: dry air with specific heat, and moisture content with latent heat.

The heating system must be designed with the capability of warming the cooler infiltrated *dry air* to the temperature of the air inside the structure. The amount of heat required to do this is referred to as the *sensible heat loss* and is expressed in Btu/h. The two methods used for calculating heat loss by air infiltration are: (1) the crack method, and (2) the air-change method.

The Crack Method

The crack method is the most accurate means of calculating heat loss by infiltration, because it is based on actual air leakage through cracks around windows and doors and takes into consideration the expected wind velocities in the area in which the structure is located. The air-change method (see below) does not consider wind velocities, which makes it a less accurate means of calculation.

Calculating heat loss by air infiltration with the crack method involves the following basic steps:

1. Determine the type of window or door (see Table 4-3).
2. Determine the wind velocity and find the air leakage from Table 4-3.

Table 4-3 Infiltration Rate Through Various Types of Windows*

Type of Window	Remarks	Wind Velocity, Miles per Hour					
		5	10	15	20	25	30
Double-Hung Wood Sash Windows (Unlocked)	Around frame in masonry wall—not caulked	3	8	14	20	27	35
	Around frame in masonry wall—caulked	1	2	3	4	5	6
	Around frame in wood-frame construction	2	63	11	17	23	30
	Total for average window, non-weather-stripped, 1⁄16-in crack and 3⁄64-in clearance; includes wood frame leakage	7	21	39	59	80	104
	Ditto, weather-stripped	4	13	24	36	49	63
	Total for poorly fitted window, non-weather-stripped, 3⁄32-in crack and 3⁄32-in clearance; includes wood frame leakage	27	69	111	154	199	249
	Ditto, weather-stripped	6	19	34	51	71	92
Double-Hung Metal Windows	Non-weather-stripped, locked	20	45	70	96	125	154
	Non-weather-stripped, unlocked	20	47	74	104	137	170
	Weather-stripped, unlocked	6	19	32	46	60	76

Rolled Section Steel Sash	Industrial pivoted, 1/16-in crack	52	108	176	244	304	372
	Architectural projected, 1/32-in crack	15	36	62	86	112	139
	Architectural projected, 3/64-in crack	20	52	88	116	152	182
	Residential casement, 1/64-in crack	6	18	33	47	60	74
	Residential casement, 1/32-in crack	14	32	52	76	100	128
	Heavy casement section, projected, 1/64-in crack	3	10	18	26	36	48
	Heavy casement section, projected, 1/32-in crack	8	24	38	54	72	92
Hollow Metal, Vertically Pivoted Window		30	88	145	186	221	242

**Expressed in cubic feet per foot of crack per hour. The infiltration rate through cracks around closed doors is generally estimated at twice that calculated for a window.*

Courtesy ASHRAE 1960 Guide

3. Calculate the lineal feet of crack.

4. Determine the design temperature difference.

The data obtained in these four steps are used in the following formula:

$$H = 0.018 \times Q(t_i - t_o) \times L$$

where

H = heat loss, or heat required to raise the temperature of air leaking into the structure to the level of the indoor temperature (t_i) expressed in Btu per hour.

Q = volume of air entering the structure, expressed in cubic feet per hour (Step 2 above).

t_i = indoor temperature.

t_o = outdoor temperature.

0.018 = specific heat of air (0.240) times density of outdoor air (approximately 0.075).

L = lineal feet of crack.

Determine the infiltration heat loss per hour through the crack of a 3 ft × 5 ft average double-hung, non-weather-stripped, wood window based on a wind velocity of 20 mph. The indoor temperature is 70°F, and the outdoor temperature 20°F.

The air leakage for a window of this type at a wind velocity of 20 mph is 59 ft^3 per foot of crack per hour. This will be the value of Q in the air infiltration formula. The lineal feet of the crack is (2 × 5) plus (3 × 3), or 19 ft (the value of L in the formula). t_i = 70°F, and t_o = 20°F. Substituting these data in the air infiltration formula gives the following results:

$$\begin{aligned} H &= 0.018 \times Q(t_i - t_o) \times L \\ &= 0.018 \times 59(70 - 20) \times 19 \\ &= 1.062 \times 50 \times 19 \\ &= 1008.9 \text{ Btu/h} \\ &= 1009 \text{ Btu/h} \end{aligned}$$

Table 4-4 represents a typical wall infiltration chart for a number of different types of wall construction. Like the air infiltration rate through windows, it is also based on wind velocity (see Table 4-3).

Table 4-4 Infiltration Through Various Types of Wall Construction

Type of Wall	*Wind Velocity, Miles per Hour*					
	5	10	15	20	25	30
Brick Wall						
8½ in Plain	2	4	8	12	19	23
Plastered	0.02	0.04	0.07	0.11	0.16	0.24
Plain	1	4	7	12	16	21
13 in Plastered	0.01	0.01	0.03	0.04	0.07	0.10
Frame Wall, Lath and Plaster	0.03	0.07	0.13	0.18	0.23	0.26

Courtesy ASHRAE 1960 Guide

Air-Change Method

In the air-change method, the amount of air leakage (i.e., infiltration) is calculated on the basis of an assumed number of air changes per hour per room. The number of air changes will depend upon the *type* of room and the number of walls exposed to the outdoors. Table 4-5 is an example of a typical air-change chart used in the air-change method.

Table 4-5 Fresh Air Requirements

Type of Building or Room	*Minimum Air Changes per Hour*	*Cubic Feet of Air per Minute per Occupant*
Attic spaces (for cooling)	12–15	
Boiler room	15–20	
Churches, auditoriums	8	20–30
College classrooms		25–30
Dining rooms (hotel)	5	
Engine rooms	4–6	
Factory buildings (ordinary manufacturing)	2–4	
Factory buildings (extreme fumes or moisture)	10–15	
Foundries	15–20	

(continued)

Table 4-5 (continued)

Type of Building or Room	*Minimum Air Changes per Hour*	*Cubic Feet of Air per Minute per Occupant*
Galvanizing plants	20–30	
Garages (repair)	20–30	
Garages (storage)	4–6	
Homes (night cooling)	9–17	
Hospitals (general)		40–50
Hospitals (children's)		35–40
Hospitals (contagious diseases)		80–90
Kitchens (hotel)	10–20	
Kitchens (restaurant)	10–20	
Libraries (public)	4	
Laundries	10–15	
Mills (paper)	15–20	
Mills (textile—general buildings)	4	
Mills (textile—dyehouses)	15–20	
Offices (public)	3	
Offices (private)	4	
Pickling plants	10–15	
Pump rooms	5	
Schools (grade)		15–25
Schools (high)		30–35
Restaurants	8–12	
Shops (machine)	5	
Shops (paint)	15–20	
Shops (railroad)	5	
Shops (woodworking)	5	
Substations (electric)	5–10	
Theaters		10–15
Turbine rooms (electric)	5–10	
Warehouses	2	
Waiting rooms (public)	4	

Ventilation Heat Loss

In larger structures (e.g., commercial or industrial buildings), ventilation is necessary for both health and comfort. In houses, the amount of infiltrated air is generally sufficient to supply the ventilation requirements.

When ventilation is provided, the heating system warms the outside air used for this purpose (i.e., ventilation) to the inside design temperature. The heat required will be equivalent to the heat lost by ventilation and can be calculated with the following formula:

$$H = 0.018 \times Q(t_i - t_o)$$

where

H = heat loss by ventilation.
Q = volume of ventilation air.
0.018 = specific heat of air (0.240) times density of outdoor air (approximately 0.075).

The Average Value Method

Heat loss from a structure can also be calculated by using average values for four basic factors. These four factors and their assigned values are:

1. Wall factor (0.32 Btu)
2. Contents factor (0.02 Btu)
3. Glass factor (1 Btu)
4. Radiation factor (240 Btu)

Each of these four factors is based on a 1°F temperature difference and must be multiplied by the *design* temperature difference (see *Design Temperature Difference* in this chapter) obtained for the structure and its geographical location.

The average value for each factor represents the amount of heat (in Btu) that will pass through 1 square foot of the material in 1 hour. Thus 0.32 Btu will pass through *a square foot* of ordinary brick wall or through a wall where the average frame construction is used *in 1 hour.* It should be stressed that these are *average* values useful in a simple and quick way of calculating heat loss.

The wall factor of 0.32 Btu varies for different kinds of walls, and heat loss calculations based on the use of average values will be only approximately accurate. More precise heat loss calculations can be made by using coefficients of heat transmission (U-factors).

The glass factor (1 Btu) is used to represent the amount of heat that will pass through 1 square foot of glass in an hour.

The radiation factor (240 Btu) was originally used to represent the amount of heat in Btu given off by an ordinary free-standing, cast-iron radiator on the basis of the square-foot area of heating surface per hour under average conditions. The average value of the radiation factor was based on repeated test observations that the amount of heat given off by ordinary cast-iron radiators per degree difference in temperature between the steam (or water) in the radiator and the air surrounding the radiator is about 1.6 Btu per square foot of heating surface. Taking this as a basis, a steam radiator under 2½ lb pressure corresponds approximately to 220°F; when surrounded by air having a 70°F inside air temperature, it gives off heat at a rate of 240 Btu ($220 - 70 \times 1.6 = 240$ Btu).

Radiators became so varied in design (with corresponding variations in the sizes of heating surface areas) that it became necessary to regard 240 Btu as a standard value.

Hot-water heating systems are planned on the basis of the Btu per hour capacity of each emitting unit. In steam heating systems, the radiation to be used is selected on the basis of the square foot EDR (equivalent direct radiation) capacity of each unit. One square foot of EDR equals 240 Btu/h. The EDR capacity of a space is expressed in square feet and may be determined by dividing the total heat required (expressed in Btu/h) by 240. For example, a room requiring 12,000 Btu/h to heat it would have a steam radiation requirement of 50 ft^2 EDR capacity.

The steps involved in determining heat loss with the average value method are:

1. Decide upon the desired inside air temperature.
2. Obtain the winter outside design temperature of the location of the structure from published lists or a local weather bureau.
3. Determine the design temperature difference (the difference between the temperatures found in Steps 1 and 2).
4. Calculate the volume of the structure or space.
5. Calculate the total area (in square feet) of window glass.
6. Calculate the total area of wall surface exposed to the outdoors.
7. Deduct the total area of window glass (Step 5) from the total area of wall surface exposed to the outdoors (Step 6).
8. Multiply the volume of the structure or space (Step 4) by the contents factor (0.02) to determine the Btu required to raise

the temperature of the volume an amount equal to the design temperature difference (Step 3).

9. Multiply the *net* wall surface area (Step 7) by the wall factor (0.32) to determine heat loss through the walls.
10. Multiply the total window glass surface area (Step 5) by the glass factor (1) to determine heat loss through glass.
11. Add the figures obtained in Steps 8 to 10 to obtain the total heat loss per hour for the structure or space (this is equal to the total heat *required* per hour).

The 11 steps outlined above might be more clearly understood if shown in a sample problem.

Let's assume that you must calculate the heat loss for a space 14 ft wide × 14 ft long × 9 ft high, and that 60 ft^2 of window glass is equally divided between two exposed walls. The inside desired temperature is 70°, and the winter outside design temperature is −10°F. Your procedure (corresponding step by step to those outlined above) is as follows:

1. 70°F inside desired temperature.
2. −10°F winter outside design temperature.
3. 80°F design temperature difference.
4. 1764 ft^3 volume (14 × 14 × 9).
5. 60 ft^2 of glass.
6. 252 ft^2 of wall surface exposed (2 sides) to the outdoors (14 + 14 × 9 = 252).
7. 192 ft^2 of *net* wall surface exposed (252 − 60 = 192).
8. 2822.4 Btu (1764 × 0.02 × 80).
9. 4915.2 Btu (192 × 0.02 × 80).
10. 4800 Btu (60 × 1 × 80).
11. 12,537.6 Btu (2822.4 + 4915.2 + 4800).

Heat Loss Tabulation Forms

Manufacturers of heating equipment will often provide heat loss tabulation forms and instructions for their use. Lists of the most commonly used *U*-values (or factors) are included with the forms. Figures 4-4 and 4-5 show examples of these forms and instructions obtained from Amana Refrigeration, Inc.

RESIDENTIAL HEAT LOSS TABULATION

Job Name: ______________________________ Date ______________________________

Location: ______________________________ Computed By ______________________________

Inside Design Temp. ______________________________

Outside Design Temp. ______________________________

Temp. Difference ______________________________

Temp. Difference +10 ______________________________ Bathroom __________

Room		Living Room		Dining Room		Kitchen		Bedroom 1		Bedroom 2						Bath		Rooms Btu Totals
1. Room Size (LxWxH)																		
2. Linear Ft. Exp. Wall																		
3. Floor Area																		
	Factor	Ar. or Ln. Ft.	BTU	Ar. or Ln. Ft.	BTU	Ar. or Ln. Ft.	BTU	Ar. or Ln. Ft.	BTU	Ar. or Ln. Ft.	BTU	Ar. or Ln. Ft.	BTU	Ar. or Ln. Ft.	BTU	Ar. or Ln. Ft.	BTU	
4. Windows																		
5. Doors																		
6. Exp. Wall																		
7. Ceiling																		
a																		
b																		
8. Infiltration c																		
d																		
9. SUB TOTALS																		
10. Sub-Total for T.D.																		
11. Floor Btu for T.D.																		
12. Partition Btu for T.D.																		
13. Room Total Btu Loss																		

14. + % Duct Loss __________

15. Grand Total Btu Load __________

Windows	U Factors
Single Pane	1.13
Double Pane, Sealed	.57
Storm Windows, Tight	.45
Storm Windows, Loose	.75
Glass Block	.50

T.D. = TEMPERATURE DIFFERENCE

Exp. Walls	U Factors		
Frame Construction With:	Clapboard or Shingle	Brick Venr.	Stone Venr.
No Insulation	.25	.28	.30
¾" Insl. Board	.19	.21	.22
2" Insl. Batts	.11	.12	.12
3⅝" Insl. Batts	.09	.10	.10

% Duct Loss Schedule

	1 Story	1½ Story	2 Story
Ducts Insul.	10	8	6
Ducts No Insul.	30	25	20

Wood Doors	U Factors	
Nom. Thickness	No Storm	Storm Door
1"	.69	.35
1½"	.52	.30
2"	.46	.28
2½"	.38	.25

Ceilings
(Vent: Attic Space Above)

No Insulation	.50
2" Insulation	.10
4" Insulation	.09
6" Insulation	.05

Infiltration (factor times lineal ft. of crack)

		Weather Stripped	Not Weather Stripped
Double Hung Window	(a)	.22	.35
Pivoting Windows	(b)	.30	.50
Doors	(c)	.60	1.00
Fixed Window	(d)	.13	.13

CONCRETE FLOOR ON GROUND
Btu Per Hr. Per Lineal Ft. of Edge

Outside Design Temp.	25	20	15	10	5	0	−5	−10	−15	−20	−25	−30
1" Vert. Insul. Extending Down 18" Below Surface	55	62	68	75	80	85	90	95	100	105	110	115
1" L Type Insul. Extending 12" Down and 12" Under	50	57	63	70	75	80	85	90	95	100	105	110
2" L Type Insul. Extending 12" Down and 12" Under	40	45	50	55	60	65	70	75	80	85	90	95
No edge insulation and no heat in slab	This construction not recommended.											

Double Wood Floors or Partition

With Air Space					
Temp. Diff.	10	20	30	40	50
Factor No Insul.	3.4	6.8	10.2	13.6	17.0
Factor 2" Insul.	1	2	3	4	5

Figure 4-4 Residential heat loss tabulation form. *(Courtesy Amana Refrigeration, Inc.)*

Customer .... Address ....

Buyer .... Installation by ....

Estimate Number .... Estimate by .... Date ....

Equipment Selected ....; Model ....; Size ....

Direction House Faces ....; Gross Floor Area .... sq. ft.; Gross Inside Volume .... cu. ft.

Design Conditions:

	Dry-Bulb Temperature (F)
Outside	....

ITEM	AREA (Sq. Ft.) Except Item 4; Linear Ft.	FACTOR (Circle the factors applicable.) DESIGN DRY-BULB TEMPERATURE (F) −30	−20	−10	0	10	20	25	30	35	BTU/HR (Area x Factor)
1. WINDOWS (total of all windows)											
Single-glass	....	113	102	90	79	68	57	51	45	40	....
Double glass or glass block	....	50	45	40	35	30	25	23	20	18	....
2. WALLS											
No insulation (brick veneer, frame, stucco, masonry)	....	25	23	20	18	15	13	11	10	9	....
1 in. insulation or 25/32 in. insulating sheathing	....	20	18	16	14	12	10	9	8	7	....
2 in. or more insulation	....	10	9	8	7	6	5	5	4	4	....

3. ROOFS											
Pitched or flat with vented air space and:											
No insulation	....	31	28	25	22	19	16	14	12	11	....
2 in. insulation	....	10	9	8	7	6	5	5	4	4	....
4 in. insulation	....	7	6	6	5	4	4	3	3	2	....
Flat with no air space and:											
No insulation	....	50	45	40	35	30	25	23	20	18	....
25/32 in. insulation	....	25	23	20	18	15	13	11	10	9	....
1 1/2 in. insulation	....	15	14	12	11	9	8	7	6	5	....
3 in. insulation	....	10	9	8	7	6	5	5	4	4	....
4. FLOORS											
Over—(sq. ft. area):											
Basement	....	0	0	0	0	0	0	0	0	0	....
Enclosed crawl space	....	23	20	18	16	14	11	10	9	8	....
Open crawl space	....	34	31	27	24	20	17	15	14	12	....
On ground—(linear feet):											
No edge insulation	....	81	73	65	57	49	41	37	32	28	....
1 in. edge insulation	....	68	61	54	48	41	34	31	27	24	....
2 in. edge insulation	....	55	50	44	39	33	28	25	22	19	....
5. OUTSIDE AIR											
Total Floor Area (sq. ft.)	....	14	13	11	10	8	7	6	6	5	....
6. TOTAL (Sum of 1 through 5)											

Figure 4-5 Residential heating load estimate form. *(Courtesy Amana Refrigeration, Inc.)*

Estimating Fuel Requirements and Heating Costs

Estimating fuel requirements and heating costs is *not* an exact science. It involves too many variables to be anything other than an estimation. Moreover, there are four different formulas used for making these calculations. Depending on which formula is used, it is possible to arrive at four different estimations of fuel requirements and heating costs for a specific situation. It is no wonder, then, that two competent engineers can submit estimates that will differ as widely as 30 percent or more. These facts are not offered to discourage anyone but to present the true picture. Any attempt to calculate fuel requirements and heating costs will not produce *precise* figures—only an estimate. The problem is to make this estimate as close an approximation to the real situation as possible.

The four formulas used in calculating fuel requirements and heating costs are:

1. Heat loss formula
2. Corrected heat loss formula
3. NEMA formula
4. Degree-day formula

If one formula is used to calculate the fuel requirements and heating costs with one type of fuel (e.g., oil) and another formula is used for a second type (e.g., natural gas), the results are practically worthless for comparison purposes. For example, the NEMA formula (created by the National Electric Manufacturers Association) will present the use of electric energy much more favorably if the results are compared with calculations for oil or natural gas based on the heat loss formula. A *true* comparison is possible only when two fuels are both calculated with the *same* formula. Each of these formulas is described in the following sections.

The Heat Loss Formula

The *heat loss formula* results in higher percentages of total requirements because it does not take into consideration internal heat gains obtained from appliances, sunlight, the body heat of the occupants, electric lights, and other sources. The corrected heat loss formula includes these factors (see the following section).

The heat loss formula will include the following data:

1. Heat loss expressed in Btu
2. Total hours in the heating season

3. Average winter temperature difference
4. Btu per unit of fuel
5. Efficiency of utilization
6. Difference between inside and outside design temperatures

The product of the first three items (1–3) is divided by the product of the last three (4–6). In other words, heat loss × total hours × average winter temperature difference/Btu per unit of fuel × efficiency of utilization × (inside design temperature – outside design temperature).

The method for calculating heat loss is described in Chapter 3 and is expressed in Btu per square foot per hour per degree Fahrenheit design temperature difference.

The total hours in the heating season will depend on the location of the house or building. If it is located in a southern state, the heating season will be much shorter than if it is located in a colder climate. Assuming that the heating season begins October 1 and ends May 1, the total number of days for which heat may be required is 212. This figure is multiplied by 24 (hours) to obtain the total hours in the heating season (5088).

The average winter temperature difference is found by subtracting the average low temperature from the average high temperature for the location in which the house or building is situated.

The Btu per unit of fuel is determined for each type, and this information can usually be obtained from the local distributor of the fuel.

Most heating equipment will burn oil or gas at an 80 percent combustion efficiency. Electric energy is generally considered to be used at 100 percent efficiency.

The outside design temperature is the lowest temperature experienced in a locality over a 3- to 4-year period. These outside design temperatures are available for a large number of localities throughout the United States. If the house or building is located in a small town or rural area, the nearest known outside design temperature is used. This inside design temperature is the temperature to be maintained on the interior of the house or building. The difference between the two design temperatures is used in the heat loss formula.

Table 4-6 illustrates the use of the heat loss formula for three types of fuel/energy:

1. Oil
2. Natural gas
3. Electricity

Table 4-6 Applying Heat Loss Formula to Three Types of Fuels

(1) No. 2 oil, 140,000 Btu/gal, 80% efficiency, 35,000 Btu/h loss, 73°F inside temperature:

$$\frac{35{,}000 \times 5088}{140{,}00 \times 0.80} \times \frac{73 - 41}{70 - 0} = 727 \text{ gallons}$$

(2) Natural gas, 100,000 Btu per therm, 80% efficiency:

$$\frac{35{,}000 \times 5088}{100{,}000 \times 0.80} \times \frac{73 - 41}{70 - 0} = 1018 \text{ therms}$$

(3) Electricity, 3413 Btu/Wh, 100% efficiency:

$$\frac{35{,}000 \times 5088}{3413} \times \frac{73 - 41}{70 - 0} = 23{,}852 \text{ kWh}$$

Courtesy National Oil Fuel Institute

The Corrected Heat Loss Formula

The *corrected heat loss formula* was devised in 1965 by Warren S. Harris and Calvin H. Fitch of the University of Illinois. It takes into consideration internal heat gains, which have shown a marked increase over the past 40 to 50 years. These internal heat gains make a significant contribution to the total required heat of a house or building, a factor that makes the original 65°F heat base established some 40-odd years ago too low for making a correct estimate. Furthermore, an inside average temperature of 73°F is probably more correct than the 70°F temperature previously used.

According to Harris and Fitch, the degree-day base (for all areas except along the Pacific Coast) must be corrected by the percentage given in Table 4-7. The calculated heat loss should then be reduced by 3.5 percent for each 1000 ft. the house or building is located above sea level. If the heat loss is below 1000 Btu per degree temperature difference, the 65°F degree-day base should be further reduced to the figures given in Table 4-8.

Table 4-9 illustrates the use of the corrected heat loss formula for determining comparative fuel requirements for No. 2 oil and natural gas.

The Degree-Day Formula

The *degree-day formula* was devised some 40-odd years ago by the American Gas Association and other groups and has since been revised (see pages 627–28 of the 1970 *ASHRAE Guide*) to reflect

Table 4-7 Reduction in Standard Degree Days for All Areas Except the Pacific Coast

Degree Days at 65°F Base	*Reduce by This Percentage*	*New Total*
10,000	3.15	9685
9500	3.20	9196
9000	3.22	8710
8500	3.23	8226
8000	3.25	7740
7500	3.80	7215
7000	3.85	6720
6500	4.35	6218
6000	4.55	5727
5500	4.60	5247
5000	4.65	4768
4500	4.70	4289
4000	5.30	3788
3500	5.35	3313
3000	6.15	2816
2500	7.15	2321
2000	8.00	1840
1500	9.50	1358
1000	11.90	881
500	15.10	425

Courtesy National Oil Fuel Institute

Table 4-8 Reduction in Degree-Day (DD) Base When Calculated Heat Loss (Btu per Degree Temperature Difference) is Less Than 1000

Calculated Heat Loss	*Revised DD Base, °F*
200	55
300	60
400	61
500	62
600	63
700	64
800	64
900	64
1000	65

Courtesy National Oil Fuel Institute

Table 4-9 Using Corrected Heat Loss Formula for Determining Comparative Fuel Requirements

Assume 500 Btu/h per degree temperature difference (35,000 Btu/h at 70°F difference); degree days (dd) with 65°F base, 5542; average indoor temperature 73°F.

Correction factor from Table 4-7 is 4.60%.

From Table 4-5, revised degree-day base is 62°F.

Reduction in base is (65 − 62) 3°F.

Multiply 0.0460 × 3 = 0.1380 and deduct from 1.0000 = 0.8620.

Multiply 5542 (dd at 65° base) by 0.8620 = 4557.

Example:

No. 2 Oil:

35 × 0.00304 × 4557 = 485 gallons per year

Natural Gas:

35 × 0.00429 × 4557 = 684 therms per year

Courtesy National Oil Fuel Institute

internal heat gains and the levels of insulation. The correction factors involved in the revision are given in Tables 4-10 and 4-11. These correction factors should also be applied to the corrected heat loss formula (see the previous section).

The degree-day formula is based on the assumption that heat for the interior of a house or building will be obtained from sources other than the heating system (e.g., sunlight and body heat of the occupants) until the outside temperature declines to 65°F. At this point the heating system begins to operate. The consumption of fuel will be directly proportionate to the difference

Table 4-10 Unit Fuel Consumption per Degree Day per 1000 Btu Design Heat Loss

	Utilization Efficiency		
Fuel	***60%***	***70%***	***80%***
Gas in therms	0.00572	0.00490	0.00429
Oil in gallons (141,000 Btu)	0.00405	0.00347	0.00304

Courtesy National Oil Fuel Institute

Table 4-11 Correction Factors for Outdoor Design Temperatures*

	Outside Design Temp., °F				
	−20	−10	0	+10	+20
Correction factor	0.778	0.875	1.000	1.167	1.400

**To be applied to Table 4-1 when design temperature is higher or lower than 0°F.*
Courtesy National Oil Fuel Institute

between the 65°F base temperature and the mean outdoor temperature. In other words, three times as much fuel will be used when the mean outdoor temperature is 35°F than when it is 55°F. The mean outdoor temperature can be determined by taking the sum of the highest and lowest outside temperatures during a 24-hour period, beginning at midnight, and dividing it by 2. Each degree in temperature below 54°F is regarded as 1 *degree day*.

The degree-day formula is applied by dividing the heat loss figure by 1000 and multiplying the result by the figure for the unit fuel consumption per degree day per 1000 Btu design heat loss, which, in turn, is multiplied by the total number of degree days in the heating season (calculated on a 65°F base) and then by the correction factors given in Tables 4-10 and 4-11. The application of this formula is illustrated in Table 4-12.

Table 4-12 Application of Degree-Day Formula

35,000 Btu/h loss; outside design temperature 0; no correction factor needed; degree days, 5542.

No. 2 oil, 140,000 Btu per therm, 80% efficiency:

$$35 \times 0.00304 \times 5542 = 590 \text{ gallons per year}$$

Natural gas, 100,000 Btu per therm, 80% efficiency:

$$35 \times 0.00429 \times 5542 = 832 \text{ therms per year}$$

Electricity, required Btu/year

$$832 \text{ (therms)} \times 100{,}000 \text{ (Bru/therm)} \times 0.80 \text{ (efficiency)}$$

$$= 66{,}560{,}000 \ \frac{66{,}560{,}000}{3413} = 19{,}502 \text{ kWh}$$

Courtesy National Oil Fuel Institute

Table 4-13 Application of NEMA Formula

kWh = heat loss ÷ 3413
× annual degree days
× constant (usually 18.5)

divided by difference between indoor and outdoor design temperature.
Example:

$$\frac{35{,}000 \text{ Btuh}}{3413} = 1030$$

$$\frac{1030 \times 5542 \times 18.5}{70 - 0} = 15{,}086 \text{ kWh}$$

Courtesy National Oil Fuel Institute

The NEMA Formula

The *NEMA formula* was created by the National Electric Manufacturers Association. A constant of 18.5 is commonly used in the formula and reflects the percentage of heat loss that must be replaced by the heating system during a 24-hour period. In other words, the heating system will provide heat 77 percent of the time during a 24-hour period (24 × 18.5), while other heat sources will supply heat during the remaining 23 percent.

The NEMA formula consists of dividing the heat loss by 3413 (the number of Btu per kilowatt-hour) and multiplying this figure by the total annual number of degree days, which figure is then multiplied by the constant (usually 18.5). The resulting product is then divided by the difference between the indoor and outdoor temperatures. The application of the NEMA formula is illustrated in Table 4-13.

A major criticism of this formula by distributors of oil and gas heating fuels is that use of the heat loss formula gives a much higher kilowatt-hour consumption rate than use of the NEMA formula.

Manual J and Related Materials Used for Sizing Heating/Cooling Systems

The calculation methods contained in the *Residential Load Calculations, Manual J,* published by the Air Conditioning Contractors of America (ACCA), is the most accurate and common ones used by HVAC contractors/installers to size heating and air conditioning systems correctly. There are also many computer programs and worksheets designed to simplify the calculation procedures found in *Manual J.*

Recommended Publications Providing Additional Information About Load Calculations and Sizing Residential Heating and Air Conditioning Systems

ANSI/AHAM RAC-1-1992, Room Air Conditioners, Association of Home Appliance Manufacturers (AHAM). Available from AHAM, 20 North Wacker Drive, Chicago, IL 60606, (312) 984-5800.

Cooling and Heating Load Calculation Manual, GRP 138, American Society of Heating, Refrigerating, and Air-Conditioning Engineers, Inc. Available from ASHRAE (see Appendix A. Professional and Trade Associations).

Cooling and Heating Load Calculation Principles, American Society of Heating, Refrigerating, and Air-Conditioning Engineers, Inc. (ASHRAE). Available from ASHRAE (see Appendix A. Professional and Trade Associations).

Heat Loss Calculation (for Calculator) (for hydronic systems). Available from Hydronics Institute. (see Appendix A. Professional and Trade Associations).

Heat Loss Software (CD-ROM), E. Hogan. Available from Hydronics Institute (see Appendix A. Professional and Trade Associations).

Residential Duct Systems, Manual D (2nd ed.), Air Conditioning Contractors of America (ACCA). Available from ACCA (see Appendix A. Professional and Trade Associations).

Residential Equipment Selection Manual, Manual S (2nd ed.), Air Conditioning Contractors of America (see Appendix A. Professional and Trade Associations).

Residential Load Calculation, Manual J (8th ed.), Air Conditioning Contractors of America. Available from ACCA (see Appendix A. Professional and Trade Associations).

Web sites:

www.heatload.com
www.buildscape.com
www.HVAC-Software.com

Other Heating Costs

Energy requirements for nonheating purposes are an important addition factor to be considered when estimating total fuel costs for a house or building. This nonheating energy is used to operate oil and gas burners, automatic washers and dryers, water heaters, stoves, refrigerators, and the variety of other appliances and work-saving devices deemed so necessary for modern living.

This nonheating energy is commonly electricity or gas. Electricity is used to operate thermostats and other automatic controls found in heating and cooling equipment, burners, water heaters, home laundry equipment, refrigerators, and blowers for air conditioners, to mention only a few. Much of this equipment (e.g., water heaters and home laundry equipment) is also designed to be operated in conjunction with gas. Each of these appliances consumes energy that will account for a certain percentage of the homeowner's utility bill. This must be accounted for when estimating the total fuel requirement and heating cost.

The most common method of determining which appliance is the most suitable one for a particular set of circumstances is by making a comparison of their energy use. For example, the estimated usage for home laundry equipment (either a washer or a dryer) is 1 hour per week per occupant. In other words, a dryer would be operated 5 hours per week on the average in a household containing five people. For purposes of comparison, an electric clothes dryer draws 5.0 kWh and a gas clothes dryer uses approximately 3.5 therms of gas per month. Although these figures appear insignificant, they gain importance when the total nonheating energy use for all the appliances is added together. The cost per fuel unit will depend upon the local utility rates (see the following section). Some appliances, such as water heaters, will have to be compared on a basis that includes a number of different energy requirements met under a variety of environmental conditions (see Tables 4-14 and 4-15).

The problem for the homeowner is to determine which type of appliance burns which type of fuel most efficiently and at the lowest cost. Probably the poorest sources of information are the distributor of the particular appliance and the rate correspondent of the utility. They will, of course, be biased in favor of their own products. This is as it should be, because this is the reason they are in business. If they made a habit of recommending the products of their competitors, they would soon be out of business.

Quite often, consumer magazines give the energy use rates of different appliances and are quite frank in their comparison. This source is probably the most objective one to be found. However, knowing the rate at which an appliance consumes energy is only half the solution. It is also necessary to determine the utility rates.

Determining Utility Rates

No estimation of heating costs is complete until the utility rates for the various fuels available are determined and included in the estimate. Utility rates depend on so many variables that we can

Table 4-14 Usage Estimates for Gas Water Heating*

Size of Unit	Hot-Water Requirements	Cold-Water Temperature: Northern Localities 40°F	North Central Localities 50°F	South Central Localities 60°F	Southern Localities 70°F
0-BR	30 Gallon/Day	12.8	11.2	10.0	8.5
1-BR	40 Gallon/Day	16.4	14.7	12.9	10.9
2-BR	50 Gallon/Day	19.7	17.5	15.2	13.2
3-BR	60 Gallon/Day	22.9	20.5	17.7	15.3
4-BR	70 Gallon/Day	26.6	23.6	20.6	17.6
5-BR	80 Gallon/Day	30.2	27.0	23.5	20.2

**Estimates are given in therms per dwelling per month. Where wholesale electricity is to be used for lighting, refrigeration, cooking, and domestic hot water, the monthly demand for all purposes may be estimated at 2.65 watts per kWh.*

Courtesy National Oil Fuel Institute

Table 4-15 Usage Estimates for Electric Water Heating*

Size of Unit	Hot-Water Requirements	Cold-Water Temperature: Northern Localities 40°F	North Central Localities 50°F	South Central Localities 60°F	Southern Localities 70°F
0-BR	30 Gallon/Day	210	185	165	140
1-BR	40 Gallon/Day	280	250	220	185
2-BR	50 Gallon/Day	350	310	270	235
3-BR	60 Gallon/Day	420	375	325	280
4-BR	70 Gallon/Day	490	435	380	325
5-BR	80 Gallon/Day	560	500	435	375

**Estimates are given in kWH per dwelling per month.*

Courtesy National Oil Fuel Institute

calculate only an estimated cost, not a precise figure. The problem is to arrive at as accurate an estimation as possible.

Among the many variables involved in determining utility rates are the following:

1. Differing rate structures
2. Fuel/energy adjustment factors
3. Variations in energy block sizes and prices
4. Rate structure qualification procedures
5. Community utility taxes
6. Maximum demand charges

Differing rate structures exist for different facilities or usage. The problem is to determine which rate structure applies to the particular situation. This information can usually be obtained by contacting the local utility and consulting with the utility rate correspondent.

Although this is a simplified explanation, a typical utility bill is computed by adding the demand charge, the energy or commodity charge, and the fuel adjustment charge, and dividing by the total amount of electricity or gas used during the billing period. This produces the average electric or gas cost. The procedure is illustrated in Table 4-16.

Table 4-16 Determining Utility Rates

Demand Charge

300 × $0.60	=	$180.00
50 × $0.45	=	$ 22.50
		$202.50

Commodity Charges

Therms		Rate	×	Amount
3,000	×	5.0¢	=	$150.00
2,000	×	4.0¢	=	90.00
5,000				$240.00

Total commodity charges: 8 × $240.00 =	$1920.00
Fuel adjustment (40,000 × $0.01819)	727.60
	$2647.60

Average gas cost— $\frac{\$1620 + \$2647.60}{40{,}000} = 10.6¢/\text{therm}$

Chapter 5

Heating Fuels

Only the principal characteristics of those heating fuels used for domestic heating purposes will be considered in any great detail in this chapter. Descriptions of *firing methods* for the more commonly used heating fuels are found in Chapter 1, "Oil Burners," Chapter 2, "Gas Burners," Chapter 3, "Burning Solid Fuels" in Volume 2, and Chapter 3, "Stoves, Fireplaces, and Chimneys" in Volume 3.

Heating fuels are measured in physical units, such as gallons (fuel oil, propane, and kerosene), cubic feet (natural gas), tons (coal and pellets), kilowatt hours (electricity), and cords or pounds (wood).

Another way to measure heating fuels is by heat content. The British thermal unit (Btu) is the value commonly used in the United States for expressing the energy value or heat content of a heating fuel. One Btu is the amount of energy required to raise the temperature of 1 pound of water 1 degree Fahrenheit (°F), when the temperature of the water is about 39°F. The average Btu contents of different heating fuels are listed in Table 5-1.

The heating fuel values listed in Table 5-1 are the gross (or higher) values commonly used in energy calculations in the United States. These values are estimated by the Energy Information Administration in the Annual Energy Review and other sources. Net (or lower) heating values may also be used in energy calculations.

Table 5-1 Average Btu Content of Common Heating Fuels

Fuel Type	*Number of Btu/Unit*
Fuel oil (No. 2)	140,000/gallon
Natural gas	1,025,000/thousand cubic feet
Propane	91,330/gallon
Coal	28,000,000/ton
Electricity	3,412/kWh
Wood (air dried)	20,000,000/cord or 8,000/pound
Kerosene	135,000/gallon
Pellets	16,500,000/ton

Courtesy Office of Energy Efficiency and Renewable Energy/U.S. Department of Energy

When a fuel is burned in an appliance (e.g., a furnace or boiler), the combustion process causes the water contained in the fuel to vaporize. The water vapor contains heat energy, which is lost when the combustion gases are vented to the exterior. The difference between the higher and lower heating values is the amount of energy required to vaporize the water contained in a fuel or created in the combustion process. This difference can range from approximately 2 to 60 percent, depending on the fuel used.

Note

The newer, high-efficiency, condensing forced-air furnaces are designed to capture much of the heat energy contained in the water vapor before it exits the furnace stack and enters the chimney. Because electricity is not burned in a combustion process, there is no difference between a gross (higher) and net (lower) value.

Heating fuels also can be classified as solid, liquid, or gaseous, depending upon their physical state. Examples of solid fuels are coal and coke. Fuel oils are classified as liquid fuels; gaseous fuels include natural gas, manufactured gas, liquefied petroleum gas, and related types.

Natural Gas

Natural gas is a generic term commonly applied to those gases found in or near deposits of crude petroleum. It is *not* connected with the production of oil and should always be regarded as a separate and independent entity. Natural gas is piped under pressure from the gas fields to the consumer centers, and these pipelines now serve a considerable portion of the United States.

Natural gas is the richest of the gases and contains from 80 to 95 percent methane with small percentages of the other hydrocarbons. The heating value of natural gas varies from 1000 to 1200 Btu/ft^3, with the majority of gases averaging about 1000 Btu/ft^3. The caloric value will depend upon the locality.

Natural gas can be divided into three basic types:

1. Associated gas
2. Nonassociated gas
3. Dissolved gas

Associated gas is a free (undissolved) natural gas found in close contact with crude petroleum. *Nonassociated gas* is also a free gas but is not found in contact with the crude petroleum deposit. As the name suggests, *dissolved gas* is found in solution in the crude petroleum.

Manufactured Gas

A *manufactured gas* is any gas made by a manufacturing process. Raw materials used in the production of manufactured gas include coal, oil, coke, natural gas, or one of the other manufactured gases. For example, coal gas is made by distilling bituminous coal in either retorts or by-product coke ovens. Often several of these raw materials are used together as a base for the production of a manufactured gas.

The types of manufactured gases commercially available include:

1. Coal gas or by-product coke-oven gas
2. Oil gas
3. Blue water gas
4. Carbureted water gas
5. Producer gas
6. Reformed natural gas
7. Liquefied petroleum gas

Most of these gases have low caloric values (generally between 500 and 100 Btu/ft^3) and are produced primarily for industrial use. Producer gas and liquefied petroleum gas are also used for domestic purposes. The former may be used alone or in combination with other gases. Of all the manufactured gases, liquefied petroleum gas enjoys the widest application in domestic heating and cooking.

Liquefied Petroleum Gas

Liquefied petroleum (LP) gas is a hydrocarbon mixture extracted primarily from wet natural gas and sold commercially as propane, butane, bottled gas, or under a variety of different brand names. The terms *dry* and *wet* natural gas refer to the gasoline content per 1000 ft^3. Natural gas is regarded as dry if it contains less than 0.1 gallon of gasoline per 1000 ft^3. and wet if it contains more than 0.1 gallon.

Propane is used extensively for domestic heating purposes. It contains 2516 Btu/ft^3 (91,547 Btu per gallon), or about two and a half times the Btu content of methane (natural gas). Table 5-2 compares the typical properties of propane and butane. Unlike natural gas, propane is heavier than air. An undetected leak can be quite dangerous because the escaping gas will accumulate in layers at floor level. An explosion can be set off if the gas reaches the level of the pilot flame in the furnace.

Table 5-2 Typical Properties of LP Gas

Property	*Butane*	*Propane*
Btu per ft^3 60°F	3280	2516
Btu per lb	21,221	21,591
Btu per gal	102,032	91,547
Ft^3 per lb	6.506	8.58
Ft^3 per gal	31.26	36.69
Lb per gal	4.81	4.24

Courtesy National LP-Gas Association

Propane is frequently delivered in bulk and stored in large, stationary tanks holding from 100 to 1000 gallons. It is then piped into the home or building in much the same manner as natural gas. For small usage (e.g., trailers or homes that use only enough gas for one or two appliances), the LP gas is delivered to 5- to 25-gallon cylinders. The suggested clearances for LP gas cylinders and storage tanks are shown in Figure 5-1. The size of the storage tank required by an installation depends upon the weather zone in which it is located. Recommend tank sizes for the various weather zones in the United States are given in Figure 5-2.

Fuel Oils

Fuel oils are hydrocarbon mixtures obtained from crude petroleum by refining processes. They may be divided in the following six classes or grades:

1. *No. 1 fuel oil* for vaporizing pot-type and other burners designed for this fuel
2. *No. 2 fuel oil* for general-purpose domestic heating not requiring a lighter No. 1 oil
3. *No. 3 fuel oil* (obsolete since 1948)
4. *No. 4 fuel oil* for installations not equipped for preheating
5. *No. 5 fuel oil* for installations equipped for preheating
6. *No. 6 fuel oil* for burner installations equipped for preheating with a high-viscosity fuel

This classification of fuel oils is based on several characteristics, including: (1) viscosity, (2) flash point, (3) pour point, (4) ash content, (5) carbon residues, and (6) water and sediment content. All

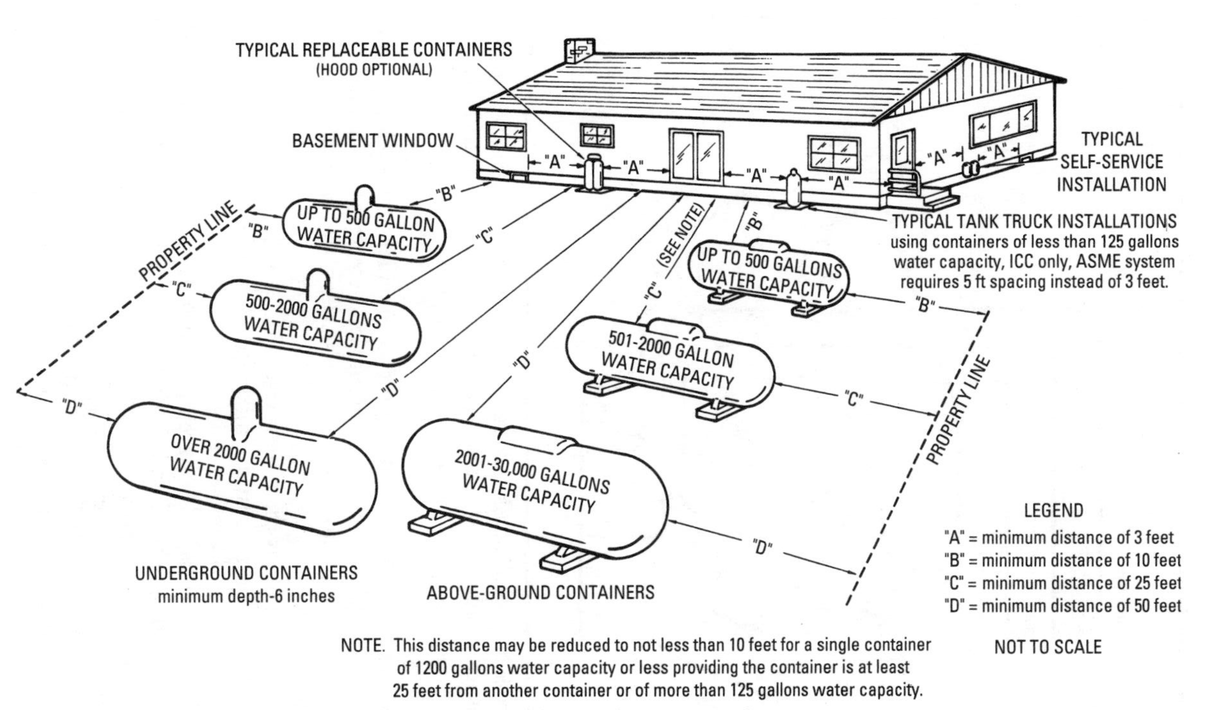

Figure 5-1 Suggested clearances for LP gas cylinders and storage tanks. *(Courtesy National LP Gas Association.)*

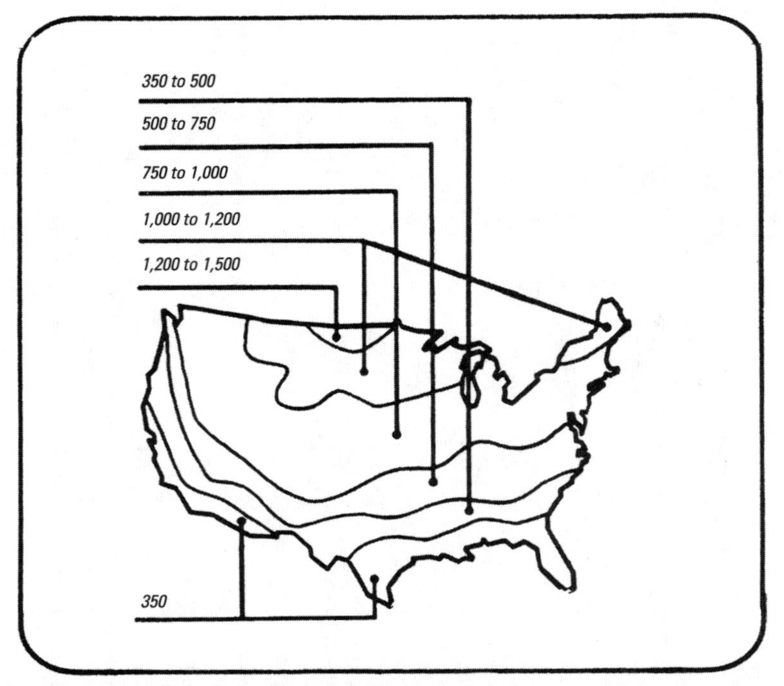

Figure 5-2 The six basic weather zones and the size of tank required to ensure an adequate supply of gas in each. *(Courtesy National LP Gas Association)*

these characteristics are important to consider when selecting a fuel oil. For example, the water content will determine its suitability for outdoor storage. A low viscosity indicates a fuel oil with good flow characteristics.

No. 1 fuel oil has the lowest viscosity (1.4–2.2). This is a distillate oil used in atomizing and similar burners because it can be easily broken down into small droplets.

No. 2 fuel oil is a distillate oil used in domestic oil burners that do not require preheating. It is slightly heavier than a No. 1 fuel oil. No. 4 fuel oil is slightly heavier than No. 2 and will work in some oil burners without preheating.

No. 2 and No. 4 fuel oils contain the highest ash content (0.10 to 0.50 percent by weight, respectively). The other grades contain little or no ash. Although slagging (i.e., the formation of noncombustible deposits) is commonly associated with solid fuels such as

Table 5-3 Btu Content of Principal Fuel Oils

Grade or Type	*Unit*	*Btu*
No. 1 oil	Gallon	137,400
No. 2 oil	Gallon	139,600
No. 3 oil	Gallon	141,800
No. 4 oil	Gallon	145,100
No. 5 oil	Gallon	148,800
No. 6 oil	Gallon	152,400
Natural gas	Cubic foot	950–1150
Propane	Cubic foot	2550
Butane	Cubic foot	3200

Courtesy Honeywell Tradeline Controls

coal, this problem also occurs in oil-burning equipment (see *Ash, Slag, and Clinker Formation* later in this chapter).

No. 5 and No. 6 (or *Bunker C*) fuel oils are heavier than the other grades and require preheating. In automatically operated oil burners, the preheating of the fuel oil must be completed before it is delivered for combustion. When the fuel oil has reached a suitable atomizing temperature, the preheating is considered completed.

PS 300 and *PS 400* are heating oils used on the West Coast. They roughly correspond to No. 5 and No. 6 fuel oils, respectively. *PS 300* is the lighter of the two and is used primarily for domestic heating purposes. *PS 400* is heavier, requires preheating, and is used for industrial purposes.

The grade of fuel oil to be used in an oil burner is specified by the manufacturer. This should also be stipulated on the label of the Underwriters Laboratories, Inc. (UL), and Underwriters' Laboratories of Canada. Table 5-3 compares the Btu content of the principal fuel oils.

Coal

Coal is a solid fuel created from deposits of ancient vegetation that have undergone a series of metamorphic changes resulting from pressure, heat, submersion, and other natural formative processes occurring over a long period of time. Because of the different combinations of these processes acting on the vegetation and the widely differing forms of vegetation involved, coal is a complex and nonuniform substance.

Coal is often classified according to the degree of metamorphic change to which it has been subjected into the following categories:

1. Anthracite
2. Bituminous coal
3. Semibituminous coal
4. Lignite

The formation of *peat* represents a precoal stage of development. After increased pressure and time, *lignite* begins to form. The next stage is *semibituminous coal,* followed by *bituminous coal* and finally *anthracite.* Each of these stages is characterized by an increase in the hardness of the coal. Anthracite is the hardest of the coals. *Graphite* represents a postcoal developmental stage and cannot be used for heating purposes.

Anthracite is clean, hard coal that burns with little or no luminous flame or smoke. It is difficult to ignite but burns with a uniform, low flame once the fire is started. Anthracite contains approximately 14,400 Btu/lb and is used for both domestic and industrial heating purposes. Its major disadvantage as a heating fuel is its cost.

Anthracite is divided by size into several different grades. Each of these grades (e.g., egg size, buckwheat size, pea size) is suitable for a specific size of firepot. They are described in greater detail in Chapter 3 of Volume 2 ("Coal-Firing Methods").

Bituminous coal is softer than anthracite and burns with a smoky, yellow flame. A great amount of smoke will result if it is improperly fired. The term *bituminous coal* actually covers a whole range of coals, many of which have widely differing combustion characteristics. Some of the coals belonging to this classification are hard, whereas others are soft.

The available heat for bituminous coal ranges from a low of 11,000 Btu/lb (*Indiana* bituminous) to a high of 14,100 Btu/lb (*Pocahontas* bituminous). The heat value of the latter approximates that of anthracite (about 14,400 Btu/lb); however, unlike anthracite coal, it is available in far greater supply, a factor that makes it a very economical solid fuel to use.

Semibituminous coal is a soft coal that ignites slowly and burns with a medium-length flame. Because it provides so little smoke, it is sometimes referred to as *smokeless coal.*

Lignite (sometime referred to as *brown coal*) ignites slowly, produces very little smoke, and contains a high degree of moisture. In structure it is midway between peat and bituminous coal. Lignite contains approximately half the available heat of anthracite (or about 7400 Btu/lb) and burns with a long flame. Its fire is almost smokeless, and it does not coke, a characteristic it shares with anthracite.

Lignite is considered a low-grade fuel, and its caloric value is low when compared with the other coals. Moreover, it is difficult to handle and store.

Coke

Coke is the infusible, solid residue remaining after the distillation of certain bituminous coal or as a by-product of petroleum distillation. Coke may also be obtained from petroleum residue, pitch, and other materials representing the residue of destructive distillation.

Coke will ignite more quickly than anthracite but less readily than bituminous coal. It burns rapidly with little draft. As a result, all openings or leaks into the ash pit must be closed tightly when coke is being burned.

Since less coke is burned per hour per square foot of grate than coal, a larger grate is required and a deep firepot is necessary to accommodate the thick bed of coal. Since coke contains very little hydrogen, the quick-flaming combustion that characterizes coal is not produced, but the fire is nearer even and regular.

The best size of coke recommended for general use, for small firepots where the fuel depth is not over 20 in, is that which passes over a 1-in screen and through a 1½-in screen. For large firepots where the fuel can be fired over 20 in deep, coke that passes over a 1-in screen and through a 3-in screen can be used, but a coke of uniform size is always more satisfactory.

Briquettes

Briquettes are a solid fuel prepared from coal dust and fines (i.e., finely crushed or powdered coal) by adding a binder and holding it under pressure.

The binder is commonly pitch or coal tar, but other substances and materials are also used. A briquette made with a pitch binder has the highest caloric value, exceeding that of any coal. The ash content is generally less than that of coal. Briquettes have been made from lignite coal dust and fines without a binder (only pressure being used).

Coal Oil

Coal oil is a heating fuel formerly derived from coal tar. Formerly it enjoyed widespread use, but it is has now been largely replaced by petroleum-based products. It is also called *paraffin oil* or *kerosene*.

Today, kerosene is obtained from petroleum through a refining process.

Wood as Fuel

A good, well-seasoned hardwood will provide approximately half as much heat value per pound as does good coal (Table 5-4). Wood is easy to ignite, burns with little smoke, and leaves comparatively little ash. On the other hand, it requires a larger storage space (commonly outdoors), and more labor is involved in its preparation.

Wood is commonly sold by the cord for heating purposes. A *standard cord* measures 4 ft high by 8 ft wide and contains wood cut to 4-ft lengths for a total of 128 ft^3. Only about 70 percent of a standard cord actually represents the wood content, however, because of the existence of air spaces between the wood.

The heat value of wood depends upon the *type* of wood being burned. Heat values per cord for a number of different types of wood are given in Tables 5-5 and 5-6. The data used to compile the tables were obtained from *Use of Wood for Fuel* (U.S. Department of Agriculture Bulletin No. 753). The wood heat

Table 5-4 Typical Firewoods

Name of Wood	*Type of Firewood*	*Combustion Characteristics*
Ash, white	Hardwood	Good firewood
Beech	Hardwood	Good firewood
Birch, yellow	Hardwood	Good firewood
Chestnut	Hardwood	Excessive sparking (can be dangerous)
Cottonwood	Hardwood	Good firewood
Elm, white	Hardwood	Difficult to split, but burns well
Hickory	Hardwood	Slow, steady fire; best firewood
Maple, sugar	Hardwood	Good firewood
Maple, red	Hardwood	Good firewood
Oak, red	Hardwood	Slow, steady fire
Oak, white	Hardwood	Slow, steady fire
Pine, yellow	Softwood	Quick, hot fire; smokier than hardwood
Pine, white	Softwood	Quick, hot fire; smokier than hardwood
Walnut, black	Hardwood	Good firewood, but difficult to find

Table 5-5 Heat Value (Million Btu per Cord) of Wood with 12% Moisture Content*

Name of Wood	*Heat Value*	*Equivalent Coal Heat Value*
Ash, white	28.3	1.09
Beech	31.1	1.20
Birch, yellow	30.4	1.17
Chestnut	20.7	0.80
Cottonwood	19.4	0.75
Elm, white	24.2	0.93
Hickory	35.3	1.36
Maple, sugar	30.4	1.17
Maple, red	26.3	1.01
Oak, red	30.4	1.17
Oak, white	32.5	1.25
Pine, yellow	26.0	1.00
Pine, white	18.1	0.70

**The equivalent coal heat values are based on 1 ton (2000 lb) of anthracite coal with a heat value of 13,000 Btu/lb.*

Table 5-6 Heat Value per Cord (million Btu) of Green Woods*

Name of Wood	*Heat Value*	*Equivalent Coal Heat Value*
Ash, white	26.0	1.00
Beech	27.1	1.04
Birch, yellow	27.2	1.05
Chestnut	19.2	0.75
Cottonwood	18.0	0.69
Elm, white	22.2	0.85
Hickory	29.0	1.12
Maple, sugar	27.4	1.05
Maple, red	23.7	0.91
Oak, red	27.5	1.06
Oak, white	28.7	1.10
Pine, yellow	23.7	0.91
Pine, white	17.3	0.67

**The equivalent coal heat values are based on 1 ton (2000 lb) of anthracite coal with a heat value of 13,000 Btu/lb.*

values are determined for cords containing approximately 90 cu ft of solid wood (i.e., about 70 percent of a standard cord). Note that the heat values for both green and dry wood (12 percent moisture content) are given in Tables 5-5 and 5-6. It is important to remember that the amount of moisture in a wood will affect its burning characteristics. A green wood will burn much more slowly than a drier one, and its fire will be more difficult to start.

All wood can be divided into either hardwood or softwood. Contrary to popular belief, the basic difference between the two groups is *not* the hardness of the wood. In other words, a wood that can be classified as a softwood is not necessarily softer than a hardwood. These terms do not refer to the physical properties of the wood but to its classification as a coniferous (needle or cone-bearing) or deciduous (broad-leafed) tree. A softwood comes from a coniferous tree, whereas a hardwood is obtained from a deciduous tree. You will find that some softwood trees produce wood that is harder than the wood obtained from some hardwood trees.

Chimney or flue fires are always a hazard when wood is used as a heating fuel. Resins and soot collect in these areas over time and can be ignited if there is a flare up of the fire. The possibility of this happening can be eliminated or greatly reduced by cleaning the chimney or flue from time to time.

Softwoods (e.g., white or yellow pine) contain greater amounts of resin than do hardwoods. This tends to give softwoods flammability, but it also results in more smoke.

Ash, Slag, and Clinker Formation

Ash is the noncombustible mineral residue that remains in a furnace or boiler after the fuel has been thoroughly burned. During the combustion process, the combustible portion of the fuel is consumed and the noncombustible ash remains in place. As the process continues, the ash residue is subjected to a certain degree of shrinkage. This shrinkage results in portions of the ash fusing together and forming more or less fluid globules, or *slag*. This process is referred to as *slagging*.

Under the proper temperature conditions, the fluid slag globules can solidify and form *clinkers* in the fuel bed or deposits on the heating surfaces of the furnace or boiler. Clinkers can obstruct the necessary air flow to the fire and result in its reduced efficiency or extinction. These deposits on the heating surfaces will reduce their heating capacity and must be removed. Suggestions for removing clinkers are given in Chapter 3 of Volume 2 ("Coal-Firing Methods").

Soot

Soot is a fine powder consisting primarily of carbon produced by the combustion process. Because of its extreme light weight, it frequently rises with the smoke from the fire and coats the interior walls of the chimney and flue. Although the heat loss from the insulating effect of a soot layer is small (generally under 6 percent), it can cause a considerable rise in the stack temperature. Soot accumulation can also clog the flues, thereby reducing the draft and resulting in improper combustion. Soot may be blasted loose from the walls of the chimney or flue with a jet of compressed air, or it may be sucked out with a vacuum cleaner. Another method is to use a brush to remove the accumulated soot layer from the walls.

Comparing Heating Fuel Costs

The cost of a heating fuel, its energy content, and the efficiency by which it is converted to useful heat are important factors for consideration when choosing a heating system. Table 5-1 lists the common heating fuels and their energy content (Btu/unit). Comparisons can be made by selecting one type of fuel (e.g., fuel oil) and a specific amount of that fuel oil (e.g., 100 gallons) as a basis for making a comparison in terms of their energy content and cost. Using the data provided in Table 5-1, one can see that 100 gallons of No. 2 fuel oil has an energy content of 14,000,000 British thermal units. That is roughly equivalent to 70 percent of a cord of air dried wood, a half ton of coal, or a little less than 14,000 cubic feet of natural gas. Multiply these figures by the cost per gallon (oil), cord (wood), ton (coal), or cubic foot (natural gas) in the area where the structure is located, and you have a cost basis for comparing heating fuels.

Another factor to consider when comparing heating fuel costs is the efficiency of the heating appliance itself. In other words, how efficient is the heating appliance (furnace, boiler, water heater, etc.) in turning a particular fuel into heat energy? Table 5-7 lists a variety of different heating appliances and their average fuel conversion efficiencies with different fuels.

In forced-warm-air heating systems and hydronic heating systems, the heated air or hot water is forced through the ducts or pipes by fans or pumps. The fans or pumps are operated by electricity, which is an additional cost when using these types of heating systems.

Table 5-7 Estimated Average Fuel Conversion Efficiency

Fuel Type	*Appliance Type*	*Fuel Efficiency*
Bituminous coal	Hand-fired central heating appliance	45.0
	Stoker-fired central heating appliance	60.0
Oil	High-efficiency central heating appliance	89.0
	Typical central heating appliance	80.0
	Water heater (50 gallon)	59.5
Gas	High-efficiency central heating furnace	97.0
	Typical central heating boiler	85.0
	Minimum efficiency central heating furnace	78.0
	Room heater (unvented)	99.0
	Room heater (vented)	65.0
	Water heater (50 gallon)	62.0
Electricity	Baseboard (resistance)	99.0
	Forced-air central heating appliance	97.0
	Heat pump central heating system	200+
	Ground source heat pump	300+
	Water heater (50 gallon)	97.0
Wood and pellets	Franklin stove	30.0–40.0
	Stoves with circulating fans	40.0–70.0
	Catalytic stoves	65.0–75.0
	Pellet stoves	85.0–90.0

Courtesy Office of Energy Efficiency and Renewable Energy/U.S. Department of Energy

Chapter 6

Warm-Air Heating Systems

Air is the medium used for conveying heat to the various rooms and spaces within a structure heated by a warm-air furnace. It is also the principal criterion for distinguishing warm-air heating systems from other types in use.

The warm-air furnace is a self-contained and self-enclosed heating unit that is usually (but not always) centrally located in the structure. Depending upon the design, any one of several fuels can be used to fire the furnace. Cool air enters the furnace and is heated as it comes in contact with the hot metal heating surfaces. As the air becomes warmer, it also becomes lighter, which causes it to rise. The warmer, lighter air continues to rise until it is either discharged directly into a room (as in the so-called pipeless gravity system) or carried through a duct system to warm-air outlets located at some distance from the furnace. After the warm air surrenders its heat, it becomes cooler and heavier. Its increased weight causes it to fall back to the furnace, where it is reheated and repeats the cycle. This is a very simplified description of the operating principles involved in warm-air heating, and it especially typifies those involved in gravity heating systems.

Classifying Warm-Air Heating Systems

A *warm-air heating system* is one in which the air is heated in a furnace and circulated through the rest of the structure either by gravity or motor-driven centrifugal fans. If the former is the case, then the system is commonly referred to as a *gravity warm-air heating system*. Any system in which air circulation depends *primarily* on mechanical means for its motive force is called a *forced-warm-air heating system*. The stress on the word "primarily" is intentional because some gravity warm-air systems use fans to supplement gravity flow, and this may prove confusing at first. In any event, one of the oldest forms of classifying warm-air heating systems has been on the basis of which method of air circulation is used: gravity or forced air.

Forced-warm-air heating systems are often classified according to the duct arrangement used. The two basic types of duct arrangements used are:

1. Perimeter duct systems
2. Extended-plenum duct systems

A *perimeter duct system* is one in which the supply outlets are located around the perimeter (i.e., the outer edge) of the structure, close to the floor of the outside wall or on the floor itself. The return grilles are generally placed near the ceiling on the inside wall. The two basic perimeter duct systems are:

1. The perimeter-loop duct system (Figure 6-1)
2. The radial-type perimeter duct system (Figure 6-2)

An *extended-plenum system* (Figure 6-3) consists of a large rectangular duct that extends straight out from the furnace plenum in a

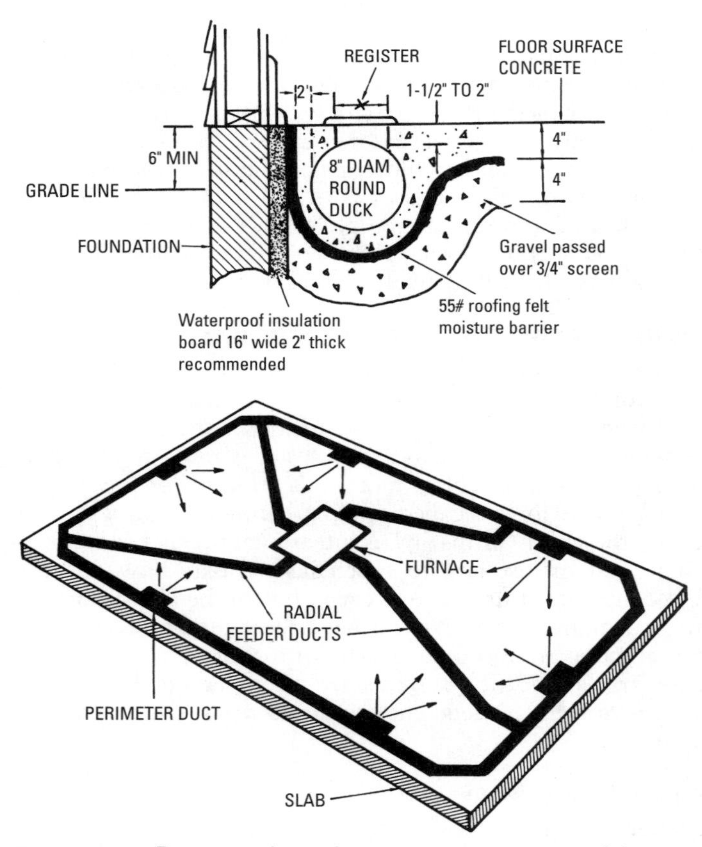

Figure 6-1 Perimeter-loop duct system in concrete slab construction.

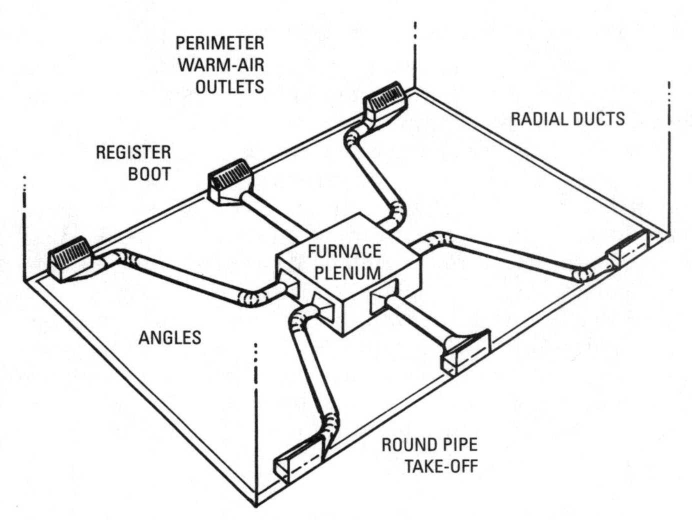

Figure 6-2 Radial perimeter duct system.

straight line down the center of the basement, attic, or ceiling. Round supply ducts connect the plenum to the heat-emitting units.

These and other modifications of duct arrangements are described in considerable detail in Chapter 7 of Volume 2, "Ducts and Duct Systems."

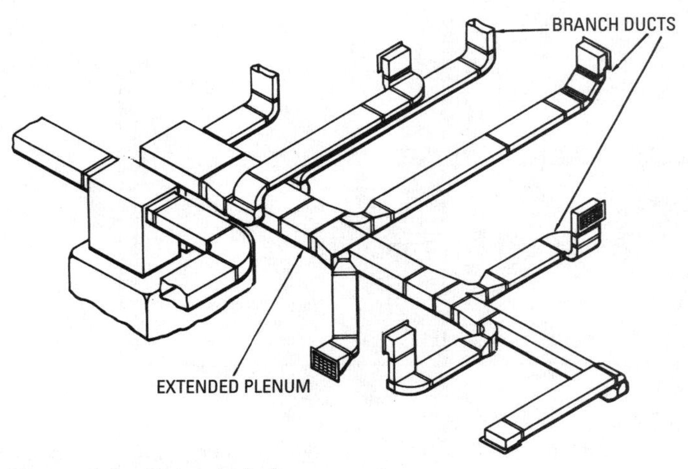

Figure 6-3 Extended-plenum system.

Gravity Warm-Air Heating Systems

A *gravity warm-air heating system* (Figure 6-4) consists of a properly designed furnace (with casing and smoke pipe), a warm-air supply delivery system, and a cool-air return system.

In a central heating system, the warm air is delivered to the rooms and spaces being heated through a system of air ducts, the exception being a gravity floor furnace or a pipeless furnace. Those ducts located in the basement and extending (or leading) from the furnace to the basement ceiling are referred to as *leaders*. The *stacks* are warm-air ducts or pipes that connect to the leaders and extend vertically within the walls and up through the building. The warm air enters a room through registers located on the walls or floor (the former will generally connect to a stack, the latter to a leader). The cooler room air exits a room by means of return grilles and travels back through return air ducts to the furnace where it is reheated and recirculated.

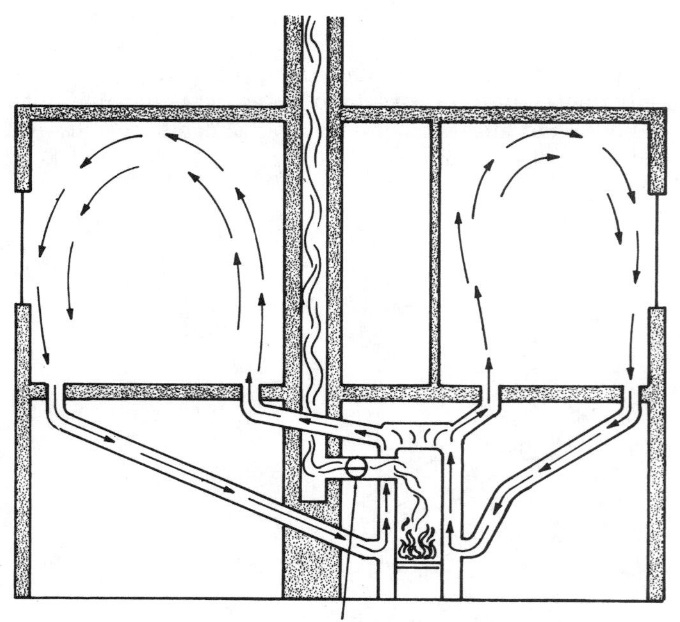

Figure 6-4 Operating principle of a gravity warm-air heating system.

A gravity warm-air heating system differs from the forced-warm-air type by relying primarily on gravity to effect air circulation. The operating principle of the gravity warm-air heating system is based on the fact that the weight per unit volume of air decreases as its temperature increases and increases as its temperature decreases. As the air is heated in the furnace, it expands and becomes lighter. Because it is lighter, it is displaced by the cooler, heavier air entering the furnace. The warmer, lighter air moves through the leaders and up through the wall stacks, where it enters the rooms through the registers. As the air cools, it becomes heavier and returns through the return air grilles and ducts to the furnace—hence the name *gravity* warm-air heating systems.

Air circulation in a gravity warm-air heating system depends upon the difference in temperature between the rising warm air and the cooler air that is falling and returning to the furnace for reheating. The greater the difference in temperature, the faster the movement of the air; however, natural conditions will place upper limits on the speed of the air movement, making it advisable to equip a gravity-type central furnace with a fan as an integral part of its construction. Such an integral fan should be powerful enough to overcome internal duct resistance to air flow and is *absolutely* necessary if air filters are used in the heating system. The motive force of air circulation in a standard gravity system is not strong enough to overcome the resistance of filters.

Planning a Gravity Warm-Air Heating System

Several trade and professional associations provide information for planning a gravity warm-air heating system. A very useful source of information is the *Gravity Code and Manual for the Design and Installation of Gravity Warm Air Heating Systems* (Manual 5) from the former National Warm-Air Heating and Air Conditioning Association. You should also check relevant publications from its successor association, the Air-Conditioning Contractors of America (ACCA).

The following recommendations for planning such a system are offered as a basic planning guide:

1. Calculate the heat loss from each room in the structure, and add these together for the *total heat loss*.
2. Plan the leaders (i.e., the horizontal ducts or pipes leading from the furnace) so that none are longer than 10 ft in length.
3. Keep the number of elbows in the leader to a minimum.

4. Locate the warm-air outlets (registers) and return air grilles as close to the floor in each room as possible.
5. Locate the warm-air outlets (registers) on the inside walls of each room and in that part of the room closest to the furnace.
6. Select warm-air and return air ducts according to their Btu carrying capacities.
7. Select a furnace capable of delivering in Btu/h a register delivery *equal* to the total heat loss calculated for the structure.
8. Locate the furnace in the *lowest* part of the structure and as near to the center as possible.

Forced-Warm-Air Heating Systems

A *forced-warm-air heating system* (Figure 6-5) consists of a furnace equipped with a blower, the necessary controls, an air-duct system, and an adequate number of suitably located warm-air

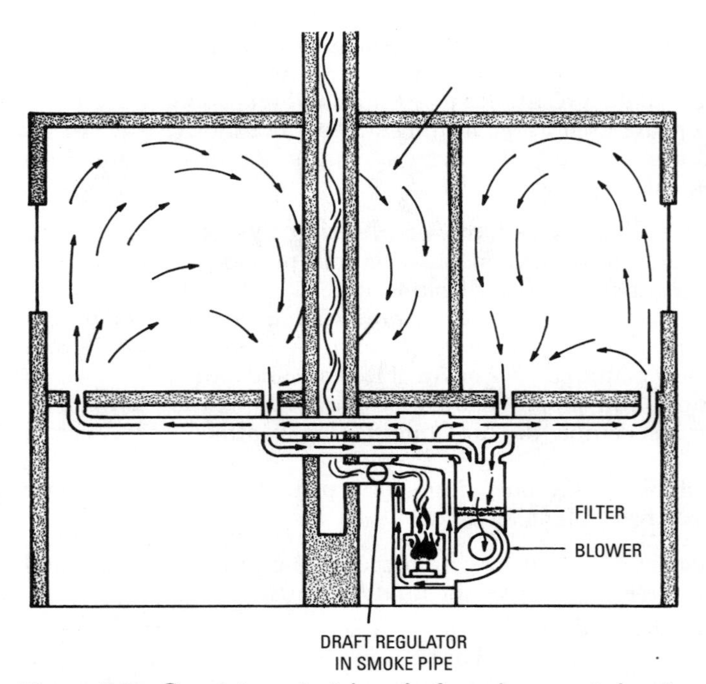

Figure 6-5 Operating principles of a forced-warm-air heating system.

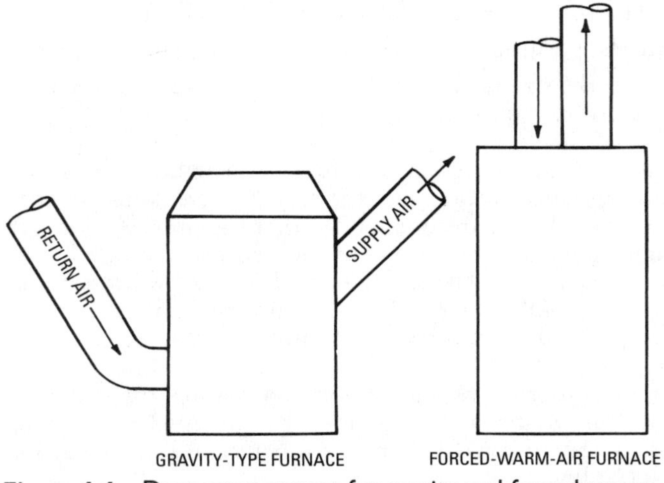

Figure 6-6 Duct arrangement for gravity and forced-warm-air furnaces.

registers and cold-air returns. The warm air travels from the furnace through the supply ducts to the room registers. The cold air returns through the return to the furnace, where it is reheated and recirculated. The return-air ducts are joined together before reaching the furnace, a feature distinguishing the duct arrangement in this type of heating system from that found in heating systems relying upon the gravity flow principle of air circulation (Figure 6-6).

Planning a Forced-Warm-Air Heating System

ASHRAE publications (such as recent editions of the *ASHRAE Guide*) summarize design procedures for planning several types of forced-warm-air heating systems. The following six elements are common to each planning procedure:

1. Calculate the heat loss for each room or space, and from this data determine the total heat loss for the structure.
2. Determine the required furnace-bonnet capacity from the total heat loss of the structure.
3. Determine the location of the diffusers (warm-air outlets) on the building plan.

4. Calculate the required Btu/h delivery of each diffuser.
5. Locate the position of the feeder ducts on the building plan.
6. Calculate the size of each feeder duct on the basis of the total Btu/h delivery it must supply.

Other elements in the design of a forced-warm-air heating system will be specifically oriented to the characteristics of each system. For example, it is recommended that the maximum length of an extended plenum (nonexistent in a forced-warm-air perimeter system) not exceed 35 ft. The locations and types of warm-air outlets and return-air inlets selected for the system will also depend upon the type of forced-warm-air heating system used in the structure.

Useful and authoritative sources of information for installing warm-air heating systems are provided by the Air-Conditioning Contractors of America (ACCA) and the National Fire Protection Association (NFPA).

Proprietary systems should always be installed according to the manufacturer's instructions. Any variation from these instructions increases the probability of error and future problems in the system. It may also void the equipment warranty.

Note

Local codes and ordinances always take precedence over installation instructions provided by manufacturers. When there is a conflict, consult the local regulatory body for a resolution.

Perimeter-Loop Warm-Air Heating Systems

The *perimeter-loop warm-air heating system* (Figure 6-1) was originally designed for use in residences built on a concrete slab rather than over a basement. The proven success of this duct arrangement in providing efficient and economical heating has resulted in its installation in all types of construction. This is rapidly becoming one of the most popular forms of warm-air heating.

In the perimeter-loop system, round ducts are embedded in the concrete slab or suspended beneath the floor. The air is heated in a warm-air furnace equipped with a blower and is forced through the ducts leading from the furnace to a continuous duct extending around the outer perimeter of the structure. The registers (diffusers) through which the heated air enters the various rooms and spaces

Figure 6-7 Floor location for a warm-air outlet.

are located along this outer perimeter duct. The most efficient operation can be achieved by placing these warm-air outlets on the floor next to the wall and *beneath a window* (Figure 6-7). By placing the warm-air outlets here, window drafts are eliminated and the colder outside walls are kept warmed. Research shows that up to 80 percent of the heat loss from a structure can occur at these locations.

The Air Conditioning Contractors of America's publication *Manual 4—Installing Techniques for Perimeter Heating and Cooling* is the most authoritative and useful source of information about this subject. (Ask for the most recent edition, because the manual has undergone a number of revisions.)

Ceiling Panel Systems

A *ceiling panel system* is a forced-warm-air heating system in which the heated air is delivered through ducts to an enclosed space above a false ceiling. There are no air supply outlets in the ceiling, and the heated air is therefore blocked from direct penetration of the occupied spaces below. Because its downward path is blocked, the heated air spreads over the entire surface area of the false ceiling. The ceiling eventually absorbs the heat, and the cooler air is returned to the furnace for reheating and recirculation. The heat is transferred from the ceiling to the occupied spaces by radiation. A more appropriate name for this type of panel heating system might be "ceiling *space* panel system." Other ceiling panel systems are described in Chapter 1 of Volume 3, "Radiant Heating."

A ceiling panel system of this type does not differ from the standard forced-warm-air heating system except in the design of the heat-emitting unit. Instead of using compact (individual heat-emitting units) a ceiling panel system uses a portion of the structure itself. As a result, considerable experience is necessary for installing this type of heating system. It is recommended only for new construction, and never as a conversion from an existing heating system.

Crawl Space Plenum Systems

A *crawl space plenum system* uses the entire underfloor crawl space as a sealed plenum chamber to distribute warm or cool air. The use of the crawl space under the house as a plenum was first investigated and developed in the 1970s with funds provided by several wood technology organizations. This system is called the *Plen-Wood System.*

In most conventionally shaped structures, air distribution is achieved without using ducts. In large, irregularly shaped structures, stub ducts are added to provide uniform air distribution. The conditioned air is supplied to the plenum by a forced-warm-air furnace, a downflow (or counterflow) furnace and cooling unit, or a heat pump with an air handler. No modifications are required to the HVAC heating and cooling unit.

The sealed underfloor crawl space in the Plen-Wood System is used as a plenum for the air supply in the winter and an air return during the summer (Figure 6-8). A damper in the HVAC unit is used to switch from one to the other. By using only the fan of the HVAC unit, comfortable temperatures can be maintained most of the year by circulating air from the plenum throughout the house.

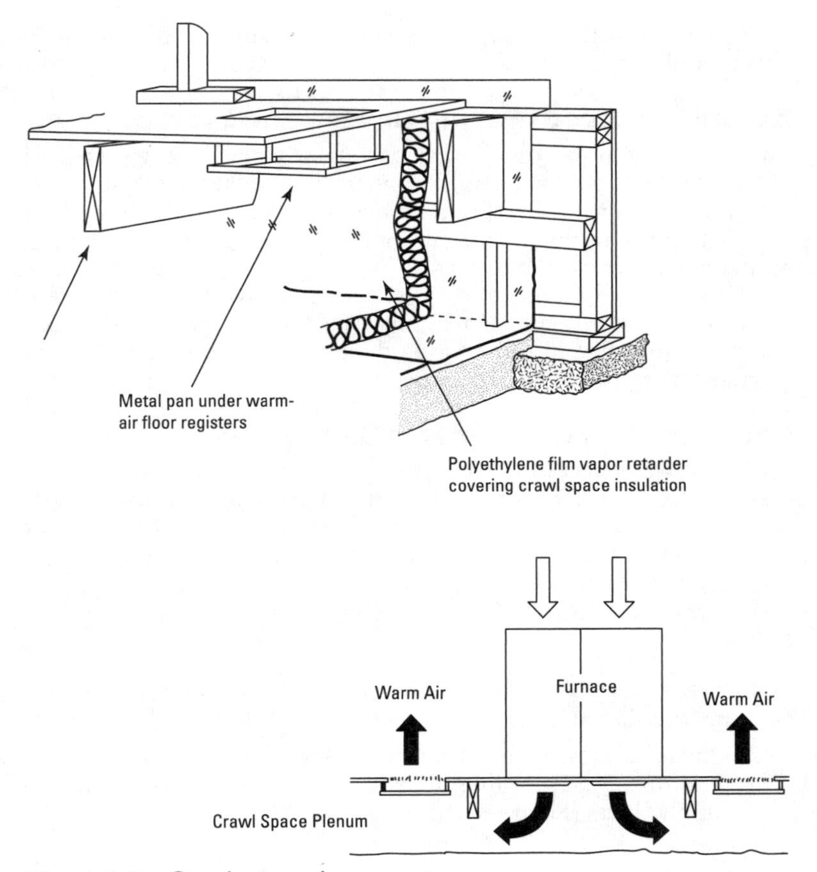

Figure 6-8 Crawl space plenum system.

Check your local building code before deciding on a Plen-Wood System. Although the Plen-Wood System is accepted by the major building codes, local building codes may place special restrictions on it. In many areas these local codes require the use of fully ducted heating supply runs. They will not permit the use of an open crawl space plenum. This system also requires the construction of a closed, tightly sealed crawl space.

Contact the American Plywood Association (P.O. Box 11700, Tacoma, WA 98411) for additional information about the Plen-Wood

System. See also the appropriate section in Chapter 7 of Volume 3, "Ducts and Duct Systems."

Zoning a Forced-Warm-Air Heating System

The design of some residences, particularly the larger split-level homes, often causes balancing problems for the heating system. As a result, certain rooms will remain much colder than the rest of the house. This is a problem that cannot be corrected simply by turning up the heat, because the rest of the house will become too hot. The most effective solution is to divide the heating system into two separate zones, each controlled by its own thermostat. A motorized damper is installed in the existing duct system and is regulated by thermostats to obtain the balance.

Balancing a Warm-Air Heating System

It is not always necessary to go to the expense of zoning to ensure that each room receives enough heat. Sometimes a heating system can be balanced by reducing the air delivery to rooms or sections of the structure that require less heat. This results in automatically diverting more air and heat to those areas that require it.

The procedure for balancing a warm-air heating system is as follows:

1. Pick a day for balancing the system when the outdoor dry-bulb temperature is 40°F or below.
2. Open the dampers in all warm-air outlets as widely as possible (Figure 6-9). The same holds true if the dampers are in the supply ducts (Figure 6-10).

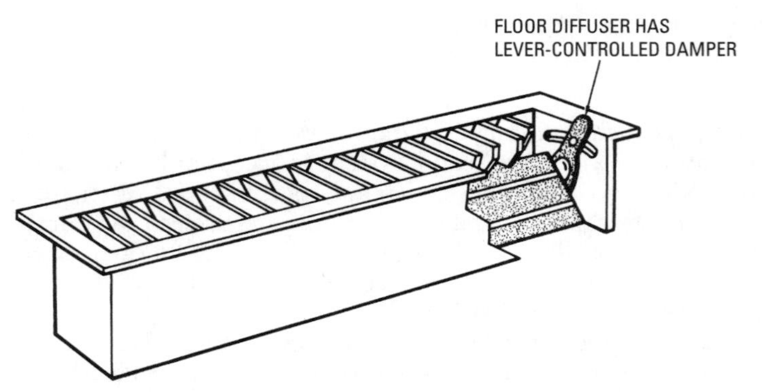

Figure 6-9 Adjusting floor diffuser to full open position.

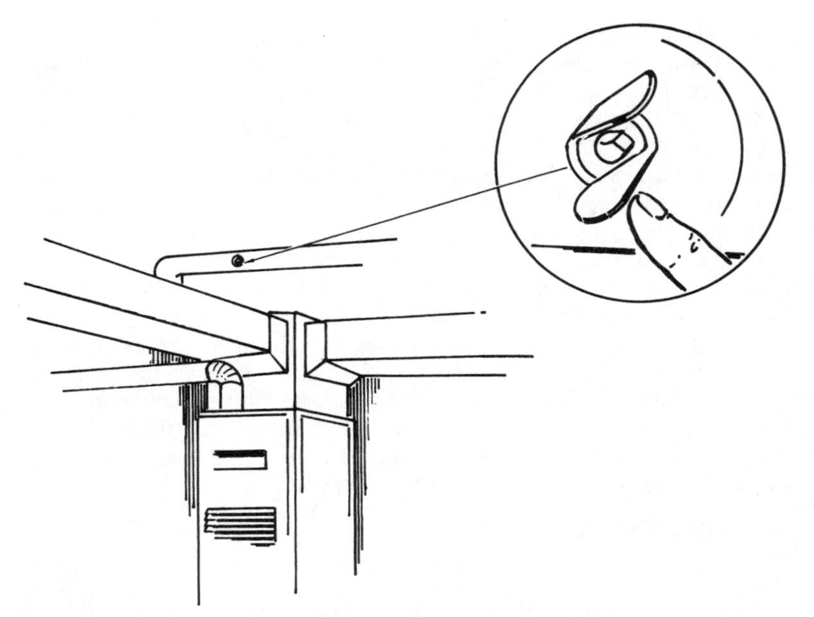

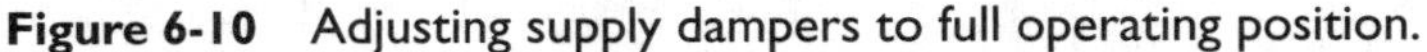

Figure 6-10 Adjusting supply dampers to full operating position.

3. Leave the thermostat at one setting for at least 3 hours, and make certain the furnace blower is running.
4. Check the temperatures in all rooms. This can be done with thermometers (if you are certain they register equally), or simply by making your own judgment.
5. Leave the dampers in the coldest rooms wide open, and adjust the dampers in the warmer rooms to obtain the desired balance. After each adjustment, allow the system to stabilize for at least 30 minutes before checking the temperature or making the next adjustment.

When balancing a heating system, do not expect immediate results. A little patience makes it well worth the effort.

Warm-Air Furnaces

The warm-air furnace is a self-contained heating unit designed to supply warm air to the interior of a structure. Warm-air furnaces used in central heating systems usually employ a system of ducts to distribute the air to the various rooms and spaces within the structure.

These furnaces can be classified according to the method of air circulation into the following two basic categories: (1) *gravity* warm-air furnaces, and (2) *forced*-warm-air furnaces. A distinct advantage of the forced-warm-air furnace is that its blower can move the air in any direction. As a result, this type of furnace can be located anywhere in the structure. Furthermore, the ducts do not have to be located above the heating unit, as is the case with gravity warm-air furnaces. These advantages of the forced-warm-air furnace have contributed to its tremendous popularity over gravity-type furnaces. Both types are described in considerable detail in Chapter 10, "Furnace Fundamentals."

Other criteria used for classifying warm-air furnaces include: (1) the type of fuel used (e.g., gas, oil, coal, or electricity), (2) the method of air distribution (ducts or pipeless), and (3) the method of firing (automatic or manual). Specific chapters in this book deal individually with some of these criteria (e.g., Chapter 11, "Gas Furnaces," or Chapter 12, "Oil Furnaces").

Control Components

The controls used in a warm-air heating system will depend upon a number of factors, including: (1) the method of air circulation used (e.g., forced or gravity), (2) the size of the structure, and (3) the method used for supplying the fuel (automatic or manual) to the fire. In other words, for all intents and purposes, a control system is custom designed to fulfill the requirements of a specific heating system.

All gravity and forced-warm-air furnaces in central heating systems are controlled by a thermostat located in one of the rooms. If the heating system is zoned, two or more room thermostats are used. In forced-warm-air systems equipped with automatic burners or strokers, the thermostat will also control the operation of these units.

Each furnace should be equipped with a high-limit control to shut off the air or gas burned when plenum air temperatures exceed the furnace manufacturer's design limits. The high-limit control is automatic and will switch on the burner again as soon as the air temperature in the plenum has returned to normal (i.e., reached a level below the manufacturer's design limits). The high-limit control is frequently designed to operate in conjunction with the fan (blower) switch in forced warm-air furnaces.

Other controls used in warm-air heating systems are described in Volume 2—Chapter 4, "Thermostats and Humidistats," Chapter 5, "Gas and Oil Controls," and Chapter 6, "Other Automatic Controls."

Ducts and Duct Sizing

Ducts are passageways used for conveying air from a furnace or cooling unit to the rooms and spaces within a structure. The ductwork is also used to return the air to its source for recirculation.

The air is distributed to the rooms or spaces by means of grilles, registers, and diffusers. The size and location of the supply-air outlets and the return-air inlets are determined by such design factors as: (1) use (heating and cooling), (2) air velocity, (3) throw, (4) drop, and (5) desired distribution pattern. These and other aspects of air outlet and air inlet design and selection are considered more thoroughly in Chapter 7 of Volume 2, "Ducts and Duct Systems."

Designing and sizing a duct system is a compromise between the requirements and limitations of both the structure and the heating and cooling system. Duct sizing will be affected by a number of variables, including (1) the architectural design of the structure, (2) space limitations, (3) the required air supply, (4) the allowable duct air velocities, (5) the desired noise level, and (6) the capacity of the blower. Adequately sized ducts represent the best possible compromise among these variables.

Accuracy in estimating the resistance to the flow of air through the duct system is important in the selection of blowers for application to such systems. Resistance should be kept as low as possible in the interest of economy. However, underestimating the resistance will result in failure of the blower to deliver the required volume of air. The various calculation methods used for sizing ducts are given in Chapter 7 of Volume 2, "Ducts and Duct Systems."

A number of precautions should be taken in the design of a duct system. For example, careful study should be made of the building drawings with consideration given to the construction of duct locations and clearances. Other recommendations that should be considered when designing a duct system can be summarized as follows:

1. Keep all duct runs as short as possible, bearing in mind that the air flow should be conducted as directly as possible between its source and delivery points, with the fewest possible changes in direction.
2. Select locations of duct outlets so as to ensure proper air distribution.
3. Provide ducts with cross-sectional areas that will permit air to flow at suitable velocities.
4. Design for moderate velocities in all ventilating work to avoid waste of power and reduce noise.

Cooling with a Warm-Air Heating System

The simplest method of cooling a house in the summer is to use the blower of a forced-warm-air furnace. Most furnaces have a manual blower switch that can be used to circulate the air.

If you plan to install a central air conditioning system capable of maintaining comfortable indoor temperatures during the warmer months, you must first examine the existing furnace and duct system and determine what changes (if any) must be made to handle the cooling load.

Forced-warm-air furnaces can often be converted to year-round air conditioning by installing a cooling coil in the furnace supply duct and connecting it to a compact compressor-condenser unit located outdoors. The existing duct system may be inadequate for air conditioning, but this can be modified to handle the cool air by increasing the size of certain ducts or by installing additional ducts.

Some furnaces (particularly gravity warm-air furnaces) cannot be fitted with cooling coils, *but* the existing ductwork is adequately sized for cooling. If this should be the case, the furnace can be bypassed with the cooling coils and air handling equipment installed in a short length of duct that feeds into the main supply- and return-air ducts at a point near the furnace. Dampers must be placed in the ducts between the furnace and the cooling coil to prevent the warm and cool air from mixing.

If both the existing furnace and the existing ductwork cannot be used, it will be necessary to install a separate and independent air conditioning system. Adding one or more window units is the most common method of accomplishing this. Slightly more expensive is the use of a through-the-wall unit, which is installed by cutting a hole in the wall. The compressor-condenser section is located on the outside to reduce the noise level. Neither window units nor through-the-wall units use ductwork to distribute the cool air.

The most expensive separate and independent air conditioning systems are the single-package-unit systems and the very popular split systems. Both use ductwork to distribute the cool air. A single-package-unit system is limited to the installation of the cooling unit in the attic or basement, with the former location offering the fewest design problems. The advantage of a split system is that the cooling unit can be located anywhere inside the structure. In both systems, the preferred location of the compressor-condenser section is outdoors.

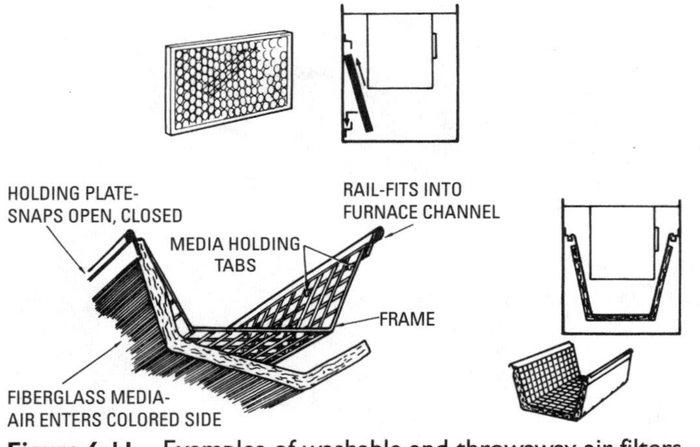

Figure 6-11 Examples of washable and throwaway air filters used in warm-air furnaces.

Air Cleaning

The air cleaning equipment used in forced-warm-air heating and cooling systems commonly takes the form of washable or disposable air filters or electronic air cleaners.

Washable and disposable air filters are installed in the return-air plenum of a forced-warm-air furnace as shown in Figure 6-11. These are dry filters consisting of cellulose fibers, steel wool, or some other suitable material set in a wire frame. These filters are effective only when dust concentrations are limited to 4 grains per 1000 ft^3 of air or lower. Consequently, they are restricted in use to residences or small buildings. Air filters are not used in gravity warm-air heating systems because they impede the rate of air flow.

Electronic air filters are designed for return-air-duct installation at the furnace, air handler (in the duct), or air conditioning unit. They generally operate on the electrostatic precipitation principle and are capable of removing up to 95 percent of all airborne particles (e.g., dust, tobacco, and smoke). The dust particles are given an electric charge when they pass through an ionizing field and are collected when they subsequently pass between collector plates having an opposite charge. Typical installations are illustrated in Figure 6-12.

The various types of air filters and electronic air cleaners used in cleaning and filtering the air are described in Chapter 12 of Volume 3, "Air Cleaners and Filters."

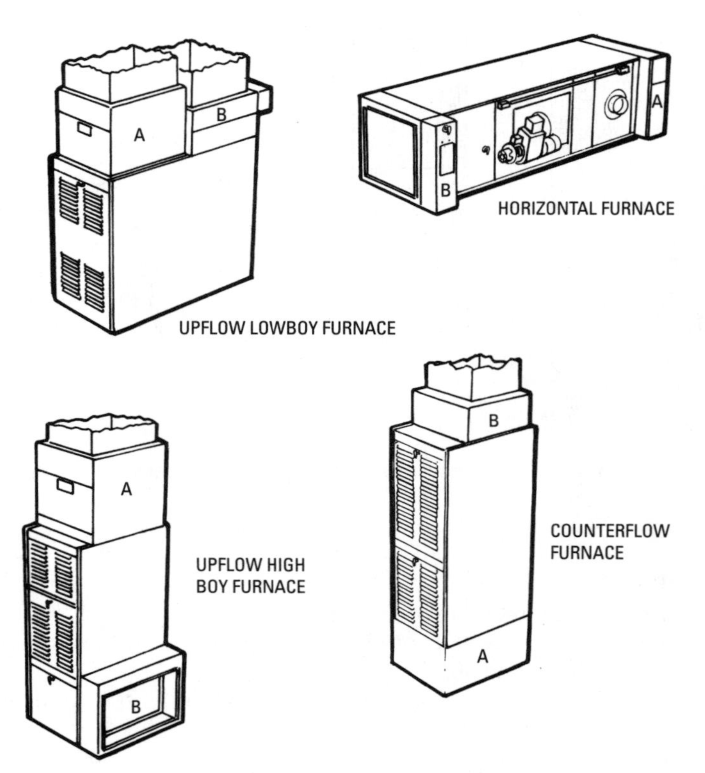

Figure 6-12 (A) Typical installations of cooling coils and (B) electronic air cleaners on forced-warm-air furnaces.

Humidifiers and Dehumidifiers

Air that has a very low level of humidity is too dry for comfort and can cause damage to walls, interior woods, and furnishings. This results from the fact that dry air absorbs moisture from other sources, including the human body. The moisture-robbing effect of dry air is particularly noticeable on the sensitive nasal and throat membranes. Moisture can be added to the air by humidification, and the device used to add moisture is called a *humidifier.*

The humidifiers used in heating and cooling systems are designed to maintain the relative humidity within the comfort zone. They are available in a number of different types, each based on a different operating principle. Pan-type humidifiers, for example, contain a reservoir of water that evaporates into the warm air flowing through the supply duct. The water level in the reservoir is controlled by a float control. The desired humidity level is determined by a humidity control setting. When the relative humidity drops below the humidity control setting, a humidifier fan is actuated and air is blown over the water in the reservoir. Moisture is then picked up by the air and blown into the space to be humidified. Other types of humidifiers include (1) spray-type air washers and (2) steam-type humidifiers (either air or electrically operated). Typical humidifier installations are illustrated in Figure 6-13.

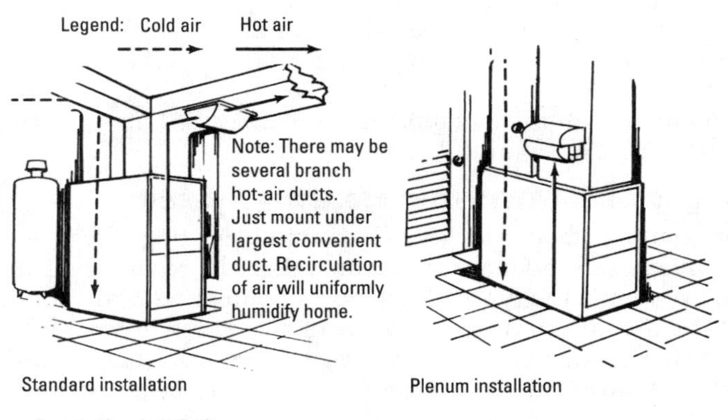

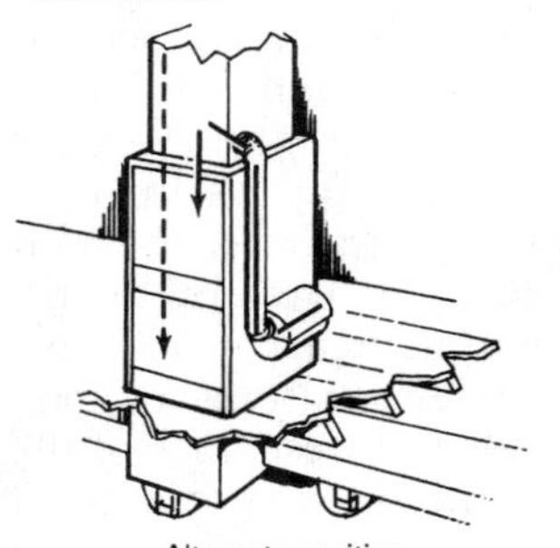

Figure 6-13 Typical installation of power humidifiers on forced-warm-air furnaces.

Sometimes air will have a humidity level that is too high. The excess moisture resulting from this condition can also cause damage to walls, interior woods, and furnishings as well as prove very uncomfortable for the occupants. Excess moisture can be removed from the air, and the device used to remove the moisture is called a *dehumidifier.*

Dehumidifiers operate either on the cooling or absorption method. The former method accomplishes dehumidification by an air washer with a water-spray temperature lower than the dew point of the air passing through the unit. Condensation occurs, and both latent and sensible heat are removed.

In the absorption method of dehumidification, sorbent materials are used for removing moisture from the air. Air to be dehumidified is drawn or blown through a screened bed of dry-solid absorbent, and the water vapor in the air is caught and retained in the pores of the absorbent.

Both humidifiers and dehumidifiers are described in greater detail in Chapter 11 of Volume 3.

Advantages of a Warm-Air Heating System

By present comfort standards, gravity warm-air heating offers no special advantage and too many disadvantages to be recommended as a heating system in the types of structures popular with the public today. The architectural design of these structures *necessitates* the use of a forced-warm-air heating system. Because the gravity system lacks a blower, air circulation (and therefore heat distribution) depends upon the temperature difference between the rising warm air and the descending cold air. Unfortunately, many of our houses and buildings are designed in such a way that a gravity system would require a considerable temperature difference to obtain the desired rate of air movement.

If you design the structure *around* a gravity heating system, then quite different results are possible. For example, a residence built by a gentleman named Wendell Thomas in western North Carolina contains a gravity-type heating system using a simple wood-burning stove that produces indoor temperatures ranging between 60 and 75°F year-round. The north and west sides of the structure were set into the ground. No windows were placed in the north wall, and the windows in the other walls were two or three panes thick.

As shown in Figure 6-14, the warm air heated by the stove rises and gives off its heat to the room. As it loses its heat, it cools and descends along the inside surfaces of the exterior walls,

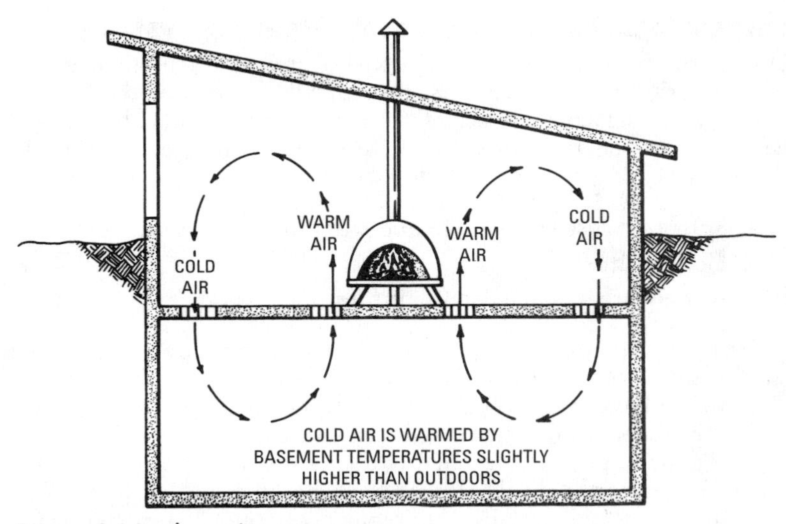

Figure 6-14 A gravity-type system. *(Courtesy Mother Earth News)*

passes through metal air vents in the floor, and then enters the basement where it is warmed somewhat by the slightly warmer indoor temperatures. It then rises through metal air vents placed in the floor around the stove where it is reheated and recirculated.

In a structure designed to take into consideration the operating principles of a gravity heating system, the operating costs will be lower than any other type system.

The advantages of a forced-warm-air heating system are as follows:

1. The installation costs are lower than those for hot-water or steam heating systems.
2. Heat delivery is generally quicker than other systems.
3. Heat delivery can be shut off immediately.
4. Air cleaning and filtering can be cheaply and easily provided.
5. The humidity level of the air can be easily controlled.
6. Air conditioning can be added without great difficulty or cost.
7. The furnace (because of its blower) can be placed in a number of different locations in the structure.

Disadvantages of a Warm-Air Heating System

One of the principal disadvantages of a *gravity* warm-air heating system is that the furnace *must* be centrally located at the lowest point in the structure. This, of course, severely limits the design flexibility of the structure itself. Other disadvantages of gravity warm-air heating systems include:

1. Slow response to heat demand from the thermostat
2. Slow air movement
3. Inadequate or nonexistent air filtering

Forced-warm-air heating systems depend upon blowers to circulate the air through a network of sheet-metal ducts. Both the operation of the blower and the expansion and contraction of the ducts as the air temperature rises or falls can cause distracting noise.

Warm-air outlets can frequently create decorating problems. In order to operate effectively, the registers should never be blocked by furniture or carpeting. This requirement, of course, restricts furniture arrangement and the placement of carpet.

The blower in a forced-warm-air heating system will create a certain amount of air turbulence. This rapid agitation of the air causes dust particles to circulate and deposit on walls, furniture, and other surfaces. Ordinary filters will remove most of the larger particles but are not very effective against smaller ones. An electronic air filter is recommended for the latter.

Warm-air heating systems require separate hot-water heaters to supply the hot water necessary for household use. This requirement for additional equipment adds to the initial installation costs for the system (in a hot-water heating system, the hot water for household use is supplied by the boiler).

Forced-warm-air heating systems supply convected heat through forced-air movement. The heat is supplied in bursts (rather than a continuous, sustained flow) when the room thermostat calls for it. This results in room temperatures varying up and down several degrees. The variation in temperature can be considerably reduced by setting the fan switch for continuous blower operation.

Troubleshooting a Warm-Air Heating System

Troubleshooting recommendations for a warm-air heating system will vary depending on the furnace make and model, and the type of installation. *Always* read and carefully follow the furnace manufacturer's troubleshooting recommendations if an operating manual is available. This troubleshooting section contains general remedies for common operating problems and their causes (Table 6-1).

Table 6-1 Troubleshooting Warm- Air Heating System

Symptom and Possible Causes	*Suggested Remedies*
No heat	
1. Power may be off. Check fuse box or circuit breaker panel for blown fuses or tripped breakers.	**1.** Replace fuses or reset breakers. If the problem continues, call an electrician or an HVAC technician.
2. Check thermostat (programmable type) for dead batteries.	**2.** Replace batteries and reset thermostat.
Insufficient heat	
1. Incorrect thermostat setting	**1.** Set thermostat above desired room temperature. Set the system switch on the thermostat to HEAT or AUTO.
2. Dirty filters	**2.** Clean or replace filters.
3. Poor air circulation	**3.** Check supply registers and return grilles for blockage.
4. Open circuit	**4.** Reset circuit.
5. Main power switch off	**5.** Turn main power switch to ON position.
6. Blocked outdoor coil (split system heating/cooling)	**6.** Clear away blockage from outdoor coil.
7. Air flow into intake and exhaust pipes of 90%-type furnace blocked	**7.** Clear debris, ice, or snow from outdoor PVC pipes.
Blower not operating	
1. Fan switch setting	**1.** Set the fan switch to AUTO if you want the blower to function only while the furnace is operating; set it to ON for continuous operation.
2. Blower door not completely closed	**2.** Close blower door to restore power.

(continued)

Table 6-1 (continued)

Symptom and Possible Causes	*Suggested Remedies*
Humidity levels too low	
1. Incorrect dehumidistat setting	**1.** Check setting and correct if necessary.
2. Incorrect humidifier setting	**2.** Check setting and correct if necessary.
3. Thermostat fan switch set on manual	**3.** Change setting to automatic.
4. Furnace ventilation switch set on continuous (low)	**4.** Change setting to automatic.
Humidity levels too high	
1. Incorrect dehumidistat setting	**1.** Check setting and correct if necessary.
2. No dehumidistat	**2.** Install dehumidistat if necessary.
3. Thermostat fan switch set on automatic	**3.** Change setting to manual.
4. Furnace ventilation switch set on automatic	**4.** Change setting to continuous (low).

Chapter 7

Hydronic Heating Systems

Hot-water heating systems use water as the medium for conveying and transmitting the heat to the various rooms and spaces within a structure. The motive force for the water in these systems is based either on the gravity flow principle or forced circulation. The latter type (referred to as *hydronic,* or *forced-hot-water heating*) is the one currently used in residential and commercial heating system. In a typical hot-water heating system, the water is heated in a boiler or water heater and circulated through pipes to baseboard convectors or radiators located in various rooms. The energy source used to heat the water may be oil, natural gas, propane, electricity, or a solid fuel, depending on the heating unit.

The hot water is circulated through pipes to baseboard convectors or radiators located along the walls of the rooms, or through radiant panels installed in the floors or ceilings. A centrally located thermostat controls room temperatures in smaller houses or buildings. When the thermostat calls for heat, the boiler or water heater heats the water and sends it to the room convectors or radiant heating panels where it is released and distributed through the rooms by natural convection. In larger houses or buildings, the heating system is zoned with individual thermostats controlling the temperatures in each zone.

Classifying Hot-Water Heating Systems

Hot-water heating systems can be classified in a number of different ways, depending on the criteria used. Three broad classification categories based on the following criteria are generally recognized:

1. Type of water circulation
2. Piping arrangement
3. Supply water temperature

In all hot-water heating systems, the water is circulated either by forcing it through the line or by allowing it to flow naturally. The latter is referred to as a *gravity hot-water heating system* because circulation results from the difference in weight (specific gravity) of the water caused by temperature differences (heavy when cold, light when hot). In a *forced-hot-water heating system,* the accelerated circulation of the water can result from several commonly employed methods, including: (1) using high pressures, (2) superheating the circulating water and condensing the steam, (3) introducing steam or air into the main riser pipe, (4) using a combination of pumps and local boosters, and (5) using pumps alone.

The four principal piping arrangements used in hot-water systems are:

1. One-pipe system
2. Two-pipe direct-return system
3. Two-pipe reverse-return system
4. Series-loop system

These piping arrangements are described and illustrated in several sections of this chapter (see, for example, the sections, *One-Pipe System* and *Two-Pipe Direct-Return System*).

If a hot-water heating system uses supply water temperatures above 250°F, it is classified as a *high-temperature system*. High-temperature systems are used in large heating installations such as commercial or industrial buildings. A *low-temperature system* is one having a supply water temperature below 250°F and is used in residences and small buildings.

One-Pipe System

A *one-pipe system* (Figure 7-1) is one in which a single main pipe is used to carry the hot water throughout the system. In other words, the same pipe that carries the hot water to the heat-emitting units (i.e., radiators and convectors) in the various rooms and spaces within the structure also returns the cooler water to the boiler for reheating. Each heat-emitting unit is connected to the supply main by two separate branch pipes (separate feed and return lines). (See Figure 7-2.)

The hot water flows from the boiler or heat exchanger (if a steam boiler is the heat source) to the first heat-emitting unit, through it to the second unit, and so on through each of the heat-emitting units in the one-pipe system until it exits the last one and returns to the boiler or heat exchanger.

One-pipe systems may be operated on either forced or gravity circulation. Special care must be taken to design the system for the temperature drop found in the heat-emitting units farthest from the boiler. This is particularly true of one-pipe systems designed for gravity circulation.

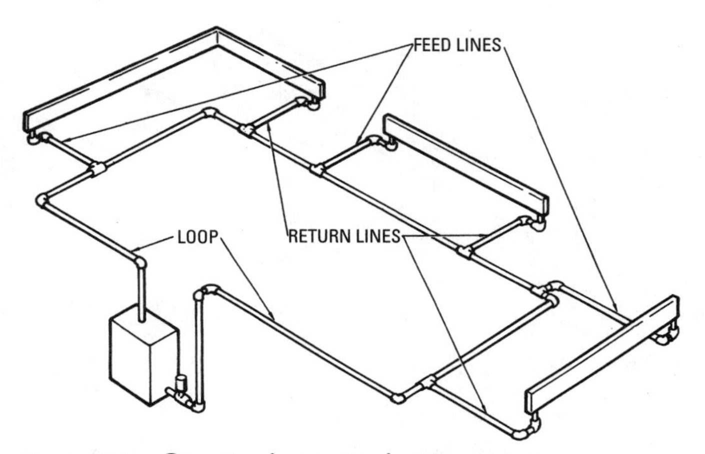

Figure 7-1 One-pipe hot-water heating system.

A principal advantage of a one-pipe system is that one or more heat-emitting units can be shut off without interfering with the flow of water to other units. This is *not* true of series-loop systems, in which the units are connected in series and form a part of the supply line.

In some large one-pipe systems, zoning can be achieved by providing more than one piping circuit from the boiler (Figure 7-3). In such cases, each piping circuit is equipped with its own thermostat and circulating pump. Sometimes these circuits are erroneously

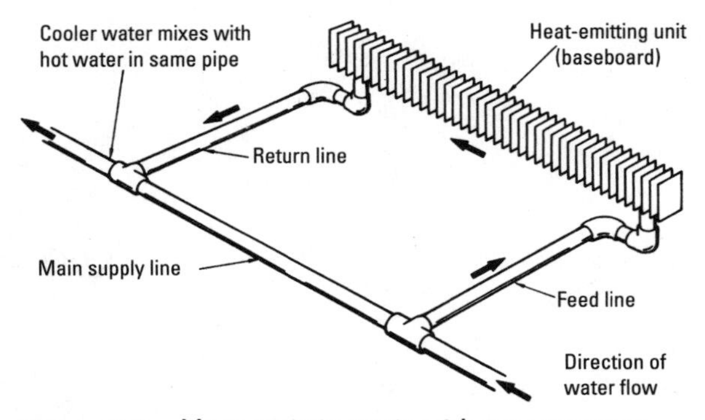

Figure 7-2 Heat-emitting unit with two separate branch pipes connected to the main supply line.

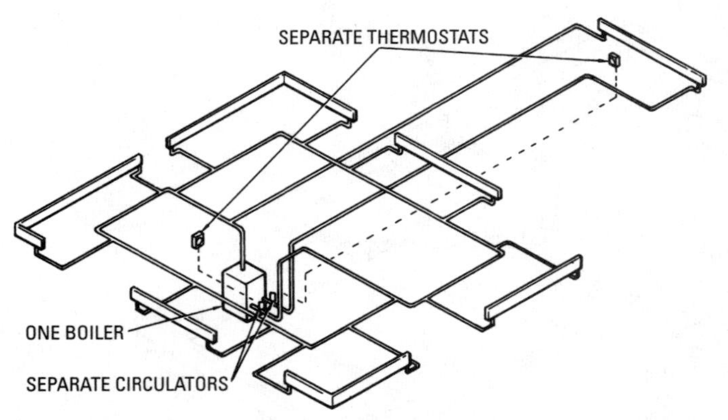

Figure 7-3 One-pipe hot-water heating system with more than one piping circuit for zoning.

referred to as "loops" and are confused with the piping arrangement in a series-loop system.

Series-Loop System

In the *series-loop system* (Figure 7-4), the heat-emitting units form a part of the piping circuit, i.e., the *loop,* which carries the hot water from the boiler around the rooms and spaces within the structure

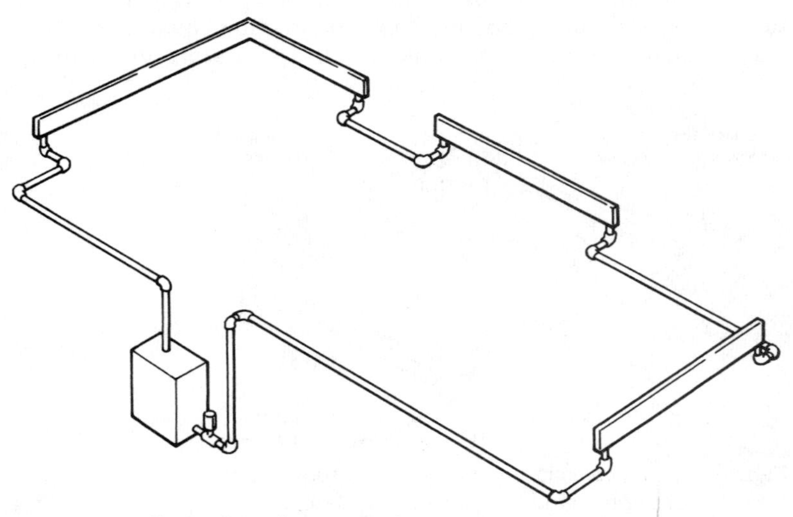

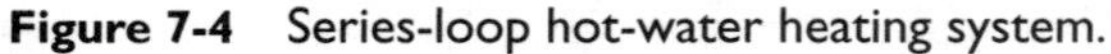
Figure 7-4 Series-loop hot-water heating system.

and back to the boiler again for reheating. In other words, there are no branch pipes connecting the main supply pipe to the heat-emitting units as in the one-pipe system. In the series-loop system, the hot water flows from the boiler through a length of the main supply pipe to the first heat-emitting unit in the circuit (loop). It then flows through the unit to a length of the main supply pipe connected at the opposite end, flows through this pipe to the second heat-emitting unit in the circuit, and so on until the entire circuit is completed.

The series-loop system is cheaper and easier to install than other piping arrangements because it eliminates the need for branch pipes and reduces the amount of pipe used in the main circuit to the comparatively short lengths connecting the heat-emitting units. There is no need for one continuous length of main pipe.

Because the heat-emitting units are connected in series and constitute a part of the main supply line, the same hot-water supply passes through each unit in succession. As a result, the heat-emitting unit closest to the boiler receives the hottest water, whereas units farther away receive water several degrees cooler. Furthermore, individual units cannot be shut off (unless there is a special bypass piping arrangement) without obstructing the flow of water to units farther along the line.

Two-Pipe, Direct-Return System

In a *two-pipe, direct-return system* (Figure 7-5), hot water returns directly to the boiler from each heat-emitting unit. In other words,

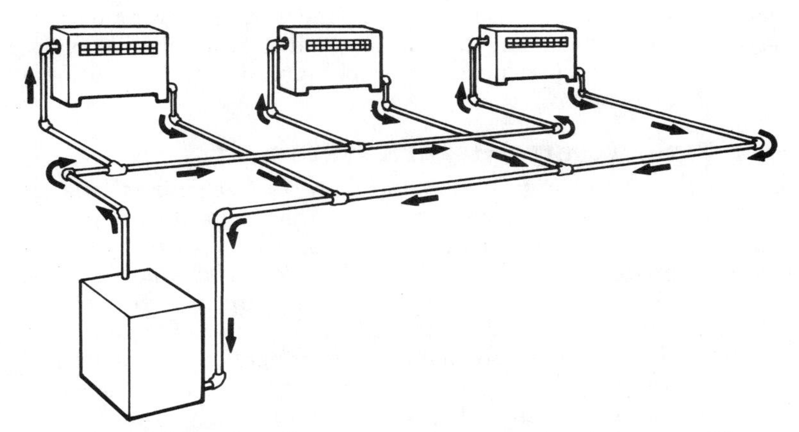

Figure 7-5 Two-pipe, direct-return, hot-water heating system. The water from each heat-emitting unit returns directly to the boiler without passing through other units in the system.

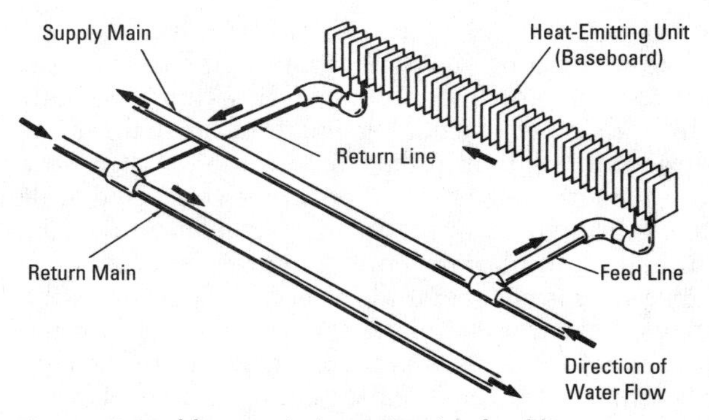

Figure 7-6 Heat-emitting unit with feed line connected to the supply main and return line connected to the return main.

the hot-water supply and return mains are separate pipes. The heat-emitting units are connected to the supply and return lines by separate branches (Figure 7-6).

Each heat-emitting unit represents the midpoint in a complete circuit within a two-pipe, direct-return system. The boiler completes the circuit. The farther a heat-emitting unit is located from the boiler, the greater the length of the piping in the circuit. Although this factor causes problems in balancing the water supply among the various circuits, balancing can be achieved in a number of ways, including (1) balancing cocks, (2) proper pipe sizing, or (3) using a reverse-return system.

Two-Pipe, Reverse-Return System

A *two-pipe, reverse-return system* (Figure 7-7) achieves a balance in the water supply by creating circuits to the radiators of approximately equal length. Instead of allowing the return supply of water to proceed directly to the boiler from each radiator, the return main carries the water in the opposite direction for a predetermined distance before turning back to the boiler. The first radiator has the shortest supply main but the longest return main. For the farthest radiator, the reverse is true. Regardless of the position of a radiator in the system, the total length of pipe within the circuit of which it forms a part will be essentially the same as that of any other circuit.

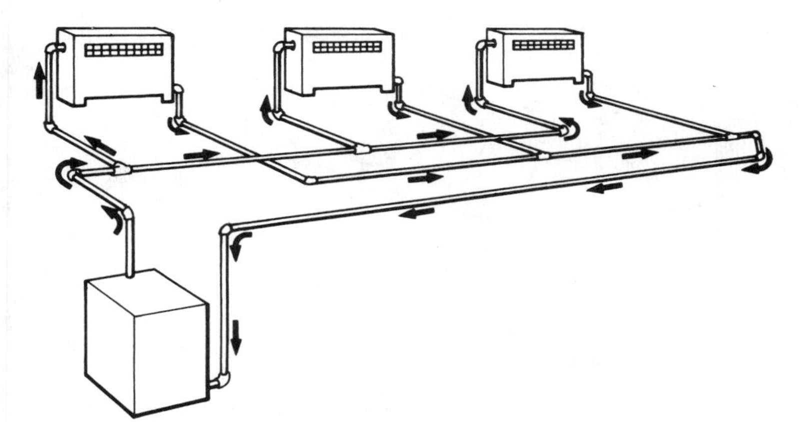

Figure 7-7 Two-pipe, reverse-return, hot-water heating system.

Combination Pipe Systems

Sometimes two or three different pipe arrangements will be combined in a single heating system. For example, it is not unusual to find a two-pipe, reverse-return system combined with a series-loop system. Combination pipe systems are usually found in commercial or industrial buildings and are fitted to specific design needs.

Zoning a Two-Pipe System

Balancing problems for two-pipe hydronic heating systems in larger houses and buildings can be solved by splitting the existing system into two or more separate zones independently controlled by their own thermostats. This can be accomplished by installing a two-position valve in the hot-water supply line. The valve is actuated by its own thermostat (Figure 7-8).

Radiant Panel Heating

Forced hot water can also be used in a radiant panel heating system (Figure 7-9). In this heating system, no radiators or convectors are used. The hot water circulates through pipes concealed in the floor, ceiling, or walls, which function as heat-emitting surfaces. Radiant baseboard systems are also used. The principles of radiant heating are described in greater detail in Chapter 1 of Volume 2, "Radiant Heating."

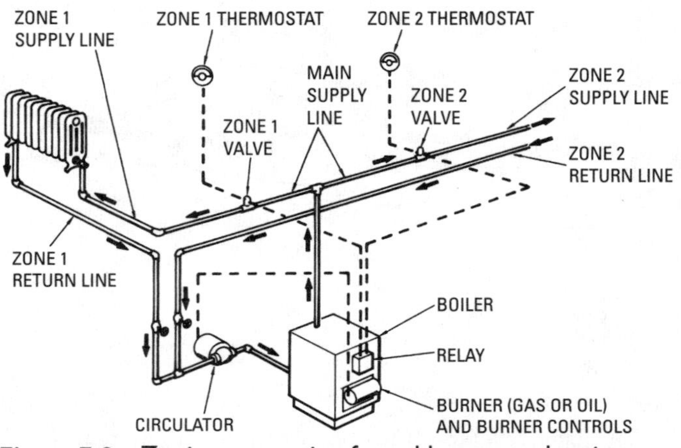

Figure 7-8 Zoning a two-pipe forced hot-water heating system. Thermostatically actuated zone valves are installed on the main water supply line.

Other Applications

It is possible to use the boiler of a hydronic heating system to supply heat for such purposes as melting snow and ice on walks and driveways, heating a swimming pool, or heating a patio. Separate circuits are created for each of these functions and are controlled by

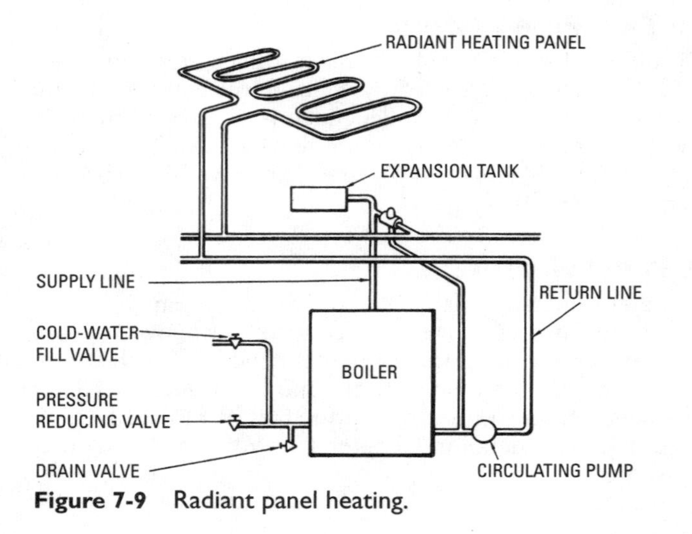

Figure 7-9 Radiant panel heating.

their own thermostats. They are designed to tap into the main heating circuit from which they receive their supply of hot water. If winter temperatures stay at or below freezing for long periods of time, the water in an outdoor circuit must be protected with an antifreezing additive. See Chapter 5, "Heating Swimming Pools" in Volume 3 for detailed coverage of this subject.

Gravity Hot-Water Heating Systems

Gravity hot-water heating systems can still be found in older buildings. They are commonly identified by their cast-iron boilers and the large-diameter wrought iron or black iron supply mains used to deliver the hot water to the rooms. These older cast-iron boilers were fired originally by coal or wood but have been converted to oil or gas use; see Chapter 16, "Boiler and Furnace Conversions" in this volume.

The difference in weight (specific gravity) of the water at different temperatures is used to circulate the water in a gravity hot-water heating system. In other words, the motive force is due to the difference in density of the water at different temperatures; heavy when cold, light when hot. For this reason, gravity systems are also referred to as *thermal, or natural, hot-water heating systems.*

A gravity hot-water heating system produces a continuous, uniform heat. The water temperature can be regulated by a manually operated draft damper located in each room. The room air temperature can be regulated by turning the radiator valve.

In order for the hot water to transmit heat from the boiler (heating unit) to the radiators or other heat-emitting devices, there must be a constant movement (circulation) of the water from the heater to the radiators and back again. As mentioned above, this circulation in gravity systems is due to the difference in density (weight per

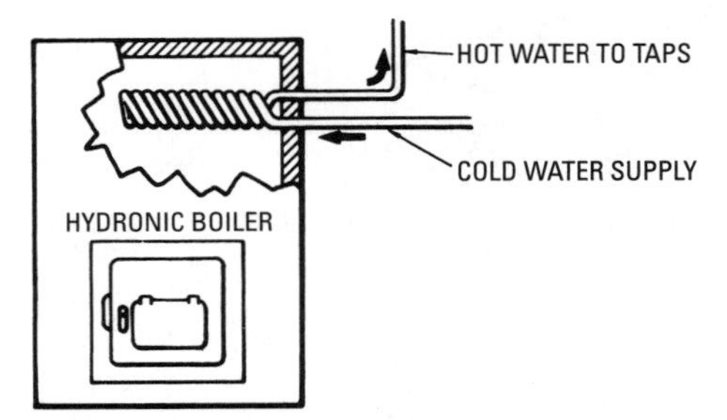

Figure 7-10 Using the heat exchanger principle to provide domestic hot water without a special tank.

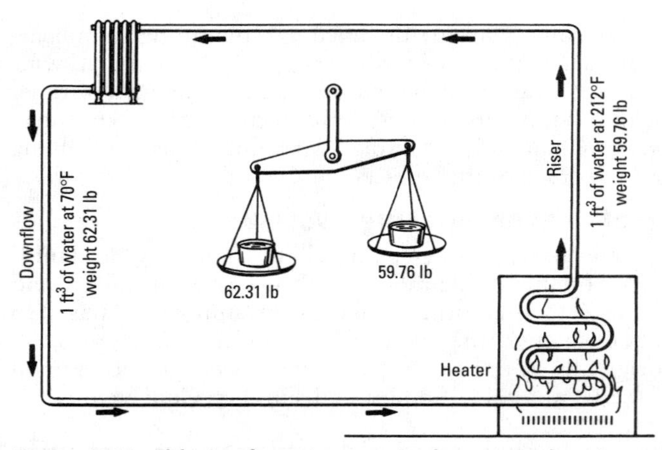

Figure 7-11 Motive force in a gravity hot-water heating system.

unit volume) of water at different temperatures. For example, 1 ft^3 of water at, say, 68°F weighs 62.31 lb; at 212°F it will weigh 59.82 lb. This difference in weight is 2.49 lb (62.31–59.82), which is available to cause circulation, as shown in Figure 7-11.

The difference in weight is caused by the expansion of water as its temperature is increased. This principle is illustrated in Figure 7-12. Note that the top surface *abcd* of the cubic foot *A* has expanded to *A*', an increase in volume represented by the linear increase *aa*'.

In the two-pipe system illustrated in Figure 7-13, the unbalanced weight between the water in the riser and downflow pipe forms a

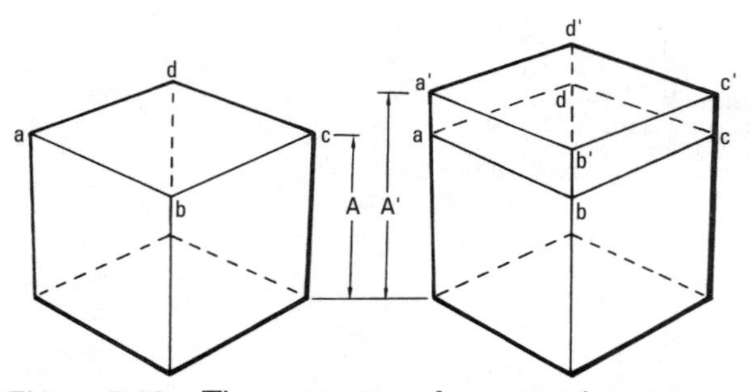

Figure 7-12 The expansion of water with rise in temperature.

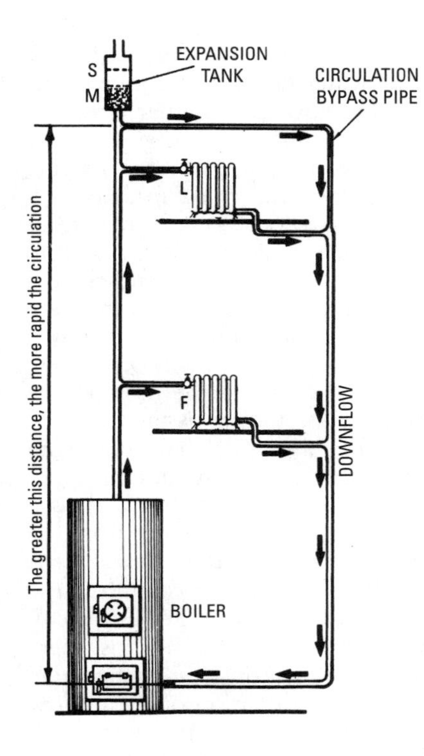

Figure 7-13 Two-pipe gravity hot-water heating system.

motive force that causes the water to circulate through the system as indicated by the arrows.

The two-pipe gravity hot-water system illustrated in Figure 7-13 is equipped with an expansion tank. In operation, after the fire is started, the temperature of the water in the boiler rises, and the water expands. This disturbs the equilibrium of the system, causing the colder and heavier water in the downflow pipe to flow downward, pushing the warmer and lighter water in the riser upward, thus starting circulation.

A circulation, or bypass, pipe is provided to form a continuous path for the flowing water in the event that all the radiators are shut off. In the absence of this provision, all the water in the upflow pipe would be pushed up into the expansion tank followed by steam.

The expansion tank provides two essential functions for this system. As the water is heated, it expands and the tank provides space for this increase in volume. Thus, as it is heated, the water will expand from, say, elevation *M* to elevation *S* in the expansion tank

(Figure 7-13). The higher the elevation of the tank, the greater the pressure that may be brought on the water in the boiler (due to the head). As a result, the system may be worked at higher temperatures when desired.

Sometimes a special piping circuit can be built into a heating system (see *Two-Pipe, Direct-Return System* and *Two-Pipe, Reverse-Return System* in this chapter). In the two-pipe gravity hot-water heating system illustrated in Figure 7-14, a single main is taken from the boiler to a high point under the basement ceiling and then is pitched along its run as much as possible to the return inlet of the boiler. The piping circuit formed by this main consists of a closed loop of extra-large pipe in the basement having a pitch of not less than ½ in per 10 ft of run. The main supplies all the risers to the radiators above. This arrangement is not as efficient as the two-pipe system illustrated in Figure 7-13, and for this reason the circuit main should be of very liberal size to reduce friction to a minimum.

In the so-called one-pipe system shown in Figure 7-15, special distribution tees are employed to deflect part of the water from the main into the radiators while letting the balance flow through the main to the next radiator. This is called a one-pipe system because one pipe serves both inlet and outlet of each radiator. In this system, there must be an upflow side *L* and a downflow side *F.* Thus, there are in reality two main pipes, but in function only *one* main pipe. The downflow may be considered as a continuation of the upflow instead of as a second pipe. The major objection to this system (as well as to the one illustrated in Figure 7-14) is a lack of uniformity

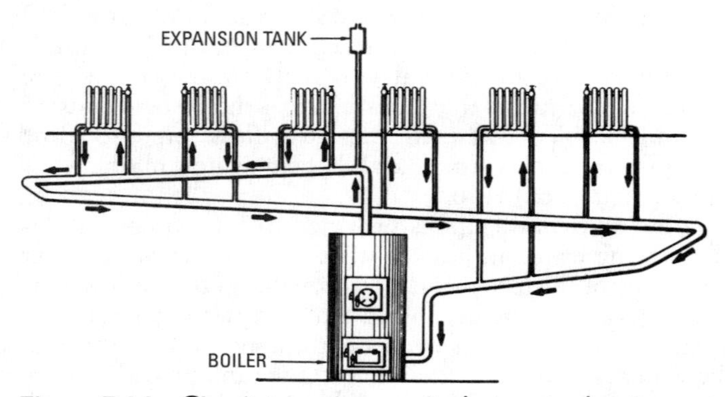

Figure 7-14 Circuit two-pipe, gravity, hot-water heating system.

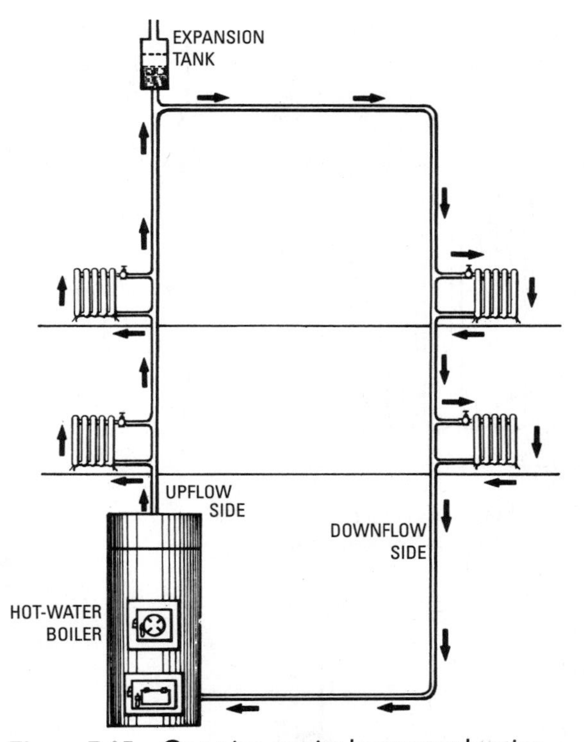

Figure 7-15 One-pipe gravity hot-water heating system.

in heat distribution, primarily because the radiators on the upflow side *L* are hotter than those on the downflow side *F.*

Distribution tees (Figure 7-16) are used in this one-pipe system. They are provided with a baffle tongue, which deflects part of the water from the main in and out of the radiator while bypassing the balance along the line as indicated by the arrows in the illustration.

Another piping system used to distribute the hot water is the *overhead arrangement* illustrated in Figure 7-17. This is practically the same as the so-called one-pipe system (Figure 7-15), except that it has a divided circuit at the high point connecting with the downflow mains. No air vents are necessary because the arrangement is such that all air works to the top and passes off into the expansion tank. One advantage of this system is a better and

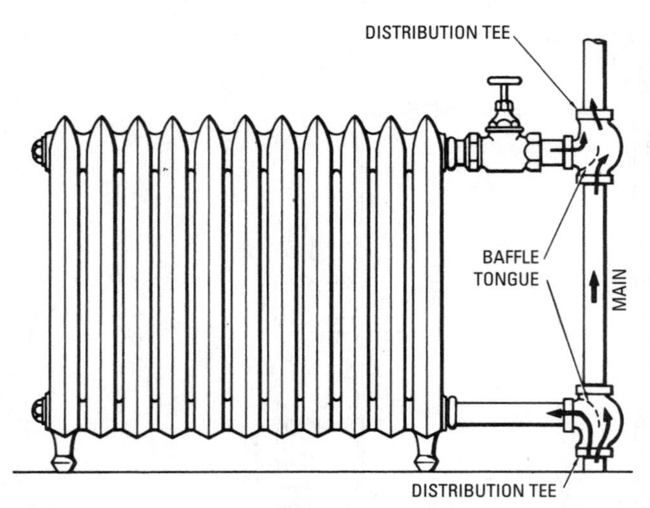

Figure 7-16 Using distribution tees to connect a cast-iron radiator to the main in a one-pipe hot-water heating system.

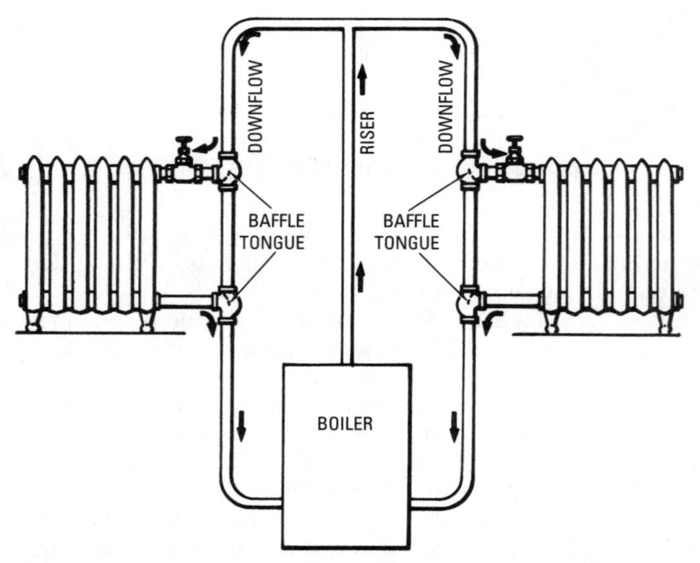

Figure 7-17 Overhead piping arrangement in a gravity hot-water heating system.

more uniform distribution of the heat. The hot water enters each downflow pipe at the same temperature, equalizing the heat given off by each pair of radiators.

Although the overhead piping arrangement is not suited to all classes of buildings, there are many buildings, such as apartments, stores, office buildings, and hotels, where the general arrangement lends itself to this system.

Forced-Hot-Water Heating Systems

Water as a medium for transmitting heat to radiators and other heat-emitting units gives up only sensible heat, as distinguished from steam systems, which heat principally by the latent heat of evaporation. The result is that the temperature of the heat-emitting units of a steam system is relatively high as compared with hot water. In a hot-water system, latent heat is *not* given off; hence more radiating surface is needed to obtain equal heating effect.

Because hot water is heavier than steam and does not rise nearly as quickly, some means must be used to accelerate the rate of flow. This is done to increase the heating capacity of the heat-emitting units, make the system more responsive to load conditions, and allow the use of smaller pipes.

Numerous methods have been introduced to accelerate the circulation of hot water. The most commonly used methods have been based on the following techniques:

1. Using radiators with nipples along both the upper and lower portion of each radiator section.
2. Introducing high pressures to gain greater temperature differences.
3. Superheating a part or all of the circulating water as it passes through the boiler and condensing the steam thus formed by mixing it with a portion of the cold circulating water of the return main.
4. Introducing steam or air into the main riser near the top of the system.
5. Forcing by pumps.
6. Combining pumps and local boosters.

The last two methods (pumps or a combination of pumps and local boosters) are generally the ones used in most modern forced-hot-water (hydronic) heating systems. (See Figure 7-18.)

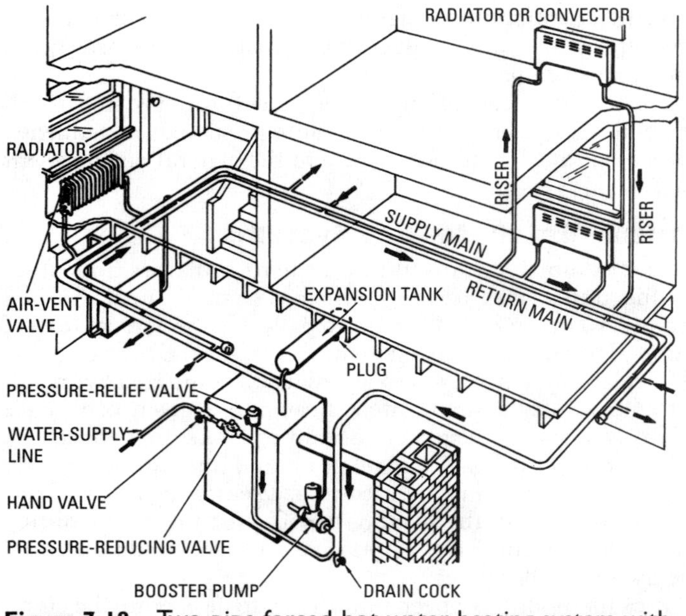

Figure 7-18 Two-pipe forced-hot-water heating system with separate supply and return mains. *(Courtesy U. S. Department of Agriculture)*

Hot-Water Boilers

The boilers used in hot-water heating systems are made of cast iron or steel (Figures 7-19 and 7-20). The cast-iron boilers generally display a greater resistance to the corrosive effects of water than the steel ones do, although the degree of corrosion can be significantly reduced by chemically treating the water.

Gas (natural or propane), oil, electricity, or solid fuels have been used as the energy source for heating the water in boilers. Multifuel boilers are designed to allow switching between two different fuels. Some manufacturers will provide conversion devices for switching from one type of gas to the other (Figure 7-21). Changing from coal to oil or gas is a frequent conversion provided for by conversion burners (see Chapter 16, "Boiler and Furnace Conversions").

Never purchase an uncertified boiler. All certified boilers are approved by one of several organizations that have assumed this responsibility. For example, many boilers are certified by either the

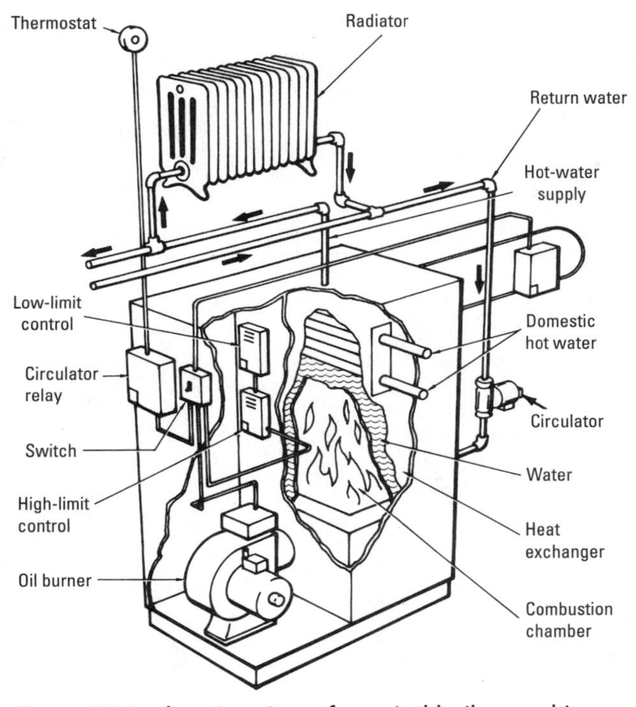

Figure 7-19 Interior view of a typical boiler used in a hot-water heating system.

Institute of Boiler and Radiator Manufacturers (cast-iron boilers) or the Steel Boiler Institute (steel boilers). They will be stamped I = B = R or SBI, respectively. Figure 7-22 illustrates some of the certifications you will encounter on boilers.

Steam boilers can also be used in hot-water heating systems, but a heat exchanger must be incorporated into the system to transfer the heat from the steam to the hot water that flows through the heat-emitting units in the rooms.

Hydronic Furnaces

In a hydronic furnace, water is first heated in a boiler or water heater and then circulated through the coils of a liquid-to-air heat exchanger connected to the furnace air handler (Figure 7-23).

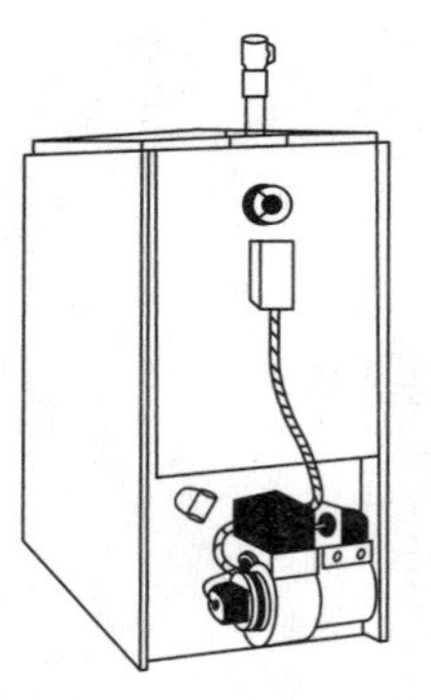

Hydrotherm residential oil-fired hydronic boiler

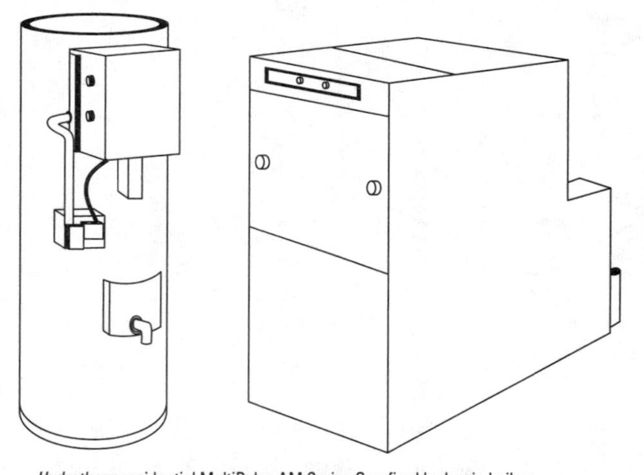

Hydrotherm residential MultiPulse AM Series Gas-fired hydronic boilers

Figure 7-20 Examples of boilers used in hot-water heating systems.

Heat is transferred from the water in the coils to the air inside the air handler compartment. A blower distributes the warm air through ducts to outlets in the different rooms of the house or building.

This type of installation is sometimes called a *hydro-air heating system*. Its operation is controlled by a centrally located thermostat working in conjunction with an aquastat or time delay. The

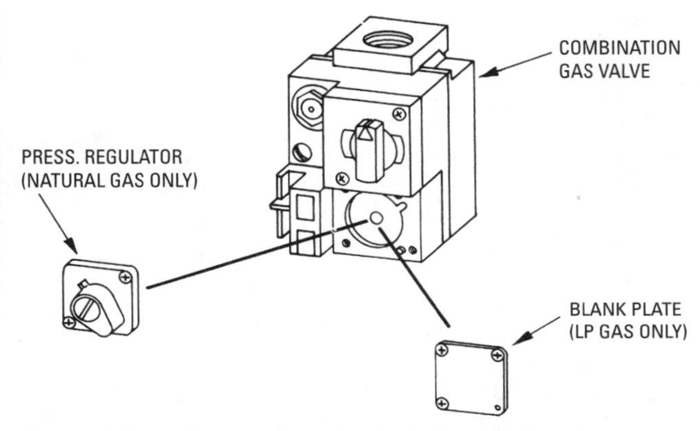

Figure 7-21 Gas conversion devices used for changing from one type of gas to another. *(Courtesy Honeywell Tradeline Controls)*

aquastat or time delay prevents blower operation until the water passing through the coils in the air handler has reached the temperature called for by the thermostat setting. In zoned systems, heat circulation is regulated by a motor-actuated damper located in or near the duct system or the air handler. The damper motor is controlled by a zone control panel.

A hydronic furnace must be equipped with an air filter to prevent dust and other matter from circulating through the ducts and passing into the rooms. The accessories used with hydro-air heating

Figure 7-22 A partial list of certification symbols for boilers.

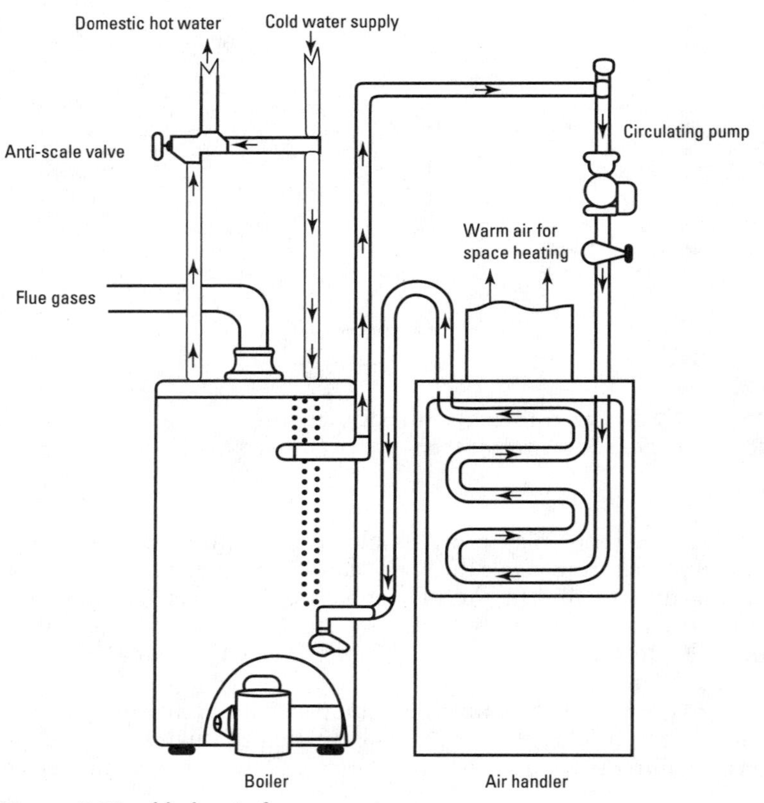

Figure 7-23 Hydronic furnace.

systems include an electronic air cleaner, a humidifier, and an air conditioning evaporator coil with a condensate pump.

Combination Water Heaters

A combination water heater is designed to produce both domestic (potable) and space heating hot water from the same source. As shown in Figure 7-24, an inner stainless steel tank containing the domestic hot water supply is immersed inside an insulated outer tank containing the prime water used for space heating. The prime water in the outer tank is heated by a gas or oil burner. The heated prime water transfers its heat by convection to the domestic water supply stored in the inner tank. The two tanks are completely separate from one another, preventing the domestic and space heating waters from

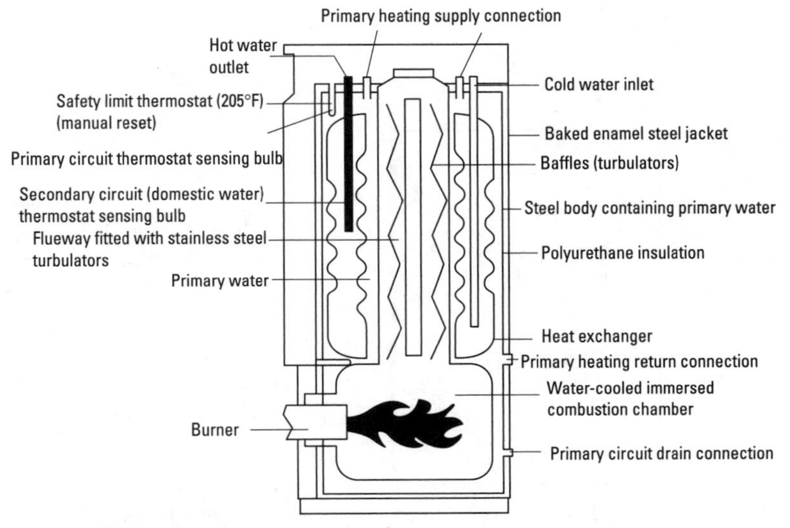

Figure 7-24 Combination water heater.

mixing. Each combination water heater operates in conjunction with a circulating pump, expansion tank, pressure-reducing fill valve, and three-way zone valve. See Chapter 4, "Water Heaters" in Volume 3 for additional information about combination water heaters.

Steam can also be used as a source of heat in a hot-water heating system. If such is the case, then a heat exchanger (Figure 7-25) must be used to convert the steam to hot water before it reaches the heat-emitting units. The heat exchanger is commonly of the steel-shell and copper-tube design. The steam flows through the shell and the water through the tubes (Figure 7-26). The hot water returns to the steam boiler, where it is reheated and changed to steam again.

Control Components

The efficient and safe operation of a hot-water heating system requires the use of a variety of different types of controls, which can be roughly divided into either system-actuating controls (e.g., the room thermostat, burner controls, and circulating pump controls) or safety controls (e.g., high-limit controls, pressure-relief valves, and pressure-reducing valves).

The safety controls prevent damage to the system by shutting it down when pressure and temperature levels become excessive. The *high-limit control,* or *aquastat,* is an example of such a control. It is

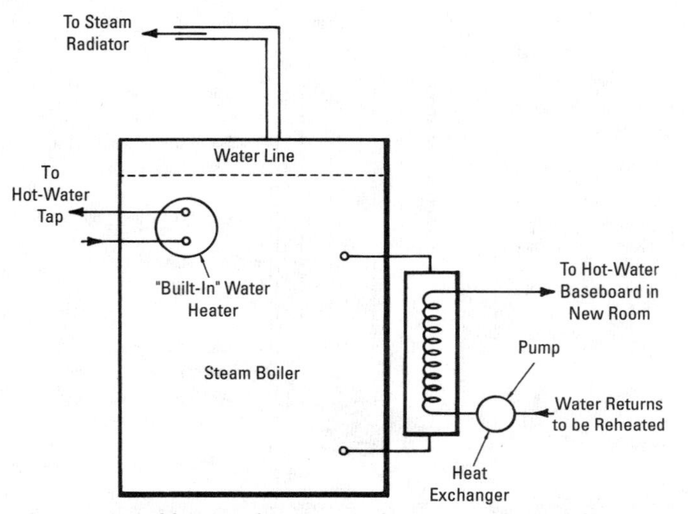

Figure 7-25 Heat exchanger used in connection with a steam boiler to provide hot water to baseboard heat-emitting units. *(Courtesy National Better Heating-Cooling Council)*

a device designed to operate in conjunction with the circulating pumps and is located on the hot-water boiler. If the pressure or temperature of the hot water exceeds the design limits of the system, the system is shut down until conditions return to an acceptable level. Some aquastat relays provide multizone control when used with a separate circulator and a relay for each zone.

Another very important safety control for hot-water heating systems are *pressure-relief valves*. These valves are designed to open when pressure in the boiler reaches a certain level and close when the pressure returns to a safe level again. Pressure-relief valves *must* be installed in accordance with current ASME or local codes.

These and other controls (e.g., main shutoff valves, pressure-reducing valves, and drain cocks) are described in considerable detail elsewhere in Chapter 15, "Steam and Hot-Water Space Heating Boilers" of this volume, and in Chapters 4, "Thermostats and Humidistats," and 9, "Valves and Valve Installation" of Volume 2.

Pipe and Pipe Sizing

The tubing used in hydronic heating systems is made of copper, steel, plastic, or rubber. The size of the pipes or tubing used in these systems depends on the following two factors: (1) the flow rate of

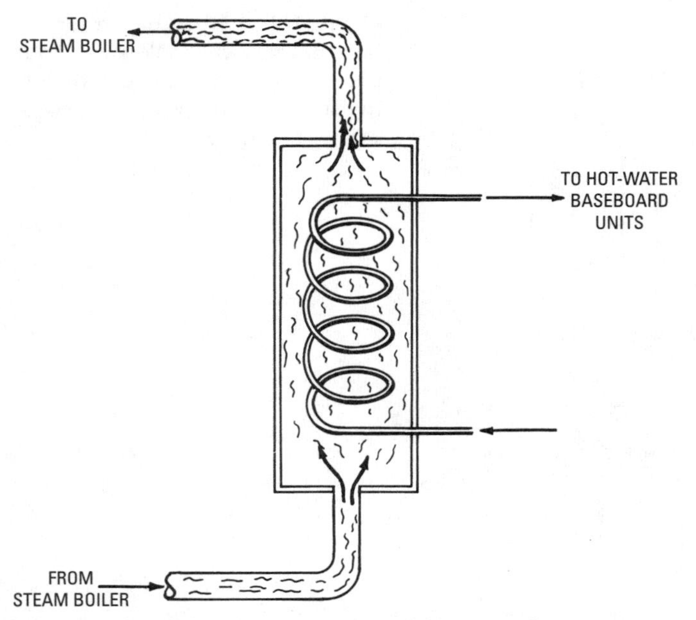

Figure 7-26 Operating principles of a heat exchanger.

the water and (2) the friction loss in the tubing. The *flow rate* of the water is measured in gallons per minute (gpm), and constant *friction loss* is expressed in thousandths of an inch for each foot of pipe length. For a description of the various types of tubing used in hydronic heating systems see the appropriate sections of Chapter 8, "Pipes, Pipe Fittings, and Piping Details" in Volume 2, which also describes the sizing methods used.

Expansion Tanks

Expansion tanks (Figures 7-27 and 7-28) are installed in hot-water heating systems to provide for the expansion and contraction of the water as it changes in temperature. Water expands with the rise of temperature, and the excess volume of the water flows into the expansion tank.

Another feature of the expansion tank is that the boiling point of the water can be increased by elevating the tank. In other words, increasing the *head* (i.e., the difference in elevation between two

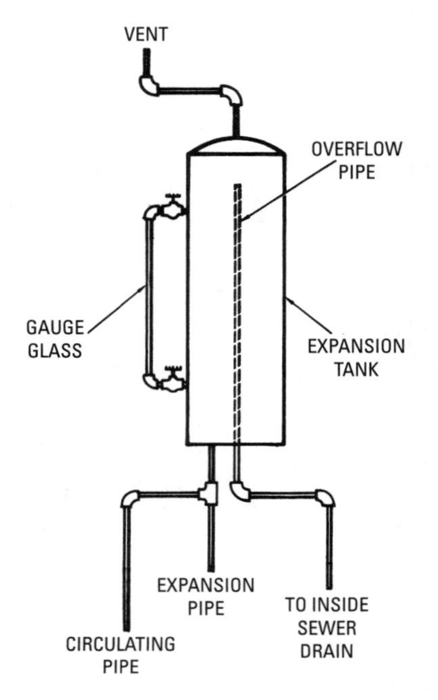

Figure 7-27 Open expansion tank. Because the tank is vented to the atmosphere, the pressure at the surface of the water is always that of the atmosphere.

points in a body of fluid) increases the pressure. As a result, the water can be heated to a higher temperature without generating steam, which, in turn, causes the radiators or other heat-emitting devices to give off more heat.

Both open and closed expansion tanks are used in hot-water heating systems. The *open* expansion tank is used on low-pressure systems, and the *closed* tank is used on high-pressure systems. Air in the tank above the water forms a cushion for increasing the pressure. As the temperature of the water rises, the water expands and flows into the tank, thus compressing the air and increasing the pressure.

The relation between pressure and volume changes of the air should be understood. According to Boyle's law, *at constant temperature the pressure of a gas varies inversely as its volume.* Thus, when the volume is reduced by one-half, the pressure is doubled. This is *not* gauge pressure, but absolute pressure (the pressure measured from true zero or point of no zero pressure).

In gravity hot-water heating systems, either closed or open piping arrangements can be used. In an *open* gravity system, the expansion tank is located at the *highest* point in the system (e.g.,

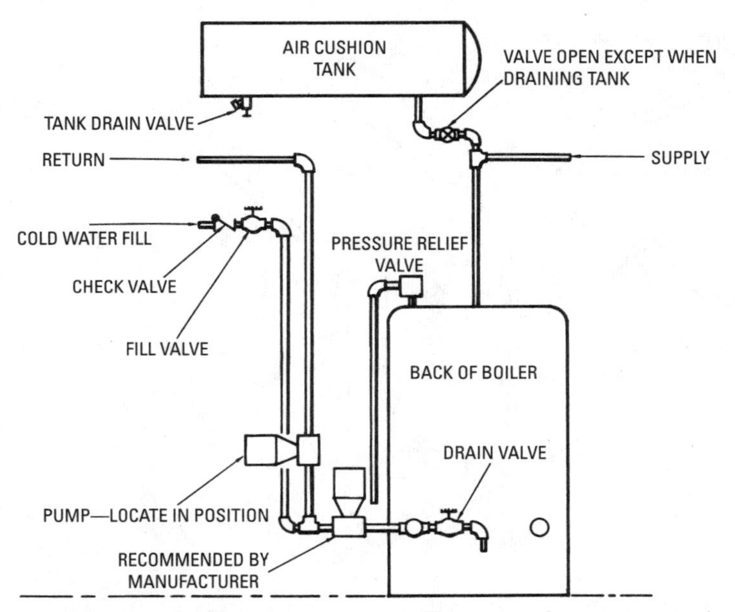

Figure 7-28 Closed expansion tank. This type of expansion tank is sealed against free venting to the atmosphere.

roof, attic, or top floor; see Figure 7-14). The expansion tank used in this piping arrangement is an open type with an overflow pipe located at the top. Provisions can be made to return the overflow water to the boiler or to discharge it through outside runoff drains.

In a *closed* gravity system, a closed, airtight expansion tank is located near the hot-water boiler (Figure 7-28). Higher pressures (and, consequently, higher water temperatures) result as pressure builds up in the system. Pressure-relief valves are installed on the main supply line to prevent the buildup of too much pressure.

Note

Hydronic heating systems are closed ones with expansion tanks located near the boiler.

Circulating Pumps (Circulators)

In forced-hot-water heating systems, circulating pumps (also sometimes called *circulators* or *booster pumps*) are used to force the hot water through the piping or tubing. Figure 7-29 illustrates a typical location of one of these pumps on a return line. The operation of

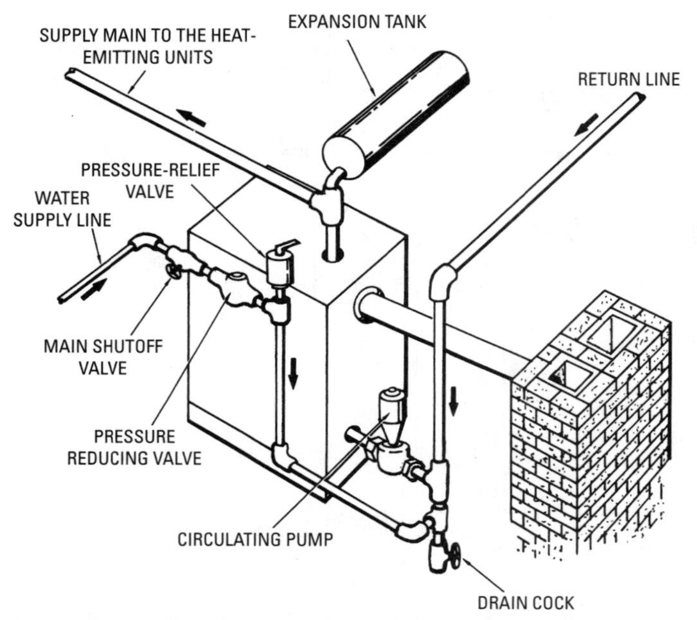

Figure 7-29 Circulating pump is located on the return water line next to the boiler.

the one or more circulating pumps in a hydronic system is controlled by a circulator relay.

Circulating pumps for hot-water heating systems are designed and built to handle a wide range of pumping capacities. They will vary in size from small booster pumps with a 5-gpm capacity to those capable of handling thousands of gallons per minute. Some are field serviceable with replaceable cartridges.

In order to select a suitable pump for a hot-water heating system (i.e., one that will produce maximum flow without overloading the pump motor), it is necessary to match the operating characteristics of the pump correctly to the requirements of the heating system.

The correct location of a pump is crucial to its successful operation. Pump manufacturers will usually give detailed information in their literature on this point. Circulating pumps with mechanical seals are very common in hot-water heating systems.

Drainage

There should be a provision for adequate drainage facilities in a hot-water heating system during shutdown periods. Pitching all pipes to a central point is an important consideration in designing the system because it is essential to effective drainage. If drainage is not provided for, there is always the danger of the unheated water freezing in the pipes during extreme cold weather. In most installations a manual drain cock is installed at the lowest point in the return line just before it reaches the circulating pump (see Figure 7-29). Chemical antifreeze solutions have also been developed for use in hot-water heating systems in an attempt to prevent the water from freezing during shut-down periods.

Heat-Emitting Units

The hot water in both gravity and forced hot-water heating systems is circulated through the pipes to the radiators or other heat-emitting units, from which the heat is transferred into the room.

Conventional cast-iron radiators (Figure 7-30) are often set on the floor or mounted to the wall. Many examples of these can still be found in older buildings. Some attempts have been made to conceal

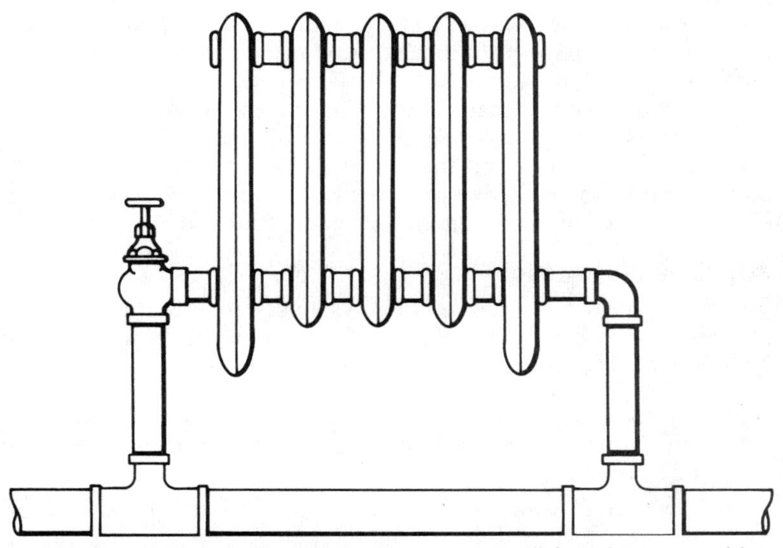

Figure 7-30 Conventional cast-iron radiator still found in many older hot-water heating systems.

radiators by recessing them in the wall of the room or enclosing them (partially or entirely) in cabinets. Recessed radiators must have some sort of insulation between them and the wall. One-inch-thick insulation board, a sheet of reflective insulation, or a combination of both is recommended for this purpose. If the radiator is partially or entirely covered with a cabinet, then openings should be provided at both the top and bottom of the cabinet for air circulation.

Convectors consist of finned tubes enclosed in a cabinet (Figure 7-31) or baseboard unit (Figure 7-32) with openings at the top and the bottom. The hot water circulates through the tubes, which radiate heat. Air enters the bottom of the cabinet or baseboard and exits through the openings in front.

Baseboard radiator units (Figure 7-33) consist of long, narrow tubes directly behind a baseboard face. The hot water flows through the tubes and heats the baseboard face, and the heat is radiated from the surface into the room. Make sure that the metal cover of the baseboard unit is not too thin (an indication of a cheaply built unit). Thin covers are easily dented, and water temperature changes in the heating element can cause a thin baseboard cover to be noisy.

Valance units along the top of the walls are also used to distribute the space-heating hot water (Figure 7-34). They were conceived as a solution to the furniture placement problem associated with baseboard units. They operate on the same principle as the baseboard units; i.e., the hot water circulates through copper tubing with attached fins, and the heat radiates out into the room.

Radiant floor systems use flexible tubing installed below the floor. The hot water from the boiler circulates through the tubes, and its heat is transferred to the floor surface. Flexible tubes or panels are also used as heat emitters in ceilings and walls.

Air Conditioning

For many years hydronic heating systems were installed without any provision for air conditioning. Two important reasons for not including cooling systems were:

- The location of the floor-mounted radiators and baseboard convectors would cause the heavier cold air to collect along the floor.
- It was impossible to capture and remove the condensation that forms when chilled water circulates through pipes /tubing embedded in floors, ceilings, or walls; or through baseboard convectors and radiators.

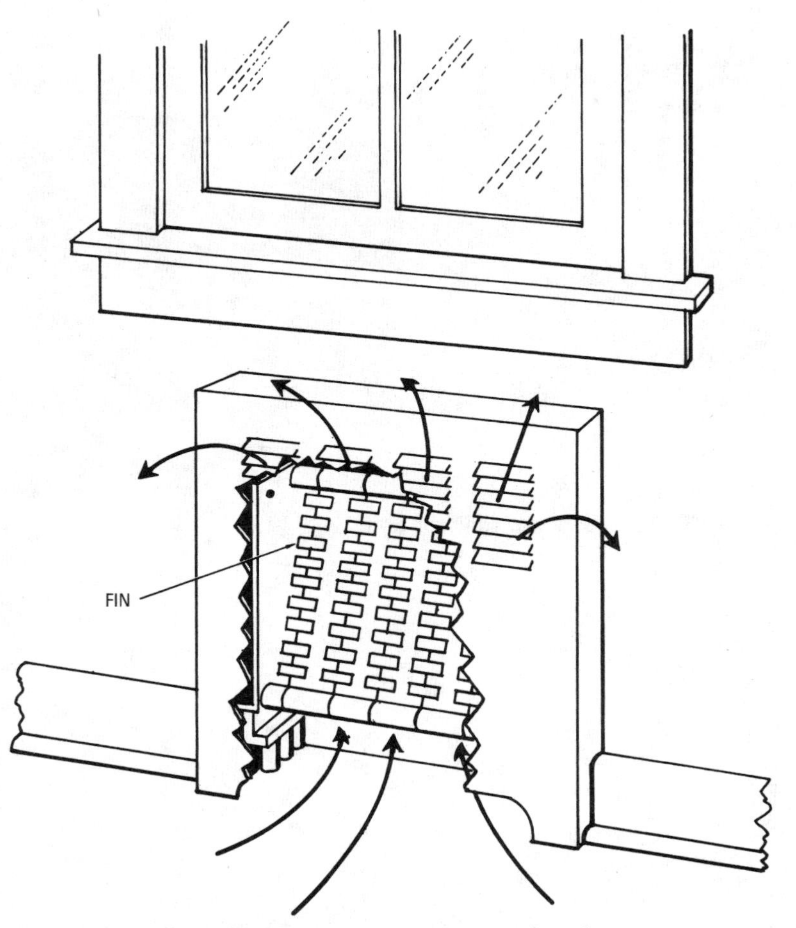

Figure 7-31 Finned-tube connector in an upright cabinet.

Until very recently these two technical problems have placed hydronic heating systems at a disadvantage when competing with the forced-warm-air systems. Consequently, hydronic systems have been limited to a much smaller share of the residential heating and cooling market. As homeowners have become more energy conscious and houses more tightly sealed against air infiltration, central air conditioning has become a necessity for houses heated by hot-water/hydronic heating systems. The methods used to add air

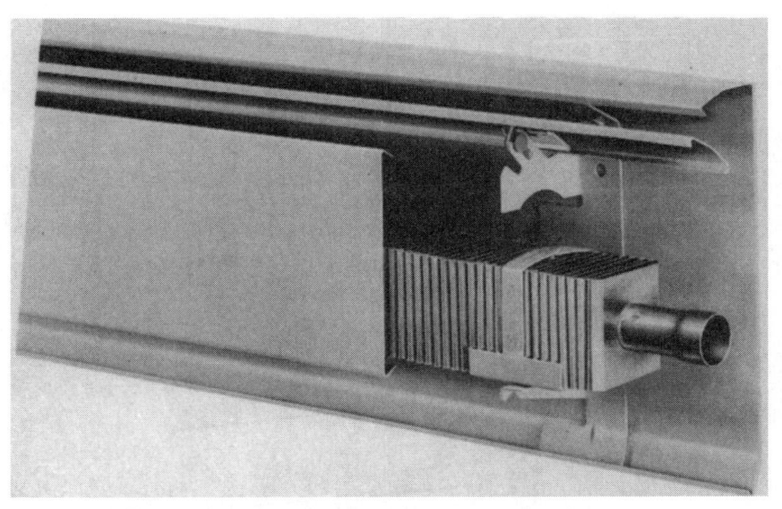

Figure 7-32 Baseboard radiator unit.

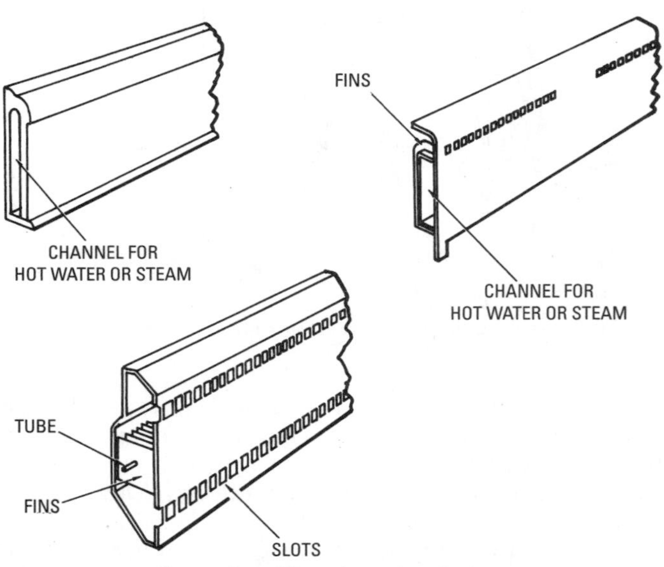

Figure 7-33 Examples of baseboard radiator units.

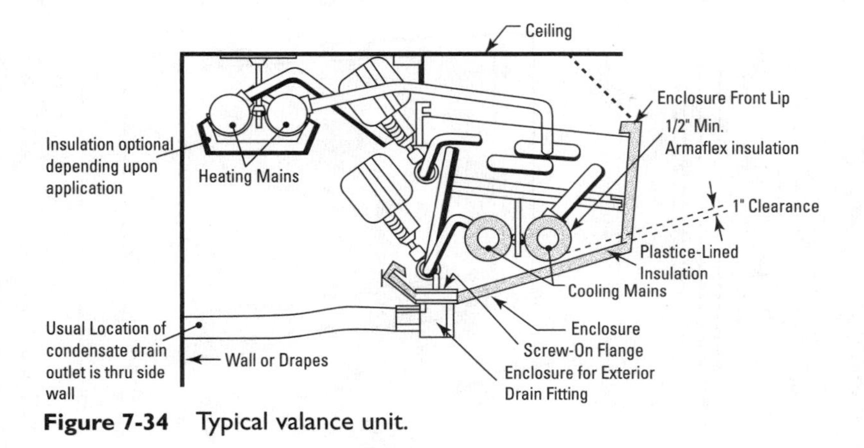

Figure 7-34 Typical valance unit.

conditioning to hot-water/hydronic heating systems are covered in Chapter 10, "Central Air Conditioning" in Volume 3.

Moisture Control

A hot-water heating system will generally maintain an adequate humidity level without the additional assistance of a separate humidifying unit. The humidifier illustrated in Figure 7-35 is an example of the type that could be used in a hydronic heating system should one be required. It is designed with a hot-water coil on the air inlet side of the unit. The heated boiler water circulates through this coil. A blower in the unit draws filtered room air over the coil and through a wetted rotating pad. The moist air is then circulated through the rooms by the blower. The humidifier is controlled by a room humidistat.

Electrically Heated Systems

Compact electric boilers, water heaters, or combination water heaters are also available for use in hydronic systems. These boilers are characterized not only by their compactness but also by their quiet operating characteristics.

The extreme compactness of an electric hydronic boiler is illustrated in Figure 7-36. The heat exchanger, expansion tank, and operating controls are all mounted on the wall in a relatively small space.

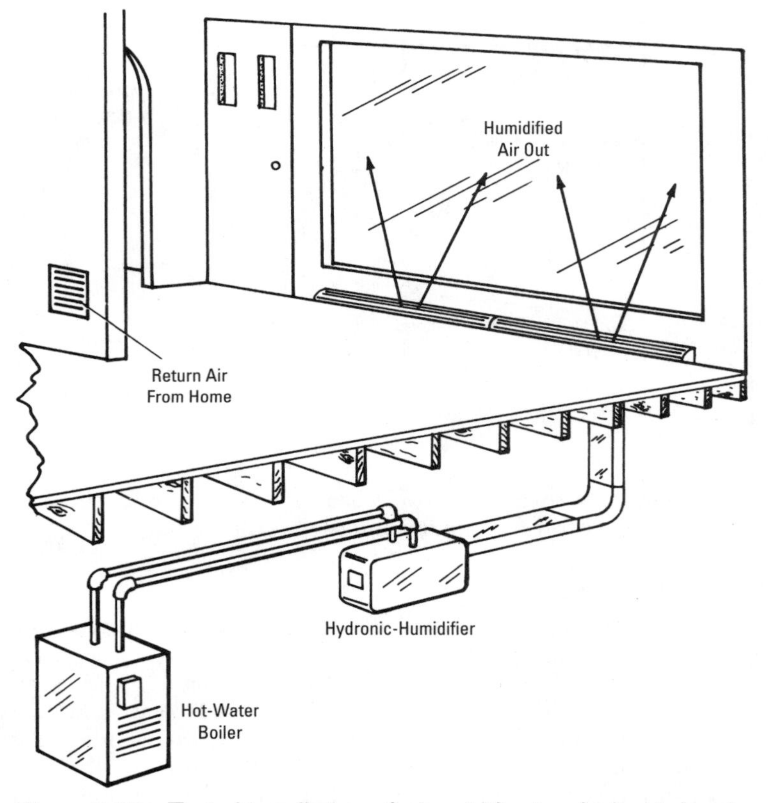

Figure 7-35 Typical installation of a humidifier in a hydronic heating system.

A central heating unit (e.g., an electric-fired boiler) can be eliminated in some systems by using thermostatically controlled electric heating components in the baseboard units. Figure 7-37 illustrates a system of this type. Note that there is no boiler. The water is circulated through the system by a pump, and a uniform temperature can be maintained by the heating element in each baseboard.

The heating system illustrated in Figure 7-37 is a single-loop installation. It is also possible to divide the whole system into a series of sealed or closed units filled with a water and antifreeze solution operating on the gravity-flow principle. Each unit could

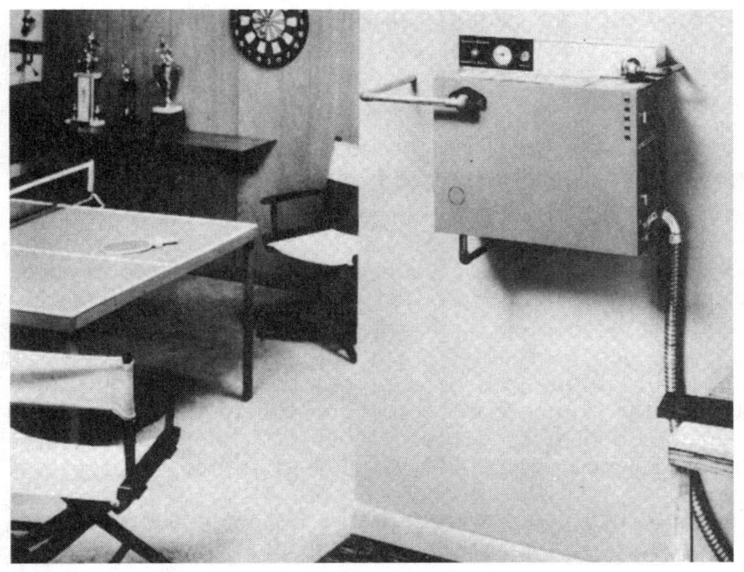

Figure 7-36 Compact electric boiler.

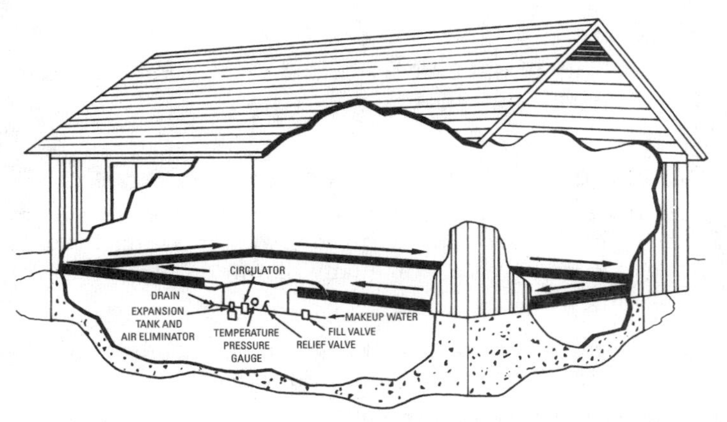

Figure 7-37 Single-loop installation consisting of electrically heated hydronic baseboard units. Water is circulated through the entire loop by a pump.

be controlled by a thermostat, or several units by a wall thermostat. The advantage of this type of installation is that the heating capacity can be easily increased if the house or building is enlarged.

Advantages of Hydronic Heating Systems

There are several distinct advantages to using hot-water heating as opposed to steam heating. For one thing, hot-water heating is more flexible than low-pressure (above-atmospheric) steam systems because the temperature may be widely varied. Due to the low working temperature of the water, the heat from a hot-water system is relatively mild, and the room atmosphere is not robbed of any of its healthful qualities. Moreover, these systems function as reservoirs for storing heat, because the radiators remain warm a considerable length of time after the fire in the boiler is extinguished. Other advantages of hydronic heating are:

- Delivery of uniform, comfortable heat without the annoying problems of cold pockets or stratification.
- Comfortable humidity levels.
- Less energy is used to circulate water than to blow air through a duct system.
- Quiet operation.
- Easy zoning.
- No possibility of dust, allergens, and unhealthful organisms being trapped in ductwork.

Disadvantages of Hydronic Heating Systems

A principal disadvantage of a hot-water (hydronic) heating system is its initial high installation cost when compared to a forced-warm-air heating system. Other commonly cited disadvantages of hydronic heating are:

- Stagnant air caused by lack of ventilation (air movement) in radiant heating systems not equipped with cooling.
- Comparatively slow heat response.
- Baseboard convectors interfere with furniture placement along walls.
- Condensation in some hydronic cooling systems.

Troubleshooting Hydronic Heating Systems

Most problems associated with hot-water heating systems involve the heating appliance (boiler, water heater, or hydro-air furnace), the water circulation pump, or the automatic controls. Table 7-1 lists some of the more general problems associated with hydronic heating systems. Specific problems involving boilers, water heaters, and automatic controls are covered in the appropriate chapters of Volumes 1 and 2 of the *Heating, Ventilating, and Air Conditioning Library*.

Table 7-1 Troubleshooting Hydronic Heating Systems

Symptoms and Possible Causes	*Suggested Remedies*
No heat	
Power may be off. Check fuse or circuit breaker panel for blown fuses or tripped breakers.	Replace fuses or reset breakers. If the problem repeats itself, call an electrician or an HVAC technician.
Check thermostat (programmable type) for dead batteries.	Replace batteries and reset thermostat.
Check automatic controls.	Turn the thermostat up or down and wait for the response. If there is no response to the changed thermostat setting after 5 minutes, the controls are malfunctioning. Test, repair, or replace as required. See troubleshooting sections in the appropriate chapters of Volume 2.
Gas burner will not operate.	See troubleshooting sections in Chapter 2, "Gas Burners" of Volume 2.
Oil burner will not operate.	Restart the oil burner by pressing the red button on the protector relay mounted on the burner. Caution: Press the red button ONLY ONE TIME. Continuing to push the button can cause an explosion. If the oil burner will not restart after the reset button is pushed once, see the troubleshooting section in Chapter 1, "Oil Burners" of Volume 2 or contact the nearest oil burner manufacturer's representative.

(continued)

Table 7-1 (continued)

Symptoms and Possible Causes	*Suggested Remedies*
Coal stoker will not operate.	See troubleshooting section in Chapter 3, "Coal Firing Methods" of Volume 2.
Circulation pump malfunctioning.	Inspect, repair, or replace as necessary.
Not enough heat	
Check thermostat setting.	Thermostat must be set several degrees higher than the desired room temperature.
Dirty baseboard convector fins.	Clean convector fins.
Dirty furnace filters.	Clean or replace as required.
Dirty fan coil (hydro-air heating system).	Clean fan coil according to manufacturer's recommended maintenance instructions.

Chapter 8

Steam Heating Systems

Steam is a very effective heating medium. Until recently, this property of steam has resulted in its being the most commonly used method of heating residential, commercial, and industrial buildings. Over the past 40 years or so, steam heating has been largely replaced in residences and small buildings by other heating systems that have often proven to be less expensive to install and operate or that operate at similar or greater levels of efficiency in small structures.

The basic operating principles of steam heating are relatively simple. A boiler is used to heat water until it turns to steam. When the steam forms, it rises through the pipes in the heating system to the heat-emitting units (radiators, convectors, etc.) located in the various rooms and spaces in the structure. The metal heat-emitting units, being cooler, cause the steam to condense and return to the boiler in the form of water (condensate, also called *condensation*) for reheating.

Classifying Steam Heating Systems

There are a number of different methods of classifying steam heating systems, but the most commonly used methods include one or more of the following features:

1. Pressure or vacuum conditions
2. Method of condensate flow to the boiler
3. Piping arrangement
4. Type of piping circuit
5. Location of condensate returns

Steam heating systems can be divided into low-pressure and high-pressure types, depending on the operating pressure of the steam used in the system. A *low-pressure system* commonly operates at a pressure of 0 to 15 psig, whereas a *high-pressure system* uses operating pressures in excess of 15 psig.

Both *vapor* and *vacuum steam heating systems* operate at low pressures (0 to 15 psig) and under vacuum conditions. The latter system uses a vacuum pump to maintain the vacuum; the vapor system does not, relying instead on the condensation of the steam to form the vacuum.

The condensate from the heat-emitting units is returned to the boiler by either gravity or some mechanical means. When the former method is used, the system is referred to as a *gravity return system.* If mechanical means of returning the condensate are employed, the system is referred to as a *mechanical return system.* The three types of mechanical devices used to return the condensation in mechanical return systems are: (1) the vacuum return pump, (2) the condensate return pump, and (3) the alternating return trap. (Each of these devices is described in the appropriate sections of this chapter.)

Using the piping arrangement as a basis for classification, a steam heating system will be either a one-pipe or a two-pipe system. A *one-pipe system* is designed with a single main that carries the steam to the heat-emitting units and the condensate back to the boiler. In other words, it functions as both a supply and a return main. In a *two-pipe system,* there is both a supply main and a return main.

The *piping circuit* may best be described as the path taken by the steam to the riser (or risers). In a *divided-circuit* installation, two or more risers are provided for the steam supply. A *one-pipe-circuit* installation, on the other hand, employs a single riser from the boiler to carry the steam supply to the heat-emitting units. A *loop-circuit* installation is used when it is necessary to operate heat-emitting units at locations *below* the water level of the boiler.

A steam heating system may also be classified according to the direction of steam flow in the risers (i.e., supply mains). An *upfeed system* is designed so that the risers are *below* the heat-emitting units. In other words, the steam supply moves from the boiler up to the heat-emitting units in the rooms and spaces within the structure. An upfeed system is also sometimes referred to as an *upflow,* or *underfeed, system.* A *downfeed system* is one in which the steam supply flows down to the heat-emitting units. In this system, the supply main is located *above* the heat-emitting units.

Sometimes the location of the condensate return is used as a basis for classifying a steam heating system. If the condensate return is located below the water level in the boiler, it is referred to as a *wet return.* A *dry return* is a condensate return located above the water-level line.

Gravity Steam Heating Systems

Gravity systems are generally limited to residences and small buildings where the heat-emitting units can be located at least 24 inches above the water-level line of the steam boiler.

A gravity steam heating system is characterized by the fact that the condensate is returned to the boiler from the heat-emitting units by means of gravity rather than mechanical means. Both one- and two-pipe installations are used, with the former being the oldest and most commonly used for a number of years (Figures 8-1 and 8-2). These one-pipe steam systems were designed to move the steam from the boiler to the heat-emitting units at no greater than 4 oz of steam pressure.

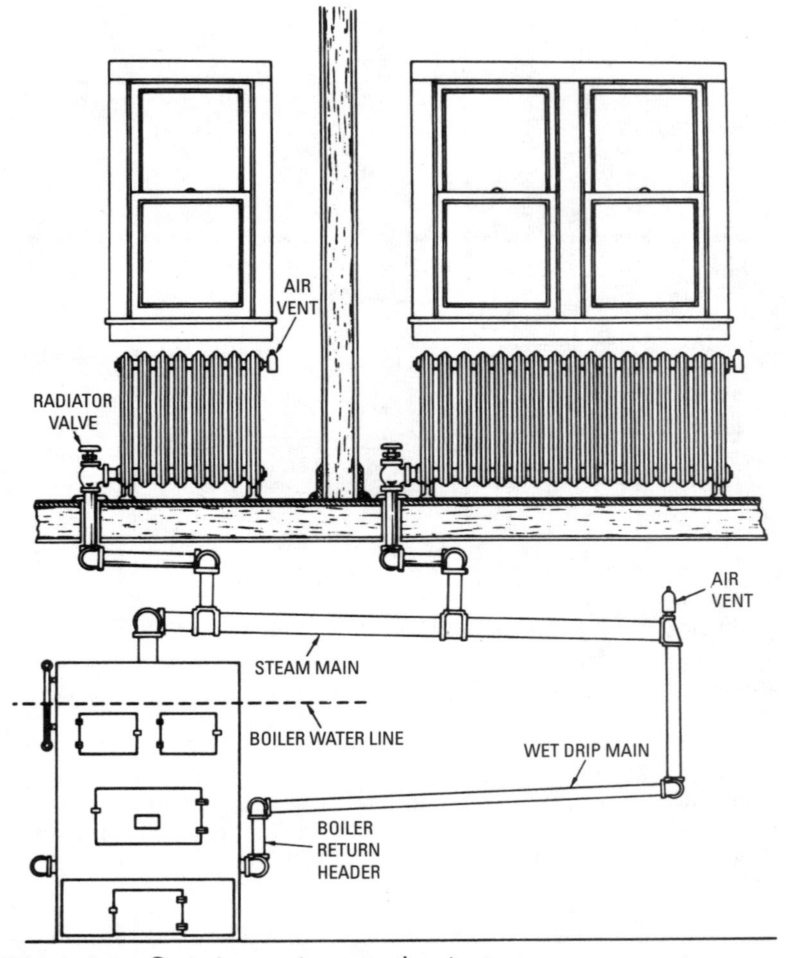

Figure 8-1 One-pipe gravity steam heating system. *(Courtesy Dunham-Bush, Inc.)*

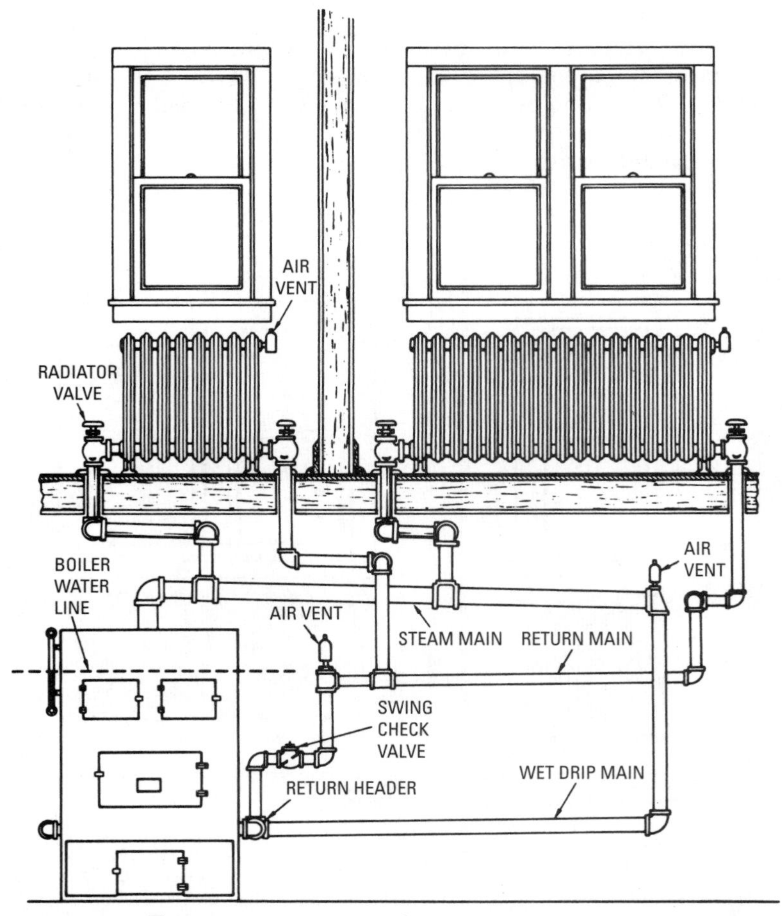

Figure 8-2 Two-pipe gravity steam heating system. *(Courtesy Dunham-Bush, Inc.)*

Gravity steam heating systems are the cheapest and easiest to install because they are adaptable to most types of structures, but they do have a number of inherent disadvantages. The principal disadvantages of gravity steam heating systems are:

1. The return lines in *one-pipe* gravity systems must be large enough to overcome the resistance offered by the steam flowing up from the boiler in the opposite direction.

2. There is the possibility of water hammer developing in one-pipe systems because the steam and condensate must flow in opposite directions in *the same pipe*.
3. Air valves (which are required) sometimes malfunction by either spurting water or failing to open. If the latter situation is the case, excess heat will build up in the system.
4. Automatic control of the steam flow from the boiler results in room-temperature fluctuations.
5. Comfortable room temperatures are possible by manually regulating the valves on individual heat-emitting units, but this results in some inconvenience.
6. In *two-pipe gravity systems,* the condensate return from each heat-emitting unit must be separately connected to a wet return or water sealed. This is expensive.

The principal reason for the development of the two-pipe gravity system is to create a means of overcoming the resistance offered by the steam flow to the condensate returning to the boiler.

One-Pipe, Reverse-Flow System

The *one-pipe, reverse-flow system* is the simplest and cheapest of this type to install. This system is easily distinguished by the absence of any wet or dry returns to the boiler. Supply mains from the boiler are inclined upward and connect with the room heat-emitting units, there being no other piping. This system is called reverse flow because the condensate flows back through the mains in a reverse direction or opposite to that of the steam flow.

The operation of a typical one-pipe, reverse-flow system is shown in Figure 8-3. Steam from the boiler flows into the main or mains (inclined upward) through the risers to the heat-emitting units at the bottom. The steam pushes the air out of the mains, risers, and heat-emitting units and escapes through air valves placed at the other end of the heat-emitting units, as shown. The condensate forming in the units flows back to the boiler through the same piping, but in the opposite or reverse direction to the steam flow.

For satisfactory operation, every precaution should be used to install the *correct* pipe sizes, especially for the mains. If the piping is too small, it will be necessary to carry excess pressure in the boiler to ensure proper operation. Without the excess pressure, operation of the remote heat-emitting units will be unsatisfactory. Satisfactory operation of these remote units can be achieved without excess boiler pressure if the pipes used in the system are the correct size.

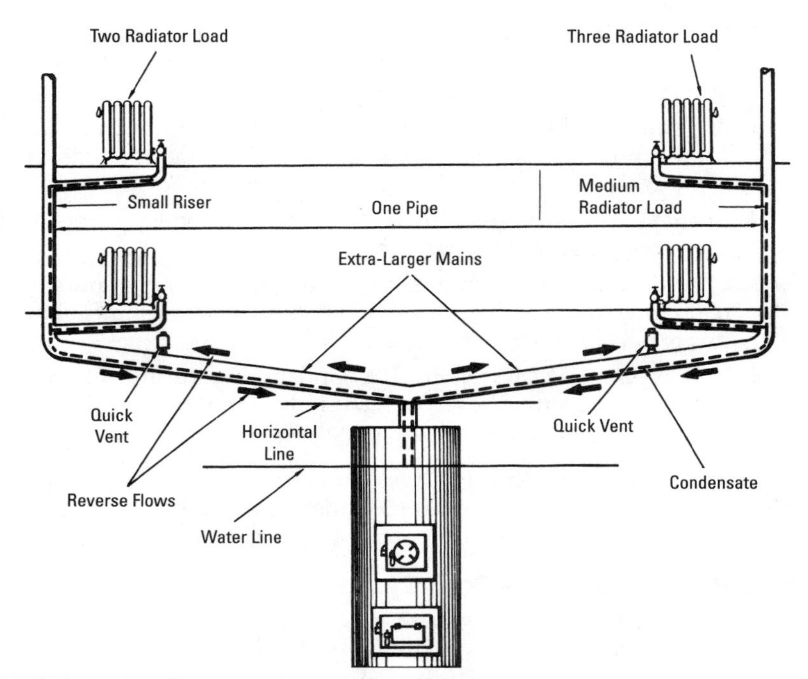

Figure 8-3 One-pipe, reverse-flow heating system.

The smaller the main, the greater the speed of the steam flow and the greater the resistance to the flow of condensate. It is very difficult for the condensate to flow back to the boiler against the onrushing steam in a long main that is in an almost horizontal position. The main should be inclined as much as conditions in the basement permit.

Leaky radiator valves can be a problem. When the valve does not close tightly, steam will work its way into the radiator and stop the condensate from coming out. The result is that the radiator soon fills with water, and when turned on again, there is difficulty getting the condensate out. This produces gurgling, hissing, and the more violent effect known as *water hammer*.

Upfeed One-Pipe System

One-pipe steam heating systems can also be of the *upfeed* type. In a standard upfeed one-pipe steam heating system, both the steam and the condensate travel through the same pipes, and the heat-emitting units are located *above* the supply mains (hence the name *upfeed system*).

An attempt to reduce the amount of condensate return in the pipes carrying the steam has resulted in a modification of the upfeed one-pipe system. In the modified system, the condensate is dripped at each radiator (and therefore from the main itself) into a wet return.

The upfeed one-pipe system shown in Figures 8-4 and 8-5 consists of a main or mains branching off from the boiler steam outlet and inclined downward, instead of upward as in the previously described one-pipe system. This arrangement causes the condensate in the mains to flow in the same direction as the steam.

The mains connect with return pipes, which carry all the condensate back to the boiler. It is only in the risers that reverse flow of the condensate takes place, and accordingly they should be large enough to take care of this reverse flow without undue turbulence.

Steam from the boiler flows through the main or mains, accruing condensate with it; the condensate returns (hence the name) to the boiler at a much lower level, as shown in Figures 8-4 and 8-5. After traversing the mains, steam flows through the risers and into the radiators. Condensation takes place as the steam warms the radiators, and the condensate drains through the risers and return pipes to the boiler.

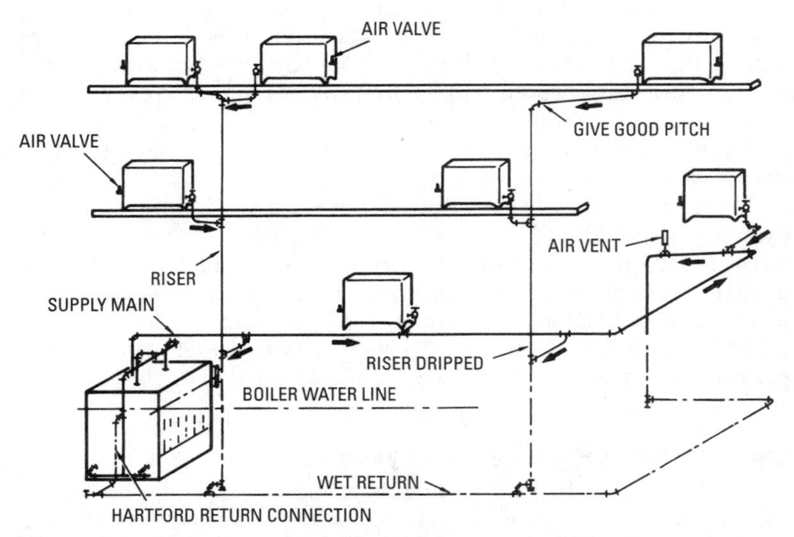

Figure 8-4 Upfeed gravity, one-pipe, air-vent steam heating system.

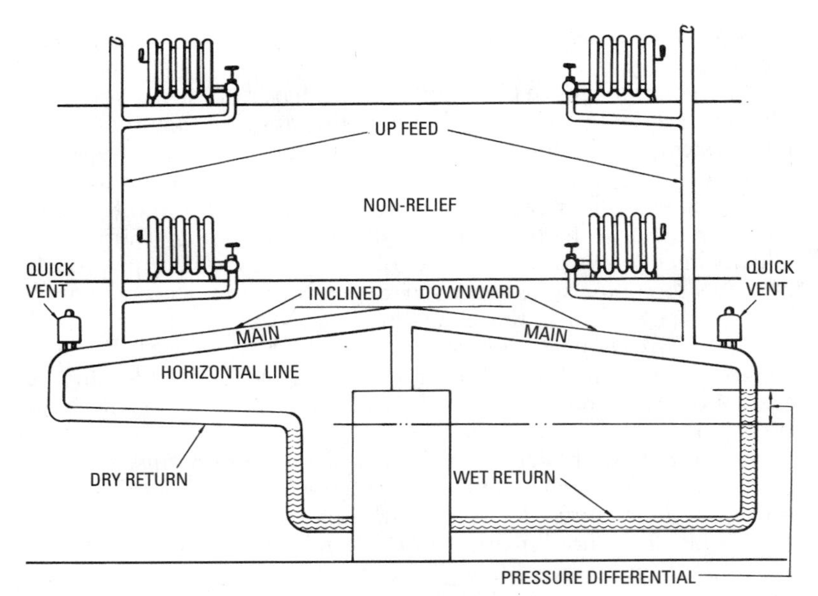

Figure 8-5 One-pipe, under- or upfeed nonrelief system.

A distinction is made between a *wet,* or *sealed, return* and a *dry return.* A wet return is below the water level in the boiler, whereas a dry return is above the water level. The advantage of a wet return is that it seals and prevents steam at a slightly higher pressure from entering the return. Under most circumstances, a wet return is preferable to a dry return. The latter may be necessary to clear doorways or other openings.

There are no disadvantages to a wet return when the system is properly installed and the valves maintained in tight condition. In most installations, these requirements are often lacking, resulting in many *noninherent* disadvantages. Among the installation problems that should be avoided are (1) pipes that are too small for the job, (2) sharp turns, (3) air pockets, (4) not enough pitch to the mains, (5) not enough air valves, and (6) air valves that are too small or cheap.

Upfeed One-Pipe Relief System

Figure 8-6 shows an *upfeed, one-pipe, relief system* applied to an eight-radiator installation. Connected to the main outlet *A* are two or more branch mains *AB* and *AC,* which supply the various risers.

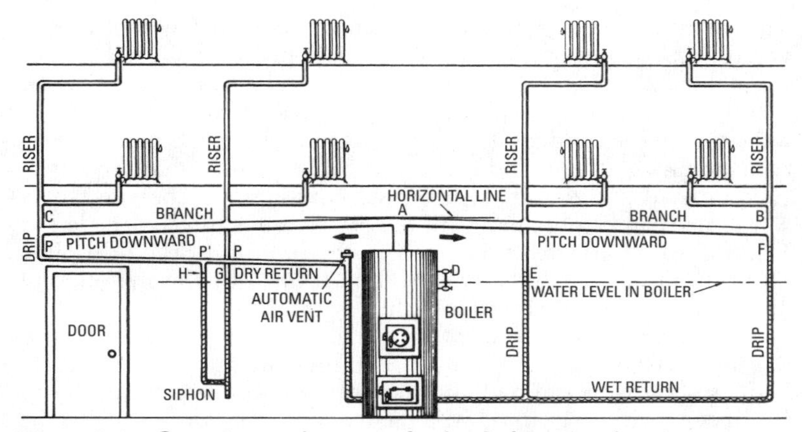

Figure 8-6 One-pipe, under- or upfeed, relief system showing locations of dry and wet returns.

Steam is supplied to each riser, and the condensate drains in the riser in reverse direction to the steam flow.

The condensate returns from the risers to the boiler by gravity through *drip pipes* (or *drop pipes*), which are virtually continuations of the return pipe, so that the condensate will flow back into the boiler.

Steam (usually at 1 to 5 psig) passes from the boiler to mains *AB* and *AC*. These branches being slightly inclined, any condensate will drain into the drip pipes. The steam passes through the risers to the radiators, where its heat is radiated to warming the rooms, thus causing condensation. Because the risers are sufficiently large, the condensate is carried by gravity in a direction reverse (that is, opposite) to the direction of the steam flow, drains down the drip pipes, and returns to the boiler through the low-level (wet) return pipe.

The condensate is forced back into the boiler by the pressure resulting from the greater head of the column of condensate in the drip pipes, as compared with the head of water in the boiler. Moreover, the water in the drip pipes, being at a lower temperature than the water in the boiler, is heavier, which upsets the equilibrium of the two columns.

This system is characterized by a slight difference in pressure (pressure differential) in the various parts of the system caused by frictional resistance offered by the pipe to the flow of the steam.

The steam flow is variable in different parts of the system due to variable condensation and badly proportioned pipe sizes. This can be explained by Figure 8-6. If the water level in the boiler is at *D,* then in operation with wet returns, the pressure difference will be balanced by the condensate standing at different levels in the different drip pipes, as at *E* and *F,* these levels being such that the difference in head and density will restore equilibrium. The effect of a wet return can be obtained with the dry return (shown at the left in Figure 8-6) by attaching a siphon to the bottom of the drip pipe. Water from the drip falls into the loop formed by the siphon; and after it is filled, it overflows into the dry return.

The water will rise to different heights, *G* and *H,* in the legs of the siphon to balance the difference in pressure at points *P* and *P′*. If the siphon were omitted and the drip pipe connected directly to the dry return, there would be a tendency for the condensate in the dry return to back up instead of draining into the boiler because the pressure in the drip pipe at *P* is greater than the pressure in the dry return. In general, the pressure varies because there is a gradual reduction in pressure as the steam flows from the boiler to the remote parts of the system. This is caused by frictional resistance offered by the pipe and fittings (Figure 8-7).

The steam flows into the pipes only when condensation is taking place. The variation in pressure exists, on the other hand, only when the steam is flowing in the pipes. The effects of pressure variation can be explained with the aid of Figure 8-8. With the plant in operation and condensation taking place in the radiators and condensate draining into the drip pipes (suppose the pressure in the boiler is 5 psig—4 psig in drip 1 and 3 psig in drip 2), then to balance these pressure differences the water will rise in drip 1 to *L* 2.3 ft above the water level in the boiler, because there is a pressure difference of 1 psig $(5 - 4)$ and the weight of a column of water 2.3 ft high is 1 lb for each square-inch of cross-section. Similarly, for drip 2, the pressure difference is 2 psig; hence, the water will rise in this column to an elevation above the water level in the boiler equal to 2.3×2, or 4.6 ft, to balance the 2-psig pressure difference. Strictly speaking, these figures are correct only when the temperature of the water in the two drip pipes is the same as the temperature of the water in the boiler. For simplicity, the difference in weight or density of the cold water in the drips and hot water in the boiler were not considered. Under these circumstances, the heavy cold water in the drips would rise to lower elevations than those shown at *L* and *F* in Figure 8-8.

Sometimes there are problems encountered on long lines with radiators located at the end of the line and near the level of the

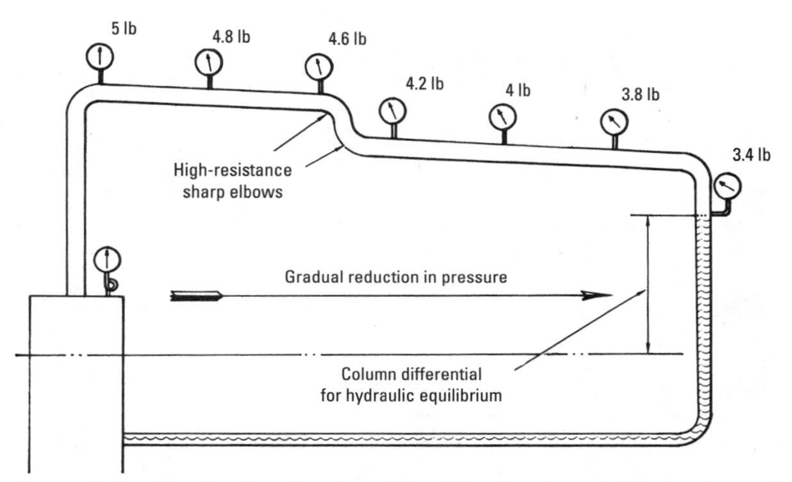

Figure 8-7 Frictional resistance offered by the pipe and fittings.

water in the boiler. On long lines where there is considerable reduction of pressure, the water sometimes backs up into the radiator, as in Figure 8-9, interfering with its operation.

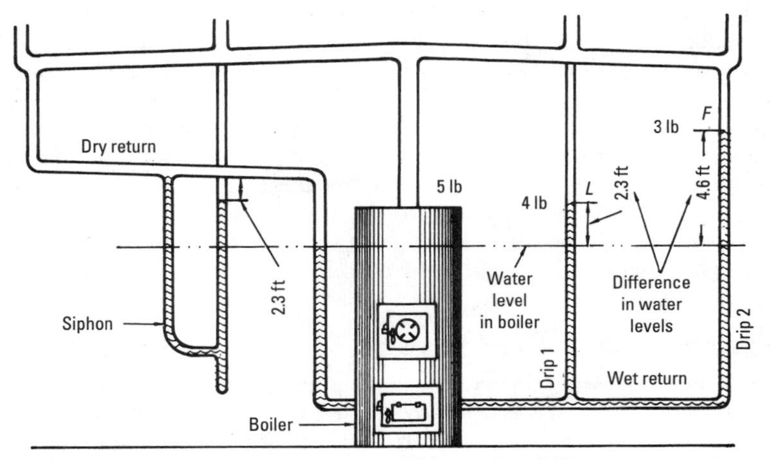

Figure 8-8 Effects of pressure variation in the different parts of a system.

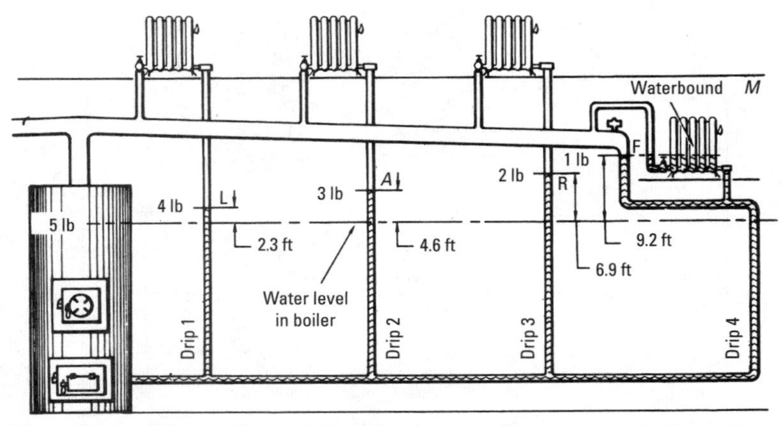

Figure 8-9 Effects of water backing up in radiator on end of long line.

Radiators located at elevations below the water level in the boiler may be operated by means of a steam loop (Figure 8-10). In the steam loop, the condenser element may consist of a pipe radiator placed on the floor above the boiler. The ample condensing surface thus provided will render the loop very active in removing the condensate, and at the same time the heat radiated from the condenser is utilized in heating. The drop leg is provided with a drain cock *D*, and the connection to the boiler is provided with a check valve. To start the system, turn on the steam at the boiler and open *D* until the steam appears. The condensation of steam in the condenser (upper radiator) will cause a rapid circulation in the riser, carrying with it the condensate from the lower radiator, which, in passing over the goose neck, cannot return but must gravitate through the upper radiator and drop leg past the check valve and into the boiler. The pipe at the bottom of the main riser, which acts as a receiver for the condensatie from the lower radiator, should be one or two sizes larger than the pipe in the main riser.

Downfeed One-Pipe System

The *downfeed one-pipe system* (also referred to as the *one-pipe oversystem* or simply the *downfeed system*) is characterized by having the heat-emitting units located *below* the supply mains. The condensate drips through the risers, thereby keeping the supply main relatively free of condensate.

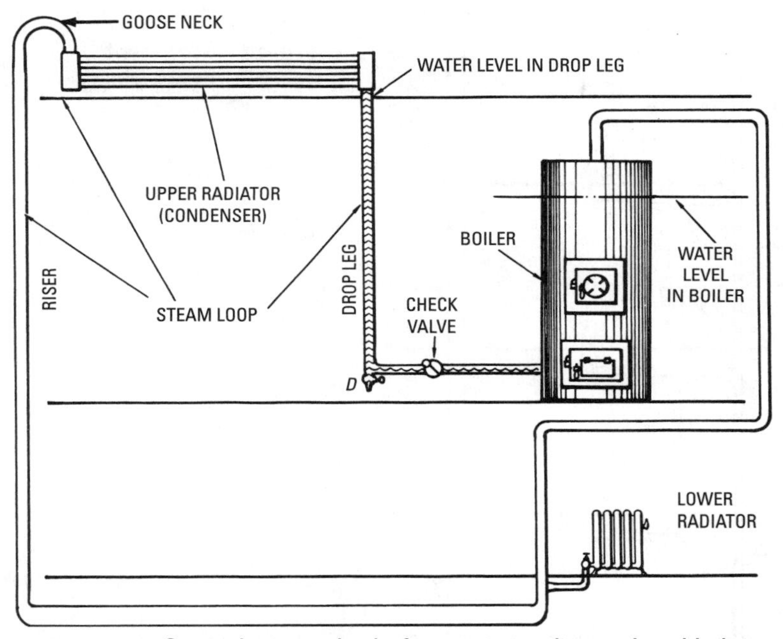

Figure 8-10 Steam-loop method of operating radiator placed below water level in boiler.

The downfeed one-pipe system is well suited for buildings 2 to 5 stories high because, if the heat-emitting units were fed from below, the risers would have to be excessively large. Instead of a steam main encircling the basement, the main is carried to the attic, forming a central riser for all the heat-emitting units. The branches in the attic correspond to the mains in the systems already described but do not carry any condensate from the heat-emitting units. These branches connect with the drops or downflow pipes that feed the heat-emitting units and drain the condensate.

As shown in Figure 8-11, steam flows from the boiler up the central upflow pipe through the branches and down through the drops or downflow pipes. Because the branches are inclined, any water or condensate in the steam drains into the upflow riser. The steam passes to the heat-emitting units through the connecting pipes. Condensate forming in the units drains through these connecting pipes into the downflow pipes. Air valves or vents are provided to rid the system of air. It should be noted that there is no *reverse* flow

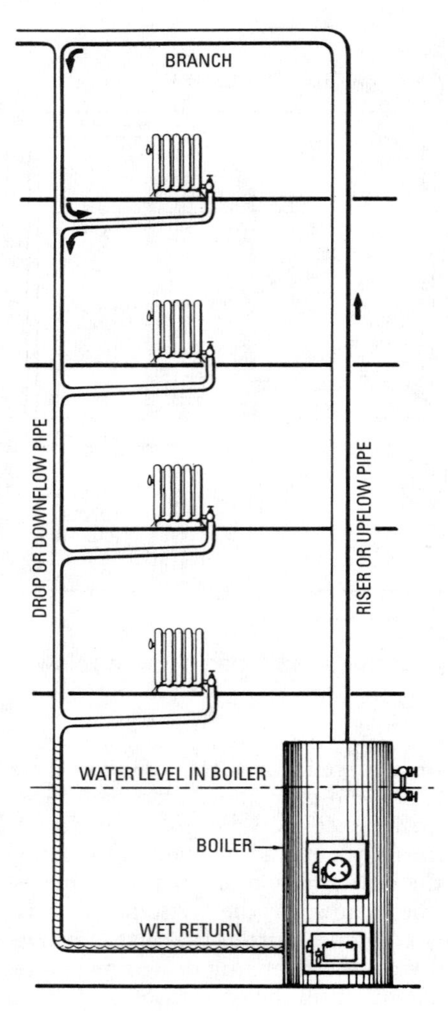

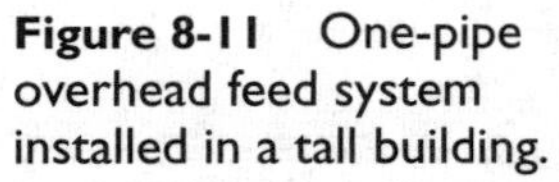
Figure 8-11 One-pipe overhead feed system installed in a tall building.

of condensate in the drop pipes. As a result, pipe sizes can be made smaller, because with parallel flow, steam velocities can be higher than in the one-pipe system.

One-Pipe Circuit System

In the *one-pipe circuit system* the steam main is carried entirely around the basement, taken off from the boiler by an elbow at the high point, as in Figure 8-12.

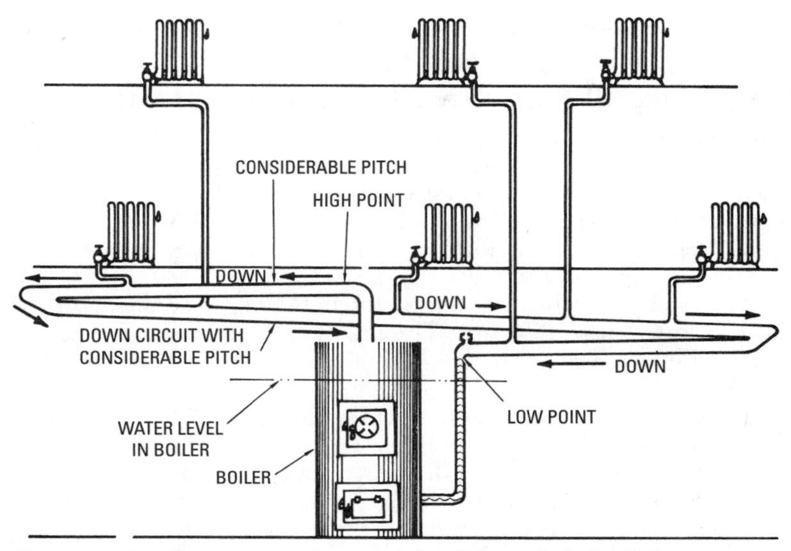

Figure 8-12 One-pipe, under- or upfeed, nonrelief, circuit system.

Note that the main *must* incline all the way from the high point to the low point. To allow for this inclination requires twice as much riser between high and low points as with the divided-circuit system, that is, where there are two mains taken off from a tee connection.

The one-pipe system is adapted for a rectangular building of low or moderate size. The size of the main (since all the steam flows through it) must be larger than in the divided-circuit system. However, especially in large installations, savings in piping may be made by installing a *tapered main.* A tapered main is one that is reduced in size along its length by connecting lengths of different-sized pipes with reducers. Eccentric reducers should be used to avoid water pockets, which would interfere with the proper drainage of the condensate. The risers are connected by being tapped from the main at various points to serve the heat-emitting units.

In a one-pipe system, the condensate drains into the main, flows in the same direction as the steam flow, and is carried to the drip pipe and then into the boiler. Since there is no return pipe as with the relief system, the circuit arrangement is less expensive to install.

Proportioning a tapered main is very important. The amount of condensate increases from the beginning to the end of the main and is considerable near the end, depending upon the number of heat-emitting units. Allowance should be made for this, and too much tapering should be avoided.

One-Pipe, Divided-Circuit Nonrelief System

The *one-pipe, divided-circuit nonrelief system* differs from the one just described in that there are two mains at the high point taken off by a tee as in Figure 8-13. These mains terminate in a U-shaped drip connection (*LF*) at the low point. Each should be vented with a quick vent as shown. Evidently each main takes care of only half the total condensate.

This steam heating system is suited to long buildings with a boiler located near the center. The end of each main is connected to a separate drip pipe connected with a common return, giving separate seals for each end.

For proper operation, these ends should not be at a lower elevation than 14 in above the boiler water line. The individual seals make the two halves of the divided circuit independent, which is desirable for unequal loads. Thus, there may be considerable difference between the pressure at *L* and *F*, each being what is necessary to balance the load.

One-Pipe Circuit System with Loop

The *one-pipe circuit system with loop* is adapted to L-shaped buildings, a circuit being used for the main building and a loop (tapped from the circuit) servicing the wing (Figure 8-14). The mains are

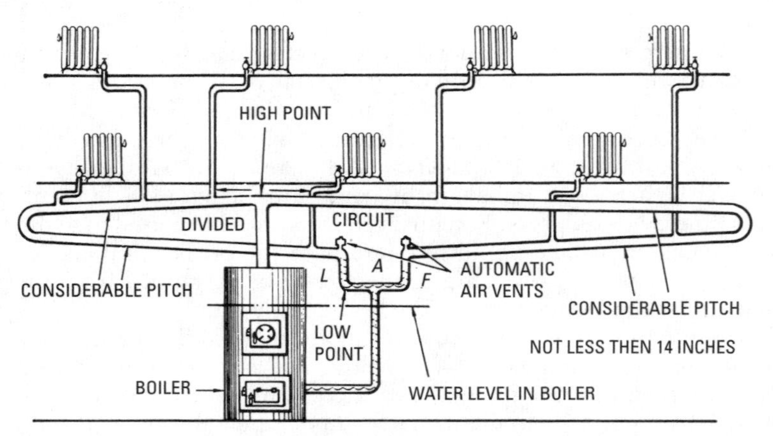

Figure 8-13 One-pipe, under- or upfeed, nonrelief, divided-circuit system.

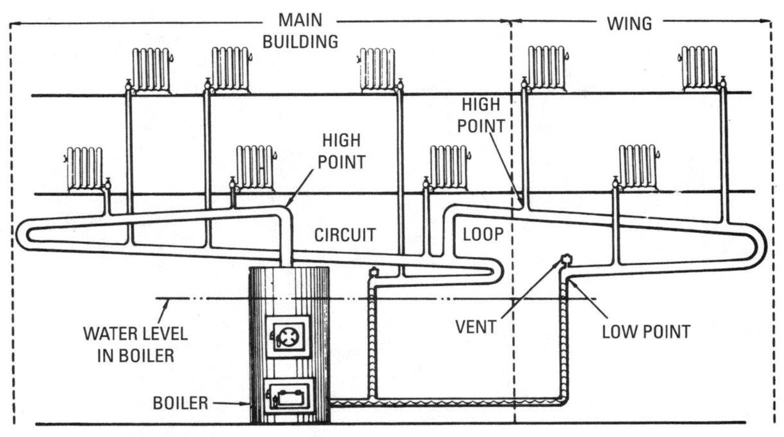

Figure 8-14 One-pipe, under- or upfeed, nonrelief, circuit system with loop for L-shaped building.

installed for the proper drainage of the condensation by providing two high points—one at the beginning of the circuit and the other at the beginning of the loop—thus giving ample margin above the boiler water line for adequate pitch in both the circuit and the loop. This is clearly shown in Figure 8-14.

The low point of the loop is higher than the low point of the circuit because the pressure at the end of the loop is less than the pressure at the end of the circuit. This requires a longer vertical drip pipe since the liquid column rises higher to balance the lower pressure.

Two-Pipe Steam Heating Systems

In two-pipe systems, separate pipes are provided for the steam and the condensate; hence they may be of smaller size than in one-pipe systems, where a single pipe must take care of both steam and condensate. Various piping arrangements are used in two-pipe systems (e.g., circuit, divided-circuit, and loop) to best meet the requirements of the building. Steam is supplied to the heat-emitting units through risers, and the condensate is returned through downflow or drip pipes.

Two-Pipe, Divided-Circuit System

A *two-pipe, divided-circuit system* is shown in Figure 8-15. In this two-pipe system, steam passes from the boiler at the high point to the mains, along which risers bring the steam to the heat-emitting

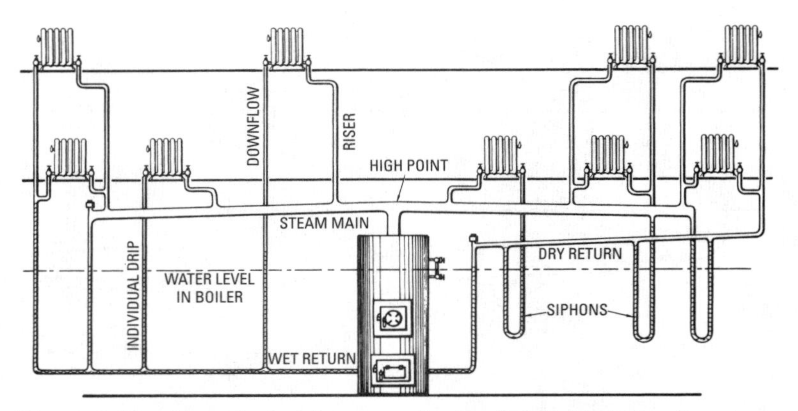

Figure 8-15 Two-pipe, under- or upfeed, relief, divided-circuit system.

units. From the opposite side of each unit is connected a drop or drip pipe. Figure 8-15 shows a wet return on the left side and a dry return on the right side.

In the wet return, the condensate is returned to the boiler by means of individual drips provided at each connection. There is a different arrangement for the dry return. The drip pipe from each heat-emitting unit terminates in a loop or siphon, which is tapped to the dry return. In operation, the condensate gradually fills the siphons and flows over into the dry return, passing into the boiler drip pipe.

A two-pipe system should also be provided with a check valve so placed as to allow water to pass into the boiler but prevent undue outflow. Under certain conditions, this prevents the water in the boiler from being driven out into the return system by the boiler pressure. The disadvantage of a check valve is that it sometimes gets stuck, a problem that can interfere with the operation of the system. An equalizing pipe with a Hartford connection loop may be used in place of a check valve to avoid this situation (see *Hartford Return Connection* in this chapter).

Vapor Steam Heating Systems

A *vapor steam heating system* is one that commonly uses steam at approximately atmospheric pressure or slightly more, and that operates under a vacuum condition without the aid of a vacuum pump (see *Vacuum Steam Heating Systems* and Figures 8-16 through 8-18).

The steam pressure at the boiler necessary to operate a vapor system is generally very low (often less than 1 lb), being no more

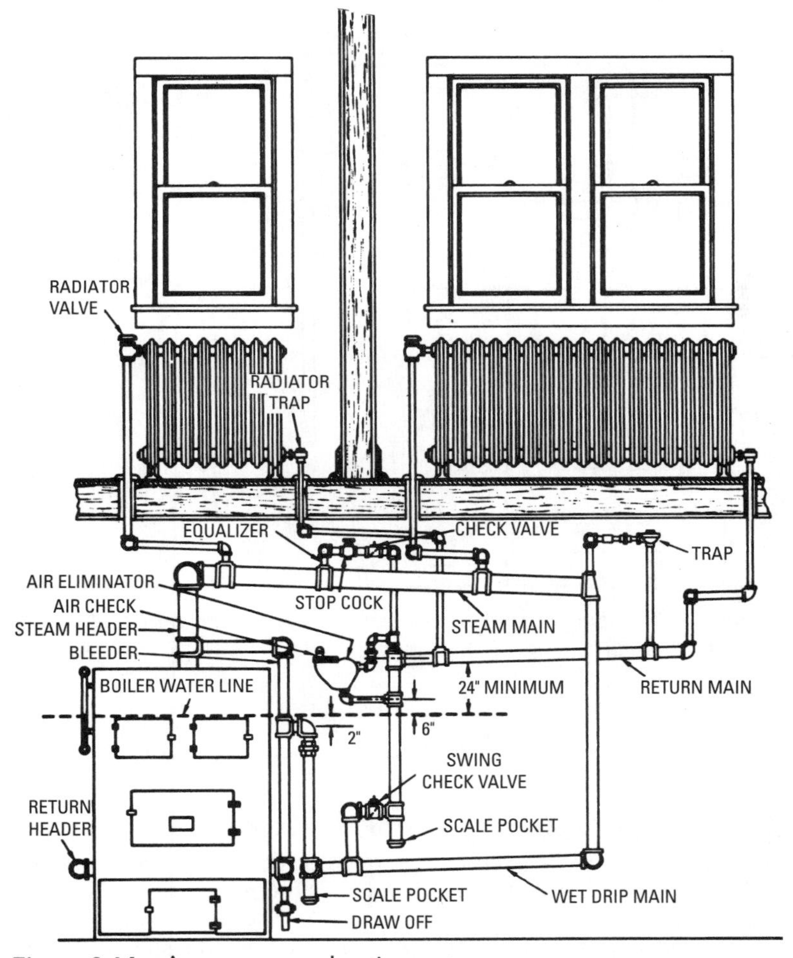

Figure 8-16 A vapor steam heating system. *(Courtesy Dunham-Bush, Inc.)*

than is required to overcome the frictional resistance of the piping system. Under most operating conditions, the pressure at the vent will be zero or atmospheric.

Vapor steam heating systems may consist of various combinations of closed or open, upfeed or downfeed, and one-pipe or two-pipe arrangements, depending on the requirements of the installation.

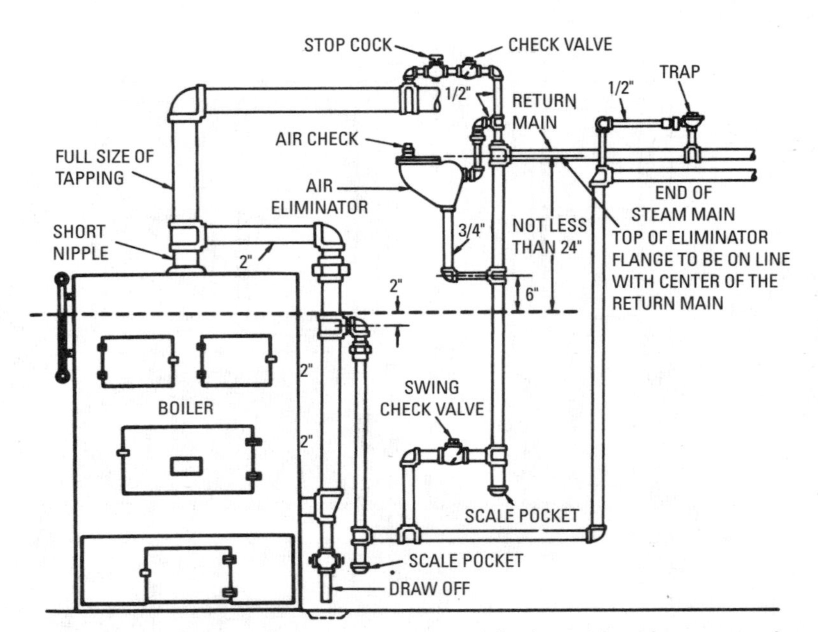

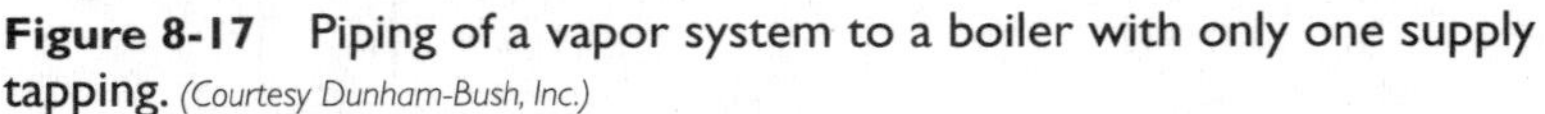
Figure 8-17 Piping of a vapor system to a boiler with only one supply tapping. *(Courtesy Dunham-Bush, Inc.)*

Any of these combinations will have certain advantages and disadvantages.

Open (Atmospheric) Vapor Systems

A vapor system with a return line open to the atmosphere *without* a check, trap, or other device to prevent the return of air is sometimes referred to as an *open,* or *atmospheric, system.*

An open vapor system is frequently used when the steam is delivered from its source under high pressure. When this is the case, pressure-reducing valves should be installed in the system to reduce the pressure of the steam to a suitable operating level. An open vapor system is also used when there is no need to return the condensate to the boiler (i.e., when it is wasted within the system). A condensate-return pump should be used when the system design requires the return of the condensate to the boiler.

In an open vapor system, the pressure at the boiler is 1 to 5 oz, or enough to overcome the frictional resistance of the piping

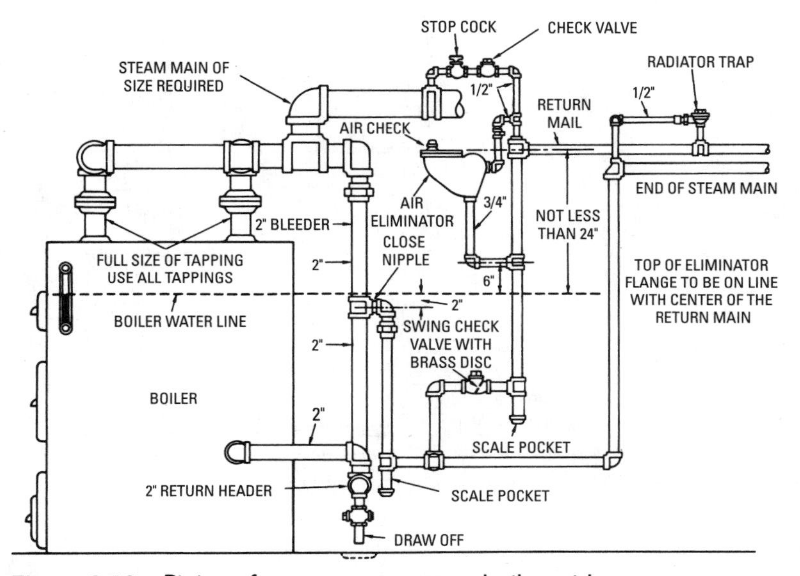

Figure 8-18 Piping of a vapor system to a boiler with two or more supply tappings. *(Courtesy Dunham-Bush, Inc.)*

system. The pressure at vent is zero gauge, or atmospheric. In operation, steam is maintained at about 5 oz pressure in the boiler by the action of the automatic damper regulator. The amount of heat desired at the radiators is regulated by the degree of opening of the supply valve. Steam enters at the top of the radiator and pushes out the air through the outlet connection, which is open to the atmosphere. The condensate returns to the boiler by gravity. This system has the advantage of heat adjustment at the radiator, but the devitalizing effect in the air is somewhat greater than in the vacuum systems because the steam entering the radiators is at a higher temperature than the steam of lower pressure in the vacuum system. It is, however, simple.

Figure 8-19 illustrates a very simplified, mechanically controlled, vapor steam heating system. Though it does not present the latest practice, it does present control principles very plainly.

The success of a vapor steam heating system depends upon the proper working of the automatic damper regulator in keeping the boiler pressure within proper limits. To accomplish pressure regulation, the dampers are controlled by a float working in a float

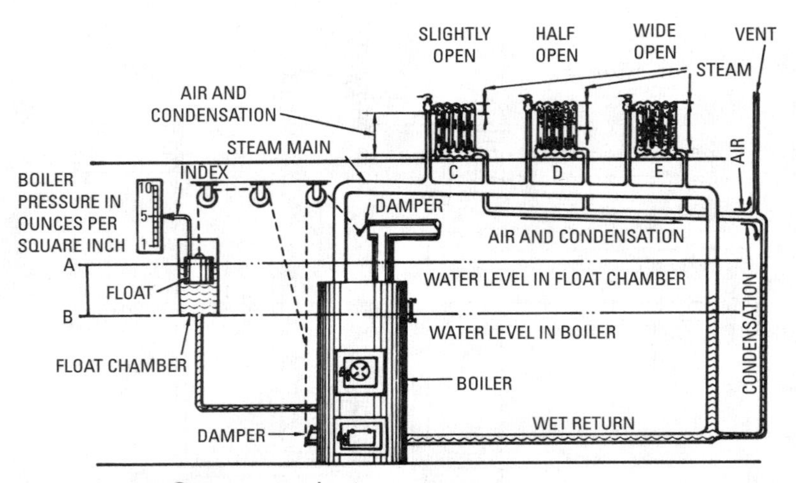

Figure 8-19 Open atmospheric vapor system.

chamber in communication with the water space in the boiler, as shown in Figure 8-19.

When the pressure in the boiler is the same as that of the atmosphere (0 psig), the water level in the float chamber (Figure 8-19) is the same as that in the boiler, and the index hand points to zero.

As steam generates, the steam pressure increases and the water level in the boiler is forced downward. The latter action causes the level in the float chamber to rise until the pressure due to the difference *AB* (Figure 8-19) of water level balances that in the boiler.

The float, in rising, connected as it is by pulleys and chains to the dampers, closes the ashpit damper, thus checking the draft and preventing a further increase of steam pressure.

In this system, the steam feed is connected to the top of the radiators and the air and condensation is taken from the bottom because steam is lighter than either air or condensation. Accordingly, when steam is admitted, it floats on top of the air, thus driving the air out through the lower connection.

The chief feature of a vapor system is that the amount of heat given off by each radiator may be regulated by the steam valve. Thus, in Figure 8-19, the valve of radiator *C* is opened just a little, which will admit only just enough steam to heat a larger portion of the radiator; with the valve wide open on *E,* the entire radiator is heated.

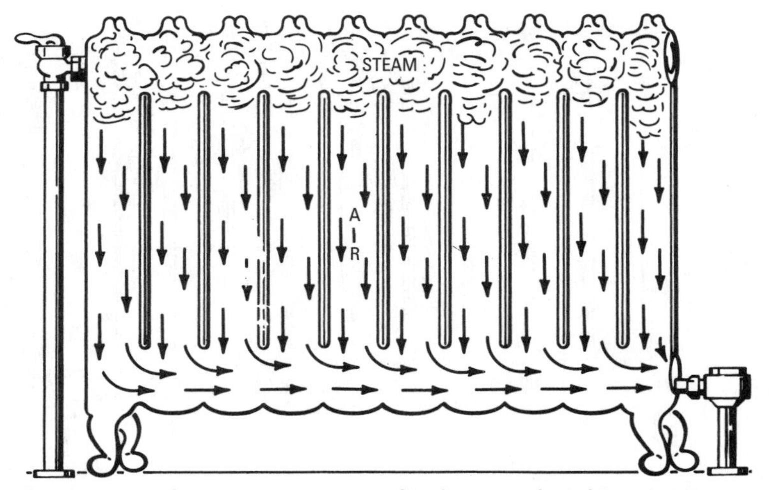

Figure 8-20 Steam entering top of radiator and pushing air out the bottom.

The kind of radiator used is the downflow type in which steam enters at one end at the top and the air and condensation pass out at the other end at the bottom.

As steam enters a cold radiator, it forces the cool air in the radiator out through the trap into the return piping. The operation of a typical downflow radiator is shown in Figures 8-20 through 8-22. Figure 8-20 shows the steam entering and air passing out through thermostatic retainer valve. Figure 8-21 shows more steam entering and condensate and the balance of the air passing out through the trap, the action progressing until (as in Figure 8-22) the radiator is full of steam.

As the radiator warms up, the steam gives off heat and condenses. The condensate, being heavier than steam, falls to the bottom of the radiator and flows to the trap through which it passes into the return piping. After the air is forced out, the steam fills the radiator and follows the condensate to the trap. The trap closes when the steam enters it because the steam is hotter than the water. This excess heat expands the valve control element, closing and holding the valve against its seat with a positive pressure, thus preventing the steam from flowing into the return piping (Figure 8-23).

The trap closes once the radiator is completely filled with steam, and heat is given off as the steam condenses. The condensate thus formed, which is cooler than the steam, flows in a steady stream to

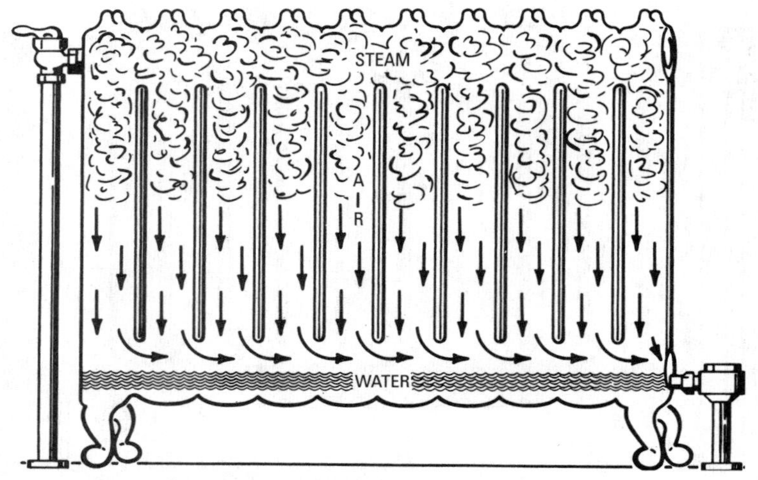

Figure 8-21 Condensate and air leaving the radiator.

the trap, which it slightly chills, causing it to open and allowing the condensate to pass out into the return piping (Figure 8-24).

When properly working, the trap adjusts itself to a position corresponding to the temperature of the condensate, just as a

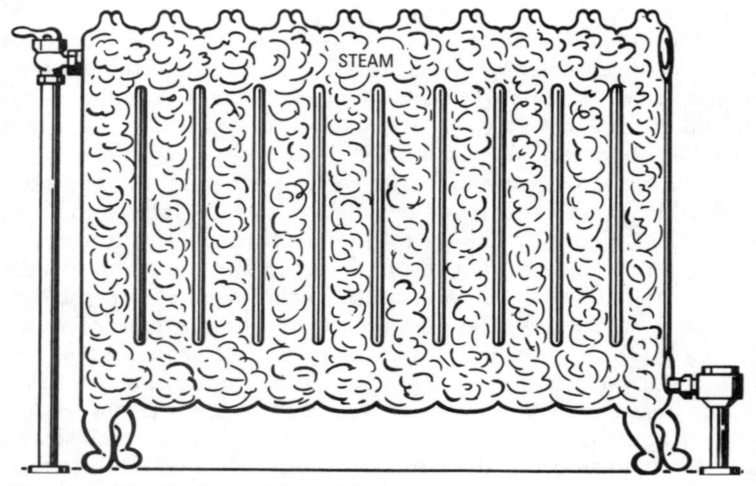

Figure 8-22 Radiator filled with steam.

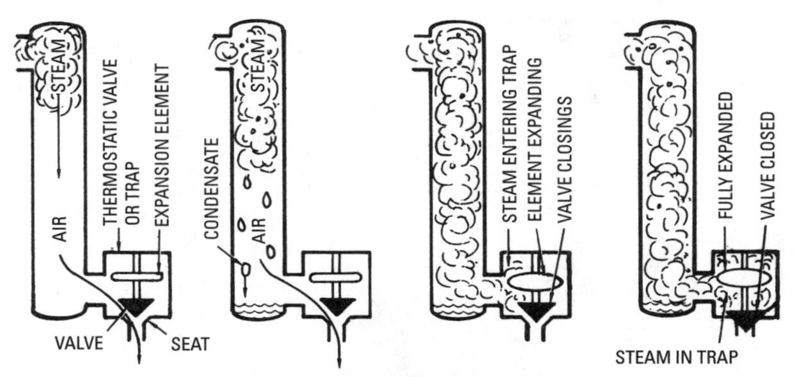

Figure 8-23 Progressive action of thermostatic check or trap from beginning of air entrance to closing of valve by expansion of actuating element.

thermometer does to the room temperature, and permits a continuous flow of condensate from the heat-emitting units (Figure 8-25).

Closed Vapor Systems

A vapor steam with a return line closed to the atmosphere is sometimes referred to as a *closed vapor* system. The condensate returns by gravity flow to a receiving device (an alternating receiver or

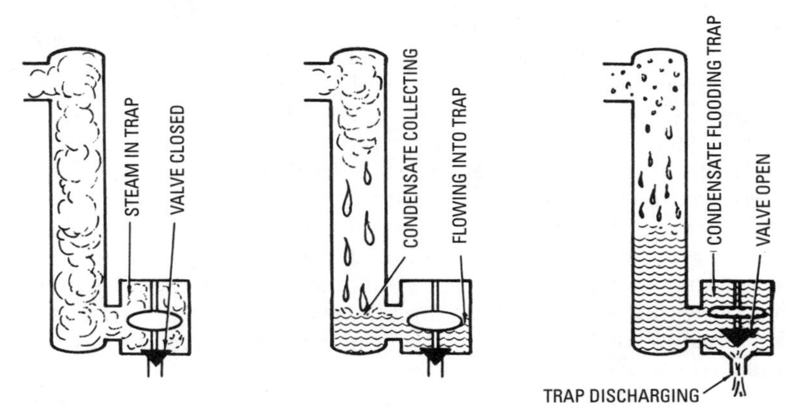

Figure 8-24 Progressive action of thermostatic check or trap closed position to open position due to contraction of actuation element when chilled by relatively cool condensate.

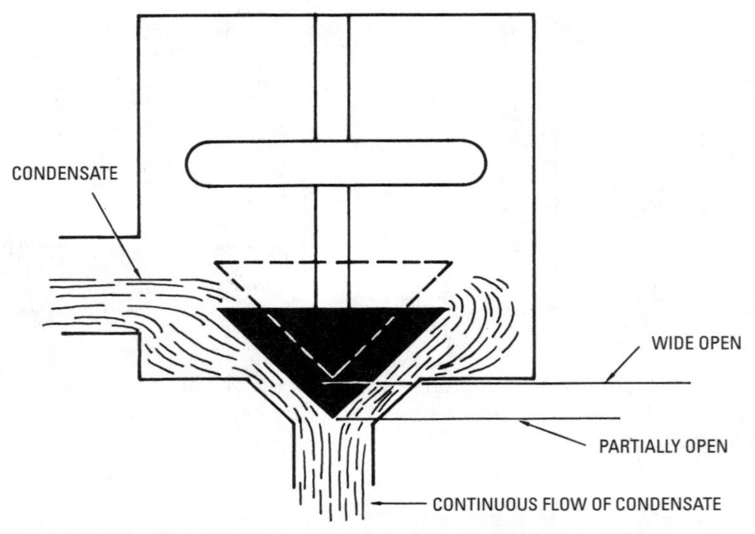

Figure 8-25 Detail of trap showing the valve in intermediate position for ideal continuation flow of condensate.

boiler-return trap), where it is discharged into the boiler. The condensate from the alternating receiver is discharged against the boiler pressure.

Because air cannot enter a closed vapor system, a moderate vacuum is created by the condensing steam. As a result, steam is produced at lower temperatures, and the system will continue to provide heat after the boiler fire has died down.

Figure 8-26 shows the arrangement of a typical upfeed, two-pipe vapor system with an automatic return trap. The heat-emitting units discharge their condensate through thermostatic traps to the dry return pipe. These systems operate at a few ounces of pressure and above, but those with mechanical condensation return devices may operate at pressures upward of 10 psi. The simplest method of venting the system consists of a ¾-in pipe with a check valve opening outward. Most systems employ various forms of vent valves, which allow air to pass and prevent its return. A dry return is provided so that the air will easily go out the vent pipe.

Vacuum Steam Heating Systems

A *vacuum heating system* is one that operates with steam at pressures less than that of the atmosphere. The object of such systems is to take advantage of low working temperatures of the steam at these low pressures, giving a mild form of heat such as is obtained with hot-water heating systems. Vacuum systems such as these, which operate at all times at pressures less than atmospheric, should not be confused with the combined atmospheric and vacuum systems described in the next section. There are distinct design differences between the two systems.

A distinction should be made between a vacuum system and a subatmospheric system. The latter differs from an ordinary vacuum system in that it maintains a controlled partial vacuum on both the supply and return sides of the system instead of only on the return side. In the vacuum system, steam pressure above that of the atmosphere exists in the supply mains and heat-emitting units practically at all times. The subatmospheric system is characterized by atmospheric pressure or higher existing in the steam supply piping and heat-emitting units only during severe weather (Figures 8-27 and 8-28).

There are a number of different methods of classifying vacuum systems (e.g., one-pipe or two-pipe and vacuum pressure or subatmospheric). For the purposes of this chapter they will be classified according to the type of vacuum: natural vacuum systems and mechanical vacuum systems.

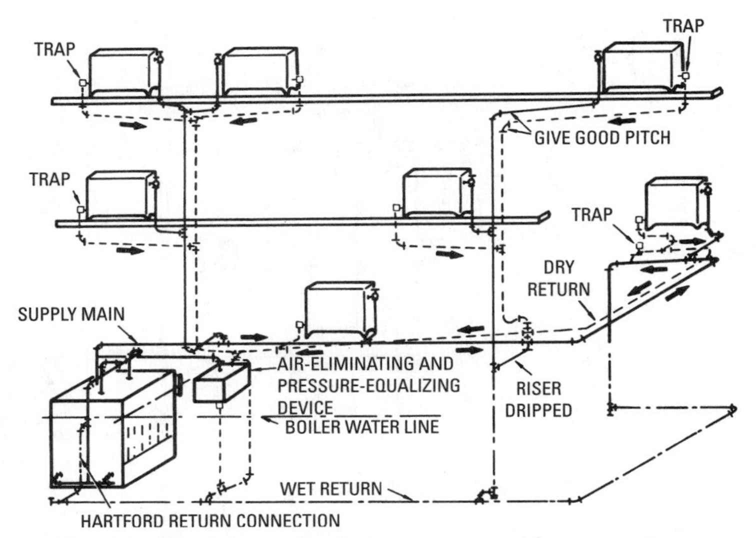

Figure 8-26 Two-pipe, upfeed vapor system with automatic return trap.

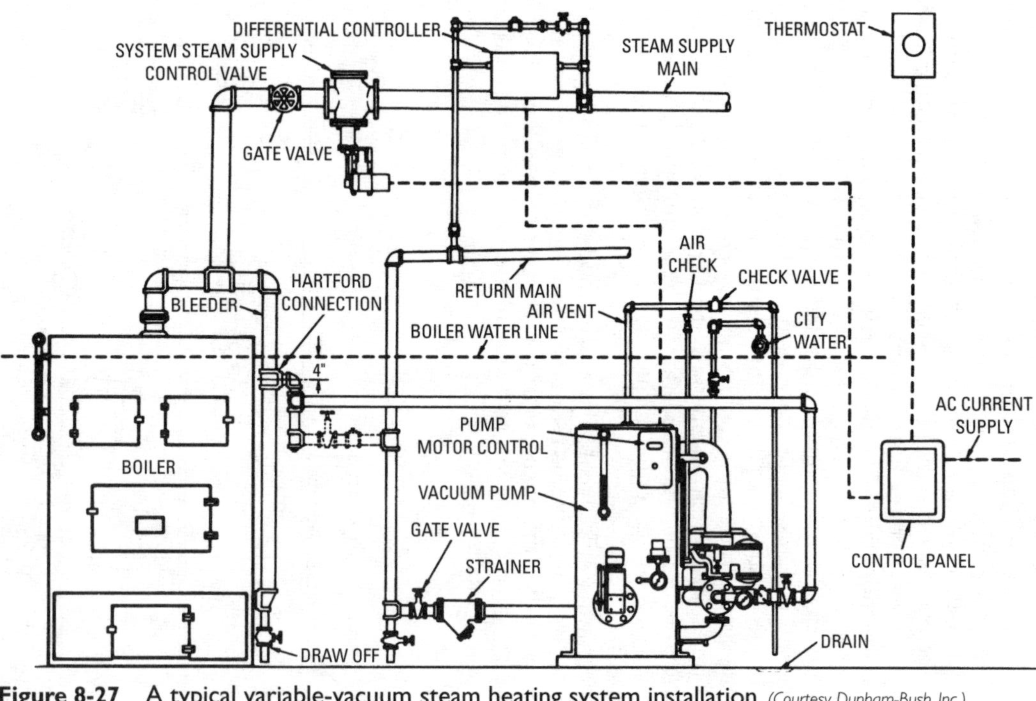

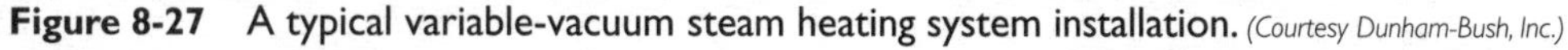
Figure 8-27 A typical variable-vacuum steam heating system installation. *(Courtesy Dunham-Bush, Inc.)*

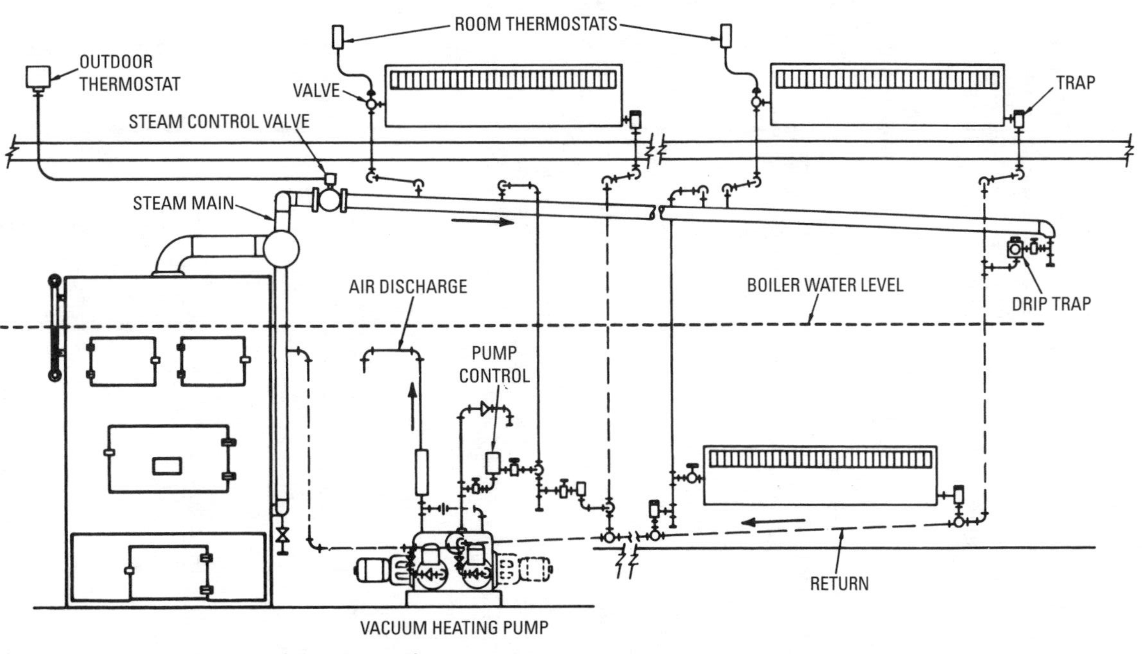

Figure 8-28 A subatmospheric steam-heating system. *(Courtesy Nash Engineering Co.)*

Natural Vacuum Systems

Any standard one- or two-pipe steam system may be converted into a *natural vacuum system* by replacing the ordinary air valve with a mercury seal or connecting thermostatic valves to the radiator return outlet on radiators and providing a damper regulator on coal-burning boilers adapted to vacuum working. The mercury-seal system is shown in Figure 8-29.

A mercury seal is virtually a barometer, consisting, as shown in Figure 8-30, of a tube (*A*) that dips just below the surface of the mercury in a cup (*B*). When the steam is raised in the boiler to pressures above atmospheric, it drives all air out of the system, the air leaving by bubbling through the mercury in cup *B*.

If the fire is allowed to go out, the steam will condense and produce a vacuum, provided all pipe fitting has been carefully done and the valve stuffing boxes are tightly packed.

In Figure 8-29 the loop at *C* prevents water from being carried over into the seal pipe when purging the system of air. If air should again enter the system, it can be expelled by raising the steam pressure above atmospheric. In very cold weather, the system can be operated at pressures above atmospheric by closing valve *D*. When fires are banked for the night, valve *D* may be opened and the system worked as a vacuum system. The flexibility of vacuum systems is in sharp contrast with low-pressure systems, where steam

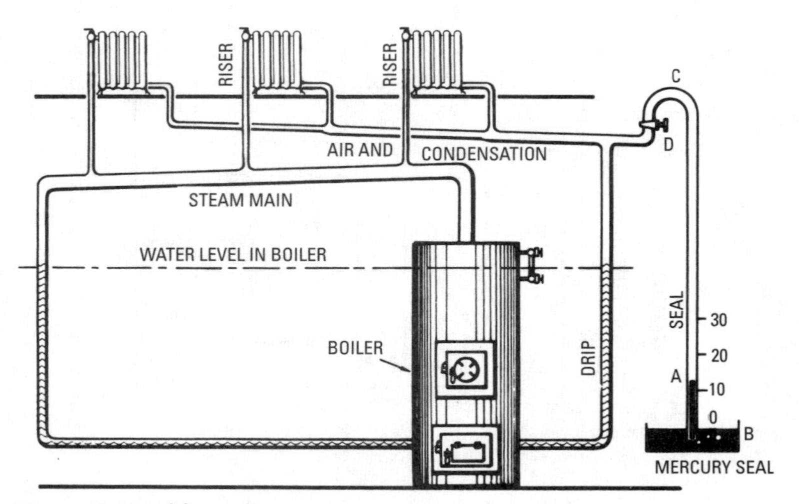

Figure 8-29 Natural vacuum, mercury-seal system.

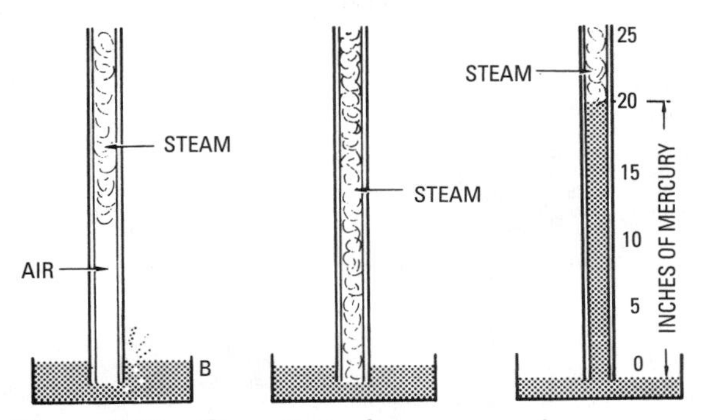

Figure 8-30 Operation of mercury seal.

disappears from the radiators as the temperature drops below 212°F. According to weather demands, the radiators may be kept at any temperature from, say, 150 to 220°F.

Another method of maintaining a natural vacuum is by using thermostatic valves instead of a mercury seal. A thermostatic valve has an expansion element that operates to close the valve when heated by hot steam and to open the valve when chilled by the relatively cold condensate.

The two kinds of thermostatic valves used are the single-unit (or retainer) valves and two-unit, or combined retainer and air-check, valves sometimes called *master thermostatic valves.*

Figure 8-31 shows the details of a master thermostatic valve, consisting of a thermostatic unit and an air check. The thermostatic unit has an expanding element, the air check, consisting of a group seat poppet check valve that is practically airtight and therefore will retain the vacuum within the system for a considerable length of time. The air check operates when excess pressure is generated in the boiler to purge the system of air, the check at other times remaining closed.

The thermostatic valve remains open while the system is being purged of air and condensation but closes when steam enters the valve chamber—it retains vacuum in the air line.

Figure 8-32 shows the operation of the natural vacuum system with retainer and master thermostatic valves. Individual thermostatic retainer valves *A, B,* and *C* are placed in the outlet of each radiator, which pass air or water but close to steam. At the end of the air line is a master thermostatic valve (*D*), which operates when the system is purged of air by excess pressure.

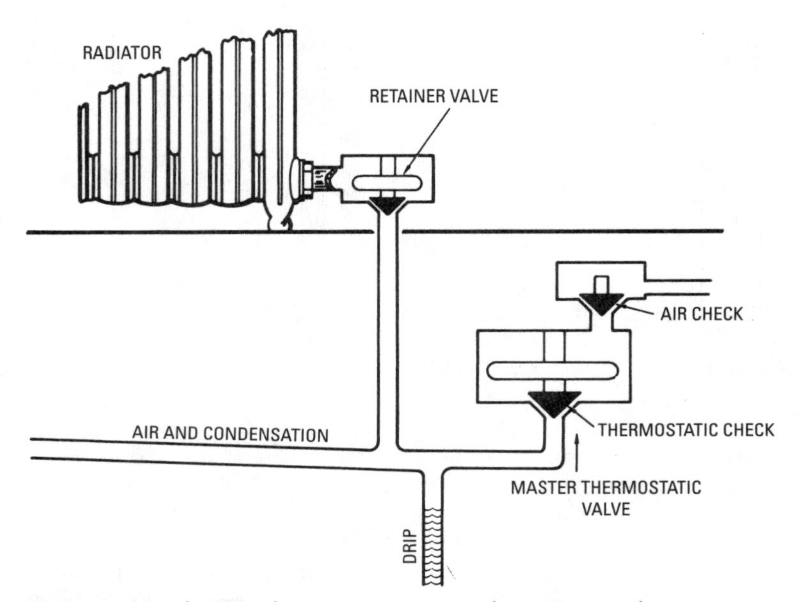

Figure 8-31 A natural vacuum system with retainer and master thermostatic valve showing sectional views of the valves.

The drip should be proportioned to prevent water entering the air line in case of high vacuum in such a manner that the vertical distance *M* between the water level in the boiler and the lowest point of the air line is not less than 2 ft for each inch of vacuum to be carried in the system.

The successful operation of natural vacuum systems depends largely on efficient damper regulators on coal-burning boilers (i.e., efficient draft control), for unless the fire is held in proper check, the pressure will rise and break the vacuum. This can waste fuel, for there may be sufficient heat in the boiler to supply steam to the system with a 5- or even 10-in vacuum and hold that heat in the system for hours.

Automatic damper regulators are designed to act by pressure, temperature, or a combination of these two. Figure 8-33 shows a regulator that acts on the pressure principle. It consists of a diaphragm connected at *B* to a lever having its fulcrum at *A* and having a weight *W* free to slide along a slot between the stops.

In starting, the weight is placed on the left side of the lever as shown (Figure 8-33). This tilts the lever (position *LF*) and opens the

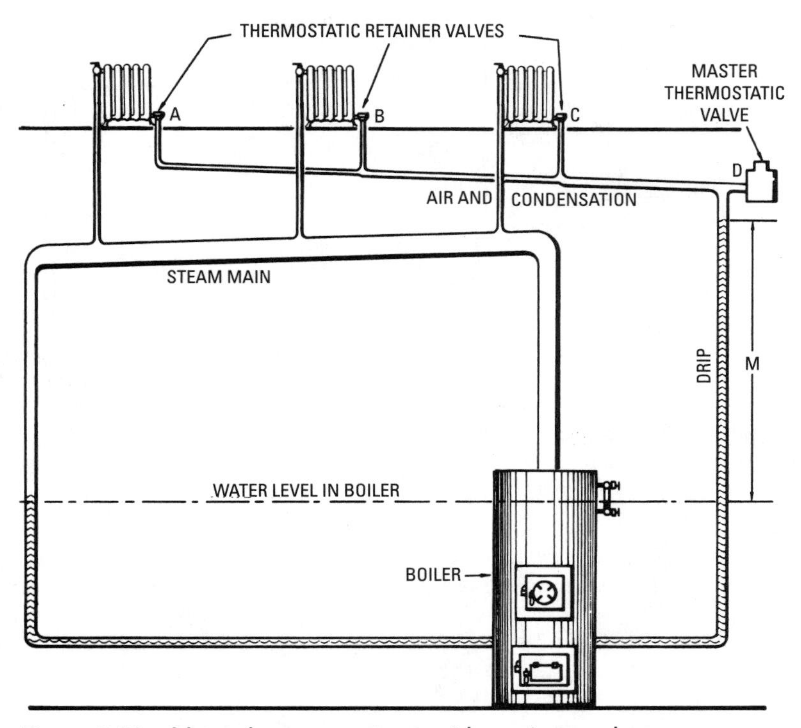

Figure 8-32 Natural vacuum system with retainer and master thermostatic valves.

damper. The weight is adjusted by the stop so that sufficient pressure is produced to clear the system of air before the regulator tips to position *L'F'* (shown in dotted lines) and closes the damper.

The regulator is gradually closing as the pressure comes on. When the regulator is entirely closed, the weight slides to the right and remains in this position until the vacuum in the system becomes strong enough to gradually open the damper—just enough to maintain a vacuum.

In the morning, the regulator may be set to the open position from the floor above by the pull chain *M*. This generates pressure and purges the system of any air that might have accumulated; then the regulator weight automatically goes to the vacuum side of the regulator and maintains the vacuum heat until more fuel is required or

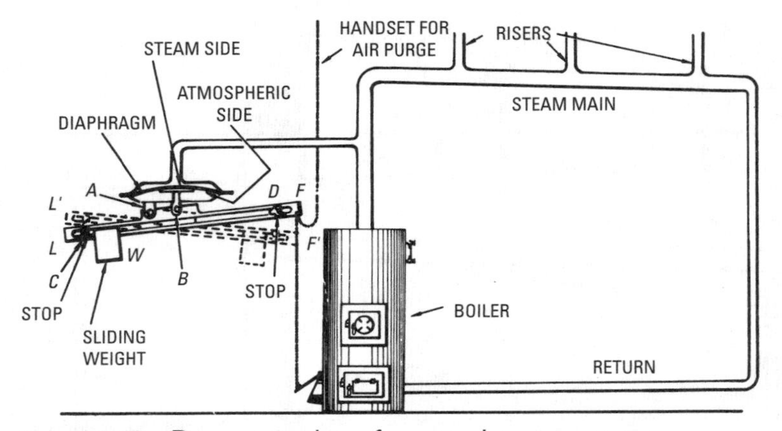

Figure 8-33 Damper regulator for natural vacuum system operating on the pressure principle.

further regulation is necessary. Temperature controls or damper regulators that depend on temperature changes for their operation may also be used (Figure 8-34). Since the temperature of steam increases with the pressure, evidently the expansion and contraction of a rod exposed to the steam can be made to operate the damper.

Figure 8-34 shows the construction of a typical thermostatic regulator. The expansion element or rod is fastened at *A* in a closed cylindrical chamber through which steam from the boiler passes to the main. The end *B* is free to move, passing out of the chamber through a stuffing box. The motion of the rod is considerably magnified by the bell crank lever, which is connected to the damper by a chain attached at *C*.

In operation, as the pressure of the steam rises, so does its temperature; the rod (which is made of a metal having a higher coefficient of expansion than that of the cylindrical chamber) expands, and its free end, *B*, moves to the right, thus causing end *C* of the lever to descend, closing the damper.

When the pressure falls, the rod contracts, and the spring that keeps the bell end in contact with the rod causes end *C* of the lever to rise and open the damper.

The lever will assume some intermediate position in actual operation, thus holding the steam at some predetermined pressure, which may be varied by means of the screw adjustment *(D)*.

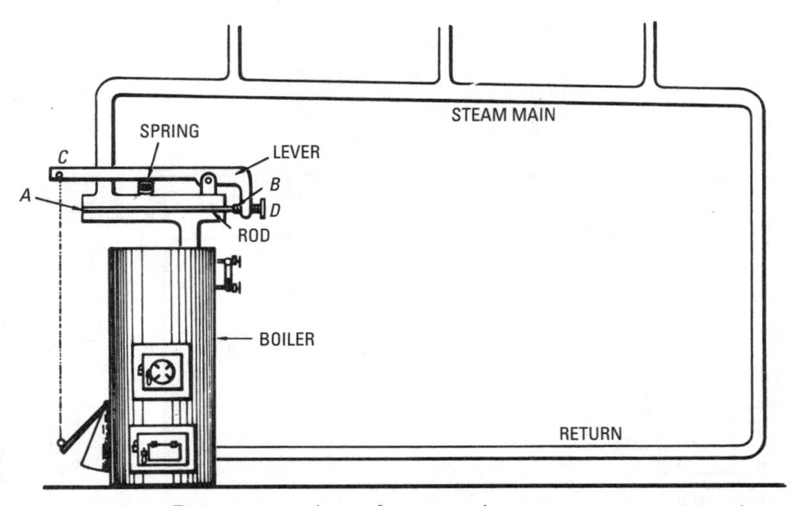

Figure 8-34 Damper regulator for natural vacuum system operating on the temperature principle.

The major objection to regulation by temperature is that there is no provision for securing excess pressure to purge the system of air in starting. This must be done by hand control of the damper. This objection can be overcome by the method of combined pressure and temperature regulation, which employs pressure for starting and temperature for running. In starting, the thermostatic portion of the regulator is closed off from the systems during which pressure is generated sufficient (about 1 pound) to purge the air from the system. After purging, the regulator automatically opens a valve to the thermostatic position, which then maintains the temperature desired, its range embracing both vacuum and low-pressure operation.

Mechanical Vacuum System

A *mechanical vacuum system* is one in which an ejector or pump is used to maintain the vacuum. The hookup of the ejector system is shown in Figure 8-35. The ejector, which may be operated either by steam or water, is started before steam is turned on in the system. Thus, after the air is removed, steam quickly fills the heat-emitting units, because the air is automatically removed as fast as it accumulates.

The system commonly used in exhaust heating employs an alleged or so-called vacuum pump, which ejects the air and condensate from

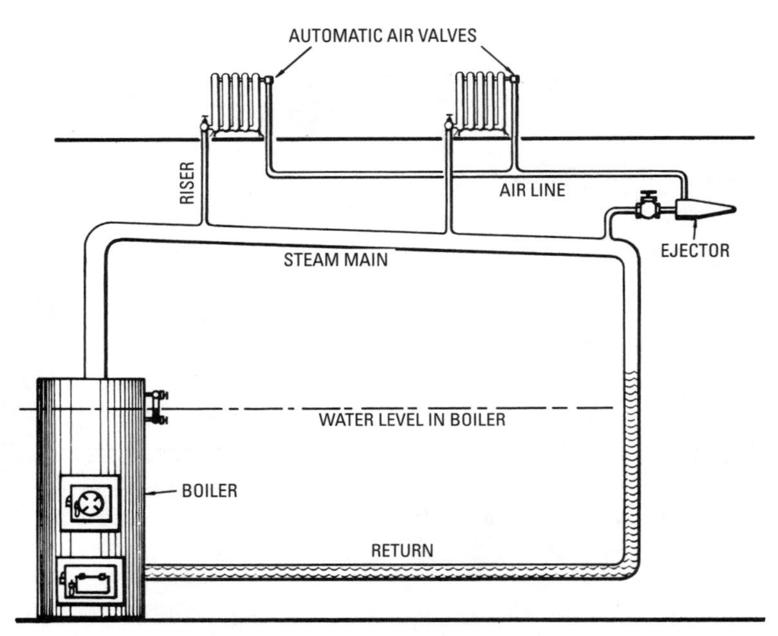

Figure 8-35 Mechanical vacuum ejector system.

the system. In operation, this device pumps out most of the air (or other gas) from the condenser, maintaining a partial vacuum. The pump cannot obtain a perfect vacuum because each stroke of the pump piston or plunger removes only a certain fraction of the air, depending on the percentage of clearance in the pump cylinder, resistance of valves, and so forth. Hence, theoretically an infinite number of strokes would be necessary to obtain a perfect vacuum (not considering resistance of the valves, clearance, etc.).

Condensation of steam creates the vacuum, and the pump that removes the air maintains a vacuum. A wet pump (that is, one that removes both air and condensate) is the type generally used. A dry pump removes only the air.

The essential features of a mechanical vacuum pump system are shown in Figure 8-36. This system is of the fractional valve distribution type. In operation, air, being heavier than steam, passes off through thermostatic retainer valves to the pump. When the steam reaches the retainer valves, they close automatically to

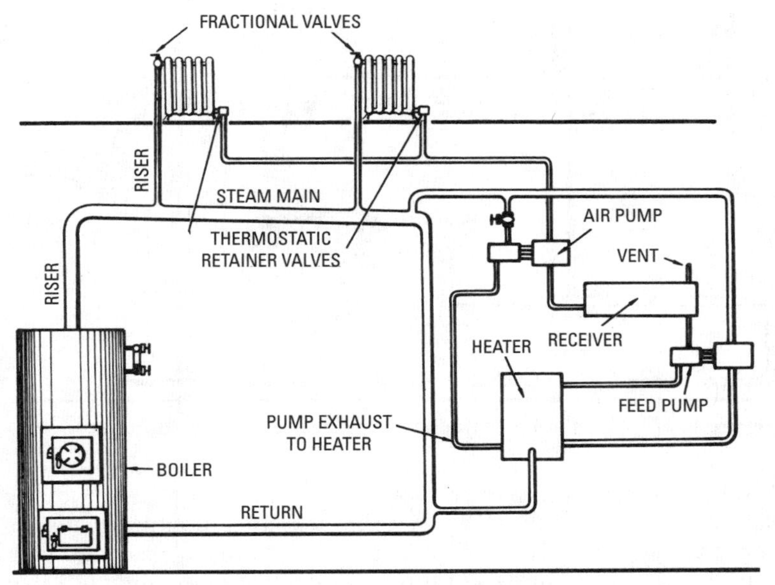

Figure 8-36 Mechanical vacuum air-pump system as applied to fractional valve distribution.

prevent the steam passing into the dry return line to the pump and breaking the vacuum. The condensate is pumped from the receiver back into the boiler by a feed pump and passes on its way through a feed water heater, where it is heated by the exhaust steam from the air and feed pumps.

Combined Atmospheric Pressure and Vacuum Systems

A combined atmospheric and vacuum system works at pressures in the boiler from 1 to 5 oz of gauge pressure (that is, above atmospheric). This pressure is needed on coal-burning installations to operate the damper regulator.

The desired vacuum in the heat-emitting units is obtained by throttling the supply steam with the unit feed valves. The working principle of this system is shown in the elementary sketch found in Figure 8-37. In operation, when steam is raised in the boiler, it passes through the steam main risers and supply valves to the heat-emitting units.

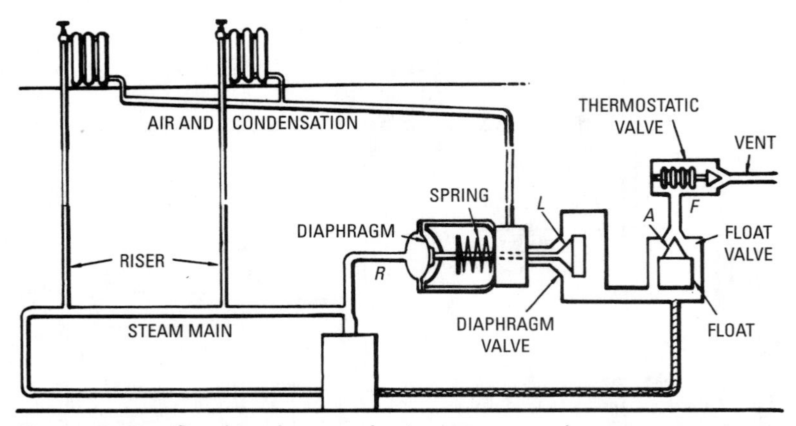

Figure 8-37 Combined atmospheric pressure and vacuum system.

The proper working of this system is obtained by an automatic device or trap that closes against the pressures of either steam or condensation and allows air, but not the steam, to pass out. The trap (Figure 8-37) consists of three elements: (1) a diaphragm valve (point *L*), (2) a float valve (point *A*), and (3) a thermostatic valve (point *F*). Connection *R*, in Figure 8-37, connects the steam outlet of the boiler to the diaphragm. When there is no pressure in the boiler, diaphragm valve *L* is held closed by the spring.

When the fire in the boiler is started and the air in the boiler expands, the diaphragm is inflated and moves the valve spring to the right (against the action of the spring), which opens valve *L*. This makes a direct opening through float valve *A* and thermostatic valve *F*, thus opening the system to the atmosphere. Valve *L* remains open as long as there is a fraction of an ounce of pressure on the boiler. When steam forms and passes through the system, it drives all the air out of the system through the three open valves (*L*, *A*, and *F*).

The heat of the steam causes the expansion element of valve *F* to expand and close the valve; thus the system is filled only with steam.

The vacuum is obtained on the principle that the steam admitted to the radiators condenses while giving off heat through the radiator walls. This causes a tremendous reduction in volume of the steam remaining in the radiators, resulting in a pressure that is less than atmospheric in radiators; i.e., a vacuum is formed. The steam condenses because of a reduction in temperature below that corresponding to the pressure of the steam.

When the radiator gives off heat in heating the room, the temperature of the steam in the radiator is lowered. This reduction in temperature causes some of the steam to condense in a sufficient amount to restore equilibrium between temperature and pressure of the steam.

The pressure falls because the temperature falls. If a closed flask containing steam and water is allowed to stand for a length of time, the atmosphere being at a lower temperature than that inside, the flask will abstract heat from the steam and water, but the heat will leave the steam more quickly than it leaves the water. The result is a continuous condensation of the steam and reevaporation of the water, during which process the temperature of the whole mass and the boiling point are gradually lowered until the temperature inside the flask is the same as that outside. This process is accomplished by a gradual decrease in pressure. Figures 8-38 through 8-40 illustrate the effect of pressure on the boiling point.

A one-pipe system may be converted into a combined system by replacing the air vent on the heat-emitting units and making the piping absolutely tight. In this conversion, a compound gauge recording both pounds of steam and vacuum inches is required (Figure 8-41).

Exhaust-Steam Heating

The term *exhaust-steam heating* relates to the source of the steam rather than to its distribution. In fact, after the exhaust steam enters the heating system, its action is no different from that of live steam

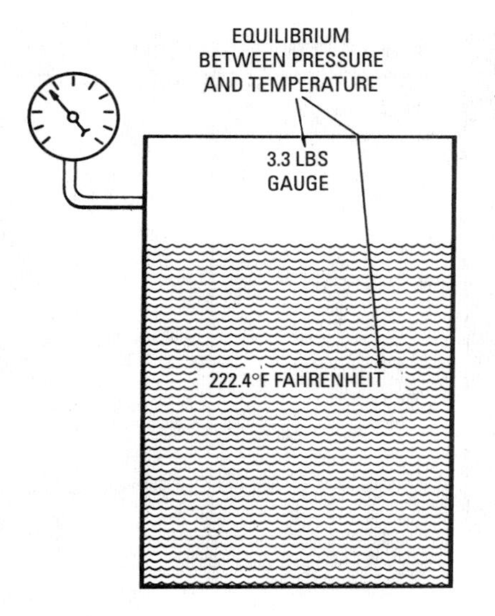

Figure 8-38 Equilibrium between the steam and the water. Equalized temperature.

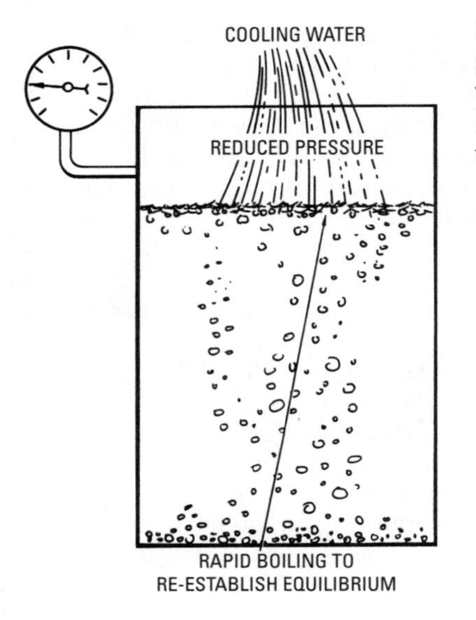

Figure 8-39 Reducing the pressure by applying cooling water to the closed vessel. Reduction of pressure causes the water to boil.

taken from a heating boiler, because it is adapted to both low-pressure and vacuum systems.

The chief difference between exhaust systems and those already described are the provisions for delivering steam from the engine to the heating system free from oil and at a constant pressure and for

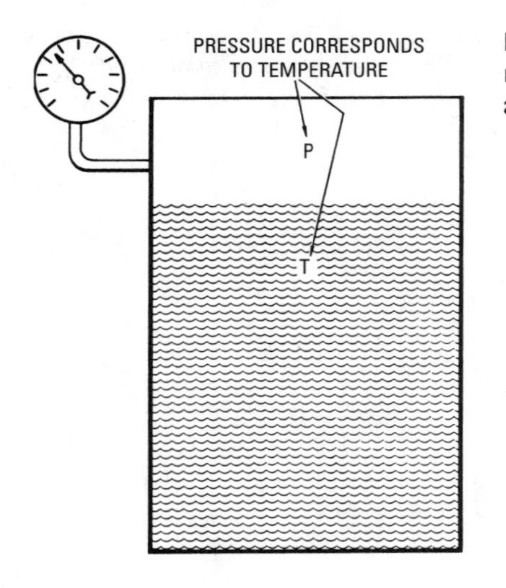

Figure 8-40 Equilibrium reestablished between pressure and temperature.

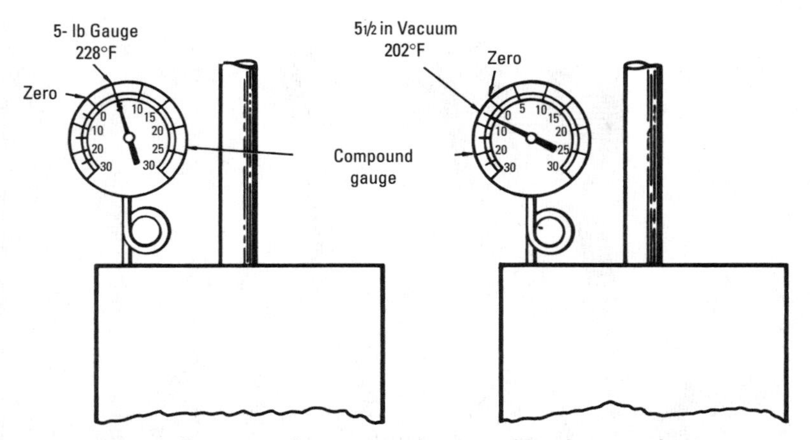

Figure 8-41 Compound gauge used on combined atmospheric-vacuum systems.

returning the condensate to the high-pressure boiler (Figure 8-42). Figure 8-43 shows the essential features of an exhaust-steam heating system having fractional control vacuum distribution.

The necessary devices between the engine and the inlet to the heating system are: (1) the oil separator, (2) the trap, (3) the back pressure valve, and (4) the pressure-regulating valve. In addition, for mechanically producing the vacuum and returning the condensation to the high-pressure boiler, an air pump, receiver with vent, and feed pump are required. A feed water heater should also be provided both for economy and to permit returning the condensation and makeup feed water to the boiler at the proper temperature.

In operation, exhaust steam from the engine first passes through the heater and then through the oil separator, which frees it from the lubrication oil, the latter passing off into the oil trap. The steam now enters the heating system at *A*, its pressure being prevented from rising above a predetermined limit by the back pressure valve (regulated by weight *B*) and maintained at a predetermined constant pressure by the pressure-regulating valve (adjusted by weight *C*).

The pressure-regulating valve is, in fact, an automatic steam "make-up" valve, which admits live steam to the heating system when the exhaust is not adequate to supply the demand, thus "making up" for this deficiency and maintaining the pressure constant.

The condensate and air are removed from the system at *D* by a wet pump (as distinguished from a dry pump, which removes only

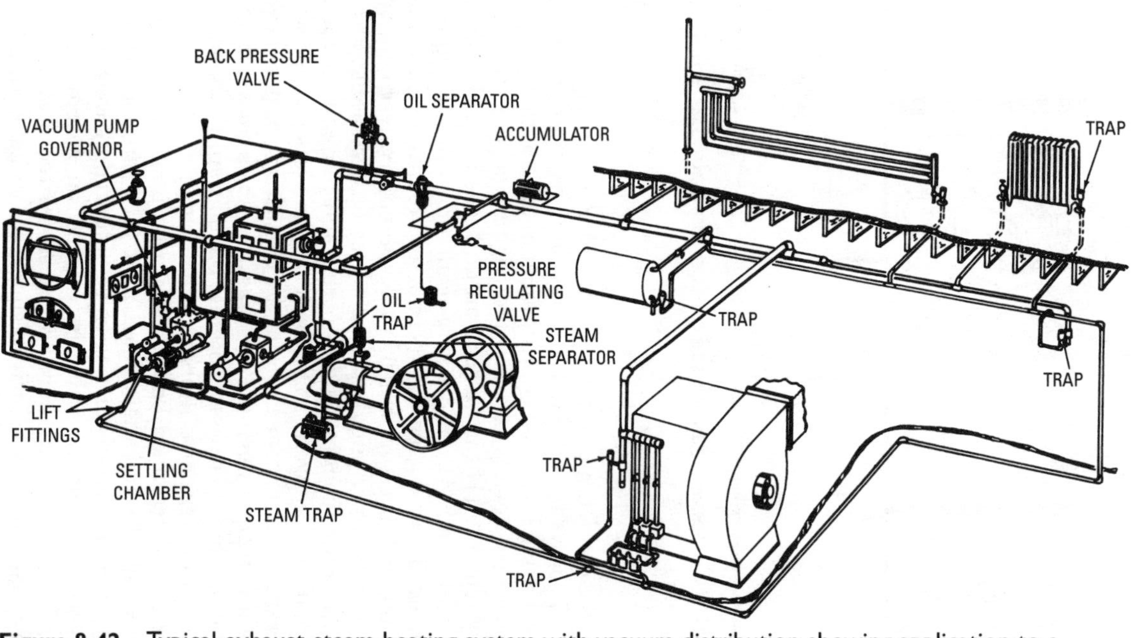

Figure 8-42 Typical exhaust-steam heating system with vacuum distribution showing application to a power plant containing an open feed water heater.

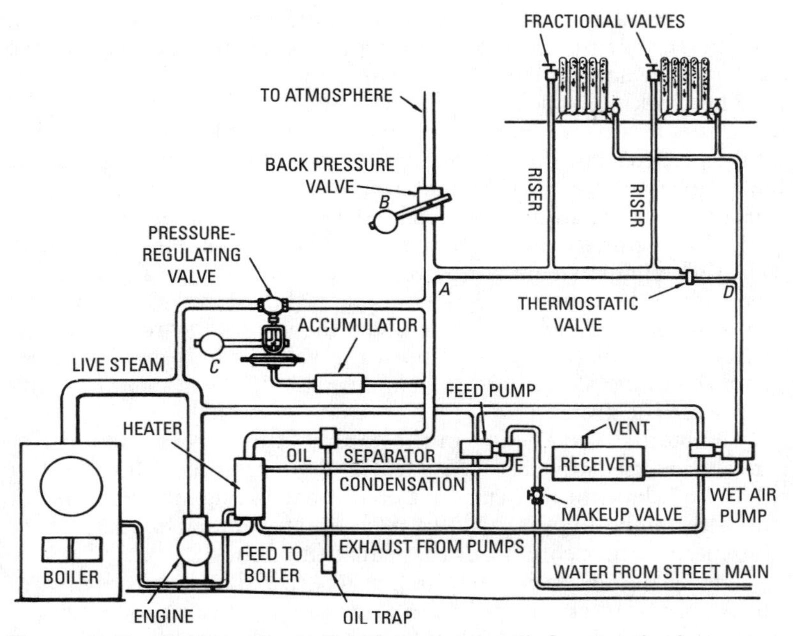

Figure 8-43 Exhaust-steam heating system with fractional valve control and automatic makeup.

the air). The condensate and air are discharged into a receiver, from which the air passes off through a vent and the condensate is pumped by a feed pump back into the boiler after passing through a feed water heater.

There is a continued loss of water through various leaks; the feed pump inlet (alleged suction) is connected at *E*, with the supply from the street main or other source, the amount entering the system being controlled by the make-up valve.

The construction of a typical regulating valve and its connections are shown in Figure 8-44. The valve is controlled by means of governing pipe *A* (Figure 8-44), connecting the diaphragm chamber to the accumulator, the latter being connected to the heating main at the point from which the pressure regulator is to be governed.

The accumulator is always half full of water, and its elevation must be such that the water line in the accumulator is level with the diaphragm so that there will not be an unbalanced column of water to exert pressure on the diaphragm.

The water is provided to protect the diaphragm from the steam, the pressure of the water being transmitted from the surface of the water in the accumulator to the diaphragm.

The working principle of the regulating valve and its connected devices is relatively simple. In operation, when the exhaust side *F* is at the predetermined pressure, this brings sufficient force against the under, or water, side of the diaphragm to overcome the downward thrust due to the adjustable weighted level and close the valve.

If the engine slows down or there is heavy demand for heat, so that the exhaust steam is not adequate to supply the demand, the pressure in the exhaust side *F* will fall, and the downward thrust of the adjustment weight will overcome the opposing pressure of the water on the diaphragm and open the valve, admitting live steam from the boiler side *L* into the exhaust side *F* in sufficient quantity to restore the pressure.

The inertia of the water in the accumulator acts as a damper to prevent oversensitiveness of the valve or hunting (i.e., the behavior of any mechanism that runs unsteadily, oscillating either too far or too little in an attempt to adjust itself to momentary fluctuations of pressure or other conditions that cause this action).

The spring under the diaphragm acts to balance the downward thrust of the lever and hold the valve in closed position when the pressure is the same on both sides of the diaphragm.

The back pressure valve is used to prevent exhaust pressure exceeding a predetermined limit. This is virtually a lever safety valve designed to work at very low pressure. Some back pressure valves are so light that they will open or close with a variation of only 2 oz. The position of the weight on the lever, whose fulcrum is at *F* (Figure 8-45), determines the exhaust pressure at which the valve will open.

Proprietary Systems

Over the years, a number of automatic heating systems have been designed and patented by manufacturers of steam heating equipment. Because these heating systems are protected by patent, they are referred to by the name of the manufacturer. Three of the most popular of these proprietary steam heating systems are:

1. The Trane vapor system
2. The Dunham differential system
3. Webster moderator systems

The *Trane vapor system,* illustrated in Figure 8-46, is a combined atmospheric and natural vacuum system installed for residential heating. In this system, the air and water return runs in

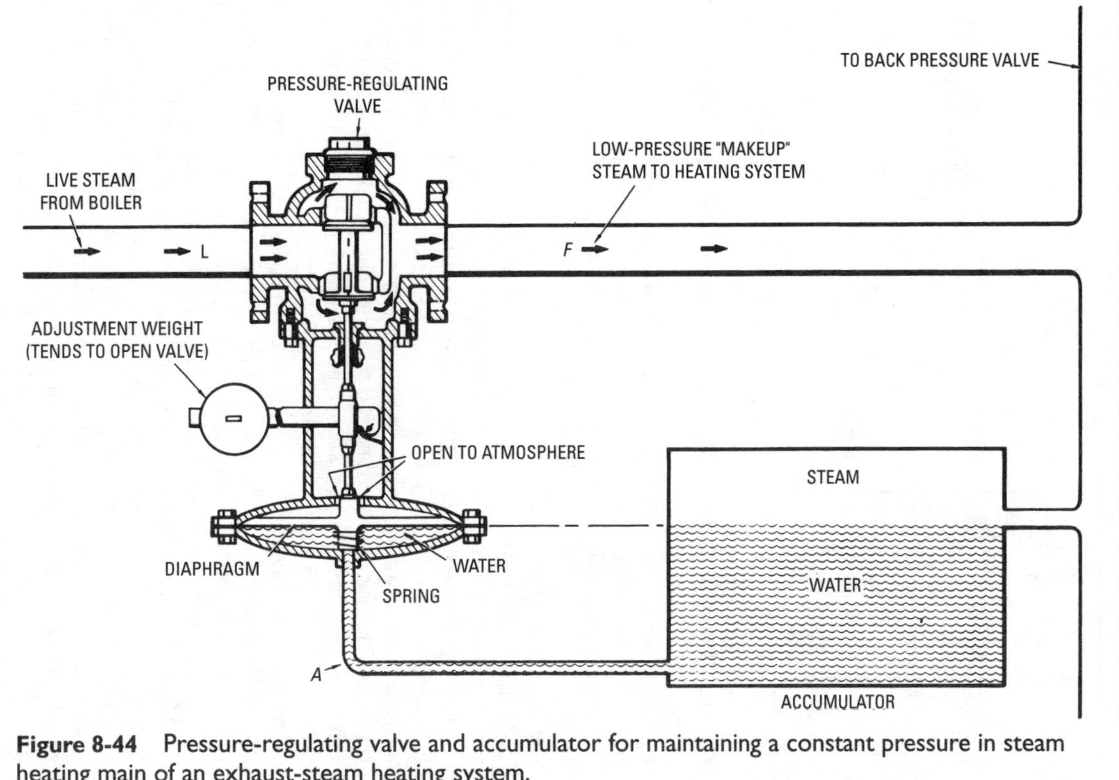

Figure 8-44 Pressure-regulating valve and accumulator for maintaining a constant pressure in steam heating main of an exhaust-steam heating system.

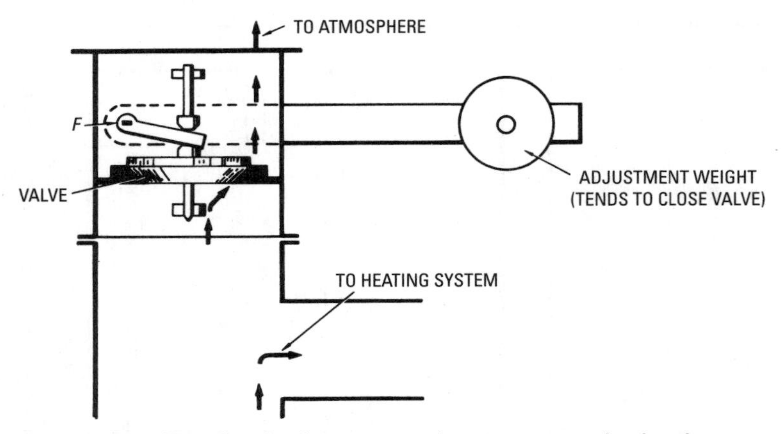

Figure 8-45 Details of exhaust-steam heating system back valve.

the same direction and is practically the same length as the supply main. When properly worked out, this feature gives the same effect as though each convector radiator were placed at an equal distance from the boiler. This tends to synchronize the heating effect of all the heat emitting units in the system.

When the fire is started in the boiler of the Trane system, the water becomes heated and steam is formed, which flows through the supply main and enters the radiators, displacing the air, which is heavier than the steam. The air and condensate drain from the radiators through radiator traps and return piping to a point near the boiler where the air is exhausted through the quick vents and float vents at the end of the steam and return mains. The water is returned to the boiler by the direct-return trap.

As the rooms become warm, less steam is condensed and the pressure in the boiler begins to rise. The rising pressure causes the damper regulator to operate, closing and opening the drafts and maintaining just the amount of pressure necessary for proper heating.

When the fire in the boiler becomes lower, condensate forms, air is prevented from entering the system, and a vacuum is created. As a result, the operation changes from atmospheric to natural vacuum, hence the name *combined atmospheric and natural vacuum system*. The reduced pressure of the vacuum allows the water to boil and furnish steam to the radiators at a lower temperature, a decided economy when the fire is low.

The boiler connections for the Trane system are shown in Figure 8-47. It is very important that the steam connection to the return

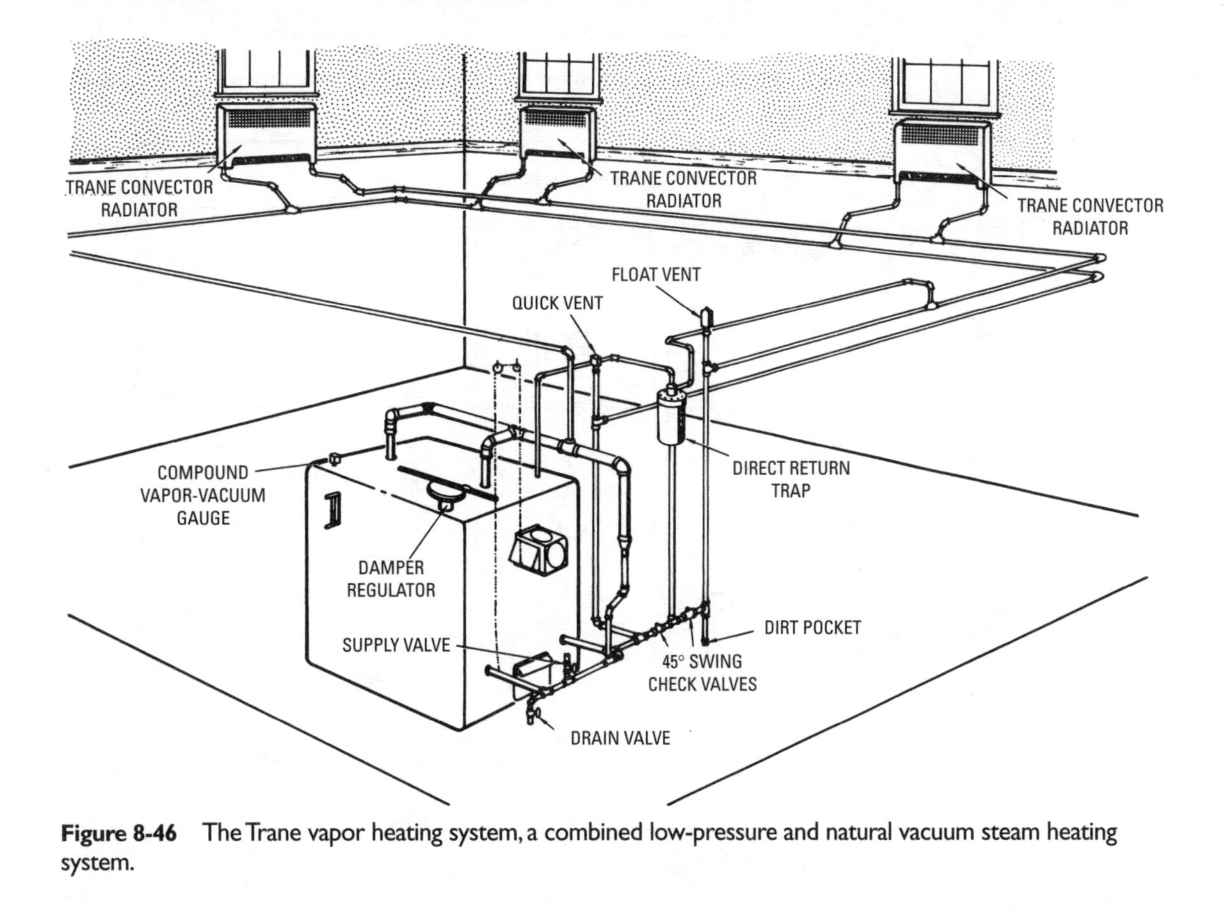

Figure 8-46 The Trane vapor heating system, a combined low-pressure and natural vacuum steam heating system.

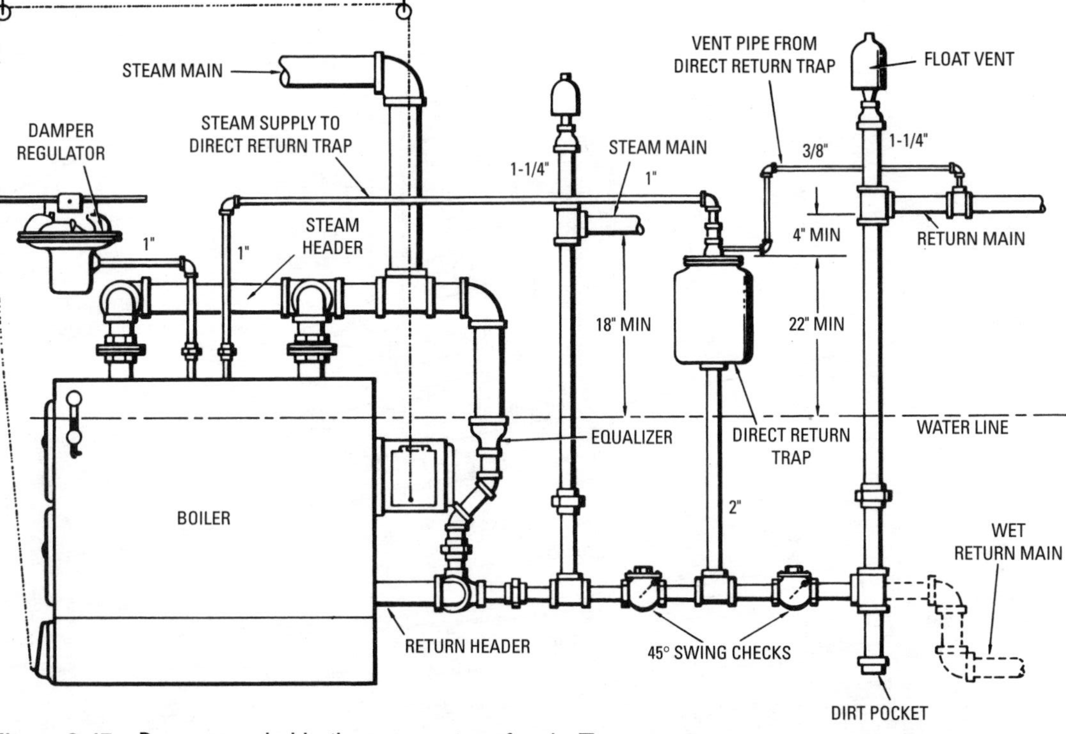

Figure 8-47 Recommended boiler connections for the Trane vapor system.

trap be taken from the steam space of the boiler and *not* from the supply piping of the header. The top of the direct-return trap must be placed at least 22 in above the water line of the boiler. In *no* case should the top of the trap be less than 4 in below the air and water return main. The dimensions from the water line to the end of the mains and the top of the return trap are the minimum allowable. Greater clearance above the water line should be employed where possible. When the ends of the steam and return mains occur in remote parts of the building away from the boiler, use the connections shown in Figure 8-48. The ends of such mains must always be vented before dropping to the wet return. The vent pipe must be installed from the piping below the trap up to the return main when a wet return is used. If a dry return is used, the vent pipe may be omitted.

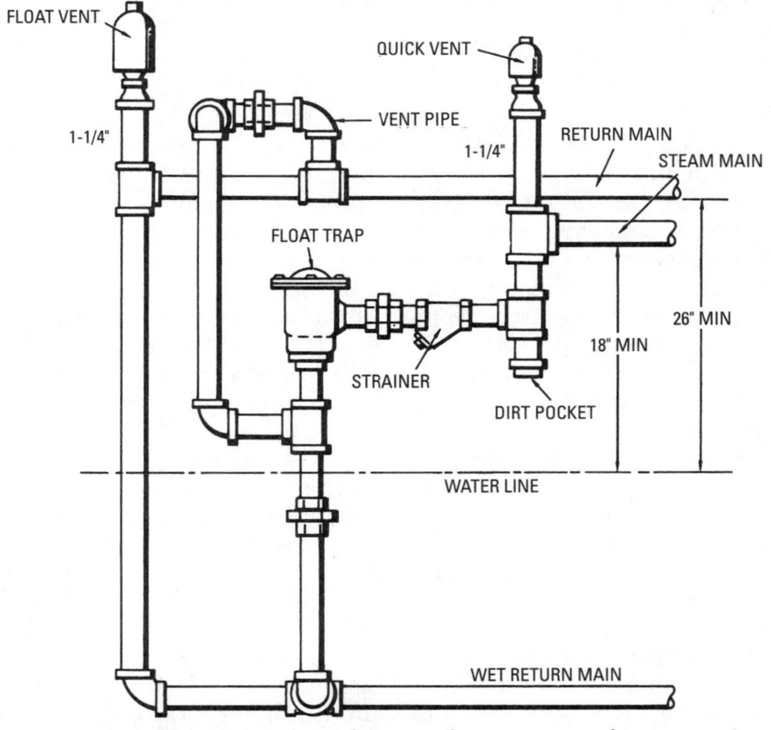

Figure 8-48 Pipe connections for use where steam and return mains drop to wet return in remote part of building.

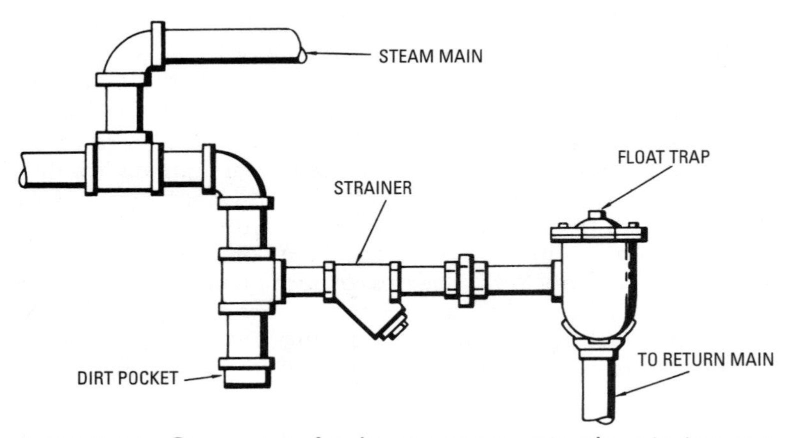

Figure 8-49 Connections for dripping steam main where it rises to a higher level.

If desired, a Hartford connection may be used between the inlet to the boiler and the return connections from the steam and return mains (see *Hartford Return Connection* in this chapter).

The recommended size for vertical piping immediately below the float vents and the quick vents is 1¼-in pipe at least 6 in long. This affords a separate chamber for air and water.

The connections for a dripping steam main that rises to a higher level are illustrated in Figure 8-49. The trap and strainer may be omitted, provided the lower main is at least 18 in above the water line and the return is connected directly into the return header of the boiler.

When the motor-operated steam valves are used on the mains, the boiler connections should be arranged as shown in Figure 8-50. An equalizer line is required to equalize between the steam main and the return main when a vacuum forms in the former after the motor-operated valve closes. A swing check valve prevents the flow of steam into the return main (Figure 8-51). Figure 8-52 shows the arrangement of boiler connections when air is eliminated from the return main through the No. 9 vent trap. Sometimes reversed circulation resulting from rapid condensation of steam will tend to create greater vacuum in the steam main than in the return main. This can be prevented by using the boiler connections illustrated in Figure 8-49.

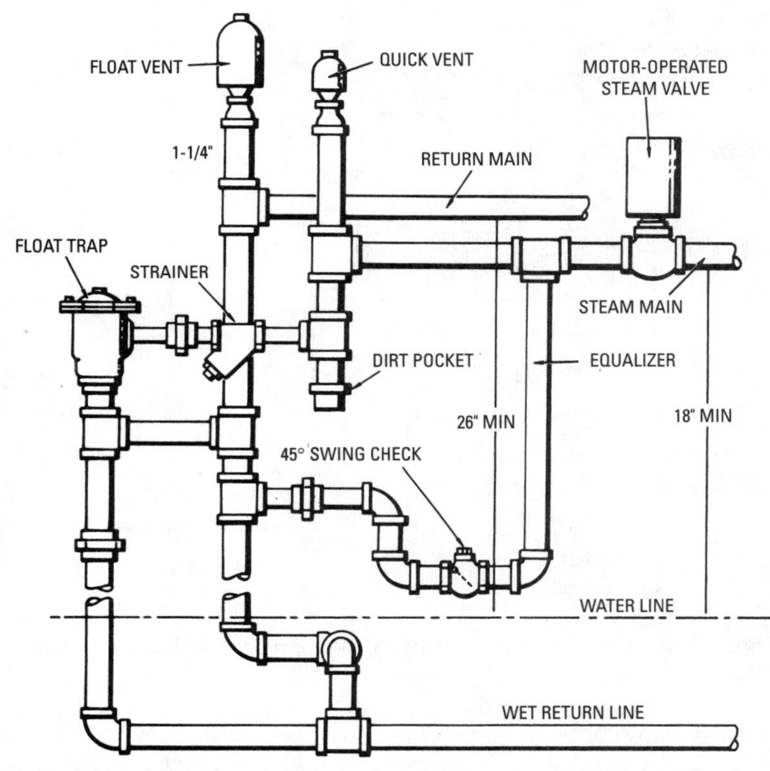

Figure 8-50 Pipe connections when motor-operated steam valve is used on mains.

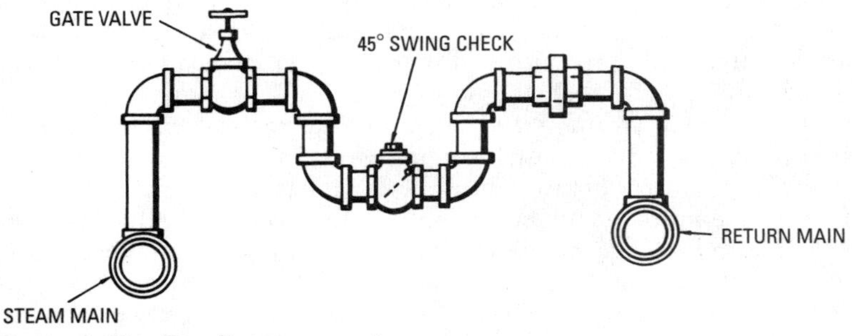

Figure 8-51 Equalizer connection.

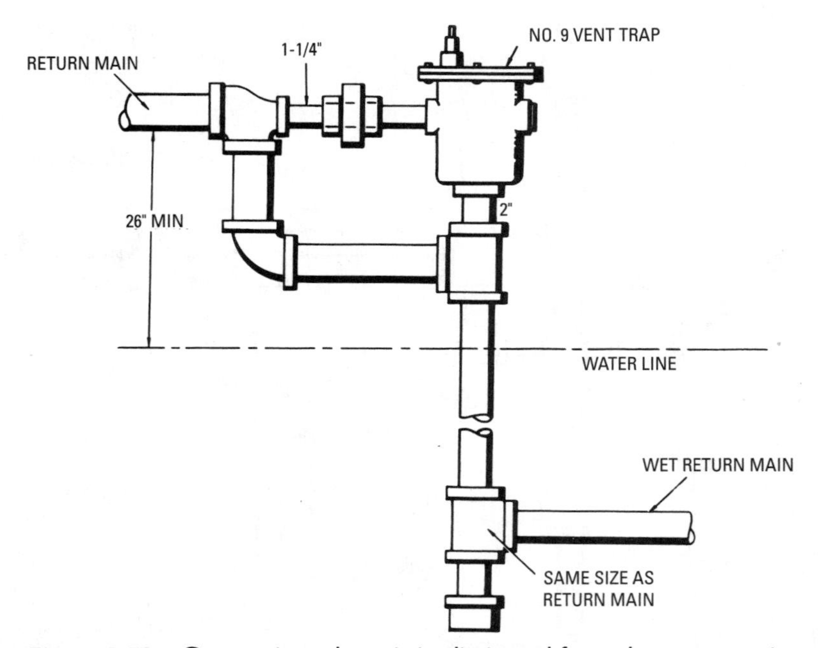

Figure 8-52 Connection when air is eliminated from the return main through the vent trap No. 9.

A typical convector radiator used in the Trane system is shown in Figure 8-53. It is equipped with an angle valve and an angle trap having horizontal laterals below the floor. Connections for Trane convector radiators are illustrated in Figure 8-54. The upper unit shows a vertical trap and a straightway valve concealed within the convector radiator. The lower unit is equipped with an angle valve and trap with a downfeed riser dripped through the angle trap.

The *Dunham differential vacuum heating system* is a simple two-pipe, power, vacuum (air-pump) return system working normally at pressures below atmospheric (subatmospheric) in the system and employing orifice supply valves on the radiators. The principal advantage of the Dunham system is that it continuously distributes heat at a variable rate equal to heat losses from the structure. This is accomplished by reducing the capacity of the system by decreasing the volume temperature of the steam.

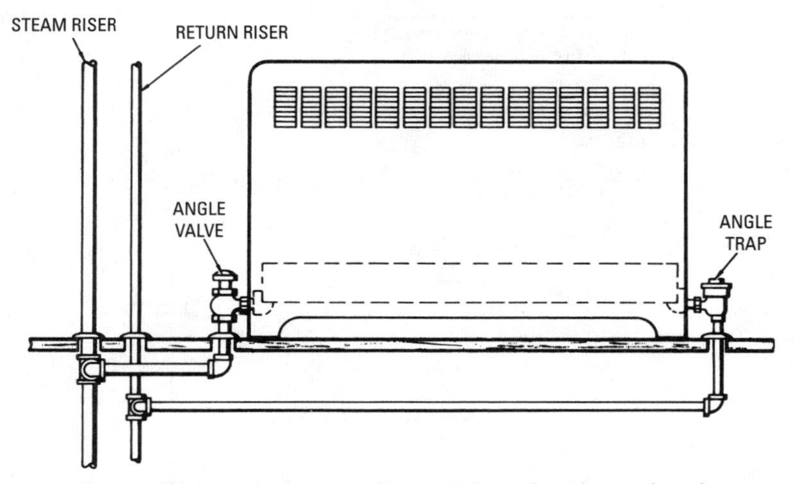

Figure 8-53 Trane convector radiator with angle valve and angle trap having horizontal laterals below floor.

Standard radiators, pipe fittings, and boiler connections are used in the Dunham system. The principal features of the Dunham system that distinguish it from other heating systems are:

1. The traps and valves
2. The condensation pump
3. The controller

Each of these components is differentially controlled; that is, each is actuated by a pressure differential (the difference in pressure in the supply piping and the pressure return piping). The Dunham system distributes the steam proportionately to all radiators, the pressure range being from 2 psig to pressures considerably lower than atmospheric. Because of the relatively constant differential in pressures between the steam and return lines, the radiators are filled with steam.

A condensate pump is connected to the differential controller and to the supply and return piping. The controller starts the pump when the pressure difference between the supply and return piping tends to fall or disappear, and stops it when the pressure differential is restored. The condensate pump is a wet air pump that handles both air and condensate.

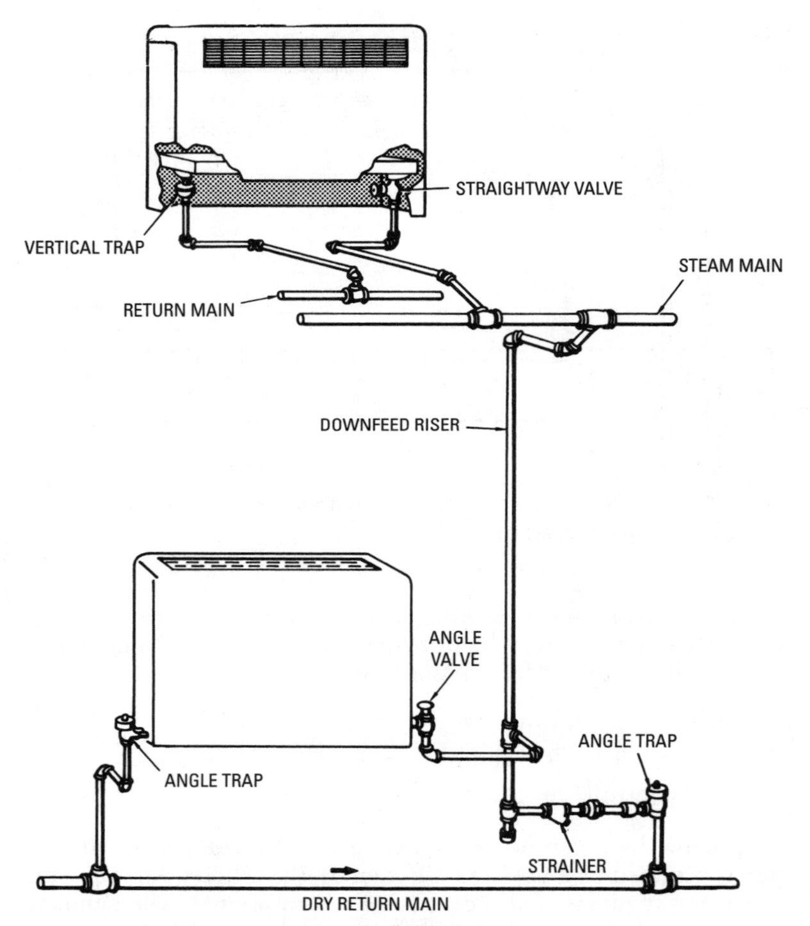

Figure 8-54 Typical connections for Trane convector radiators.

The thermostatic radiator traps are actuated by temperature changes within the radiator. Drip traps are installed at drip points to which large volumes of condensate flow. These are combined thermostatic and float traps. A control valve regulates the admission of a *continuous* flow of steam into the heating main.

Webster moderator systems are special control systems used primarily with two-pipe vapor, vacuum, and vented return systems or in modified form with some one-pipe steam heating systems. These

Webster moderator systems are controlled by the weather with an automatic outdoor thermostat. One or more hand-operated *variators* can be used to adjust the outdoor thermostat. The two moderator control systems produced by Webster are based on the following operating principles:

1. Continuous steam flow
2. Pulsating steam flow

The continuous steam flow arrangement is called the *Webster electronic moderator system* (Figure 8-55) and is suitable for medium to large buildings requiring one or more control valves. This system consists of the following basic components:

1. Outdoor thermostat
2. Main steam control valve
3. Variator

The steam supply may be taken from a high- or low-pressure boiler or any other source. The initial pressure should not be over 15 psig, using a reducing valve if necessary. The return piping may be either open or closed.

The moderator control regulates the pressure difference and will function equally well regardless of whether the pressure in the return piping is atmospheric or below. The main steam-control valve is adjusted automatically by the moderator control, which acts to reverse the direction of the motor, causing it to move the valve in the closing direction when *less* steam is required and in the opening direction when *more* steam is required. The outdoor thermostat automatically varies the steam flow in accordance with changes in outdoor temperature. Depending upon the outdoor temperature, the thermostat automatically selects the position of the main steam-control valve. Its position may be advanced or reduced by the variator to give more or less steam than is called for by the outdoor temperature.

Changes in pressure difference in the heating system are automatically compensated for by a pressure-actuated mercury tube in the control cabinet. One end of the tube is connected to the steam supply mains and the other end to the return main. If the supply pressure is unduly increased, mercury rises in the tube to unbalance resistances contained therein and the main steam valve begins to close in amount sufficient to balance resistances. A reverse action takes place when the pressure difference falls below that called for by the control equipment.

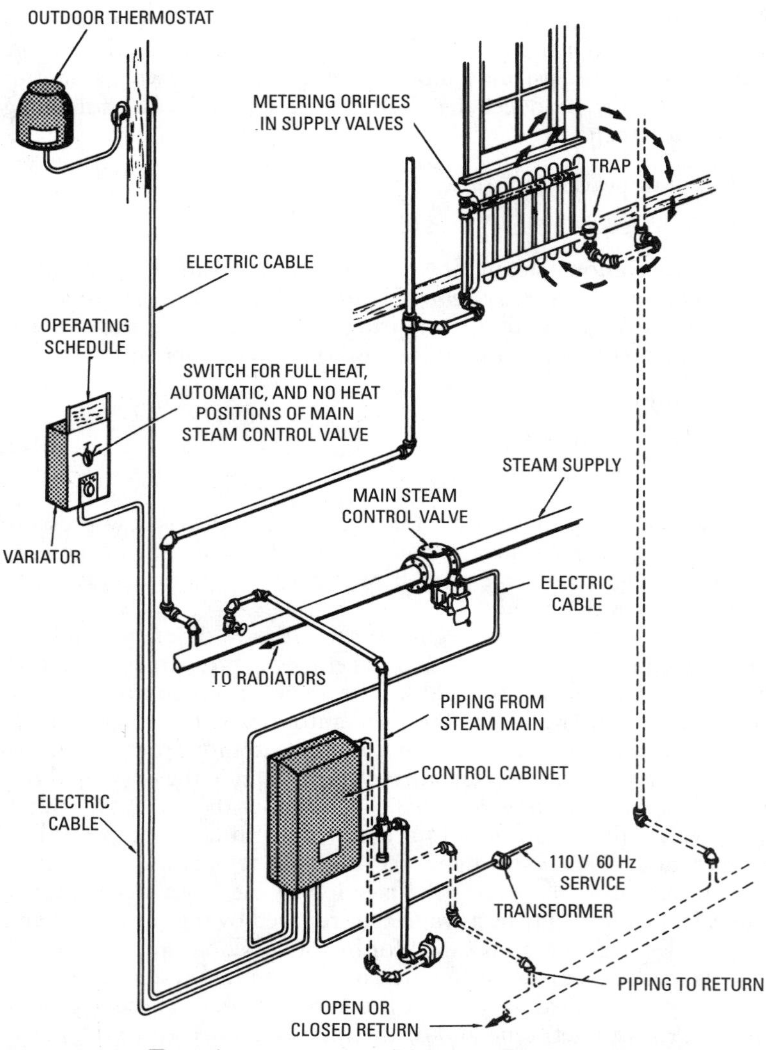

Figure 8-55 Typical arrangement of Webster E-5 electronic moderator system.

The pulsating steam flow arrangement is called the *Webster moderator system*. This system comprises a central heat control of the pulsating flow type for new or existing steam or hot-water systems. It is designed for small- and medium-size buildings and for zoning of large buildings. It directly controls the operation of burner, stoker, blower, or draft damper motors. The basic components of this system are:

1. Outdoor thermostat
2. Pressure difference controller
3. Control cabinet
4. Capillary tubing

These four components work together to open and close a valve in the steam main or to start and stop the automatic firing device at the boiler, generally through a relay. Figure 8-56 shows the general arrangement of a Webster EH-10 moderator system for a building served by a central station or street steam. A Webster EH-10 moderator system used for controlling a burner on a steam boiler is illustrated in Figure 8-57.

The pressure difference controller maintains the correct pressure difference between supply and return piping. In combination with metering orifices, this device gives an even distribution of the steam to all radiators in the system, thereby preventing over- or underheating.

Control is accomplished by varying the length of intervals during which steam is delivered to the radiators. These intervals are longest in cold weather and shortest in mild weather. The timing is such that the longest *off* interval is comparatively short so that heat output from radiators is practically continuous.

The timing mechanism inside the control cabinet is powered by a synchronous motor, which turns a cam. Timing gears between motor and cam set the length of the operating cycle. Rising on the cam is a roller connected to the arm of a switch. When the roller is on the high part of the cam, the switch is in the position for opening the control valve or starting the firing equipment. When the roller is on the low part of the cam, the switch is on the position for closing the control valve or stopping the firing equipment. The length of the *on* interval is changed automatically by the outdoor thermostat or by adjusting the variator by hand. The average length of the cycle is 30 minutes. Other gears can be furnished for cycle lengths from 12 to 60 minutes.

A variator is included in the control cabinet for manual adjustment of the rate of heat delivery to the building. The variator

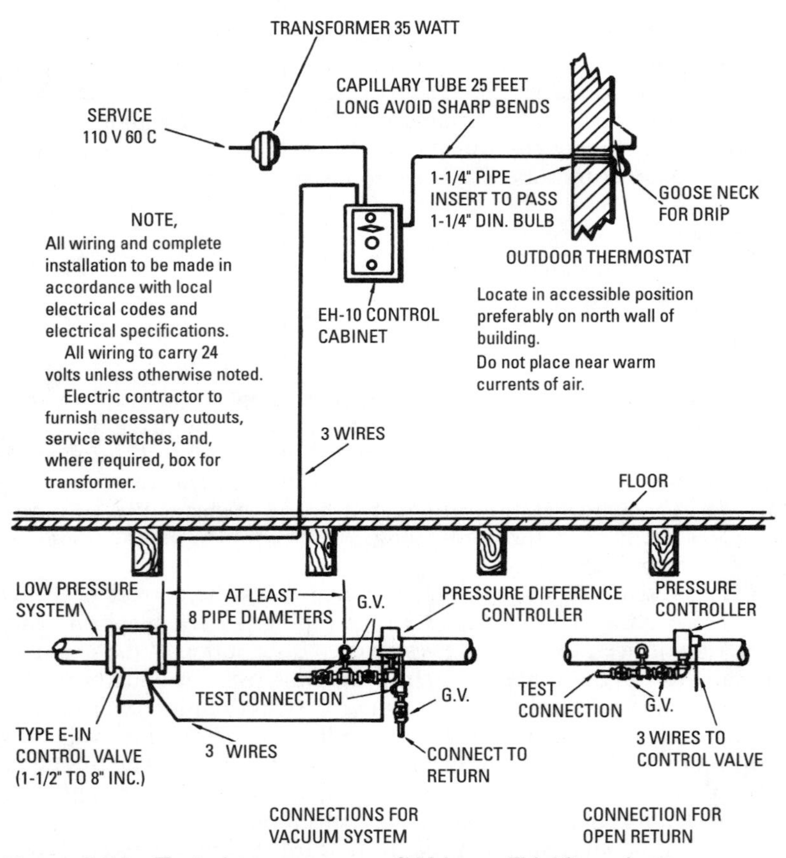

Figure 8-56 Typical arrangement of Webster EH-10 moderator system in a building served by a central station or street steam.

changes the relationship of the switch lever and roller to the cam by moving the cam itself. This is accomplished by mounting the motor, cam, and gear train on a movable carriage.

High-Pressure Steam Heating Systems

High-pressure steam heating systems operate at pressures above 15 psig (generally in the 25- to 150-psig range) and are usually found in large industrial buildings in which space heating or steam process

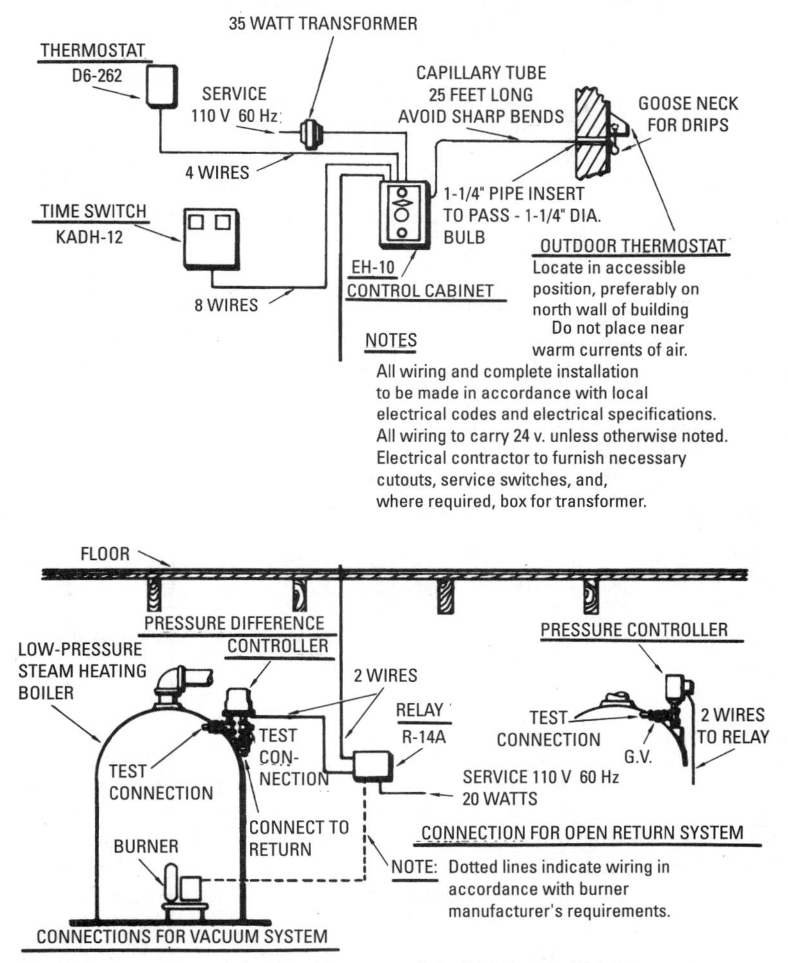

Figure 8-57 General arrangement of the Webster EH-10 moderator system controlling a burner of a steam boiler.

equipment (e.g., water heaters and dryers) is used. An example of a typical two-pipe, high-pressure steam heating system is shown in Figure 8-58. This system is also referred to as a *medium-pressure steam heating system* when the steam pressures are in the lower ranges.

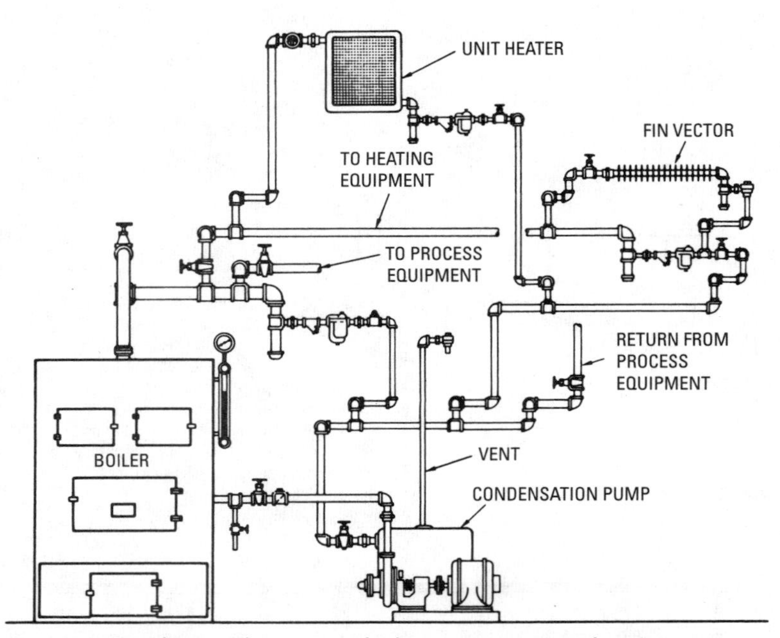

Figure 8-58 A typical two-pipe, high-pressure, steam heating system.
(Courtesy Dunham-Bush, Inc.)

One advantage of this system is that the high pressure of the steam permits the use of smaller pipe sizes. The high steam pressure also makes possible the elevation of the condensate return lines above the heating units, because the return water can be lifted into the return mains. High-pressure condensate pumps and thermostatic traps are commonly used in these systems to handle the condensate and return it to the boiler.

High-pressure steam heating systems are more expensive to operate and maintain than low-pressure systems. In most cases, a licensed stationary engineer must be hired for the installation—a factor that tends to increase the operating cost.

Steam Boilers

The boiler is the source of heat for a steam heating system, and it will operate on a number of different fuels; however, regardless of the fuel used, the operating principle will be the same. Water is heated until it boils and changes to steam. The steam is then distributed to the heat-emitting units throughout the structure either by natural or mechanical means.

Low-pressure steam heating boilers are used in residences and small buildings. The design and construction of these boilers are very similar to the boilers used in hot-water heating systems in the same size. Both boilers are described in considerable detail in Chapter 15 ("Boilers and Boiler Fittings").

To size a steam boiler, first measure the height in inches of one cast-iron radiator in the system. Then, count the number of sections and the number of tubes or columns (Figure 8-59). The sections are the divisions or separations of a cast-iron radiator as seen when standing directly in front of it. When you look at the radiator from its narrow end, you can see that each section consists of one or more vertical pipes.

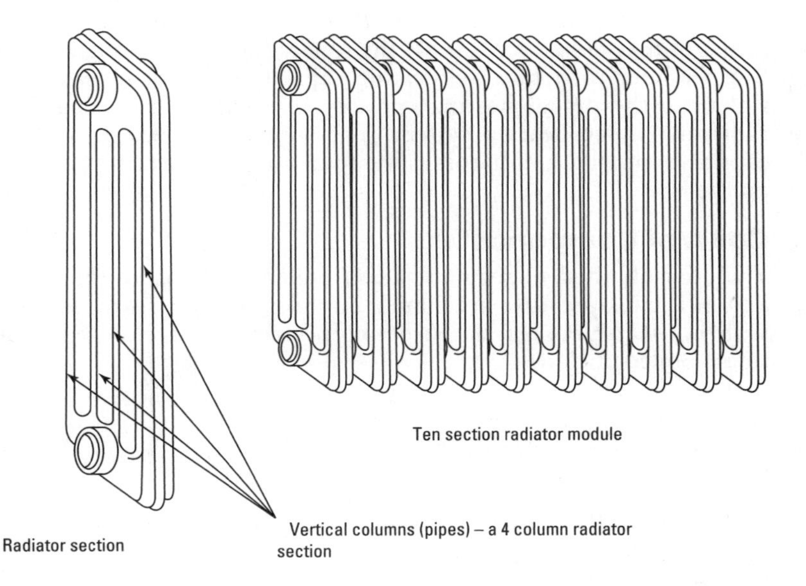

Every cast-iron radiator is made up of individual sections, which when joined together, form a single radiator module.

Figure 8-59 Steam boiler sizing.

Table 8-1 Abbreviated Column-Type Radiator EDR Chart (Number of Columns and Height)

	18"	*20"*	*22"*	*23"*	*26"*	*30"*
1 Column		1.50		1.67	2.00	
2 Columns		2.00		2.33	2.67	
3 Columns	2.25		3.00		3.75	
4 Columns	3.00		4.00		5.00	

Note

These vertical pipes are called *columns* in the traditional cast-iron radiators and are 2½ inches wide. In newer radiators, they are called *tubes* and are only 1½ inches wide.

Find the square foot Equivalent Direct Radiation (EDR) of one section of the radiator from Table 8-1. Multiply that figure by the number of sections in the radiator to arrive at the square foot EDR rating of that radiator. Multiply the square foot EDR rating by 240 Btu per hour to obtain the heating capacity of that one radiator. Calculate the design heating capacity for each of the remaining radiators in the heating system. The sum of the design heating capacities of all the radiators is the total radiation heating demand on the boiler.

Control Components

The controls used to regulate steam boilers and ensure their safe and efficient operation are similar in most respects to those used with hot-water boilers. A list of the principal controls used with steam boilers includes:

1. ASME safety valve
2. Steam pressure gauge
3. Low-water cutoff
4. Boiler water feeder
5. High-limit pressure control
6. Water gauge glass
7. Water cocks
8. Primary control (burner mounted)
9. Operating control (with tankless heater)

These and other boiler controls are described in considerable detail in Chapter 15 ("Boilers and Boiler Fittings"), Chapter 4 of Volume 2 ("Thermostats and Humidistats"), and Chapter 9 of Volume 2 ("Valves and Valve Installation").

Hartford Return Connection

A well-designed steam heating system will have a *Hartford return connection* (Figure 8-60) in the condensate return line. The purpose of the Hartford return connection (or *Hartford loop,* as it is sometimes called) is to prevent excessive loss of water for the boiler when a breakdown (such as a water leak) occurs in the return line.

As shown in Figure 8-60, an equalizer connects the lower outlet to the steam outlet. The Hartford connection is taken off from the equalizer an inch or two below the normal water level. Evidently low pressure can only draw water out of the boiler connection. This gives a low water level, and it cannot recede further because steam will flow into the connection leg as the water recedes in the leg due to pressure difference.

More details about different boiler connections are found in Chapter 15, "Steam and Hot Water Space Heating Boilers," and in Chapter 9, "Valves and Valve Installations" in Volume 2.

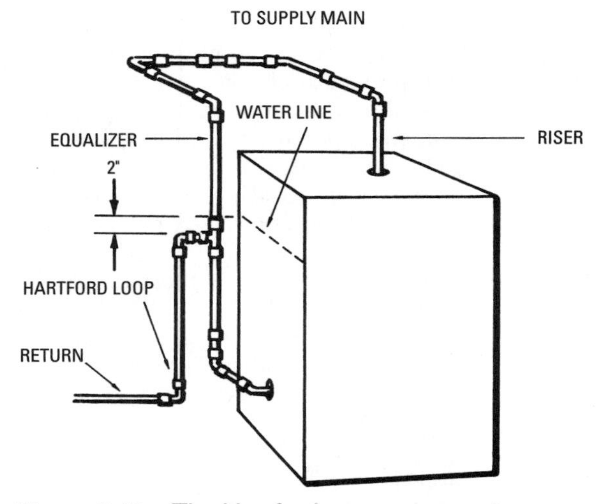

Figure 8-60 The Hartford return connection.

Pipes and Piping Details

A steam heating system requires careful planning of the piping to ensure both an efficient and a safe operation. For example, the pipe material used (e.g., wrought iron or Schedule 40 black steel) is important because the *capacity* of a particular weight pipe will depend on both its size and the material from which it is constructed.

The expansion of pipes when they become heated is another factor that must be considered when designing a steam heating system. Sufficient flexibility in the piping can be provided for by correctly designed offsets, slip joints or bellows, radiators and riser runouts, U-bends, or other expansion loops. The pitch of connections from risers must be sufficient to prevent the formation of water pockets when pipe expansion occurs.

Pipe materials, pipe sizes (and pipe-sizing methods), pipe expansion rates, pipe fittings, and piping details, such as wet and dry returns, drips, and connections to heat-emitting units, are covered in Chapter 8 of Volume 2 ("Pipes, Pipe Fittings, and Piping Details").

Steam Traps

A *steam trap* is an automatic device installed in a steam line to control the flow of steam, air, and condensate. In operation, it opens to expel air and condensation and closes to prevent the escape of steam. All steam traps operate on the principle that the pressure within the trap at the time of discharge will be slightly in excess of the pressure against which the trap must discharge.

The principal functions of steam traps in steam heating systems include (1) *draining condensate* from the piping system, radiators, and steam processing equipment; (2) *returning condensate* to the boiler; (3) *lifting condensate* to a higher elevation in the heating system; and (4) *handling condensate* from one pressure to another.

Steam traps may be classified on the basis of their operating principles as follows:

1. Float traps
2. Bucket traps
3. Thermostatic traps
4. Float and thermostatic traps

5. Flash traps
6. Impulse traps
7. Lifting traps
8. Boiler return traps

Pumps

Pumps are used in steam heating systems to dispose of condensate or return it to the boiler and to discharge excess air and noncondensable gases to the atmosphere. The specific function (or functions) of a pump depends on the type of steam heating system in which it is used. The two basic steam heating pumps are:

1. Condensate return pumps
2. Vacuum heating pumps

Heat-Emitting Units

A *heat-emitting unit* is a device that transmits heat to the interior of a room or space. The two heat-emitting units used in steam heating systems are radiators and convectors. Simply defined, a *radiator* is a heat-emitting unit that transmits heat from a direct heating surface principally by means of radiation. A *convector* may be defined as a heat-emitting unit that transmits heat from a heating surface principally by means of convection. The heating surface of a convector is usually of the extended finned tube construction.

A detailed description of the heat-emitting units used in steam heating systems is contained in Chapter 2 of Volume 3, "Radiators, Convectors, and Unit Heaters."

Air Supply and Venting

Two types of venting occur in heating systems. One deals with passing smoke and gases resulting from the burning of combustible fuels (e.g., coal, oil, and gas) to the outdoors and is described in the various chapters on furnaces and boilers. Another type of venting deals with the relief of pressure in steam heating systems by allowing a certain amount of air to escape (i.e., be vented) from the radiators. Venting radiators is described in Chapter 2 of Volume 3, "Radiators, Convectors, and Unit Heaters."

Unit Heaters

A *unit heater* is essentially a forced draft convector. A centrifugal fan or propeller is used to force the air over the heating surface and into the room or space through deflector vanes. The principal operating components of a steam unit heater are shown in Figure 8-61. Unit heaters find their widest application in industrial and commercial buildings, gymnasiums, field houses, auditoriums, and other types of large buildings.

All unit heaters are described in considerable detail in Chapter 2 in Volume 3, "Radiators, Convectors, and Unit Heaters."

Air Conditioning

Central air conditioning can be added to a steam heated structure by installing a water chiller or a separate forced-air cooling system (i.e., a split system).

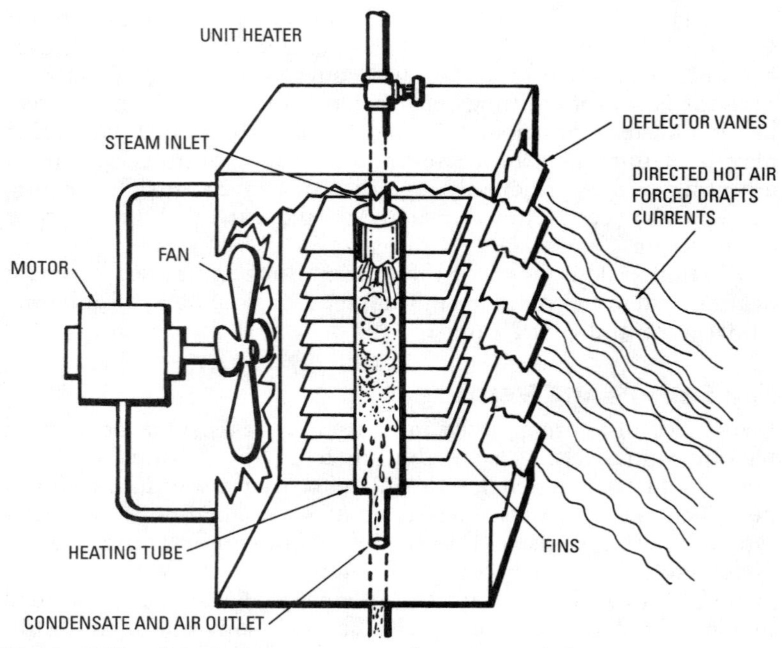

Figure 8-61 Principal components of a steam unit heater.

The water chiller and steam boiler may be installed as separate units, or a complete package containing both units may be used. Water chiller installations are very rarely used in residential air conditioning. Split-system air conditioning dominates this field (i.e., with respect to steam heated residences).

Troubleshooting Steam Heating Systems

Steam heating systems are easy to operate but require much higher levels of maintenance than other types. It is especially important to keep a watchful eye on the control system. If the controls malfunction, excessive steam pressure and overheating could cause the boiler to rupture, resulting in serious injury to the occupants of the house and expensive damage to the structure.

If the steam heating system is in good condition, it can last for years without any problems. When a problem does occur, however, the homeowner may experience some difficulty in finding local help qualified in servicing and repairing boilers and steam heating systems. This is especially true in rural areas and small towns.

The first step the homeowner must take is to determine the cause of the problem in the steam heating system. Once that is determined, the homeowner then can decide whether to make the necessary repairs, replace the boiler with a new one, or replace the entire system with another type of space heating system. An example of a problem and possible remedies are listed in Table 8-2.

One of the single best sources of information for troubleshooting steam heating systems is Dan Holohan's *Pocketful of Steam Problems (with Solutions!)*. It contains easy to understand descriptions of hundreds of steam heating problems, their possible causes, and suggested remedies. It covers in depth all types of steam heating systems. The book can be ordered from

Dan Holohan Associates, Inc.
63 North Oakdale Ave.
Bethpage, NY 11714
800-853-8882
mailroom@heatinghelp.com

Table 8-2 Troubleshooting Steam Heating Systems

Symptom and Possible Cause	*Possible Remedy*
Water hammer	
(a) Condensed water trapped in a section of horizontal steam piping	(a) Provide correct pitch of piping in steam lines down and away from boiler to drip traps; install drip traps ahead of risers, at end of steam main, and every 300 to 400 feet along steam piping; mount screen and dirt pocket of Y-type strainers horizontally; provide gravity condensate drainage from steam traps if there is a modulating steam regulator in the steam supply line.
(b) Lift installed in return line after the trap	(b) Install a flash tank on the drip trap discharge, and direct the cooled condensate into a common return line.
(c) Steam bubbles trapped and imploding in water-filled wet return lines or pump discharge piping	(c) Use properly sized gravity return lines to allow sufficient space in the piping to allow condensate (water) to flow in the bottom and steam to flow in the top without mixing.

Chapter 9

Electric Heating Systems

Different heating systems use electricity as a source of heat. Each of them employs one or more of the following methods of heat transfer:

1. Radiation
2. Convection
3. Forced air

Radiation is the transmission of heat energy by means of electromagnetic waves (infrared light). In other words, the heat is transferred directly from the surface of the heating element to the people, objects, and furnishings in the room *without* significantly heating the air.

Convection takes place after the temperature of air is increased by contact with a heated surface. As the air becomes warmer, it becomes lighter and rises. When it touches a cooler object, it gives up some of its heat to that object, becomes cooler and heavier, and sinks, completing the circulation of the air in the space. Now note the difference between radiation and convection. In the radiation method of heat transmission, the heat energy passes through the air to the individual or object in much the same way that the sun transfers its heat to earth through space and through the atmosphere. The air is *not* used as the medium of heat transfer. The surfaces are warmed, and the air picks up heat from these surfaces (convection), not directly from the infrared radiation passing through it from the heating element.

In the *forced-air* method of heat transmission, the heated or conditioned air is circulated by motor-driven fans integral to the heating unit or included elsewhere in the system. The heated air is forced directly into the room (as in the case of a wall heater) or through a duct system from a centrally located furnace (central forced-warm-air heating systems).

Central Hot-Water Systems

Some central hydronic systems (i.e., forced-hot-water heating systems) use electric-fired boilers as the heat source. These boilers (Figure 9-1) are compact units consisting of an insulated steel or

Figure 9-1 Electric-fired boiler. *(Courtesy National Better Heating-Cooling Council)*

cast-iron generator with replaceable immersion heating elements. The expansion tank, circulating pump, valves, and prewired controls are included within the boiler package, and the entire unit is assembled at the factory (Figure 9-2).

An electric-fired boiler is compact, generally having a water capacity of only 2 or 3 gal. Some electric-fired boilers are small enough to be hung on a wall in the home. The small size is possible because water has a much higher heat-carrying capacity than air, and the electric elements are immersed directly in the water. This not only saves space but provides very rapid heating of the water because almost all the heat produced by the elements is transferred directly to the water. The heated water is piped directly to the room convectors, where its stored heat is delivered to the air. Distribution of the water can be easily controlled by zone valves under the control of separate zone thermostats.

The controls for an electric boiler are similar to those found on boilers (e.g., oil- and gas-fired) in other central heating systems. The basic controls are:

1. A low-voltage thermostat
2. Pressure and temperature limit controls

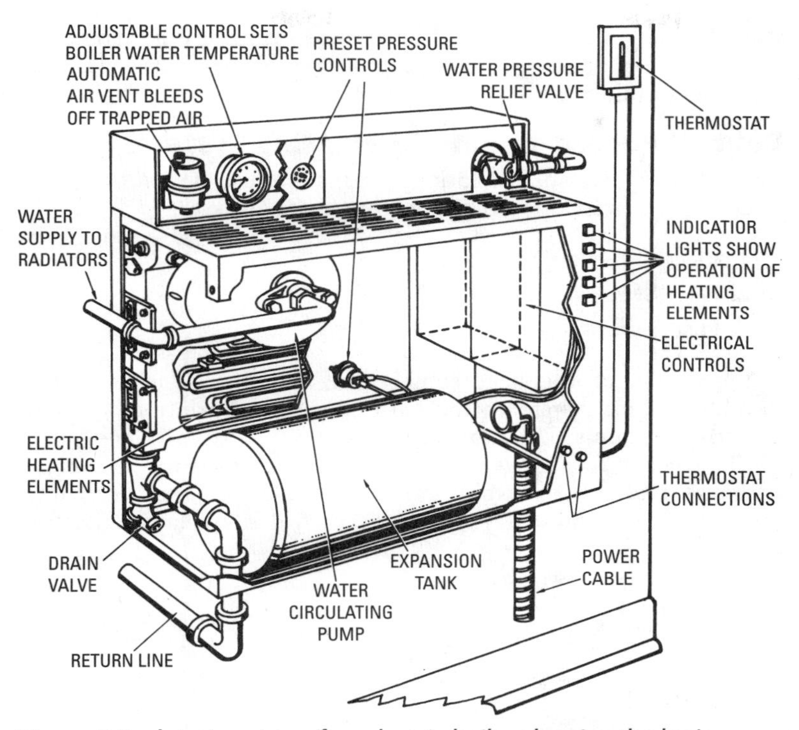

Figure 9-2 Interior view of an electric boiler showing the basic components.

3. Relay and sequence controls
4. Circulating pump controls

The circulating pump is activated by the low-voltage thermostat. Boiler overheating and pressure buildup is controlled by the boiler high-limit controls and the pressure-temperature relief valve.

More detailed descriptions of electric-fired boilers and their controls are found in other chapters of this book (see, for example, the section on electric-fired boilers in Chapter 15, "Boilers and Boiler Fittings").

For information about the piping system used in electric hydronic heating, read the appropriate sections in Chapter 7, "Hot-Water Heating Systems" and in Chapters 8 and 9 of Volume 2.

Cooling for an electric hydronic system is generally provided by individual room air conditioners (i.e., through-the-wall units) or by chilled water pumped from a central refrigeration unit. The latter installation is usually found in multifamily structures, such as apartment buildings.

Central Forced-Warm-Air Heating Systems

Central forced-warm-air heating systems that use electricity to heat or cool the air rely upon the following heat sources:

1. Electric furnaces
2. Duct heaters
3. Heat pumps

Regardless of the heat source, the air is circulated by fans through a duct system to the various rooms. Variations of these ducted air systems are shown in Figure 9-3.

The electric furnace is a complete unit designed for zero clearance and available in several models suitable for either horizontal or vertical installation. The heat is provided by fast-activating,

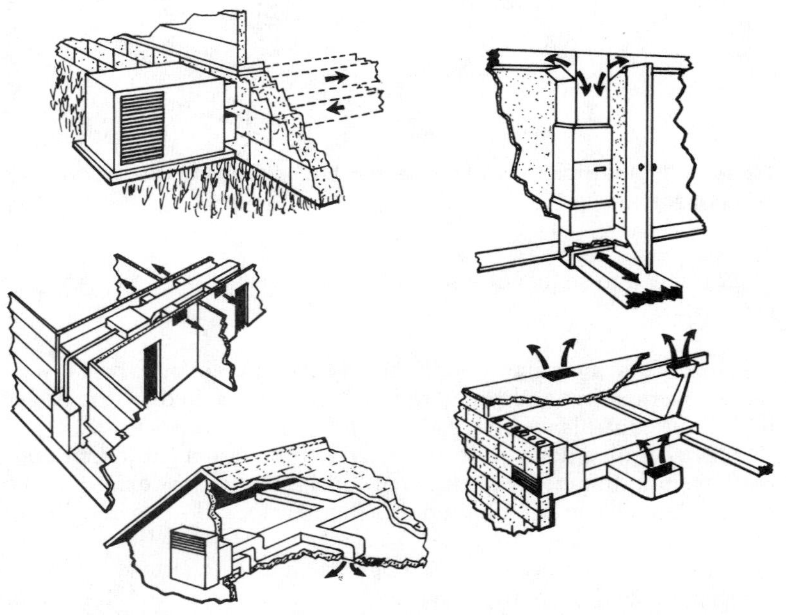

Figure 9-3 Various duct systems for electric heating and cooling.
(Courtesy Electric Energy Association)

coiled-resistance-wire heating elements. When there is more than one coil, the individual coils are energized at intervals in sequence to prevent the type of current overload possible if all were energized simultaneously. The heating coil sequencer is activated and controlled by a low-voltage thermostat mounted on a wall in some convenient location. Overheating of the furnace is prevented by the same type of high-limit control found in other types of furnaces. The warm air is forced through the duct system to outlets in the room by a blower. Evaporator coils can be added to the furnace for cooling. Additional modifications can provide humidity control (humidification and dehumidification), ventilation, and air filtration.

A more detailed description of electric-fired furnaces is found in Chapter 14, "Electric-Fired Furnaces." For further information on forced-warm-air heating systems, read the appropriate sections in Chapter 6, "Warm-Air Heating Systems." Controls for this heating system are described in the section *Electric Heating and Cooling Controls* in this chapter and in Chapter 4 of Volume 2, "Thermostats and Humidistats."

Duct heaters (Figure 9-4) are factory-assembled units installed in the main (primary) or branching ducts leading from the blower unit. Heat is provided by parallel rows of resistance wire formed in the shape of spirals, which may be sheathed or left open. The blower is housed in a specially constructed and insulated cabinet to which the ducts are connected. This unit functions as the power source for circulating the air through the duct heater and the rest of the system. Cooling can be provided by adding a cooling unit to the blower. Humidity control is possible by first supercooling the air in

Figure 9-4 Duct heaters. *(Courtesy Vulcan Radiator Co.)*

the coil and then reheating it to the desired temperature and moisture content when it passes through the duct heater.

Some duct heaters are designed to be inserted into a portion of the duct through a hole cut in its side and are more or less permanent units. Other duct heaters are assembled at the factory in a portion of duct flanged for easy installation at the site.

Duct heaters may be installed in more than one duct to provide zoned heating. When this is the case, each duct heater is provided with means for independent temperature control. Another common method is to install a single duct heater in the main (primary) duct leading from the blower unit. Controls for either installation are similar to those used on electric-fired furnaces (see earlier paragraphs in this section).

The third type of heat source used in ducted, central forced-warm-air heating systems is the air-to-air or water-to-air heat pump. These are described briefly in another section of this chapter (see *Heat Pumps*), and in greater detail in Chapter 10 of Volume 3, "Heat Pumps."

Radiant Heating Systems

Any electrical conductor that offers resistance to the flow of electricity generates a certain amount of heat; the amount of heat generated is in direct proportion to the degree of resistance. This method of generating heat is employed in *radiant heating systems*.

The conductor commonly used in radiant heating systems is an electric heating cable embedded in the floors, walls, or ceilings (Figures 9-5 and 9-6). The cables may be installed at the site (as is often the case with new construction), or they may be obtained in the form of prewired, factory-assembled, panel-type units. The heat generated by the cables is transferred to the occupants and surfaces in the room by low-intensity radiation.

Site-installed heating cables or prewired and assembled panel units are used in the following radiant heating systems:

1. Radiant ceiling panel systems
2. Radiant wall panel systems
3. Radiant floor panel systems

Radiant ceiling heating systems are by far the most commonly used type. The other two have certain disadvantages inherent in their construction. All three radiant heating systems are described in greater detail in Chapter 1 of Volume 3, "Radiant Heating."

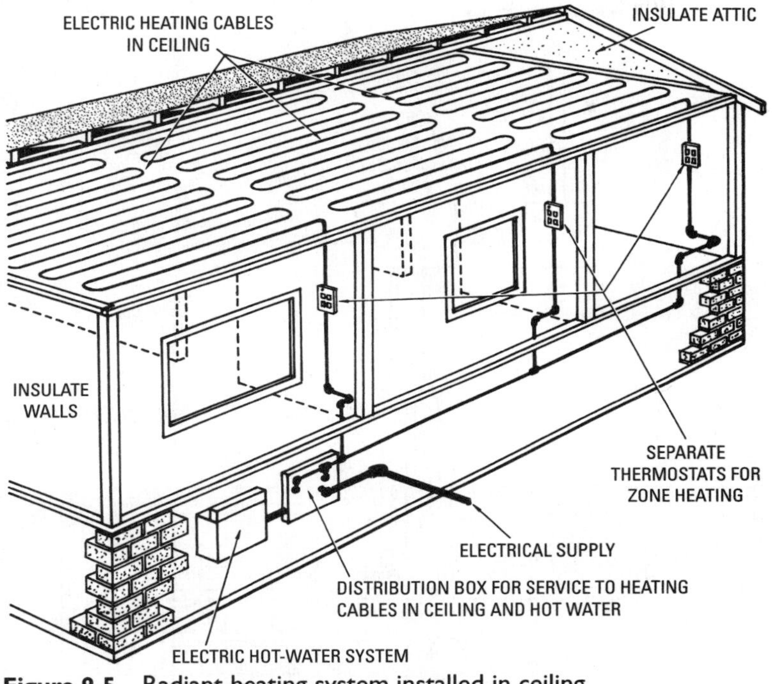

Figure 9-5 Radiant heating system installed in ceiling.

The electric heating cables are activated and controlled by wall-mounted, low-voltage or line thermostats. Cooling can be accomplished only by adding a separate and independent system.

Baseboard Heating Systems

Baseboard heating systems use electric baseboard units located at floor level around the perimeter of each room (Figure 9-7). Each unit consists of a heating element enclosed in a thin metal housing. Zoning is possible within a single wall-mounted line- or low-voltage thermostat located in each room. Heat is provided to the room *primarily* by convection (some radiation is involved) as the room air moves across the heated elements in the baseboard unit. These baseboard units are described more fully in Chapter 2 of Volume 3, "Radiators, Convectors, and Unit Heaters."

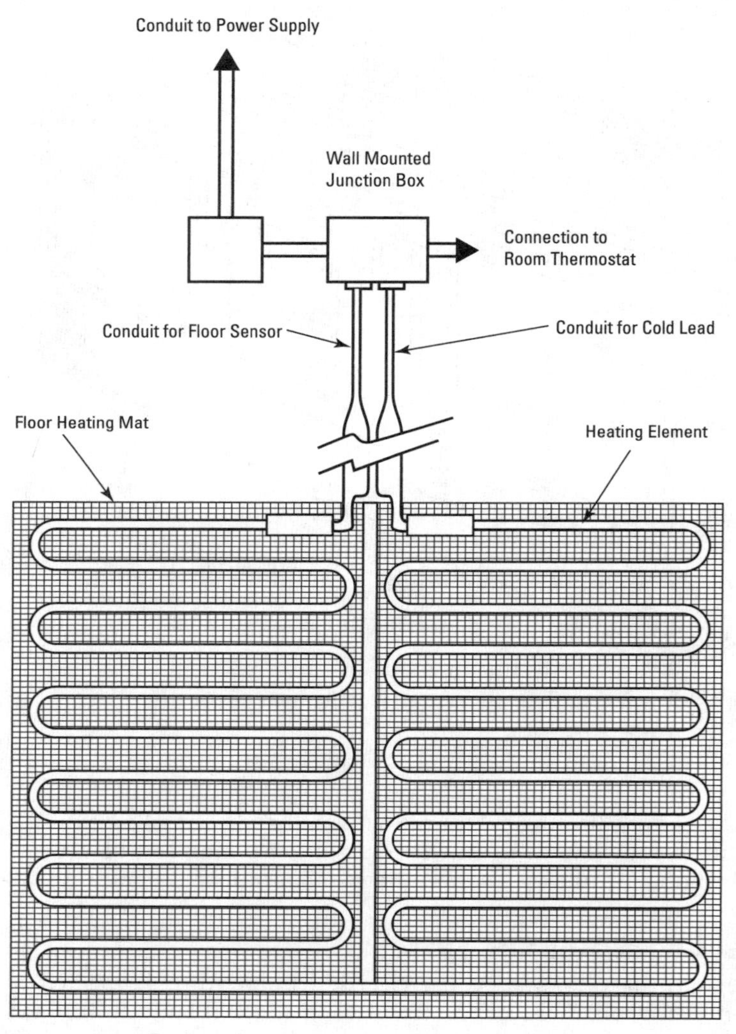

Figure 9-6 Radiant heating system installed in floor.

As is the case with radiant heating systems, cooling can be provided by installing a separate cooling system (central or room air conditioners). The baseboard heating system represents the most widely used form of electric heating.

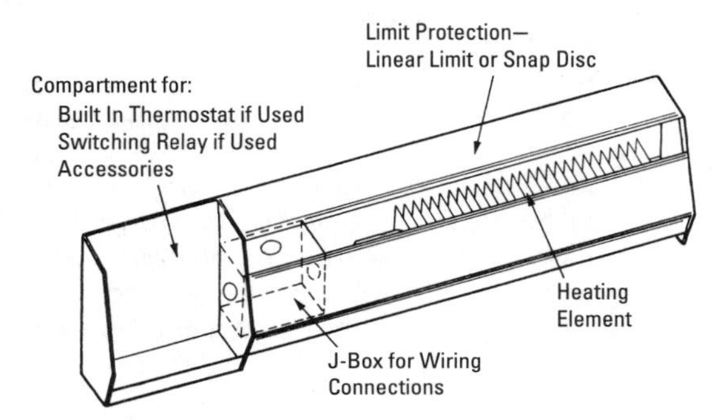

Figure 9-7 Typical electric baseboard heating unit. *(Courtesy Honeywell, Inc.)*

Electric Unit Ventilators

Electric unit ventilators (similar in appearance to electric unit heaters, shown in the next section) are used to heat, ventilate, and cool large spaces that are by nature subject to periods in which there are high densities of occupancy. They are frequently found in offices, schools, auditoriums, and similar structures. The basic components of a typical unit ventilator are:

1. The housing
2. Motor and fans
3. Heating element
4. Dampers
5. Filters
6. Grilles or diffusers

Automatic controls activate the unit ventilator and vary the temperature of the air discharged into the room in accordance with room requirements. Outdoor air is drawn through louvers in the wall and into the unit ventilator before being discharged into the interior of the structure. These ventilators are usually floor- or ceiling-mounted, depending on the design of the room.

In addition to electricity, unit ventilators may also be gas-fired or use steam or hot water as the heat medium. Unit ventilators are described in greater detail in Chapter 2 of Volume 3, "Radiators, Convectors, and Unit Heaters."

Electric Unit Heaters

Electric unit heaters (Figure 9-8) are used primarily for heating large spaces, such as offices, garages, warehouses, and similar commercial and industrial structures.

As is the case with unit ventilators, the heat-conveying medium and combustion source may be other than electricity. For example, the air may be heated by either gas-fired or oil-fired units. Steam or hot water can be substituted for air as the heat conveying medium. Selecting electricity as a medium will depend on such factors as:

1. The availability of cheap electrical power
2. The need for supplementary housing
3. The scarcity of other heat-conveying mediums

The typical unit heater contains the following basic components:

1. Fan and motor
2. Heating element
3. Directional outlet
4. Casing or housing

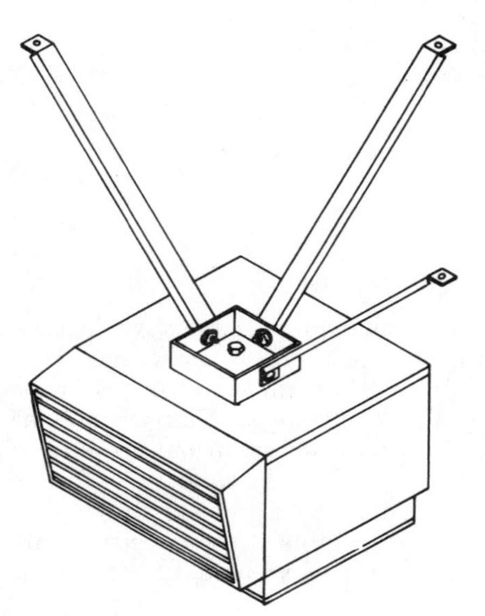

Figure 9-8 Wall- or ceiling-mounted electric unit heater.

Unit heaters may be floor-mounted or suspended from the ceiling, depending upon the design requirements of the structure.

Electric Space Heaters

Smaller unit heaters designed and constructed for domestic heating purposes are commonly referred to as *space heaters*. Like the larger units, they are also designed to operate with other heat-conveying mediums. However, the advantage of using electric space heaters over the fuel-burning types is that they do not have to be vented and require no flue.

Space heaters are available as either portable or permanent types. The latter can be fitted into ducts or mounted on walls or the ceiling.

For further information on space heaters see the appropriate sections in Chapter 2 of Volume 3, "Radiators, Convectors, and Unit Heaters."

Heat Pumps

A *heat pump* (Figures 9-9 and 9-10) is an electrically powered, reversible-cycle refrigeration unit capable of both heating and cooling the interior of a structure.

The heat source is commonly either outside air (in the air-to-air heat pump) or a closed loop of circulating water (in the water-to-air heat pump). The former is the most popular heat pump used for single-family dwellings. The water-to-air heat pump system is most often found in multifamily structures. These are essentially central heating systems, with the heat pump replacing the central furnace or boiler as the heat-generating unit.

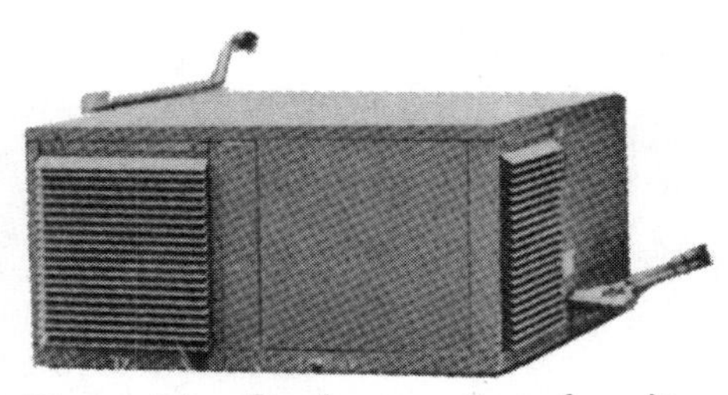

Figure 9-9 Outdoor section of a split-unit heat pump. It is connected to the inside section by refrigerant tubing and electrical wiring. *(Courtesy Electric Energy Association)*

Figure 9-10 Outside-wall-mounted single-package heat pump. *(Courtesy Electric Energy Association)*

Operating valves in the heat pump unit control the reversal of the refrigeration cycle. It removes heat from the interior of the structure, discharges it outside during hot weather, and supplies heat to the interior spaces during periods of cold weather.

The basic components of a heat pump installation are (1) the compressor, (2) the condenser, (3) the evaporator, and (4) a low-voltage thermostat. These and other aspects of heat pumps are described in greater detail in Chapter 12 of Volume 3.

Electric Heating and Cooling Controls

In an electric heating and cooling system, the best results are obtained by finding a comfortable thermostat setting and leaving it there. Constantly changing the thermostat setting results in consuming more energy (resulting in increased operating costs) and creates additional wear and tear on the equipment. Experts estimate that for each degree the thermostat is raised above the normal setting, there is a 3 percent increase in heating costs. A comfortable setting for the thermostat depends upon a number of variables, including (1) the type of system, (2) environmental conditions, and (3) your personal requirements. For example, 70°F is usually adequate for a radiant heating system (i.e., baseboard, panel-type heating installations, etc.) in a properly insulated structure. A comfortable indoor temperature setting for air conditioning is usually 76 to 78°F.

A variety of control methods are used in electric heating and cooling systems to maintain kilowatt demand at a level both economical in operation and suitable in performance. The basic controls used for these purposes are:

1. Thermostats
2. Sequence switching devices
3. Load-limiting controls
4. Time-delay relays

Thermostats can be located in each room or on each heating unit to provide decentralized control. A manual switch can be provided with each thermostat to shut off the current if the room is to be unoccupied for long periods of time.

Sequence switching devices are used to switch electrical current in sequence (rotation) from one room or circuit to the next. A *load-limiting control* is designed to shut off the current to one or more circuits when total electric demand exceeds a preset value. *Time-delays* are used to restore service after an interruption *nonsimultaneously* over a period of several minutes so that overloading is avoided. Each of these controls is examined in considerable detail in Chapter 14, "Electric Furnaces."

Insulation for Electrically Heated and Cooled Structures

Any structure electrically heated and cooled *must* be properly insulated, or operation costs will be unacceptable. Converting an older structure to electric heat is therefore not recommended unless you have no objection to the additional expense of improving the insulation or you feel you can live with the higher operating costs. Consequently, electric heating is more often considered for new construction.

Insulating a structure that is to be electrically heated and cooled requires greater attention to construction details than do other types of systems; however, the results are well worth the efforts because reduced operating costs and other gains (e.g., the reduction of outside noise and quiet operating characteristics) are soon readily apparent. Because electric heating and cooling systems require an especially well-insulated structure for efficient operation, recommendations for the required minimum levels of insulation are available from the Electric Energy Association (e.g., its publication *Electric Space Conditioning in Residential Structures*).

The recommendations found in the following paragraphs will contain references to *U*-factors and *R*-values. The *U-factor* is the overall coefficient of heat transfer and is expressed in Btu per hour per square foot of surface per degree Fahrenheit difference between air on the inside and air on the outside of a structural section. The *R-value* is the term used to express thermal resistance of the insulation. The *U*-factor is the reciprocal of the sum of the thermal resistance values (*R*-values) of each element of the structural section. Table 9-1 gives the insulation recommendations for electrically heated and cooled structures both in terms of *R*-values and in terms of *U*-factors (overall coefficients of heat transfer).

Table 9-2 lists heat loss limits for electrically heated structures as recommended by the Electric Energy Association. The values that are listed are expressed in watts and Btu/h (1 watt = 3.413 Btu/h) per square foot of floor space measured to the exterior walls. The assumed infiltration rate on which this table is based is approximately three-quarters air change per hour.

The maximum summer-heat-gain limits for electrically air conditioned homes are indicated in Figure 9-11. These figures are adapted from the FHA Minimum Property Standards for single- and multifamily structures.

Table 9-1 Maximum Winter Heat Loss for Electrically Heated Homes

	Maximum Heat Loss Values	
Degree Days	***Watts/ft²***	***Btu/h/ft²***
Over 8000	10.0	34
7001 to 8000	9.4	32
6001 to 7000	8.8	30
4501 to 6000	8.2	28
3001 to 4500	7.6	26
2000 to 3000	7.0	24

Courtesy Electric Energy Association

Table 9-2 Insulation Recommendations (*R*-Values) and Overall Coefficients of Heat Transfer (*U*-Factors) for Major Areas of Heat Loss and Heat Gain in Residential Structures

Type of Construction	Opaque Sections Adjacent to Unheated Spaces			Opaque Sections Adjacent to Separately Heated Dwelling Units		
	Walls	*Floors*	*Ceilings*	*Walls*	*Floors*	*Ceilings*
Frame	R-11 *U*:0.07	R-11 *U*:0.07	R-19 *U*:0.05	R-11 *U*:0.07	R-11 *U*:0.07	R-11 *U*:0.07
Masonry	*R*-7 *U*:0.11	*R*-7 *U*:0.11	*R*-11 *U*:0.07	*R*-7 *U*:0.11	*R*-7 *U*:0.11	*R*-11 *U*:0.07
Metal section	*R*-11 *U*:0.07	Not applicable	Not applicable	*R*-11 *U*:0.07	Not applicable	Not applicable
Sandwich	*R*-[a] *U*:0.07	*R*-[a] *U*:0.07	*R*-[a] *U*:0.05	*R*-[a] *U*:0.07	*R*-[a] *U*:07	*R*-[a] *U*:07
Heated basement or unvented crawl space	*R*-7 *U*:0.11	Insulation not required	Insulation not required			
Unheated basement or vented crawl space	Insulation not required	See "Floors"	Insulation not required			

Courtesy Electric Energy Association

Note: Insulation R-values refer to the resistance of the insulation only.

[a] *Since the thermal resistance of sandwich construction depends upon its composition and thickness, the amount of additional insulation required to obtain the maximum U-factor must be calculated in each case.*

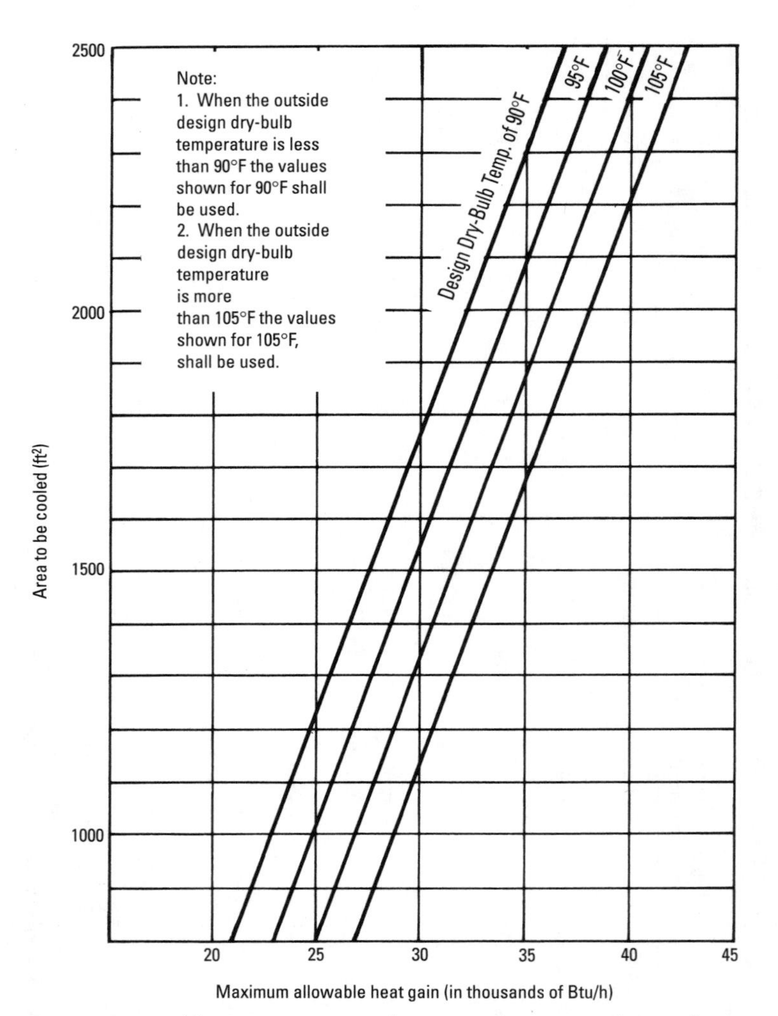

Figure 9-11 Maximum summer heat-gain for air conditioned homes. *(Courtesy Electric Energy Association)*

Note

The insulation details, recommendations, and illustrations included in Chapter 3, "Insulating and Ventilating Structures" of this volume meet the minimum requirements for an electrically heated house.

Further Information

Further and more detailed information about insulation, the problems of heat loss and gain, and methods for calculating heat loss can be found by reading the appropriate sections of Chapter 2, "Heating Fundamentals"; Chapter 3, "Insulating and Ventilating Structures"; and Chapter 4, "Sizing Residential Heating and Air Conditioning Systems."

Advantages of Electric Heating and Cooling

Among the principal advantages of using electric heating and cooling are:

1. Greater safety
2. Quiet operation
3. Economy of space
4. Reduction of drafts
5. Reduction of outside noise
6. Uniform temperature
7. Structural design flexibility

Although the chances of an explosion in gas- or oil-fired heating systems have become very small because of the safety features built into these systems, an explosion simply *cannot* occur in total electric heating. A properly installed electric heating and cooling system will last for years without any problems. If a problem occurs (e.g., a defect in the heating unit or wiring), resistance builds up in the line to the fuse or circuit breaker box. At a certain point, the fuse will blow or the circuit breaker will trip, which will automatically shut off the electricity before any damage occurs.

Electric heating units are very compact and therefore utilize very little space. In baseboard systems, there is no need for ducts or pipes to carry a heat medium from its source to the space being heated. These factors offer a great degree of structural design flexibility because duct and pipe arrangements do not have to be taken into consideration. In addition, no chimney or flue arrangement is required.

In a structure properly insulated for electric heating and cooling, there is a marked reduction of drafts and the degree of outside noise penetration. Moreover, uniform temperatures will prevail.

Finally, an electric heating system is generally quieter than other types because fewer mechanical parts are involved. This quietness of operation is particularly characteristic of baseboard-type installations.

Disadvantages of Electric Heating and Cooling

An electrically heated and cooled structure offers certain inherent disadvantages when compared with other types of heating and cooling systems. Of course, this is true of *any* system regardless of the energy source. When choosing an energy source for a heating and cooling system, it is always necessary to weigh the advantages and disadvantages of the system and its energy source carefully against the requirements you demand from them.

An electrically heated or cooled structure must be well insulated against heat gain or loss. If it isn't, the cost of heating or cooling it will be extremely high. For that reason, electric systems are seldom installed in existing structures, except in situations where a room is added or a basement is finished.

Electric heating and cooling systems generally have higher operating costs. These energy costs are dependent upon a number of variables, such as the amount of insulation, the orientation of the structure, the number of windows (total glass area), the cost of electricity where the structure is located, and the energy use habits of the occupants.

One disadvantage of electric furnaces is that they are frequently oversized. An oversized electric furnace heats up and cools down too rapidly to maintain acceptable comfort levels in the rooms of a structure. However, oversizing is not a problem limited to electric furnaces. See Chapter 4, "Sizing Residential Heating and Air Conditioning Systems" for additional comments about the problem of oversizing.

Troubleshooting Electric Heating Systems

Table 9-3 suggests remedies for troubleshooting electric heating systems.

Caution

> Never attempt to service or repair the electric controls inside an electric furnace cabinet unless you have the qualifications and experience to work with electricity. Potentially deadly high-voltage conditions exist inside these furnace cabinets.

Table 9-3 Troubleshooting Electric Heating Systems

Symptom and Possible Causes	*Suggested Remedies*
No heat	
Power may be off. Check fuse or circuit breaker panel for blown fuses or tripped breakers.	Replace fuses or reset breakers. If the problem repeats itself, call an electrician or an HVAC technician.
Check thermostat (programmable type) for dead batteries.	Replace batteries and reset thermostat.
Not enough heat	
Check thermostat setting.	Thermostats in electric heating systems must be set several degrees higher than the desired room temperature.

Chapter 10

Furnace Fundamentals

The American Society of Heating, Refrigerating, and Air-Conditioning Engineers (ASHRAE) defines a furnace as "a complete heating unit for transferring heat from fuel being burned to the air supplied to a heating system." The *Standard Handbook for Mechanical Engineers* (Baumeister and Marks, seventh edition) provides a definition that differs only slightly from the one offered by the ASHRAE: "a self-enclosed, fuel-burning unit for heating air by transfer of combustion through metal directly to the air." Contained within these closely similar definitions are the two basic operating principles of a furnace: (1) Some sort of fuel is used to produce combustion, and (2) the heat resulting from this combustion is transferred to the air within the structure. Note that *air*—not steam, water, or some other fluid—is used as the heat-conveying medium. This feature distinguishes warm-air heating systems from the other types; see Chapter 6, "Warm-Air Heating Systems."

Most modern furnaces are used in warm-air heating systems in which the furnace is a centralized unit, and the heat produced in the furnace is forced or rises by means of gravity through a system of ducts or pipes to the various rooms in the structure. This is what one commonly refers to as a *central heating system*. In other words, the furnace is generally in a centralized location within the heating system in order to obtain the most economical and efficient distribution of heat (although this is not an absolute necessity when a forced-warm-air furnace is used).

Ductless or pipeless furnaces are also used in some heating applications but are limited in the size of the area that they can effectively heat. They are installed in the room or area to be heated but are provided with no means for distributing the heat beyond the immediately adjacent area. This is a far less efficient and economical method of heating than the central heating system, but it is found to be adequate for a room, an addition to an existing structure, or a small house or building.

Classifying Furnaces

There are several different ways in which furnaces can be classified. One of the more popular methods is based on the fuel used to fire the furnace. Using this method, the following four types of furnaces are recognized:

1. Gas-fired furnaces
2. Oil-fired furnaces
3. Coal, wood, and multi-fuel furnaces
4. Electric furnaces

The first three categories of furnaces use a fossil fuel to produce the combustion necessary for heat transfer. The last one, electric furnaces, uses electricity. Whether or not electricity can be justifiably called a fuel is not of great importance here, because in this particular instance it functions in the *same* manner as the three fossil fuels: It heats the air being distributed.

Furnaces can also be classified by the means with which the heated air is distributed to the room (or rooms) in the structure. This method of classifying furnaces establishes two broad categories: (1) gravity warm-air furnaces and (2) forced-warm-air furnaces. The gravity warm-air furnaces rely *primarily* upon the principle of gravity for circulating the heated air. Because warm air is lighter than cold air, it will rise and pass through ducts or directly into the rooms to be heated. After passing off its heat, the air, now cooler and heavier, descends through returns to the furnace, where it is reheated. Gravity-type furnaces represent the earliest designs used in warm-air heating systems. They were sometimes equipped with fans (integral or booster) to increase the rate of air flow. They have been largely replaced in popularity by forced-warm-air furnaces, which are equipped with integral fans.

Forced-warm-air furnaces are often divided into three principal classes, based primarily upon the location on the furnace of the warm-air discharge outlet and the return-air inlet. Furthermore, additional design considerations dependent upon the planned location of the furnace also enter into the classification of warm-air furnaces, resulting in three types, or classes, of warm-air furnaces:

1. Upflow furnaces
    a. Upflow highboy furnaces
    b. Upflow lowboy furnaces
2. Downflow furnaces
3. Horizontal furnaces

Upflow Highboy Furnace

A typical *upflow highboy furnace* (also referred to as an *upflow furnace* or a *highboy furnace*) is shown in Figure 10-1. These are compact heating units that stand no higher than 5 or 6 ft and occupy a floorspace of approximately 4 to 6 ft^2 (2 ft × 2 ft or 2 ft × 3 ft).

The heated air is discharged through the top of the upflow furnace (hence the name), and the return air enters the furnace

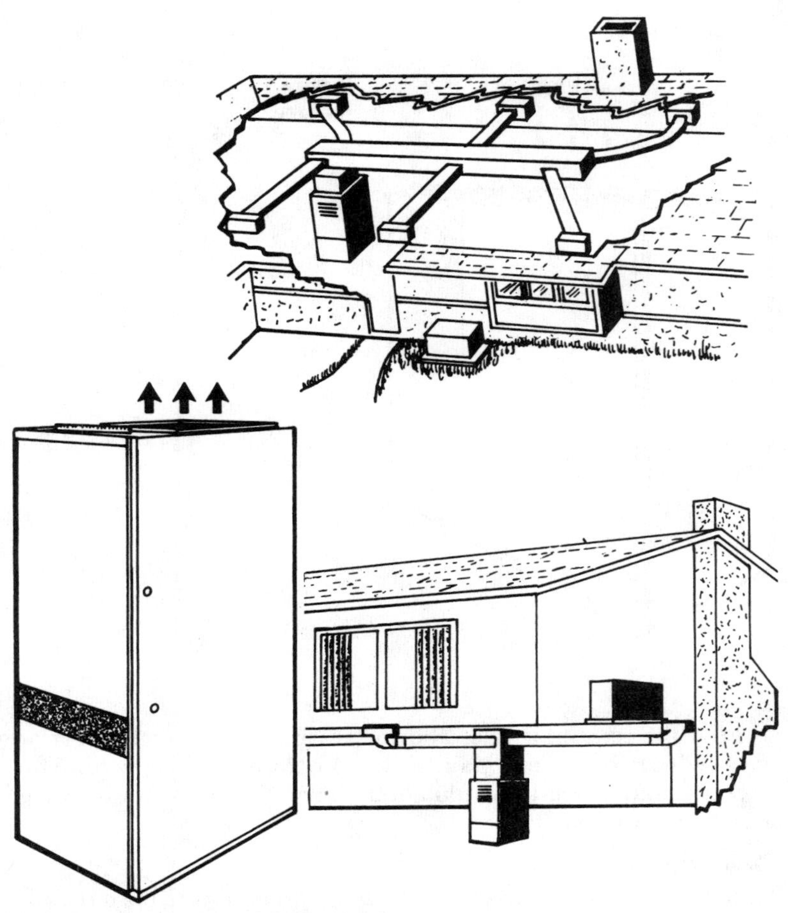

Figure 10-1 Upflow (highboy) furnace.

through air intakes in the bottom or sides. Cooling coils can be easily added to the top of the furnace or in the duct system.

Upflow Lowboy Furnace

The *upflow lowboy furnace* (also referred to as an *upflow furnace* or *lowboy furnace*) (Figure 10-2) is designed for low clearances and stands only about 4 to 4½ ft high. Although shorter than either the upflow highboy or downflow types (see below), it is longer from front to back.

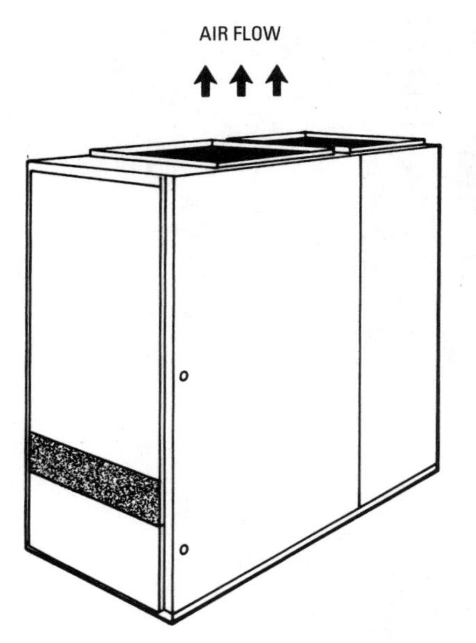

Figure 10-2 Upflow (lowboy) furnace.

Both the return-air inlet and the warm-air discharge outlet are usually located on the top of the lowboy furnace. The lowboy furnace is found in heating installations where the ductwork is located above the furnace.

Downflow Furnace

Figure 10-3 shows an example of a *downflow furnace* (also referred to as a *counterflow furnace* or a *downdraft furnace*). It is very similar in size and shape to the upflow furnace, but it discharges its

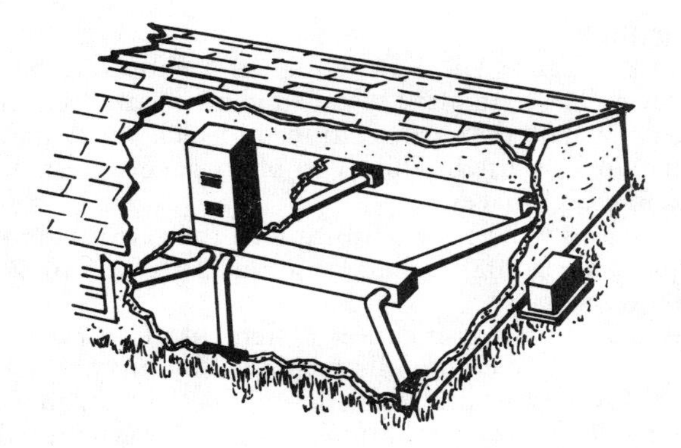

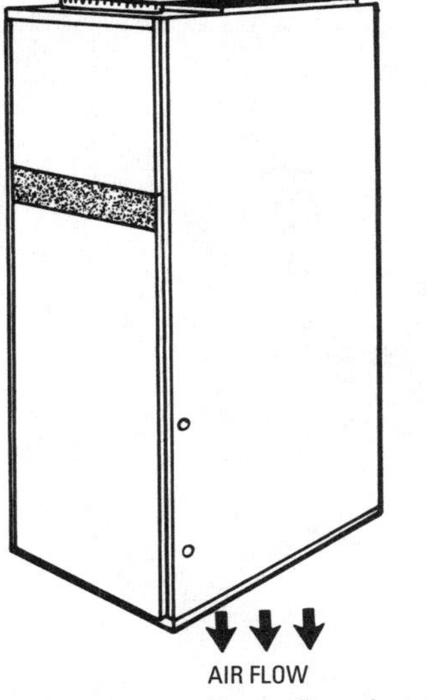

Figure 10-3 Downflow furnace.

warm air at the bottom rather than the top. The return-air intake is located at the top.

A downflow furnace is used primarily in heating installations where the duct system is either embedded in a poured concrete slab or suspended beneath the floor in a crawl space.

Horizontal Furnace

Horizontal furnaces (Figure 10-4) are designed for installation in low, cramped spaces. They are often installed in attics (and referred to as *attic furnaces*), where they are positioned in such a way that a minimum of ductwork is used. This type of furnace is also frequently installed in crawl spaces.

Although dimensions will vary slightly among the various manufactures, the typical horizontal furnace is about 2 ft wide by 2 ft high and 4½ to 5 ft long.

The terms used in this classification system (*upflow furnace, downflow furnace,* etc.) are commonly employed by furnace manufacturers in the advertising literature describing their products. This is equally true of their installation and operation manuals. Because of the widespread usage of these terms, they will be employed in the more detailed description of furnaces in the following chapters.

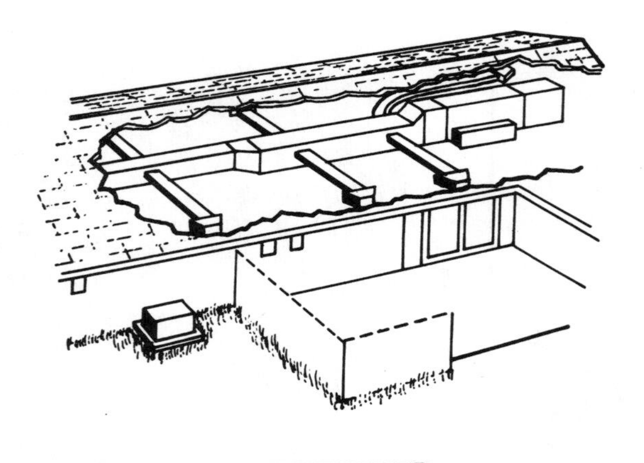

Figure 10-4 Horizontal furnace.

Gravity Warm-Air Furnaces

Gravity warm-air furnaces rely upon the fact that warm air is lighter than cold air. As a result, the warmer air rises through the ducts or pipes in the structure, gives off its heat to the rooms, and descends to the furnace as it becomes cooler and heavier. It is then reheated and rises once again to the rooms. A continuous circulation path is thus established through the heating and cooling of the air. Sometimes a fan is added to increase the rate of flow, but the primary emphasis is still the effect of gravity on the differing weights of air.

Depending upon the design, a gravity warm-air furnace will fall into one of the following categories:

1. A gravity warm-air furnace without a fan
2. A gravity warm-air furnace with an *integral* fan
3. A gravity warm-air furnace with a *booster* fan

Each of these three categories represents a different type of warm-air furnace used in *central* heating systems.

Any gravity warm-air furnace not equipped with a fan relies entirely upon gravity for air circulation. The flow rate is very slow, and extreme care must be taken in the design and placement of the ducts of pipes. Sometimes an integral fan is added to reduce the internal resistance to airflow and thereby speed up air circulation. A booster fan provides the same function but is designed not to interfere with air circulation when it is not use.

The round-cased, gravity warm-air furnace illustrated in Figure 10-5 is a coal-fired unit that can be converted to gas or oil; see Chapter 16, "Boiler and Furnace Conversions." Depending upon the model, these furnaces are capable of developing up to 108,266 Btu at register and up to 144,319 Btu at bonnet.

Floor, wall, pipeless furnaces, and some unit (space) heaters also operate on the principle of the gravity warm-air furnace. They are distinguished by the fact that the warm air is discharged directly into the room (or rooms) without the use of ducts or pipes.

Selecting a Furnace for a New House

One of the many decisions involved in building a new house is the selection of the heating, ventilating, and air conditioning system. This is probably one of the most important decisions, because the cost of operating and maintaining the system and its efficiency will be a daily fact of life for as long as you live there. It is therefore in your best interest to select the most efficient heating system that you possibly can.

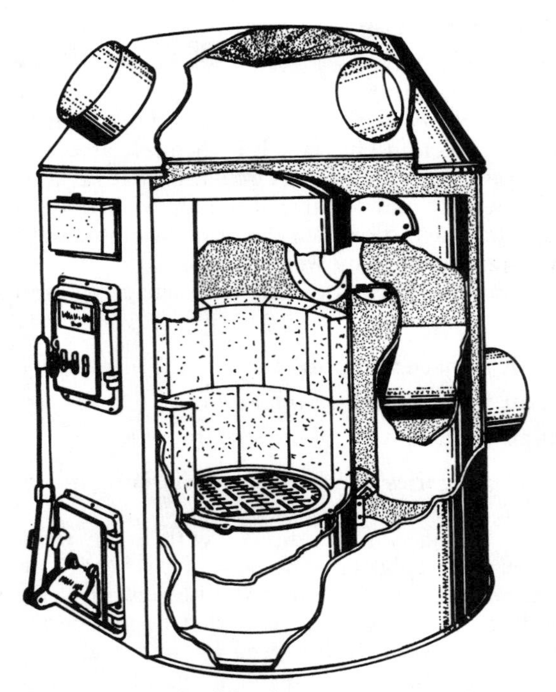

Figure 10-5 Coal-fired gravity warm-air furnace.
(Courtesy Oneida Heater Co., Inc.)

The type of heating system selected (forced-warm-air, hydronic, steam, electric baseboard, etc.) will depend upon such factors as the design of the house, the climate where the house is located, fuel costs, and personal preference. Comments on the advantages and disadvantages of the various types of heating systems are found in Chapters 6 to 9.

Assuming that, like the majority of home owners, you have decided to install a *forced*-warm-air heating system, you should pay particular attention to selecting a suitable furnace. Should this decision be left entirely to your building contractor? Certainly not. For convenience, friendship, or other reasons, he may prefer to subcontract the work to a heating firm with which he customarily does business. As a result, you may not be getting the best furnace for your needs. For example, John Doe Heating, Ventilating, and Air Conditioning Company may deal in furnace Brand X (or Brands X,

Y, and Z). It may be that none of these furnaces is as efficient and economical to operate as the one you might wish to use.

There are many manufacturers of forced-warm-air furnaces doing business in the country. The furnaces produced by these manufacturers differ from one another in a variety of ways, including:

1. Installation cost
2. Design factors
3. Type of fuel used
4. Cost of furnace
5. Furnace gross output (Btu/h)
6. Furnace net output (Btu/h)
7. Furnace efficiency

The *installation cost* depends upon the area in which you are living and can vary widely for the same furnace model from the same manufacturer. It is not something you can easily control, because it depends upon local labor costs, delivery charges, and availability of equipment.

Design factors involve developments in the heating equipment that improve (or detract from) their performance. For example, some electric-fired furnaces use a direct-drive blower motor instead of a belt-driven one. The direct-drive motor results in a smaller, more compact furnace. The belt-driven blower motor has the advantage of being more easily maintained (it is simply a matter of replacing the standard motor used in the unit). Other design developments include improvement in flame retention (e.g., in oil-fired furnaces), more efficient combustion chambers, or improvements in other components. It may take a little research, but several professional journals in the heating field carry articles reviewing new products. Consumer magazines also provide this service. Do not take the manufacturer's (or the sales representative's) claim at face value. Go to your local library and read what the experts have to say.

The type of fuel used will be an important factor in determining future operating costs. What is the cost of the various fuels in your area? Remember, the cheapest may not necessarily be the best for your purposes. Operating costs depend on the total amount of fuel necessary to heat your home on a month-to-month basis. As a general rule, electricity is more expensive than the fossil fuels, and gas slightly more expensive than oil. Read Chapter 5, "Heating Fuels," for additional information.

In considering the *cost of a furnace,* remember that the most expensive furnace is not necessarily the best one, and the cheapest is not always that bargain you expect it to be. Your *first* consideration should be furnace performance. Although the initial cost of a suitable furnace may be high relative to others, it will soon pay for itself with lower operating costs.

Furnace efficiency is expressed as a percentage and is found by subtracting *furnace net output* (the actual amount of heat delivered to the rooms) from *furnace gross output.* For example, an oil-fired furnace with a gross output of 150,000 Btu/h and a net output of 120,000 Btu/h will have an approximate efficiency of 80 percent.

By way of review, the three basic steps involved in selecting a suitable furnace for a new house are as follows:

1. Determine the *type* of heating system you want for the house.
2. Estimate the amount of heat loss from the structure; see Chapter 4, "Sizing Residential Heating and Air Conditioning Systems."
3. Select a furnace with a heating capacity capable of replacing the lost heat.

Selecting a Furnace for an Older House

Sometimes an older house or building will have a furnace that needs replacing. If you are planning to install a newer model of the same kind of furnace, you should have no serious difficulties. However, be sure that the newer model develops a similar Btu rating at both register and bonnet. In a forced-warm-air heating system, the ducts are sized in accordance with the overall requirements of the system, and a furnace is selected with a capacity to meet these requirements.

Switching from coal to gas or oil also should not present any great difficulties. Conversion burners have been designed and manufactured for just this purpose; see Chapter 16, "Boiler and Furnace Conversions."

Installing an electric-fired furnace in an older structure is not generally recommended unless the construction is particularly tight and well insulated.

Furnace Components and Controls

Most forced-warm-air furnaces have the following basic components and controls:

1. Heat exchanger
2. Blower

3. Air filter
4. Primary control
5. Fan and limit controls
6. Thermostat control

The *heat exchanger* (Figures 10-6 and 10-7) is a metal surface (0.05- to 0.06-in steel) located between the burning fuel and the circulating air in the furnace. The metal becomes hot and transfers its heat to the air above, which is then circulated through the ducts by the *blower.*

Figure 10-6 Heat exchanger for a horizontal gas-fired furnace.
(Courtesy Meyer Furnace Co.)

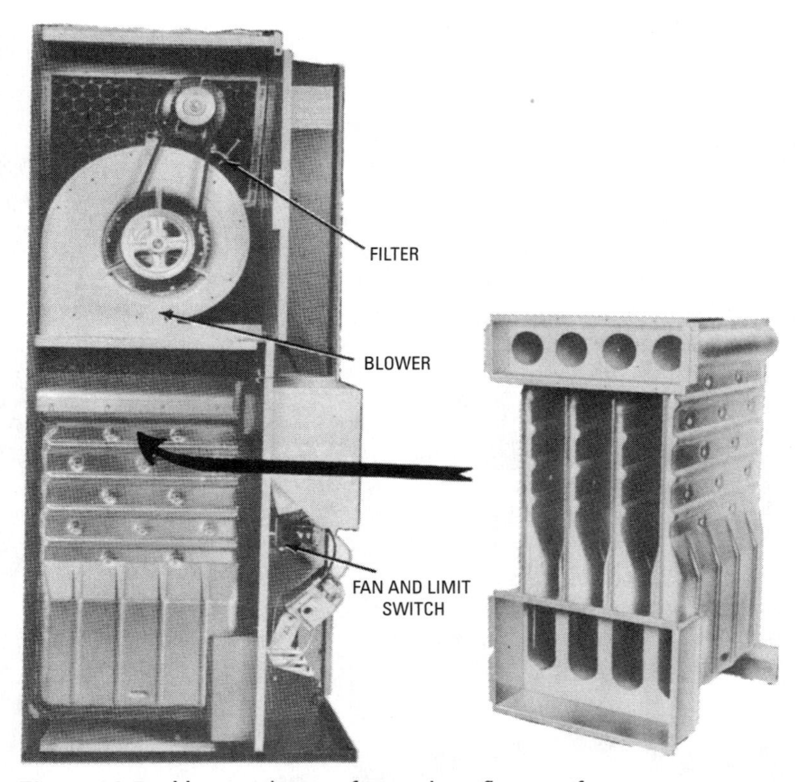

Figure 10-7 Heat exchanger for an downflow gas furnace.
(Courtesy Meyer Furnace Co.)

Most forced-warm-air furnaces are equipped with either a disposable or permanent (and washable) *air filter* to clean the circulating air. An air filter is not recommend for a gravity warm-air furnace, because it tends to restrict the air flow.

Most warm-air furnaces are thermostatically controlled. In addition, there will be a primary control as well as fan and limit controls. Among the functions of the primary control is the regulation or stoppage of the flow of fuel to the combustion chamber when the fire is out or the thermostat indicates that no further fuel is needed. The fan and limit controls are also actuated by the thermostat and are designed to start or stop the fan when the temperature in the furnace bonnet reaches predetermined and preset temperatures. In warm-air furnaces, it is suggested that the limit-control switch be placed in the warm-air plenum, with a recommended setting of 200°F for a forced-warm-air furnace and 300°F for a gravity warm-air furnace.

These and other controls are offered by furnace manufacturers in a variety of different combinations. Their primary function is to provide safe, smooth, and automatic operation of the warm-air furnace. More detailed descriptions of the controls used in furnaces and heating systems are found in Chapter 4, "Thermostats and Humidistats"; Chapter 5, "Gas and Oil Controls"; Chapter 6, "Other Automatic Controls"; and Chapter 9 of Volume 2, "Valves and Valve Installation." (Other chapters also include sections on automatic controls. Check the Index).

Most warm-air furnaces (except hand-fired coal furnaces and electric-fired furnaces) are provided with devices that automatically prepare the fuel for combustion or to feed it directly to the fire. These devices (oil burners, gas burners, and automatic coal stokers) are described in the appropriate chapters. Electric-fired warm-air furnaces utilize electric resistance heaters.

Cooling coils, electronic air cleaners, and humidifiers are examples of optional equipment that can be added to most forced-warm-air furnaces to provide total environmental control. It is best to include this optional equipment when the furnace is installed rather than add it at a later date. By doing so, you avoid complications that might arise in the overall design of your heating system. For example, adding air conditioning capability is not simply a matter of installing cooling coils. Duct sizes must also be considered, and they are not necessarily the same as those used for heating. This optional equipment is considered in detail in Chapters 7 and 8 in Volume 3.

Pipeless Floor and Wall Furnaces

A *pipeless furnace* (Figure 10-8) is commonly a gravity warm-air furnace installed in a central location beneath the floor. A single grille for the warm air and a return is provided for air circulation. This type of pipeless furnace is sometimes referred to as a *floor furnace,* although the latter is actually a permanently installed room heater and should be distinguished as such. *Wall furnaces* also belong to this category.

Both gas- and oil-fired furnaces are manufactured for installation in recesses cut from the floor. They are available in a wide range of Btu ratings for different types of installations. Some are thermostatically equipped for automatic temperature control and with blowers for forced-air circulation (although those that operate on the gravity principle of air circulation are more common). There

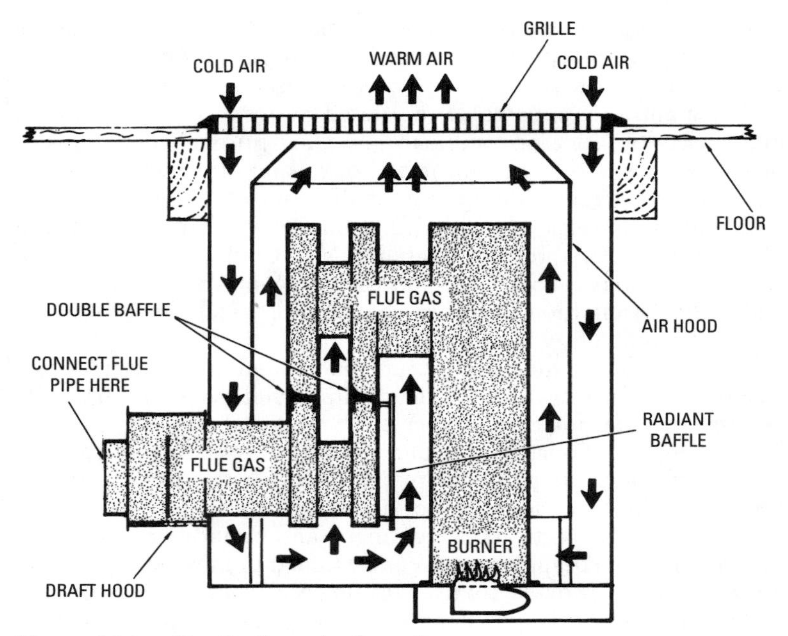

Figure 10-8 Gas-fired gravity floor furnace.

is also a choice between electric ignition and pilot flame. All gas- or oil-fired floor furnaces *must* be vented.

Another type of pipeless furnace is the gas- or oil-fired vertical furnace installed in a closet or a wall recess. The counterflow types discharge the warm air from grilles located at the bottom of the furnace, as shown in Figure 10-9.

Duct Furnaces

A *duct furnace* (Figure 10-10) is a unit heater designed for installation in a duct system where a blower (or blowers) is used to circulate the air. It is commonly designed to operate on natural or propane gas, although electric duct heaters are also available (see Chapter 7 of Volume 2).

Geothermal Furnaces

Calling this heating appliance a furnace is a bit of a misnomer. A geothermal furnace is actually a heat pump, not a furnace. It is designed to use water instead of air to deliver the heat. Geothermal

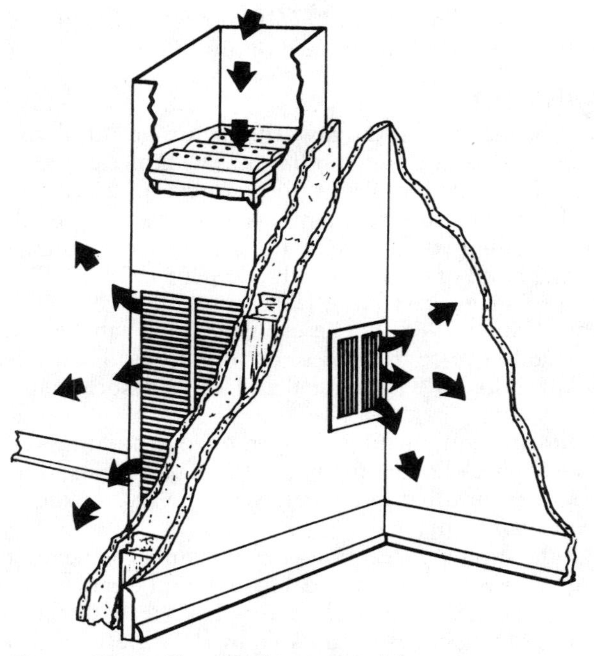

Figure 10-9 Counterflow vertical furnace. *(Courtesy U.S. Department of Agriculture)*

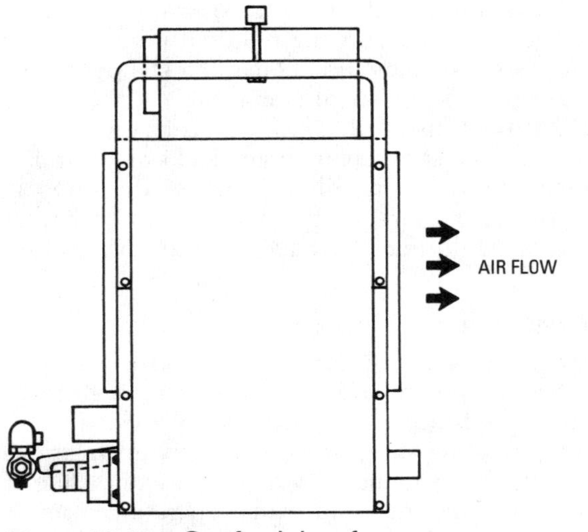

Figure 10-10 Gas-fired duct furnace. *(Courtesy Janitrol)*

furnaces are described in Chapter 12, "Heat Pumps" in Volume 3 of the *Heating, Ventilating, and Air Conditioning Library.*

Furnace Installation

No attempt should be made to install a warm-air furnace until you have consulted the local codes and standards. The American Gas Association and the National Fire Protection Association have also established codes for installing warm-air furnaces; these are available to the public through their publications (see, for example, NFPA No. 31, "Installation of Oil Burning Equipment 1972"; NFPA No. 90B, "Residence Warm Air Heating 1971"; and NFPA No. 204, "Smoke and Heat Venting 1968"). These publications can be obtained free or at a modest price from these organizations. Their addresses appear in Appendix A, "Professional and Trade Associations" of this book.

As soon as the heating equipment is shipped to your building site or existing structure, check it for missing or damaged parts. The manufacturer should be notified immediately of any discrepancies so that replacements can be made.

Follow the manufacturer's instructions for installing the heating equipment, but give precedence to local codes and regulations should any conflict arise. Most manufacturers base their installation instructions on existing codes formulated by the American Gas Association and the National Fire Protection Association.

Locate the furnace so that it has the shortest possible flue run containing as few elbows (turns in the run) as possible. Clearances around the furnace should be kept to the required minimum but within the regulations established by local codes and standards. Air ventilation must be adequate for efficient operation.

After the furnace has been installed, check all gas or oil lines for possible leaks. Make the necessary repairs if any leaks are found.

More detailed instructions for installing furnaces are found in Chapter 11, "Gas Furnaces"; Chapter 12, "Oil Furnaces"; Chapter 13, "Coal, Wood, and Multi-Fuel Furnaces"; and Chapter 14, "Electric Furnaces."

Furnace Maintenance

Furnace maintenance is a very important part of the efficient operation of a warm-air heating system and should never be neglected.

The manufacturer of the heating equipment will provide recommendations for proper maintenance, and these recommendations should be carefully followed. By doing so, you will extend the life of the equipment, improve its efficiency, and reduce operating costs.

Maintenance recommendations specific to gas, oil, electric, coal, wood, and multi-fuel furnaces are found at the end of Chapters 11 through 14. *Always* read and closely follow the furnace manufacturer's service and maintenance instructions. Maintain a periodic service and maintenance schedule and post the schedule near the furnace. The schedule should include the dates of any service, maintenance, and repairs performed, and the telephone numbers of the local furnace manufacturer's representative and/or a local technician qualified to work on the furnace.

Troubleshooting Furnaces

The automatic controls of all furnaces, as well as the heating elements in electric furnaces, are operated by electricity. If there is no electricity, the furnace will not operate and there will be no heat. Many times the problem of a furnace failing to operate can be traced to a blown fuse or a tripped circuit breaker caused by an electrical surge in the main power supply line. The fuses or circuit breakers protect the furnace (and other equipment on the circuit) from potentially damaging high voltages and currents by shutting off the power supply before the excess can harm the equipment. Replace a blown fuse or reset a tripped circuit breaker and restart the furnace. If the problem persists and you do not have the training or experience to make furnace repairs, call the local representative of the furnace manufacturer or a qualified HVAC technician for a service call.

Chapters 11 through 14 contain troubleshooting recommendations specific to gas, oil, electric, coal, wood, and multi-fuel furnaces, respectively.

Chapter 11

Gas Furnaces

This chapter is concerned primarily with introducing the basic procedures recommended for installing, operating, servicing, and repairing gas-fired forced-warm-air furnaces. Additional information about furnaces is contained in Chapter 10, "Furnace Fundamentals."

Only gas-fired furnaces approved by the American Gas Association (AGA) should be used in a heating installation. Not only does AGA certification ensure a certain standard of quality, it is often required by most local codes and ordinances.

Types of Gas Furnaces

There are a variety of different ways to classify residential gas furnaces. One method is to classify them by their shape, which is commonly determined by the direction the warm air flows out of the furnace into the spaces being heated. Furnace manufacturers manufacture upflow, downflow (counterflow), lowboy, and horizontal flow models for installation in attics, basements, closet spaces, or at floor level (Figures 11-1 through 11-3).

Furnaces are also available in different capacities. The furnace capacity (Btu output) is determined by the heating requirements of the structure—in other words, the amount of heat the furnace must provide to replace heat loss and maintain the desired comfort level.

Note

> It is very important to match the furnace capacity as closely as possible to the sizing requirements of the heating system. An oversized or undersized furnace will not heat efficiently.

Furnaces can also be distinguished by their heating efficiency. In recent years, the government and furnace manufacturers have made enormous strides in improving the heating efficiency of residential gas furnaces. Two widely used methods, the AFUE measurement and the Energy Star program of the United States Environmental Protection Agency (EPA), are described below. A furnace that meets the minimum efficiency requirements of these ratings is referred to as a high-efficiency gas furnace.

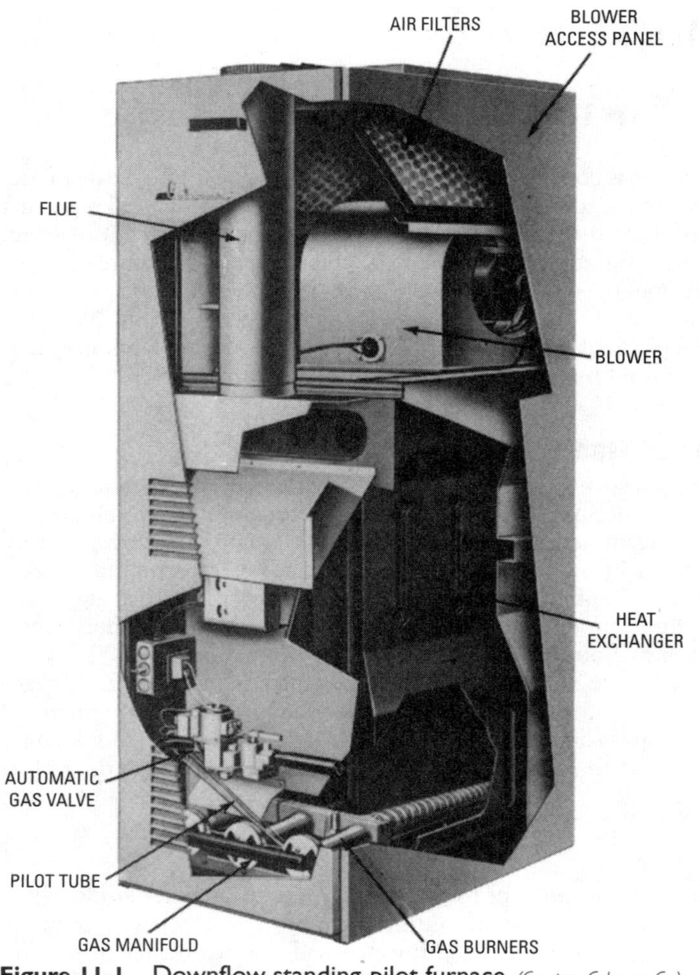

Figure 11-1 Downflow standing-pilot furnace. *(Courtesy Coleman Co.)*

Gas Furnace Energy Efficiency Ratings

AFUE Rating

The energy efficiency of a natural gas furnace is measured by its annual fuel utilization capacity (AFUE). The AFUE ratings for furnaces manufactured today are listed in the furnace manufacturer's literature. Look for the EnerGuide emblem for

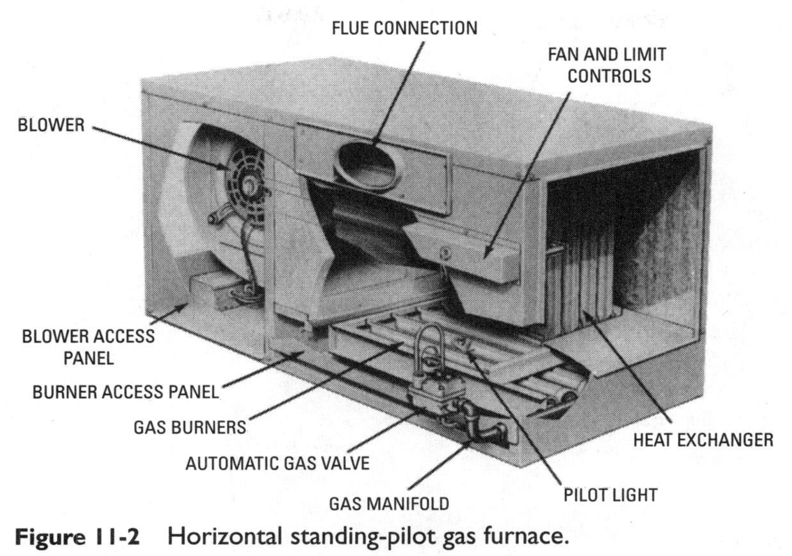

Figure 11-2 Horizontal standing-pilot gas furnace.
(Courtesy Lennox Industries Inc.)

the efficiency rating of that particular model. The higher the rating, the more efficient the furnace. The government has established a minimum rating for furnaces of 78 percent. Mid-efficiency furnaces have AFUE ratings ranging from 78 percent to 82 percent. High-efficiency furnaces have AFUE ratings ranging from 88 percent to 97 percent. Traditional standing-pilot gas furnaces have AFUE ratings of approximately 60 to 65 percent.

Energy Star Certification

Energy Star is an energy performance rating system created in 1992 by the U.S. Environmental Protection Agency (EPA) to identify and certify certain energy-efficient appliances. The goal is to give special recognition to companies that manufacture products that help reduce greenhouse gas emissions. This voluntary labeling program was expanded by 1995 to include furnaces, boilers, heat pumps, and other HVAC equipment. Both the Energy Star label and an AFUE rating above 80 percent will identify the gas furnace as an energy-efficient appliance.

Finally, gas furnaces can be classified according to the method used to light the gas burner. The traditional method was to use a

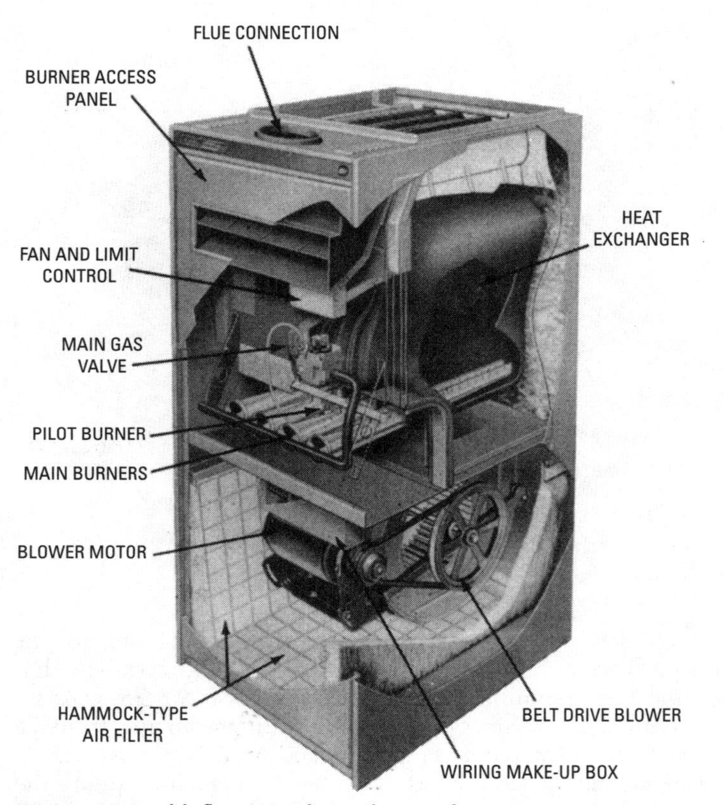

Figure 11-3 Upflow standing-pilot gas furnace. *(Courtesy Lennox Industries Inc.)*

pilot light, which was always burning. These types of furnace are called standing-pilot furnaces. They are gradually being replaced by the more energy efficient electronic ignition furnaces (see *High-Efficiency Furnaces*).

Standing-Pilot Gas Furnaces

The majority of residential gas furnaces in use today are still the conventional standing-pilot types, although they are gradually being replaced by the more efficient mid-efficiency and high-efficiency furnaces (Figures 11-2 and 11-3).

A conventional standing-pilot gas furnace consists of a naturally aspirating gas burner, draft hood, a solenoid-operated main gas valve, a continuously operating pilot light, a thermocouple safety device, a 24-volt AC transformer, a heat exchanger, a blower and motor assembly, and one or more air filters. The furnace is vented through a flue connection to a chimney.

The main gas valve used on a conventional standing-pilot gas furnace has a valve knob with a pilot position for lighting the pilot, and both a pilot tube and a thermocouple connection. A transformer provides 24-volt AC power to operate the main gas valve. When the pilot light is lit, the thermocouple sends an electrical current to hold the pilot valve open. If the pilot goes out, the thermocouple closes the valve.

Some standing-pilot gas furnaces use a millivolt generator instead of a 24-volt transformer to open the main gas valve. The main gas valve in these furnaces is specifically designed for use with the millivolt generator.

The energy efficiency of conventional gas furnaces has been improved by equipping them with a direct-drive blower and motor. Adding the direct-drive blower and motor creates a dual-capacity, variable-speed furnace that can match the heat output and circulating fan speed of the furnace to the actual heating requirements of the structure. The furnace is able to maintain the desired temperature comfort levels while, at the same time, saving energy and reducing energy costs.

Mid-Efficiency Gas Furnaces

Mid-efficiency gas furnaces are equipped with a naturally aspirating gas burner and a pilot light, but the pilot does not run continuously (Figure 11-4). The pilot is shut off when the furnace is idle (that is, when the thermostat is not calling for heat).

These furnaces have a more efficiently designed heat exchanger than the one found in conventional furnaces. The heat exchanger provides greater resistance to the flow of the combustion gases up through the flue and chimney. This allows the slowed gases to cool and condense. The heat normally trapped in these gases can be extracted from the condensate instead of being lost up the chimney. There is no draft hood as in a conventional standing-pilot gas furnace. It is replaced by a small fan in the flue pipe. The induced draft produced by the fan overcomes the cooler (heavier) exhaust gases and the resistance created by the more efficient heat exchanger. Furnaces equipped with these fans are sometimes called induced-draft furnaces. Other components (automatic controls, blower and blower motor assembly, venting, and so on) are the same as those found in a conventional standing-pilot furnace. Mid-efficiency gas furnaces are approximately 20 percent more energy efficient than a conventional gas furnace. The AFUE ratings are in the 78 to 82 percent range.

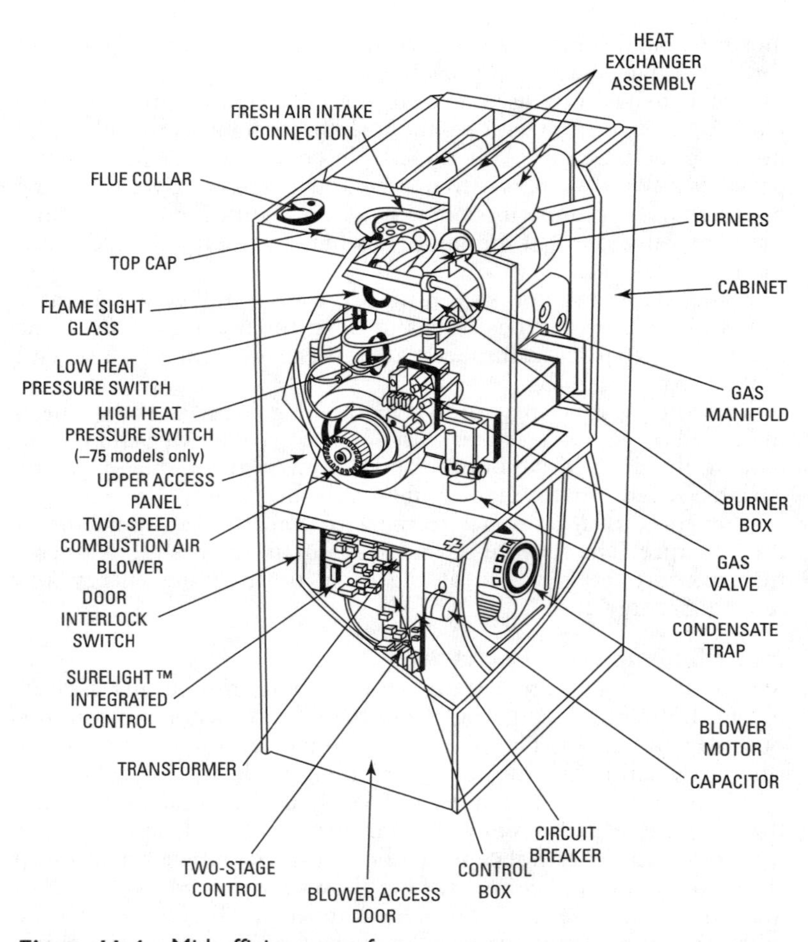

Figure 11-4 Mid-efficiency gas furnace. *(Courtesy Lennox Industries Inc.)*

Note

Mid-efficiency furnaces with AFUE ratings higher than 85 percent often experience condensation problems in the furnace and the exhaust venting.

Some manufacturers of mid-efficiency furnaces install a motorized damper in the exhaust flue pipe. The damper automatically closes when the furnace is idle to prevent heat from escaping up the chimney.

High-Efficiency Gas Furnaces

High-efficiency gas furnaces (also sometimes called condensing furnaces or electronic ignition furnaces) have AFUE ratings of 90 percent or more (Figure 11-5). The ignition process in these furnaces is controlled by a solid-state circuit board. There is no continuously burning pilot flame.

High-efficiency furnaces are equipped with two and sometimes three heat exchangers to extract the maximum amount of heat from the flue gases. The condensate resulting from the heat extraction process is then passed out of the bottom of the furnace to a nearby drain. The flue gas temperatures are low enough for the remaining gases to vent outdoors through a narrow PVC pipe.

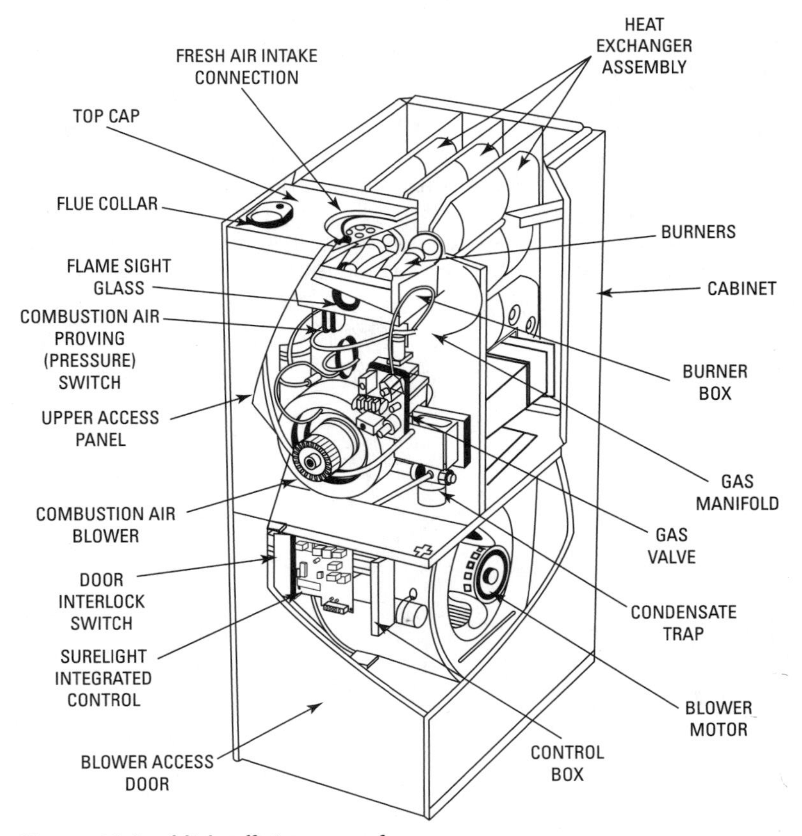

Figure 11-5 High-efficiency gas furnace. *(Courtesy Lennox Industries Inc.)*

There is no need to vent the gases up a chimney. There are two types of high-efficiency furnace: the intermittent-pilot furnace and the hot-surface ignition (HSI) furnace.

Intermittent-Pilot Furnace

An intermittent-pilot furnace (also sometimes called a spark ignition furnace) has many of the same components as a conventional standing-pilot type, including a pilot light assembly, except for a more efficient heat exchanger and an electronic ignition system. When the thermostat setting calls for heat, there begins a short ignition trial period during which a high-voltage spark is generated. The spark ignites the pilot. If the attempt to light the pilot was successful, the pilot flame must be proven through the flame rectification process. If the flame is proven, the solid-state circuit board in the furnace control box sends a signal to the main gas valve, opening it to permit gas to flow to the burner (Figure 11-6). The pilot flame then lights the gas burner, which continues to burn until the heat has reached the desired level inside the structure, whereupon the thermostat signals the electronic module or circuit board to stop the ignition process and shut off the pilot and burner.

Some intermittent-pilot furnaces use a pilot relight module and a mercury flame sensor instead of a control module to operate the pilot and gas valve. The gas valve is identical to the one used in fur-

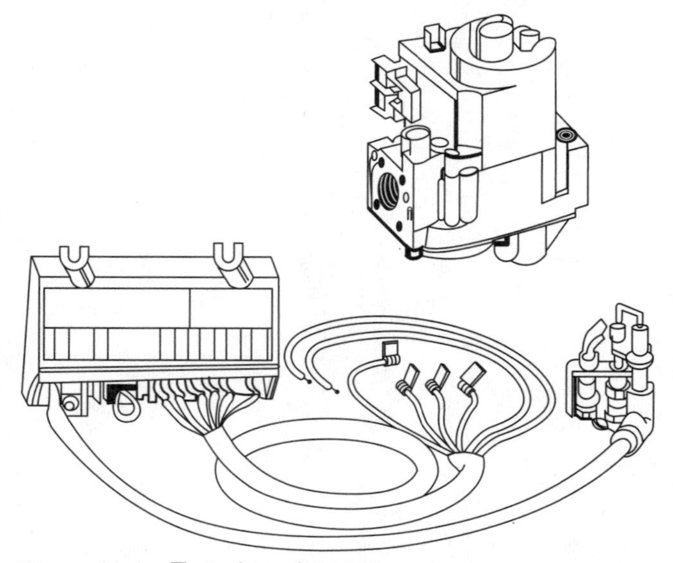

Figure 11-6 Typical spark ignition system. *(Courtesy Lennox Industries Inc.)*

naces with solid-state control modules. These furnaces are sometimes called cycle pilot furnaces.

Hot-Surface Ignition (HSI) Furnace

The hot-surface ignition (HSI) furnace is an electronically controlled furnace with two or more heat exchangers. An electronic ignition device called a hot surface igniter (also sometimes called a glow stick or glow plug) is used instead of a pilot light to start the gas burner (Figure 11-7). The tip of the igniter is positioned directly above the gas burner opening. When the thermostat setting calls for heat, the furnace begins a sequence of pre-ignition steps including a purge cycle. After the purge cycle, an electric current is sent to the igniter and heats the surface of the element until it reaches a high enough temperature to ignite the burner. At that point, the main gas valve is opened and gas flows to the burner and is ignited. A flame-sensing rod connected to a solid-state control module is used to detect (prove) the gas flame at the burner nozzle through flame rectification. If the flame can be proven, the electrical current to the igniter is shut off. In these furnaces, the solid-state ignition unit provides direct ignition and burner control.

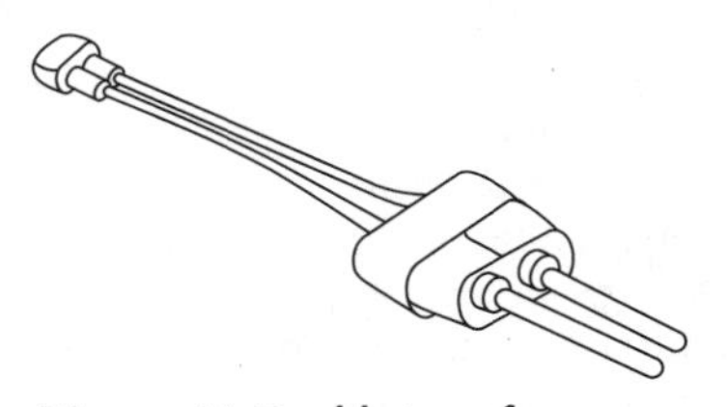

Figure 11-7 Hot-surface ignition igniter. *(Courtesy Lennox Industries Inc.)*

Note

> Some of these furnaces ignite the gas with a high-voltage direct spark instead of a red-hot igniter element. These furnaces are sometimes called direct-spark furnaces. All the other components are the same as those used in a conventional HSI furnace.

Gas Furnace Components

This section contains a brief description of the principal components of a gas-fired, forced-air furnace. Additional and more detailed information about these components is found in Chapter 2, "Gas Burners"; Chapter 4, "Thermostats and Humidistats"; Chapter 5, "Oil and Gas Controls"; and Chapter 6, "Other Automatic Controls" in Volume 2. The basic furnace components are as follows:

1. Furnace controls
2. Heat exchanger
3. Gas burners

4. Gas pilot assembly
5. Blower and motor
6. Air filter(s)

Furnace Controls

Each gas furnace is equipped with a variety of different controls used to ensure its safe and efficient operation. Not every gas furnace will have all of the controls listed here. For example, a thermocouple is a safety device used in standing-pilot gas furnaces, but not in those equipped with electronic ignition systems. Check the furnace specifications and manuals for the controls used with a specific model. Detailed descriptions of these devices are covered in Chapter 2, "Gas Burners"; Chapter 4, "Thermostats and Humidistats"; Chapter 5, "Gas and Oil Controls"; and Chapter 6, "Other Automatic Controls."

Thermostat

A room thermostat controls the operation of the furnace (Figure 11-8). The thermostat senses air-temperature changes in the space or spaces being heated and sends an electrical signal to open or close the automatic main gas valve. The thermostat is generally wired in series with the pilot safety valve, the automatic main gas valve, and

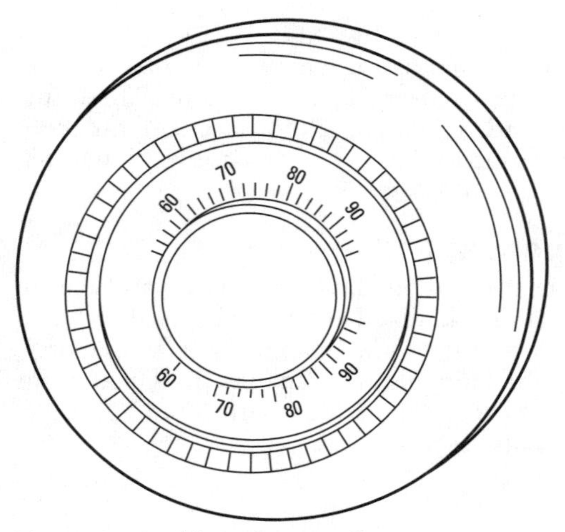

Figure 11-8 Typical room thermostat.

the limit control. See Chapter 4 of Volume 2, "Thermostats and Humidistats" for a detailed description.

Main Gas Valve

The principal function of the main gas valve is to control the flow of gas from the outside supply line (natural gas) or storage tank (propane) to the burner manifold (in the case of a burner assembly) or to the gas burner. These valves are manufactured in a variety of different shapes and sizes. They are commonly located behind the front panel of the furnace. Several different types of gas valves are illustrated in Figure 11-9. These valves are often combined with

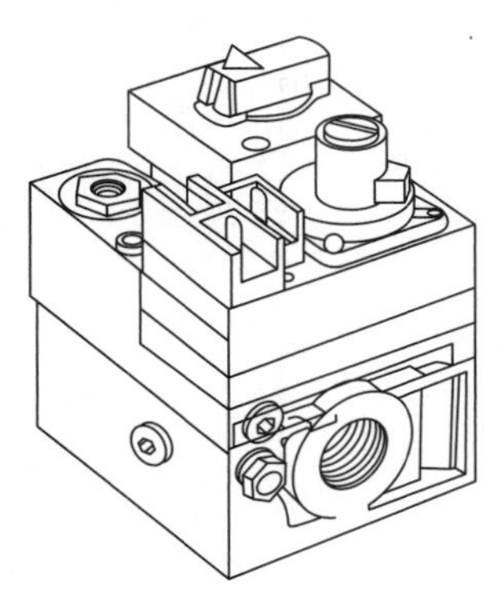

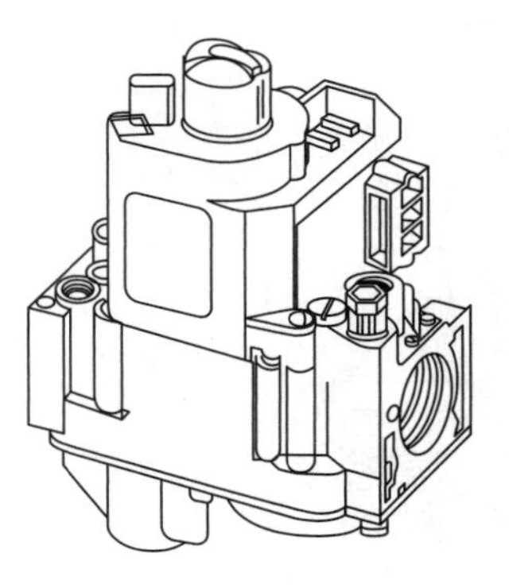

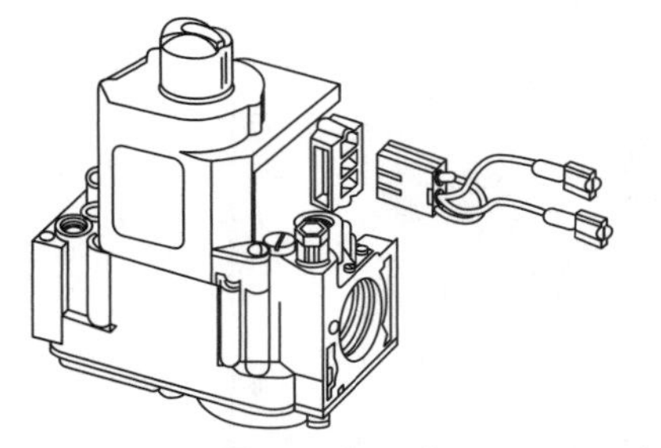

Figure 11-9 Examples of main gas valves.

safety shutoff devices that turn off both the main gas and the pilot gas, and/or pressure regulators that regulate the flow of gas. Such valves are commonly called combination gas valves, combination main gas valves, and intermittent-pilot dual-valve combination gas valves. The main gas valves in modern furnaces where there is no standing-pilot are sometimes called electronic ignition gas valves. The various types of gas valves are covered in greater detail in Chapter 6, "Gas and Oil Valves" of Volume 2.

Thermocouple

A thermocouple is a heat-sensing safety device used in a standing-pilot gas furnace to determine whether the pilot is lit before the main gas valve is opened to supply gas to the burners (Figure 11-10). The heat of the pilot flame is converted by the thermocouple into an electric current. The current is strong enough to open the pilot portion of the main gas valve, which then supplies gas to the pilot light. If the thermocouple does not detect a pilot flame, it will turn off the gas supply to the pilot. The main gas valve is operated by a current from a 24-volt AC transformer. The thermocouple is described in greater detail in Chapter 5, "Gas and Oil Controls" in Volume 2.

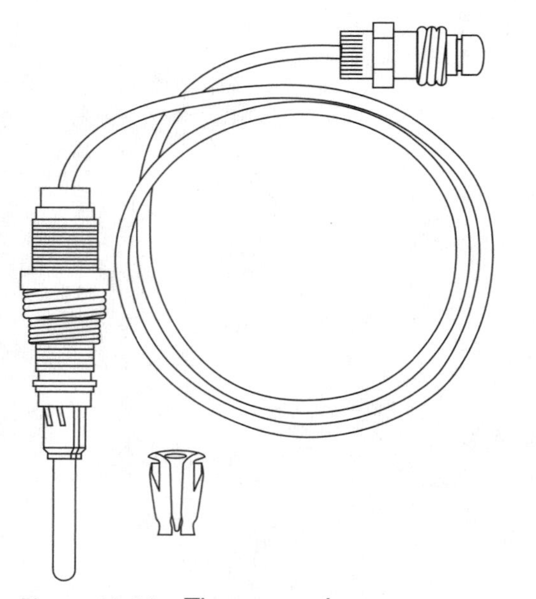

Figure 11-10 Thermocouple. *(Courtesy Lennox Industries Inc.)*

Thermopile

A thermopile (sometimes called a thermopile generator or a millivolt generator) is a pilot flame, heat-sensing safety device used in some standing-pilot gas furnaces instead of a thermocouple (Figure 11-11). It is larger than a thermocouple, delivers more electricity, and operates both the main gas valve and the pilot light. No transformer is required in gas furnaces equipped with a thermopile. The thermopile is described in greater detail in Chapter 5, "Gas and Oil Controls" in Volume 2.

Mercury Flame Sensor

A mercury flame sensor used in an electronic ignition system is illustrated in Figure 11-12. It consists of a mercury-filled sensor end, a capillary tube, and an SPDT switch assembly on the main gas valve. The sensor end, which is filled with mercury, is directly heated by the burner flame. The mercury vaporizes and expands, forcing the remaining nonvaporized mercury in its liquid state down the capillary tube to the switch on the gas valve. The weight of the liquid mercury triggers the switch and opens the gas valve.

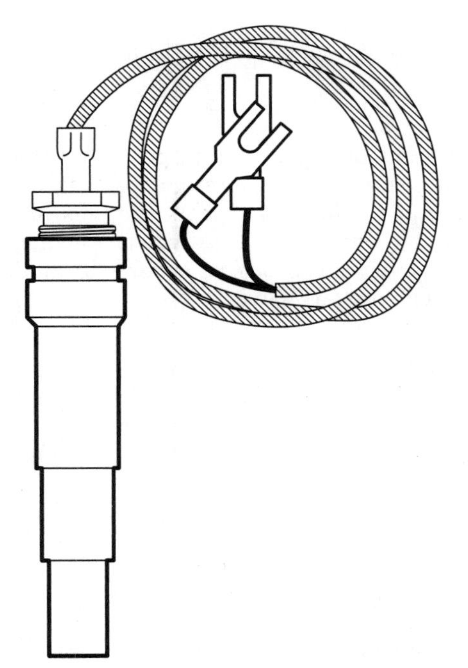

Figure 11-11 Thermopile.
(Courtesy Lennox Industries Inc.)

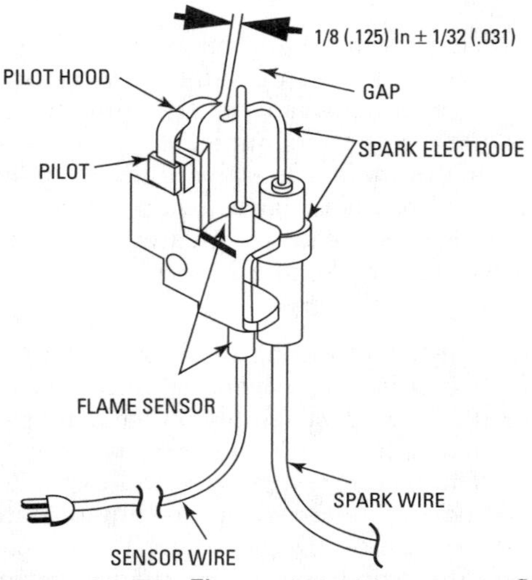

Figure 11-12 Electronic ignition system flame sensor.
(Courtesy Lennox Industries Inc.)

The mercury flame sensor is described in greater detail in Chapter 5, "Gas and Oil Controls" in Volume 2.

Gas Pressure Regulator

The gas in the supply main is frequently subjected to pressure variations. A gas pressure regulator ensures a constant gas pressure in the burner manifold. In a natural gas heating system, the pressure regulator is located in the furnace on the main gas valve. The gas pressure regulator in an LP (propane) gas heating system is located between the supply tank and the main gas valve.

Fan and Limit Control

The fan and limit control is a safety device that operates on a thermostatic principle (Figure 11-13). This combined control is designed to govern the operation of the furnace within a specified temperature range (usually 80 to 150°F), and is located in the furnace plenum, where it responds to outlet air temperatures.

One of the functions of this control is to prevent the furnace from overheating by shutting off the gas supply when the plenum air temperature exceeds the upper temperature setting (180°F) on the control. The limit-control switch (and pilot safety relay) responds to excessive plenum temperatures by closing the automatic main gas valve and thereby shutting off the flow of gas to the burner or burner assembly.

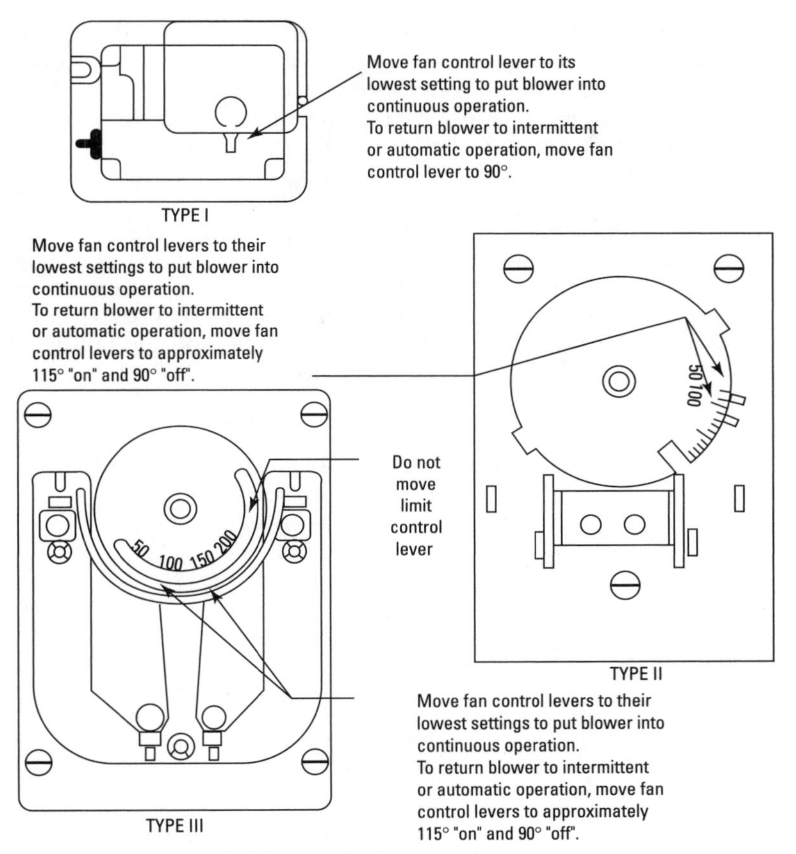

Figure 11-13 Typical fan and limit controls. *(Courtesy Lennox Industries Inc.)*

Another function of the fan and limit control is to stop the fan when the gas burner or burner assembly has been turned off. When the plenum air drops below the lower setting (80°F) on the control, the fan is automatically shut off. For additional information about fan and limit controls, read the appropriate sections in Chapter 6 of Volume 2, "Other Automatic Controls."

Heat Exchanger

A traditional standing-pilot gas furnace has a burner compartment containing the gas burner and pilot. The gas and air necessary for combustion are mixed in the venturi or mixing tube prior to ignition.

This mixture is then ignited by the pilot flame as it leaves the burner. The burner flame is encased in a combustion chamber or firebox located directly below the furnace heat exchanger. The heat of the combustion process is transferred to the metal walls of the heat exchanger and then to the air rising through it (Figure 11-14).

The air flowing through a heat exchanger is subject to temperatures that frequently approach the 700 to 800°F range. Because of these high temperatures, the airflow must be uniform across all sections of the heat exchanger (as shown in Figure 11-15) and must move at a sufficient rate of speed to keep these temperatures from building any higher. A low volume of air flowing across one section of the heat exchanger will result in overheating and the failure of the entire heat exchanger assembly (Figure 11-16).

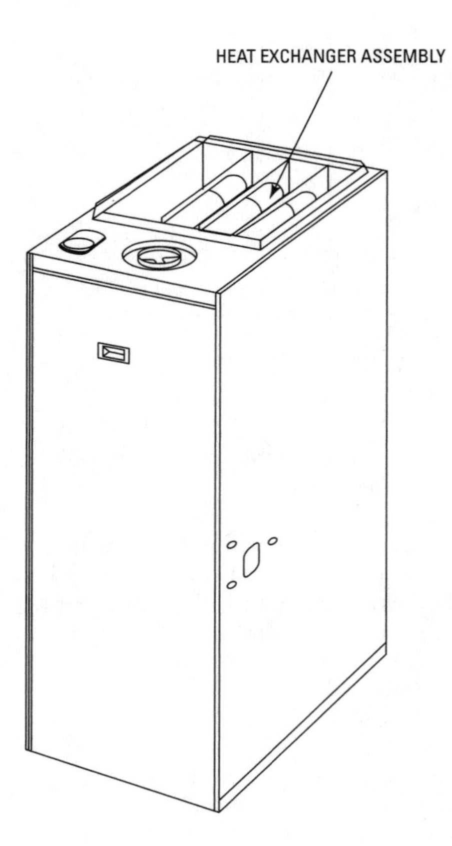

Figure 11-14 Heat exchangers. *(Courtesy Lennox Industries Inc.)*

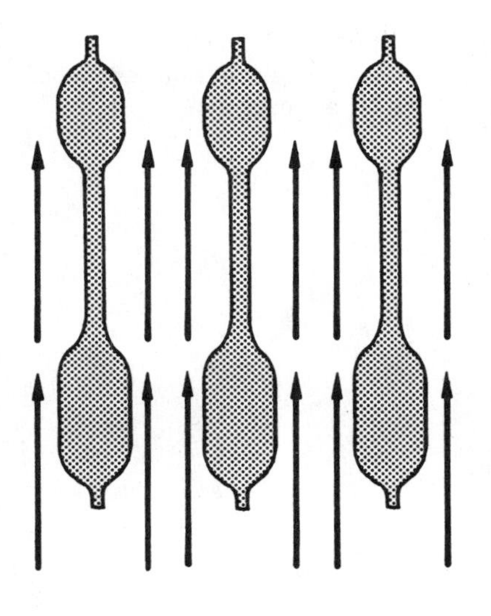

Figure 11-15 Uniform air flow across all sections of the heat exchanger. *(Courtesy Trane Co.)*

Overheating can also be caused by a portion of the gas flame impinging on the inner surface of the heat exchanger (Figure 11-17). This results in the generation of extremely high temperatures. Under normal operating conditions, the gas flame should be directed upward through the center of each section.

Another cause of heat exchanger failure is overfiring. This condition is usually traced to an excessively high manifold pressure or

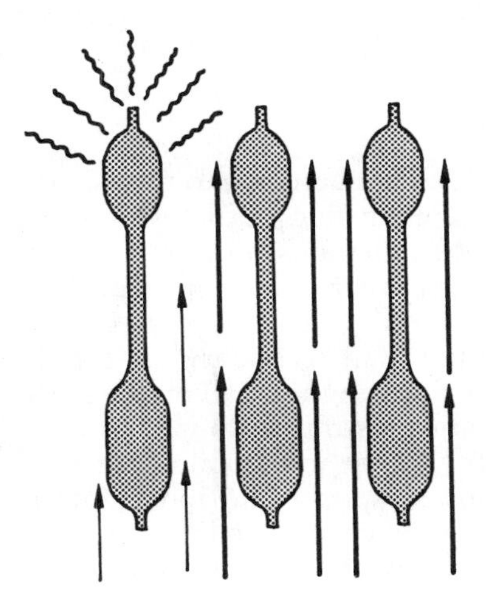

Figure 11-16 Low air flow across one heat exchanger section may result in premature failure due to overheating. *(Courtesy Trane Co.)*

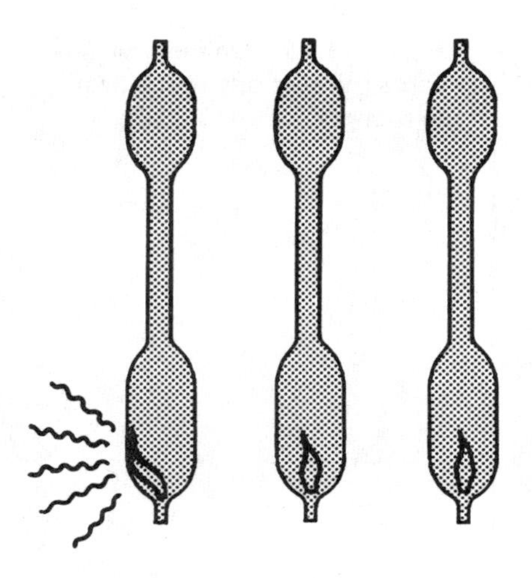

Figure 11-17 Flame impingement in one section of the heat exchanger.

an orifice that is too large. The result is a firing rate that exceeds the design limits.

If the heat exchanger becomes so covered with soot, carbon deposits, and other contaminants that it reduces the operating efficiency of the furnace, it should be thoroughly cleaned. Gaining access to the heat exchanger on some furnaces is not easy because it may require removing the burners and manifold assembly, vent connector, draft hood (diverter), and flue baffles. For this type of furnace, you are advised to call a local service repairman if you want to clean the heat exchanger. Many furnace manufacturers now provide convenient and accessible cleanout openings to make this maintenance chore easier.

Gas Burners

Gas burners and gas burner assemblies are designed to provide the proper mixture of gas and air for combustion purposes. The gas burners used in residential heating installations operate primarily on the same principle as the Bunsen burner. See Chapter 2, "Gas Burners" in Volume 2.

Gas furnaces require both primary and secondary air for the combustion process. The gas passes through the small orifice in the mixer head, which is shaped to produce a straight-flowing jet moving at high velocity (Figure 11-18). As the gas stream enters the throat of the mixing tube, it tends to spread and induce air in through the opening of the adjustable shutter. This is the primary air, which mixes with the gas before the air-gas mixture is forced

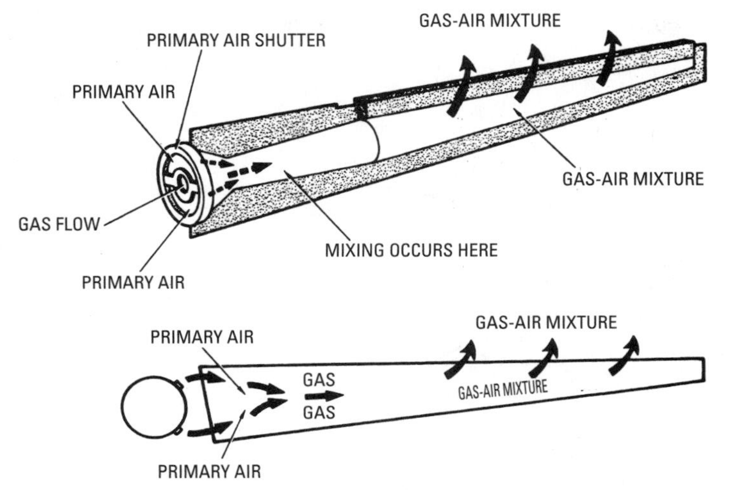

Figure 11-18 Schematic of air movements in a typical gas burner.

through the burner ports (Figure 11-18). The pressure in the gas stream forces the mixture through the mixing tube into the burner manifold casting, from which it issues through ports where additional air must be added to the flame to complete combustion. This secondary air supply flows into the heat exchanger and around the burner. Its function is to mix with unburned gas in the heat exchanger in order to complete the combustion process.

The primary air is admitted at a ration of about five parts primary air to one part gas for manufactured gas, and at a 10:1 ratio for natural gas. These ratios are generally used as theoretical values of air for purposes of complete combustion. Most burners used in gas-fired furnaces operate efficiently on 40 to 60 percent of the theoretical value. Primary air is regulated by means of an adjustable shutter. When burning natural gas, the air adjustment is generally made to secure as blue a flame as possible. Burner manufacturers provide a number of different ways to field adjust primary air. For example, shutters for primary air control are part of the main burners. Factory adjusted, angled orifices are also used on some burners.

The secondary air is drawn into the burner by natural draft. No provision is made by manufacturers to field adjust secondary air. Secondary air is controlled primarily by the design of the flue restrictors and draft diverter. The general configurations of the heat exchanger and burners are also important factors in secondary air

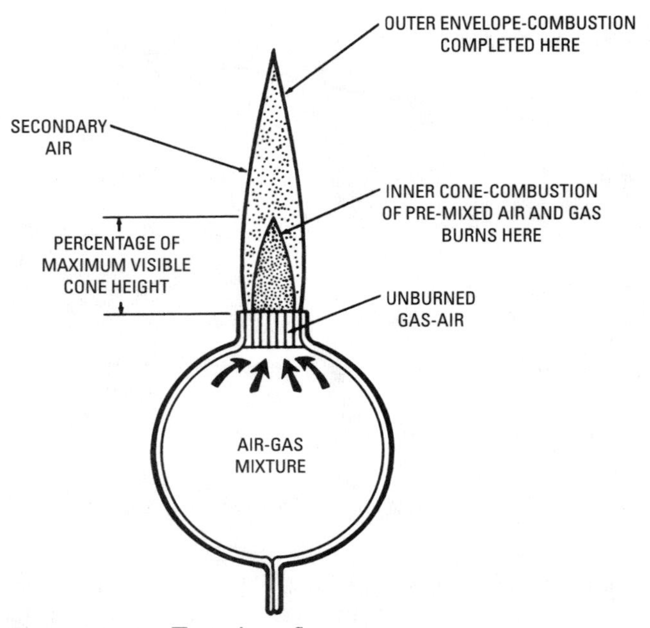

Figure 11-19 Typical gas flame.

control. Excess secondary air constitutes a loss and should be reduced to a proper minimum (usually not less than 25 to 35 percent).

As shown in Figure 11-19, a typical gas flame consists of a well-defined inner cone surrounded by an outer envelope. Combustion of the primary air and gas mixture occurs along the outer surface of the inner cone. Any unburned gases combine with the secondary air and complete the combustion process in the outer envelope of the flame.

As Figure 11-19 illustrates, the major portion of the inner cone consists of an unburned air-gas mixture. It is along the outer surface of the inner cone that combustion takes place. The height of the inner cone can be reduced by increasing the amount of primary air contained in the air-gas mixture. The extra oxygen increases the burning rate, which causes the inner cone to decrease in size.

A burner is generally considered to be properly adjusted when the height of the inner cone is approximately 70 percent of the maximum visible cone height. The gas should flow out of the burner ports fast enough so that the flame cannot travel or flash back into the burner head. The velocity must not be so high that it blows the flame away from the port(s).

Gas Pilot Assembly

The gas pilot assembly in a standing-pilot gas furnace consists of a pilot burner and a pilot safety device (Figure 11-20). The most commonly used pilot safety is the thermocouple lead operating in conjunction with a companion valve or switch. The thermocouple itself is actually a miniature generator that converts heat (from the pilot flame) into a small electrical current. The upper half of the thermocouple is located in the pilot flame. As long as the pilot flame is operating, an electrical circuit is maintained between the pilot burner and the gas valve. If the pilot flame goes out, the circuit between the thermocouple and the gas valve is broken. As a result,

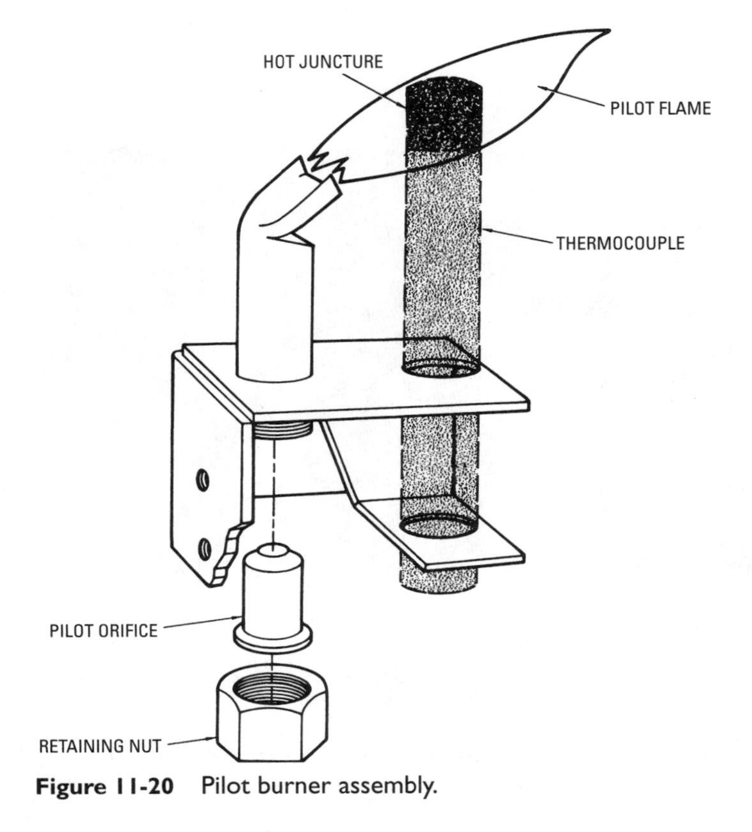

Figure 11-20 Pilot burner assembly.

the gas valve closes and shuts off the gas flow to the pilot burner. Other pilot safety devices used with gas-fired furnaces are described in Chapter 5 of Volume 2, "Gas and Oil Controls."

Smaller gas furnaces, such as those used in RV furnaces, mobile homes, and swimming pool heaters, may use a millivolt system instead of a thermocouple. The pilot heats the millivolt device until it generates enough electricity to operate the main gas valve. There is no need to incorporate a 24-volt transformer, because with a millivolt system the burner will not operate if the pilot is not lit.

Many gas furnaces contain several individual heat exchanger sections with a gas burner inserted in each section. In this arrangement, all burners must fire almost simultaneously, or there will be danger of unburned gas flowing into the combustion chamber. Simultaneous ignition can be provided by installing a crossover burner across the top of the main burners near the port area. Figures 11-21 and 11-22 show the types of crossovers in Trane gas-fired furnaces.

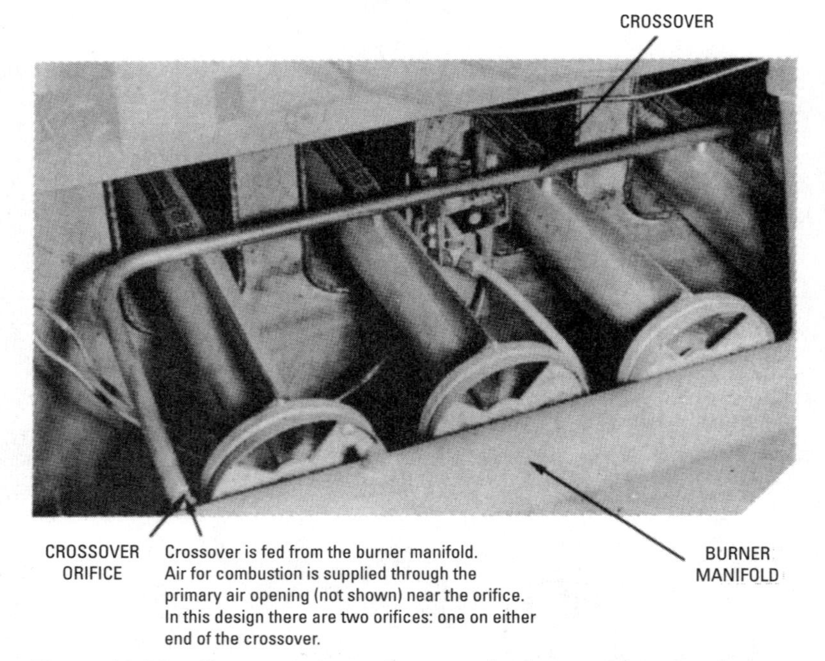

Figure 11-21 Crossover located perpendicular to and across the top of the main burners near the port area. *(Courtesy Trane Co.)*

Figure 11-22 Crossover arrangement in which each burner feeds a portion of the gas-air mixture into the crossover, where it is ignited by the pilot flame as it leaves the crossover part. *(Courtesy Trane Co.)*

High-efficiency gas furnaces do not have a continuously burning pilot light. In these furnaces, an electronic pilot assembly takes the place of the standing-pilot and thermocouple. These pilot assemblies contain a pilot, a spark ignition, and a flame sensor (Figure 11-23).

Blower and Motor

Forced-warm-air furnaces are equipped with variable-speed blowers designed for continuous duty. Blower motors are available as either direct-drive or belt-driven units (Figures 11-24 and 11-25). Most modern furnaces are equipped with direct-drive motors. See the section *Blowers and Motors* in this chapter for a detailed description.

Air Filter

A forced-warm-air, gas-fired furnace is supplied with either a disposable air filter or a permanent (washable) one. Typical gas furnace air filters are illustrated in Figure 11-26. See the section *Air Filters* in this chapter for a detailed description.

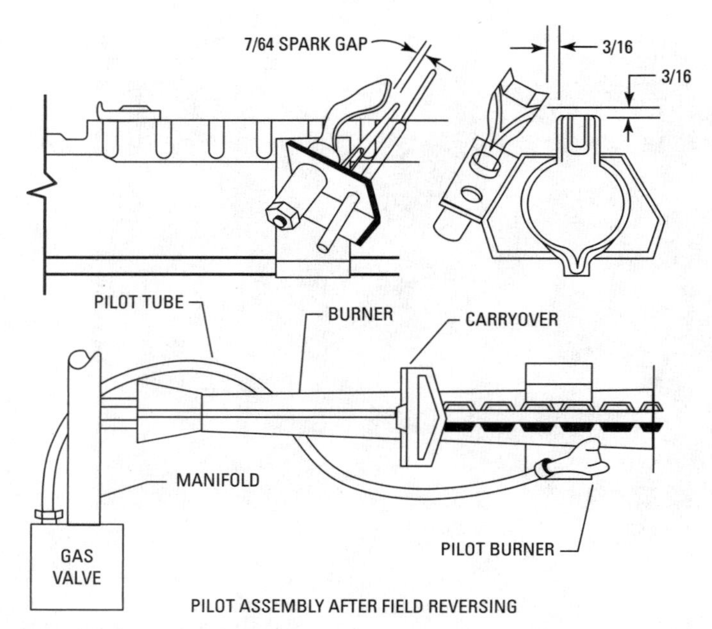

Figure 11-23 Electronic ignition pilot assembly. *(Courtesy Lennox Industries Inc.)*

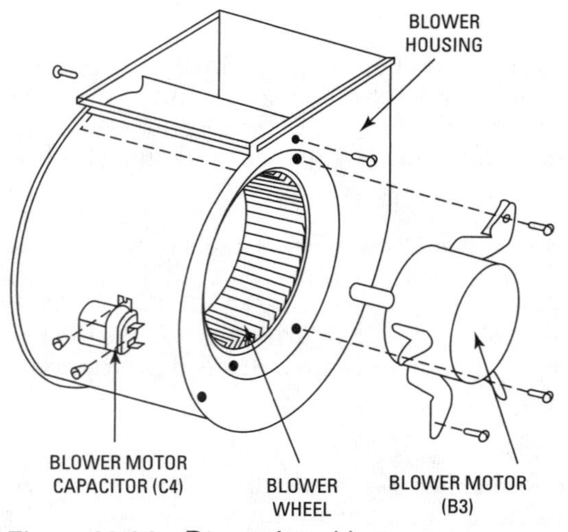

Figure 11-24 Direct-drive blower. *(Courtesy Lennox Industries Inc.)*

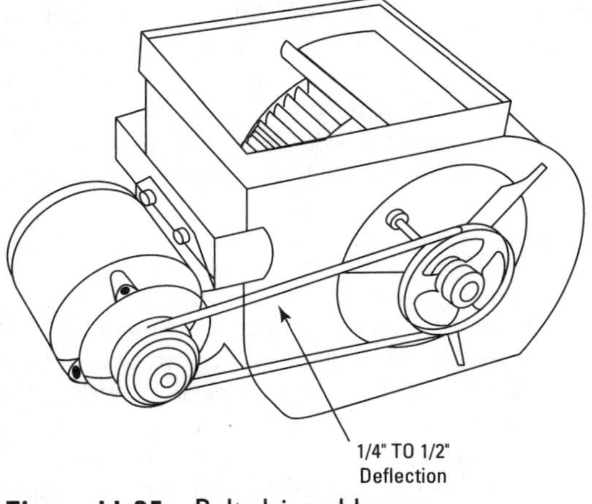

Figure 11-25 Belt-driven blower.
(Courtesy Lennox Industries, Inc.)

The first step in planning a heating system is to calculate the maximum heat loss for the structure. This should be done in accordance with procedures described in the manuals published by the Air Conditioning Contractors of America (ACCA), especially their *Manual J, Residential Load Calculations,* or by a comparable method. Correctly calculating the maximum heat loss for

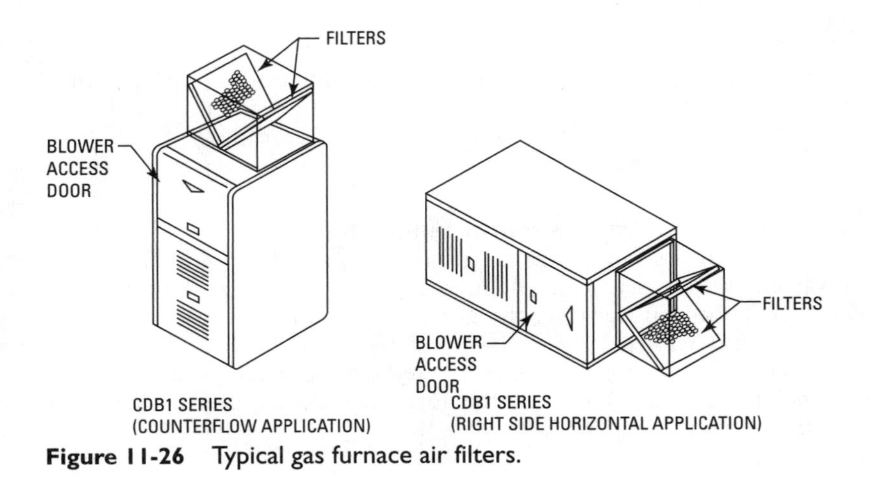

Figure 11-26 Typical gas furnace air filters.

the structure is very important because the data will be used to determine the size (capacity) of the furnace selected for the heating system. The ACCA's *Manual J* contains the most accurate calculation methods for sizing heating and air conditioning equipment. Computer programs and worksheets designed to simplify the calculation procedures found in *Manual J* are also available for purchase.

Note

High-efficiency (condensing) gas furnaces must be correctly sized. Oversizing will not lead to increased energy costs, as is the case with the traditional standing-pilot gas furnaces, but it will cause short firing cycles. If the furnace operates with short firing cycles, it will never get hot enough to dry out the condensate extracted from the combustion process. The condensate accumulates, becomes gradually more and more acidic, and begins to corrode parts of the furnace. Oversizing high-efficiency furnaces may also result in uncomfortable temperature swings in the living areas and excessive air flow from the room registers. Oversized furnaces require larger blowers than correctly sized ones.

Caution

A gas furnace must be properly set up and installed by a certified HVAC gas heat technician, a representative of the gas furnace manufacturer, or someone with equivalent experience.

Warning

If the heating or heating/cooling installation is to be approved by either the FHA or VA, heat loss and heat gain calculations must be made in accordance with the procedures described in the ACCA's *Manual J*.

Some Installation Recommendations

New furnaces for residential installations are shipped preassembled from the factory with all internal wiring completed. In order to install the new furnace, the gas piping from the supply main, the electrical service from the line voltage main, and the low-voltage thermostat must be connected. Directions for making these connections are to be found in the furnace manufacturer's installation instructions.

Caution

> A gas furnace must be properly set up and installed by a certified gas heat technician, a representative of the gas furnace manufacturer, or an HVAC professional with equivalent experience.

Make certain you have familiarized yourself with all local codes and regulations that govern the installation of a gas-fired furnace. Local codes and regulations take precedence over national standards. Your furnace installation must comply with the local codes and regulations.

Always check a new furnace for damages as soon as it arrives from the factory. If shipping damages are found, the carrier (not the factory) should be notified and a claim filed immediately. Heating equipment is shipped FOB from the factory and it is the responsibility of the carrier to see that it arrives undamaged.

Always place the furnace on a solid and level base. Doing so will reduce vibrations from the equipment and keep operating noise to a minimum. In a crawl-space installation, the furnace is either supported on a slab or on concrete blocks, or it is suspended from floor joists with ⅜-in hanger rods.

A furnace installed in an attic should be placed on a fiberboard sheeting base to absorb vibrations. Many furnaces designed for attic installations are certified for installation directly on combustible material such as ceiling joists or attic floors. In an attic installation, the furnace should be placed over a load-bearing partition for additional support.

Locations and Clearances

The proper location of a gas furnace depends on a careful consideration of the following three factors:

1. Length of heat runs
2. Chimney location
3. Clearances

A gas forced-air furnace should be located as near as possible to the center of the heat-distribution system and chimney. Centralizing the furnace reduces the need for one or more long supply ducts, which tend to lose a certain amount of heat. The number of elbows should also be kept to a minimum for the same reason. Installing the furnace near the chimney will reduce the length of the horizontal run of flue pipe required for the traditional, standing-pilot type gas furnace. The flue pipe should always be kept as short as possible.

Sufficient clearance should be provided for access to the draft hood and flue pipe. It is particularly important to locate a gas-fired furnace so that the draft hood is at least 6 in from any combustible material.

Allowing access for lighting the furnace and servicing it is also an important consideration. Most furnace manufacturers recommend a clearance of 24 to 30 inches in front of the unit for access to the burner and controls.

Clearances between the furnace and any combustible materials are usually provided by the manufacturer in the furnace specifications.

Wiring

All of the electrical wiring inside the furnace is completed and inspected at the factory before the unit is shipped. Wiring instructions will accompany each furnace and the internal (factory installed) wiring will be clearly marked. The wiring to controls and to the electrical power supply must be completed by the installer, and the wiring should be done in accordance with the diagram supplied by the manufacturer. This wiring is usually indicated on wiring diagrams by broken lines.

All wiring connected to the furnace must comply with the latest edition of the *National Electric Code* and any local codes and ordinances. Local codes and ordinances will always take precedence over national ones when there is any conflict. In order to validate the furnace warranty, many manufacturers require that a local electrical authority approve all electrical service and connections.

The two types of electrical connections required to field wire a furnace are (1) line-voltage field wiring, and (2) control-voltage field wiring. Both are illustrated in the wiring diagram for the gas furnace shown in Figure 11-27.

The line-voltage wiring connects the furnace to the building power supply. It runs directly from the building power panel to a fused disconnect switch. From there, the wiring runs to terminals L1 and L2 on the power-supply terminal block (that is, the furnace junction box).

The unit must be properly grounded either by attaching a ground conduit for the supply conductors or by connecting a separate wire from the furnace ground lug to a suitable ground. The furnace must be electrically grounded in accordance with the *National Electrical Code* (ANSI-CI-1971).

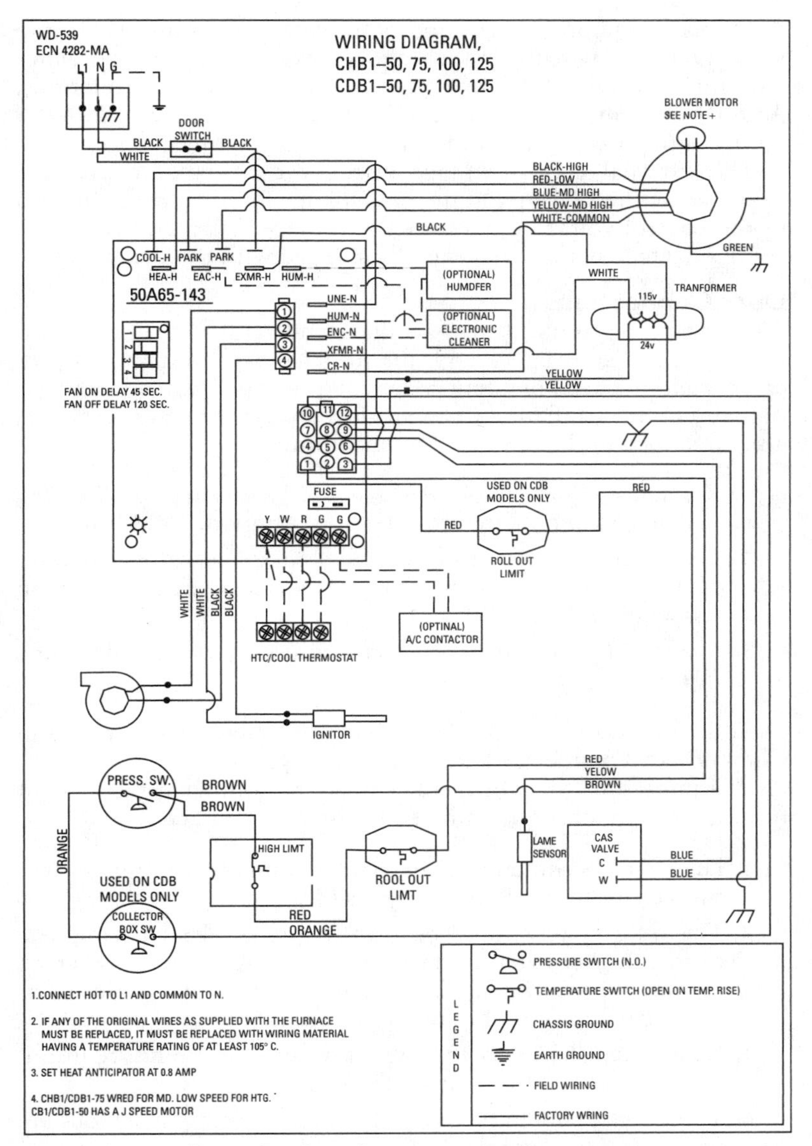

Figure 11-27 Example of gas furnace wiring diagram.

Furnaces equipped with motors in excess of ¾-hp or 12-amp ratings must be wired to a separate 240-volt service in accordance with the *National Electrical Code* and local codes and regulations. A 240-volt transformer should be installed whenever 240-volt service is required and/or a motor larger than ¾ hp is used.

The external control-voltage circuitry consists of the wiring between the thermostat and the low-voltage terminal block located in the control-voltage section of the furnace. Instructions for the control-voltage wiring are generally shipped with the thermostat.

Duct Connections

The warm-air plenum, warm-air duct, and return-air duct should be installed in accordance with the furnace manufacturer's specifications and the local building code requirements. Detailed information about the installation of an air-duct system for a gas furnace is also contained in the following publications:

1. *Residence Type Warm Air Heating and Air Conditioning System* (National Fire Protection Association, No. 90B)
2. *Installation of Air Conditioning and Ventilating Systems of Other than Residence Type* (National Fire Protection Association, No. 90A)
3. *Manual D, Residential Duct Systems* (Air Conditioning Contractors of America)

Additional information about duct connections can be found in Chapter 7 of Volume 2, "Ducts and Duct Systems."

It is important for you to remember the following facts about furnace duct connections and the air distribution ducts:

1. Duct lengths should be kept to a minimum by centralizing the furnace location as much as possible.
2. The duct system must be properly sized. Undersizing will cause a higher external static pressure than the one for which the furnace was designed. This may result in a noisy blower or insufficient air distribution.
3. Duct sizing should include an allowance for the future installation of air conditioning equipment.
4. The furnace should be leveled before any duct connections are made.
5. Seal around the base of the furnace with a caulking compound to prevent air leakage if a bottom air return is used.

6. Ducts should be connected to the furnace plenums with tight fittings.
7. Warm-air-plenum and return-air-plenum connections should be the same size as the openings on the furnace.
8. Use tapered fittings or starter collars between the ducts and the furnace plenum.
9. Ductwork located in unconditioned spaces (for example, unheated attics, basements, and crawl spaces) should be insulated if air conditioning is planned for some future date.
10. Install canvas connectors between air plenums and the casing of the furnace to ensure quiet operation. Check local codes and regulations to make certain canvas connectors are in compliance.
11. Line the first 10 ft or so of the supply and return ducts with acoustical material when extremely quiet operation is necessary.
12. Install locking-type dampers in each warm-air run to facilitate balancing the system.

Ventilation and Combustion Air

A gas furnace requires an unobstructed supply of air for combustion. In the older standing-pilot furnaces, this combustion air supply was commonly supplied through natural air infiltration (cracks in the walls and around windows and doors) or through ventilating ducts. Newer furnaces include a separate air combustion fan next to the burners to provide a sufficient supply of combustion air to the furnace. The houses are now too tightly constructed and well insulated for natural ventilation to provide enough air. These small air combustion fans or blowers are in addition to the large indoor blower on the furnace.

If a gas-fired furnace is located in an open area (basement or utility room) and the ventilation is relatively unrestricted, there should be a sufficient supply of air for combustion and draft hood dilution. On the other hand, if the heating unit is located in an enclosed furnace room or if normal air infiltration is effectively reduced by storm windows or doors, then certain provisions must be made to correct this situation. Figure 11-28 illustrates one type of modification that can be made to provide an adequate supply of combustion and ventilation air to a furnace room. As shown, two permanent grilles are installed in the walls of the furnace room, each of a size equal to 1 in^2 of free area per 1000 Btu per hour of burner output. One grille should be located approximately 6 in

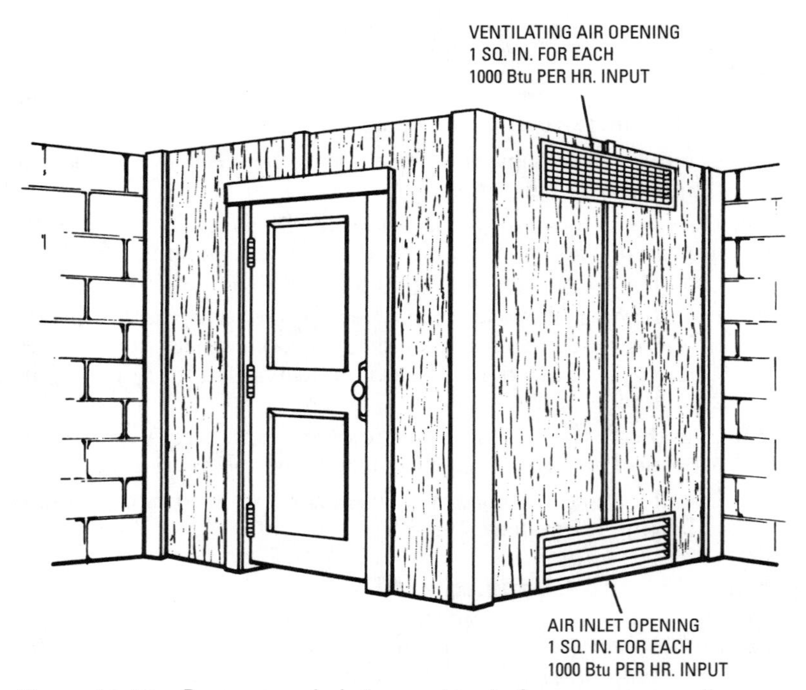

Figure 11-28 Recommended air openings in furnace room wall.

from the ceiling and the other one near the floor. A possible alternative is to provide openings in the door, ceiling, or floor as shown in Figure 11-29.

If the furnace is located in a tightly constructed building, it should be directly connected to an outside source of air. A permanently open grille sized for at least 1 in^2 of free area per 5000 Btu per hour of burner output should also be provided. Connection to an outside source of air is also recommended if the building contains a large exhaust fan.

Provisions for ventilation and combustion air are described in greater detail in *Installation of Gas Appliances and Gas Piping* (ASA Z21.30 and NFPA No. 54). These standards were adopted and approved by the National Fire Protection Association and the National Board of Fire Underwriters.

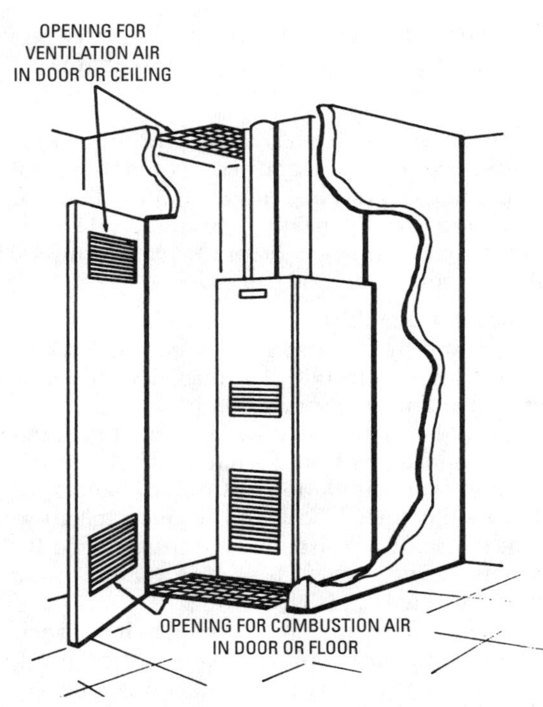

Figure 11-29 Alternative method of providing air supply openings.

Venting

Provisions must be made for venting the products of combustion to the outside in order to avoid contamination of the air in the living or working spaces of the structure. The four basic methods of venting the products of combustion to the outside are:

1. Masonry chimneys
2. Low-heat Type A prefabricated chimneys
3. Type B gas vents
4. Type C vents
5. Wall venting
6. PVC pipe venting

Masonry and prefabricated chimneys are described in later sections (see *Chimney* and *Chimney Troubleshooting* in this chapter).

Warning

Any furnace that produces heat from the combustion process (that is, burning gas, oil, coal, or wood) can potentially produce carbon monoxide gas as well. Carbon monoxide gas is odorless, invisible, and deadly toxic. The only safeguard against the accumulation of this gas before it reaches unsafe levels is to install a carbon monoxide detector.

Low-heat Type A flues are metal, prefabricated chimneys that have been tested and approved by Underwriters Laboratories, Inc. (UL). These chimneys are easier to install than masonry chimneys and are generally safer under abnormal firing conditions.

Type B gas vents (Figure 11-30) are listed by UL and are recommended for venting all standard gas-fired furnaces. They are not recommended for incinerators, combination gas-oil appliances, or appliances convertible to solid fuel. Check the furnace specifications for the type of vent to use. It will usually be AGA certified for use with a Type B gas vent. A Type C vent is used to vent gas-fired furnaces in attic installations (Figure 11-31).

Wall venting involves gas-fired appliances, which have their combustion process, combustion air supply, and combustion products isolated from the space that is being heated. These appliances are generally vented through the wall.

High-efficiency gas furnaces vent the gases and other byproducts of the combustion process through PVC (plastic) pipe, not the metal pipe used with standing-pilot gas furnaces. The PVC exhaust pipe extends vertically through the roof or horizontally through a sidewall to the outdoors. The latter is the typical connection to a replacement furnace.

The PVC exhaust pipe must be firmly and securely supported. Weak or inadequate support can cause the piping to sag. A sagging horizontal exhaust vent pipe will cause condensate (water) to collect in low spots. The condensate cannot drain back down into the furnace and eventually builds up to a point where it blocks the exhaust vent. A blocked exhaust vent will cause the furnace to shut down.

Note

Provide at least a ¼-inch per foot upward pitch for horizontal exhaust vent piping to allow the condensate to drain back down into the furnace. Never pitch the exhaust pipe downward toward the outdoors.

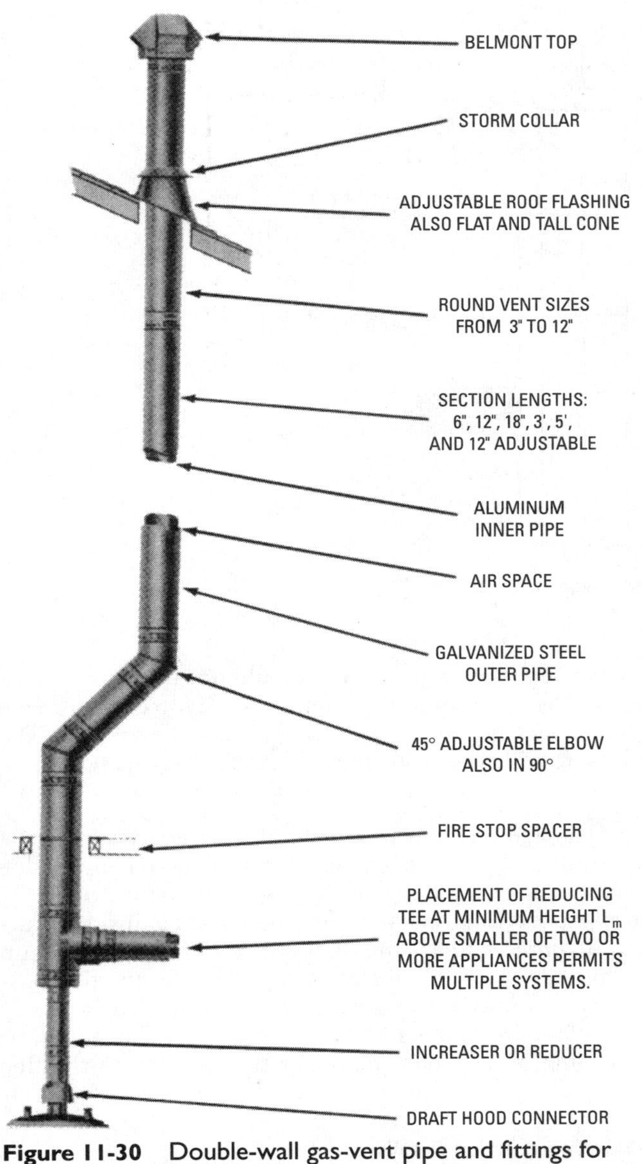

Figure 11-30 Double-wall gas-vent pipe and fittings for single or multiple Type B gas-vent system.

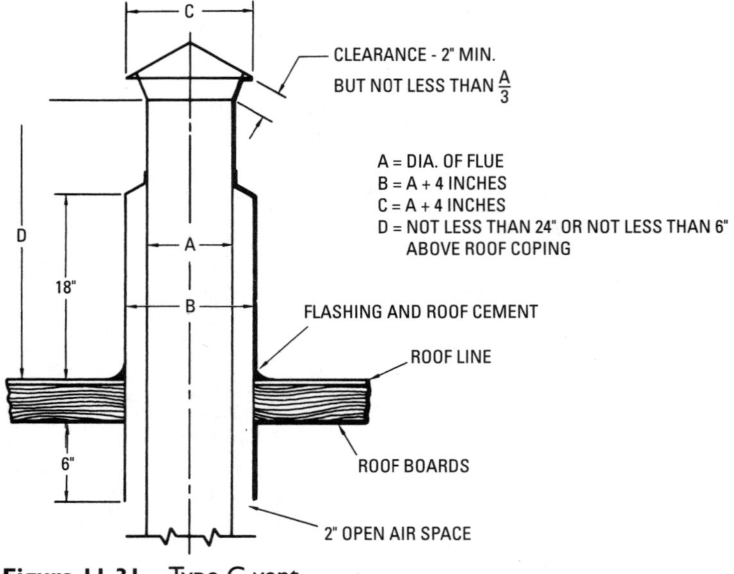

Figure 11-31 Type C vent.

High-efficiency (condensing) furnaces should not be vented into a chimney. The exhaust gases from these furnaces are too cool to create a chimney draft. Unable to rise, they condense inside the chimney. The trapped condensate will eventually damage the chimney materials.

Flue Pipe

The flue is the passage through which the flue gases pass from the combustion chamber of the furnace (or boiler) to the outside. A flue is also referred to as a flue pipe, vent pipe, or vent connector.

The term *appliance flue* refers to the flue passages inside a furnace or boiler. A *chimney flue* is the vertical flue passage running up through the chimney. A *vent connector* is specifically the flue passage between the heating unit and the chimney. It is also variously referred to as a *chimney connector* or *smoke pipe*. A *flue outlet,* or *vent,* is the opening in a furnace or boiler through which the flue gases pass.

Conventional Flue Construction Details

The flue pipe (vent pipe) connects the smoke outlet of the furnace with the chimney. It should never extend beyond the inner liner of the chimney and should never be connected to the flue of an open fireplace. Furthermore, flue connections from two or more sources should never enter the chimney at the same level from opposing sides.

The horizontal run of flue pipe should be pitched toward the chimney with a rise of at least ¼ in per running foot. Some furnace manufacturers recommend as much as a 1-in minimum rise per running foot. Do not allow the horizontal run to exceed, in length, 75 percent of the vertical run. Vent pipe crossovers in an attic must not extend at an angle of less than 60° from the vertical.

At intervals, fasten the horizontal run of flue pipe securely with sheet-metal screws and support the pipe with straps or hangers to prevent sagging. Sagging can cause cracks to develop at the joints, which may result in releasing toxic flue gases into the living and working spaces of the structure.

The point at which the flue pipe enters the chimney should be at least 2 ft above the cleanout opening of the chimney (Figure 11-32). The flue pipe must be the same size as the outlet of the flue collector

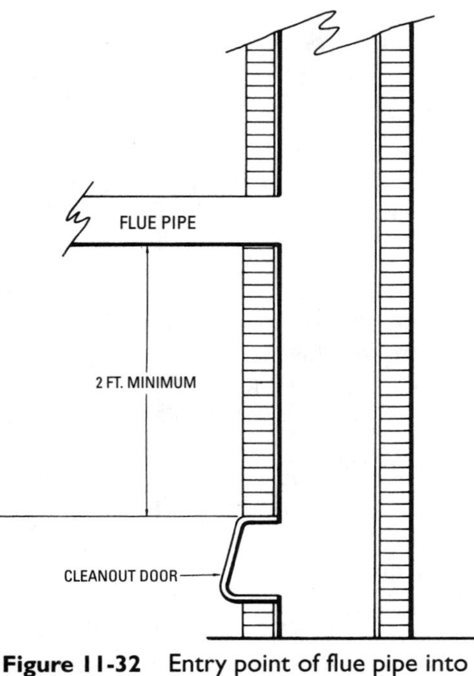

Figure 11-32 Entry point of flue pipe into chimney. *(Courtesy Dunham-Bush)*

(furnace flue collar). Never install a damper in a flue pipe or reduce the size of the flue pipe. Run the flue pipe from the draft hood to the chimney in as short a distance as possible. All joints in the flue pipe should be made with the length nearest the furnace overlapping the other.

If excessive condensation is encountered, install a drip tee in place of an elbow. Never use dampers or other types of restrictors in the vent pipe. A minimum distance should be maintained between the vent pipe and the nearest combustible material.

When more than one appliance is vented into a common flue, the area of the common flue should be equal to the area of the largest flue plus 50 percent of the area of the additional flue.

PVC Vent Construction Details

High-efficiency furnaces vent the byproducts of the combustion process through horizontal PVC piping extending from the furnace to an opening in an exterior sidewall. PVC piping is also used to bring combustion air from the outdoors to the furnace. Vent outlet and combustion air inlet details are illustrated in Figure 11-33. Figure 11-34 illustrates typical terminations for PVC piping on the outside of the sidewall. Note that the maximum wall thickness through which both the vent and combustion air pipes can pass is 18 inches. The minimum thickness is 2 inches.

Carefully follow the local codes and ordinances plus the furnace manufacturer's requirements when installing PVC vent and air intake piping. Note the following requirements:

- Install the bottom edge of the vent outlet termination elbow a minimum of 12 inches above the outlet of the combustion air termination elbow, as shown in Figure 11-32.
- Maintain a minimum horizontal distance of 4 ft between the outlet and inlet vents and any electric meters, gas meters, regulators, and relief devices.
- Insulate PVC vent pipe passing through an unheated space with 1.0-in-thick foil faced with fiberglass insulation.
- Cut PVC pipe at right angles with a hacksaw. Remove all burrs before installing the pipe.
- Seal PVC joints with a silicone caulk. Do not butt and glue cut pipe ends together. Do not use caulk to seal the PVC sleeve or coupling to the metal air intake collar on the furnace burner box.

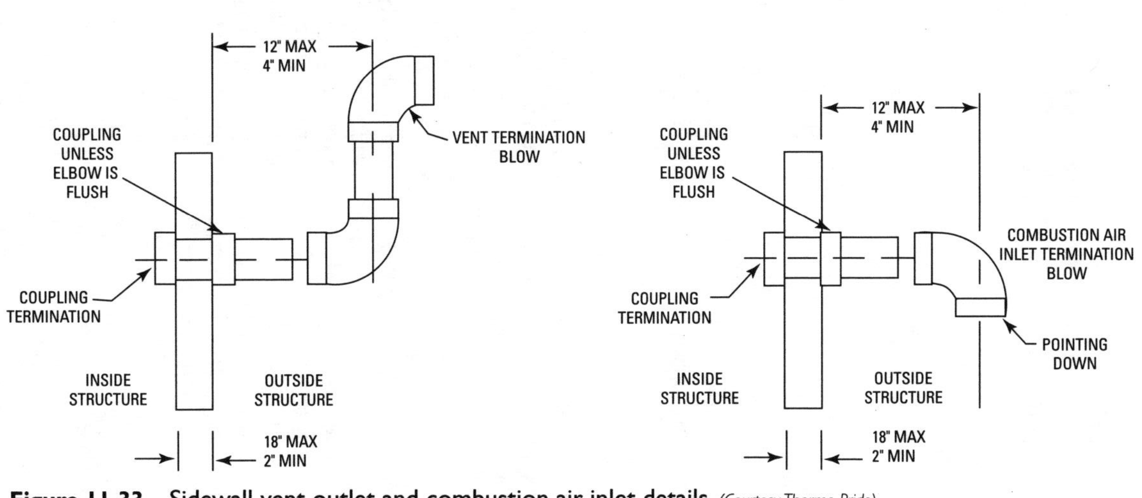

Figure 11-33 Sidewall vent outlet and combustion air inlet details. *(Courtesy Thermo Pride)*

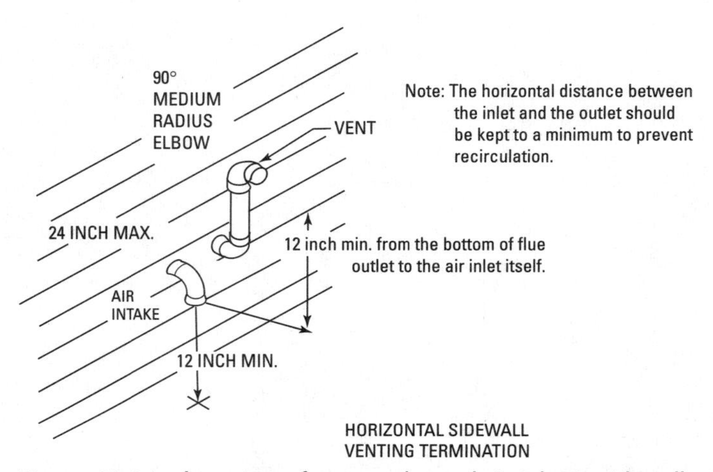

Figure 11-34 Location of vent outlet and air inlet in sidewalls.
(Courtesy Thermo Pride)

Chimney

Most chimneys are constructed of brick or metal. If a brick chimney is used, it should be lined with a protective material to prevent damage from water vapor. The most commonly used liner is smooth faced tile. Prefabricated factory-built chimneys are also used, but only those listed by Underwriters Laboratories are suitable for use with fuel-burning equipment.

The standard chimney must be at least 3 ft higher than the roof or 2 ft higher than any portion of the structure within 10 ft of the chimney flashing in order to avoid downdrafts (Figure 11-35).

An existing chimney should always be checked to make sure it is smoketight and clean. Any dirt or debris must be cleaned out before the furnace or boiler is used. See Chapter 3, "Fireplaces, Stoves, and Chimneys" in Volume 3 for additional information about chimney construction, maintenance, and repair.

Caution

> If an original furnace has been replaced by a mid-efficiency or high-efficiency furnace, the chimney cannot be used to vent the combustion gases unless it has been rebuilt or modified. The cooler gases of these more efficient furnaces lack the buoyancy of those produced in the conventional standing-pilot furnaces. Consequently, they do not rise quickly up the chimney. The condensate of these cool, slower moving gases will collect on the chimney walls and eventually destroy the masonry.

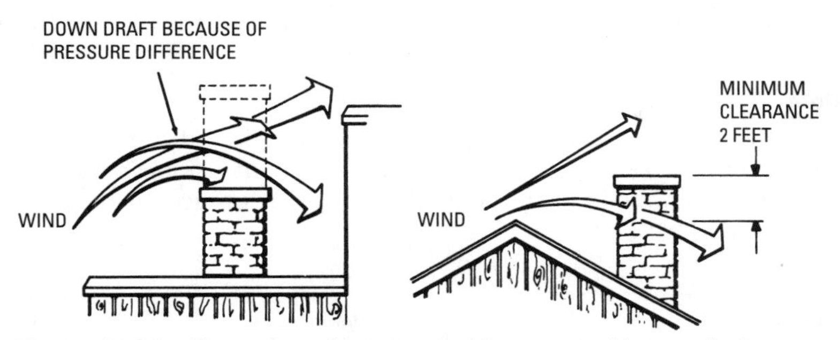

Figure 11-35 Examples of correct and incorrect chimney designs.

Chimney Troubleshooting

The chimney must give sufficient draft for combustion or the furnace will not operate efficiently. It must also provide a means for venting the products of combustion to the outside. Although a chimney-produced draft is not as important for the combustion process in gas-fired appliances as it is in other types of fuel-burning equipment, the venting capacity of the chimney is extremely important. The chimney must be of a suitable area and height to vent all the products of the combustion process.

Figure 11-36 illustrates some of the common chimney problems that can cause insufficient draft and improper venting. Some of these problems are detectable by observation; others require the use of a draft gauge.

Sometimes a chimney will be either too small or too large for the installation. When this is the case, the chimney should be rebuilt with the necessary corrections made in its design.

Draft Hood

A standing-pilot (continuously burning pilot) gas furnace is equipped with a draft hood attached to the flue outlet of the appliance. The draft hood used on the appliance should be certified by the American Gas Association. Only gas conversion furnaces equipped with power-type burners and conversion burner installations in large steel boilers with inputs in excess of 400,000 Btu/h are not required to have draft hoods.

A draft hood is a device used to ensure the maintenance of constant low draft conditions in the combustion chamber. By this action, it contributes to the stability of the air supply for the combustion process. A draft hood will also prevent excessive chimney draft and downdrafts that tend to extinguish the gas burner flame. Because of this last function, a draft hood is often referred to as a draft diverter.

Note

Mid-efficiency furnaces (the so-called induced-draft furnaces) do not have a draft hood. They use a small fan located in the flue to induce a draft. High-efficiency gas furnaces also do not have draft hoods.

Problem	***Remedy***
Top of chimney is lower than surrounding objects.	Extend chimney above all objects within 20 ft.
Chimney cap or ventilator obstruction.	Remove.
Coping restricts opening.	Make opening as large as inside of chimney.
Obstruction in chimney.	Use rod or weight on string or wire to break and dislodge.
Joist projecting into chimney.	Change support for joist so that chimney will be clear. Should be handled by a competent brick contractor.
Break in chimney lining.	Rebuild chimney with a course of brick between flue tiles.
Collection of soot at narrow space in flue opening.	Clean out with weighted brush or bag of loose gravel on end of line. May be necessary to open chimney.
Offset.	Change to straight or long offset.
Two or more openings in same chimney.	Least important opening must be closed using some other chimney flue.
Smoke pipe extends into chimney.	Shorten pipe so that end is flush with inside of tile.
Loose-fitted cleanout door.	Leaks should be eliminated by cementing all pipe openings.
Failure to extend the length of flue partition down to the floor.	Extend partition to floor level.
Loosely fitted cleanout door.	Close all leaks with cement.

Common chimney problems and their corrections. *(continued)*

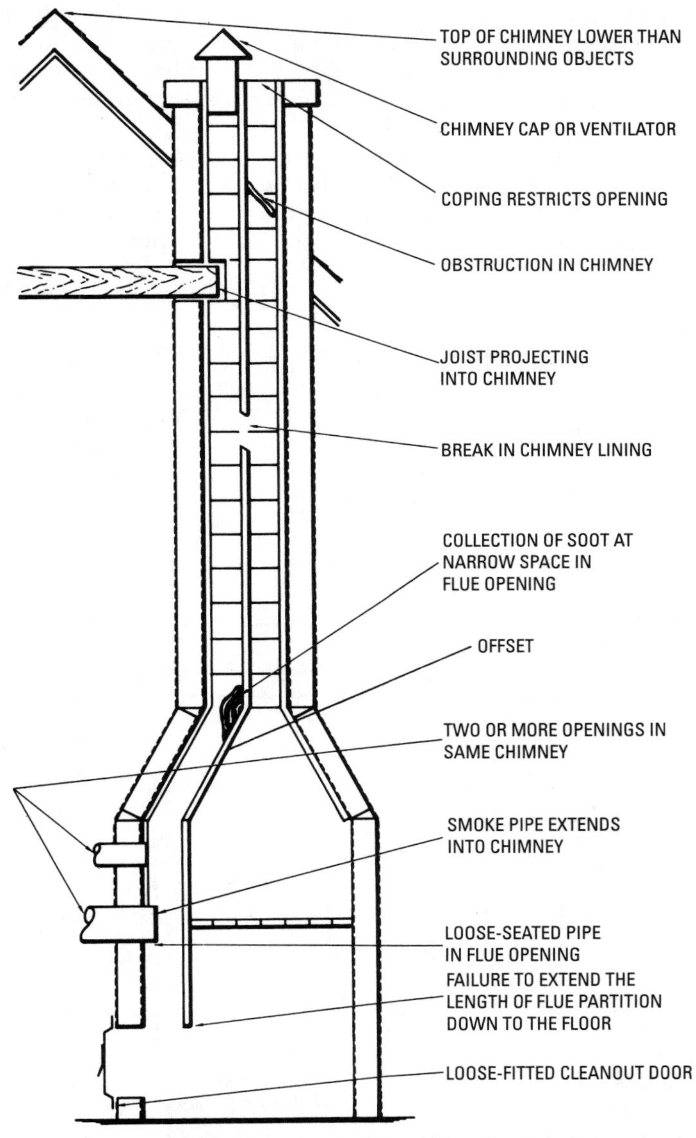

Figure 11-36 Common chimney problems and their corrections.

Draft hoods may be either internally or externally mounted, depending on the design of the furnace. Never use an external draft hood with a furnace already equipped with an internal draft hood. Either vertical or horizontal discharge from the draft hood is possible (Figure 11-37). Locations of draft hoods in conversion burner installations are illustrated in Figure 11-38.

In some installations, a neutral pressure-point adjuster is installed in the flue pipe between the furnace and the draft hood. The procedure for making a neutral pressure-point adjuster is illustrated in Figure 11-39. Always leave the neutral pressure-point

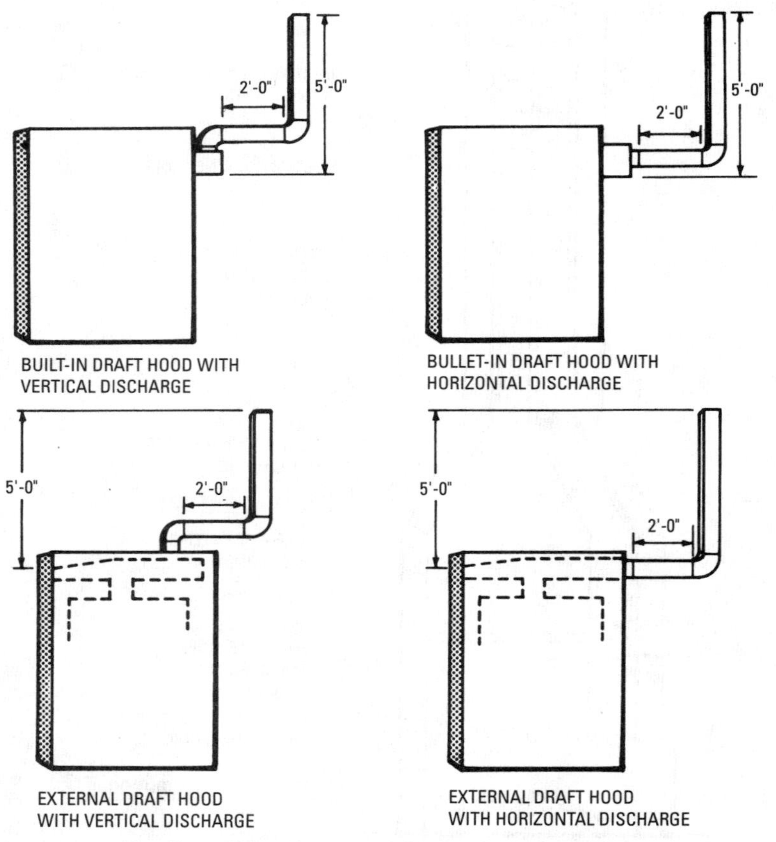

Figure 11-37 External and internal (built-in) draft hoods with vertical and horizontal discharge. *(Courtesy American Gas Association)*

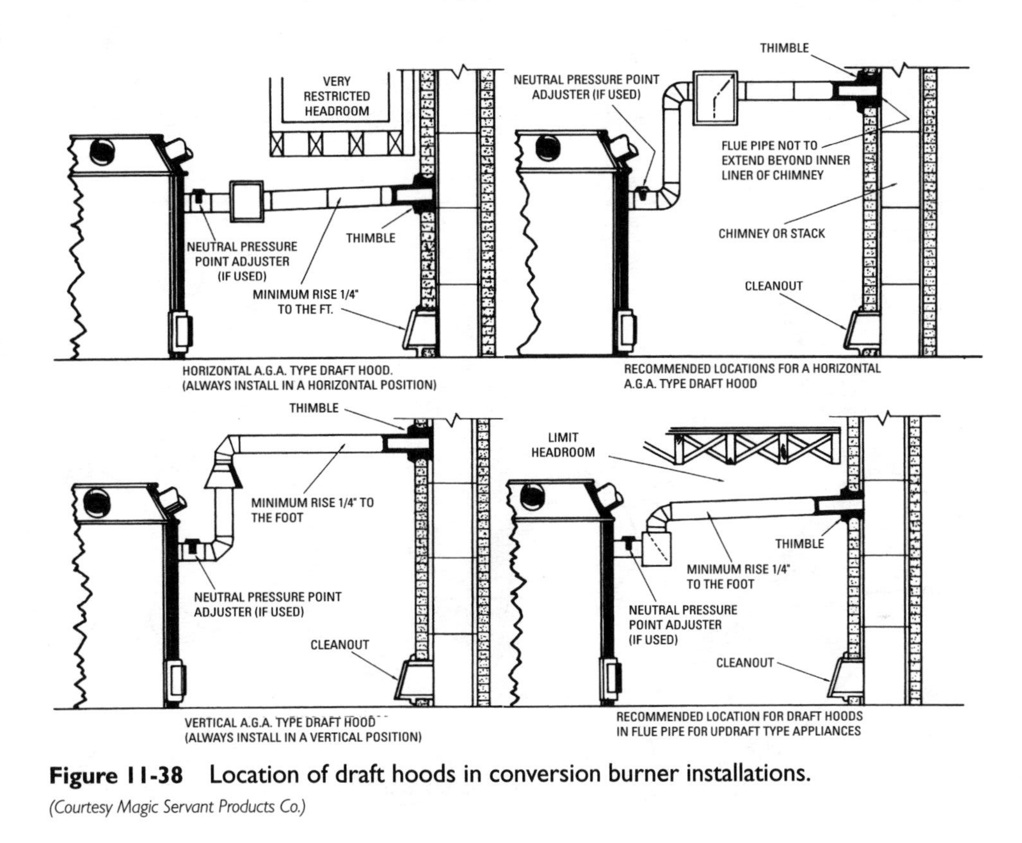

Figure 11-38 Location of draft hoods in conversion burner installations.

(Courtesy Magic Servant Products Co.)

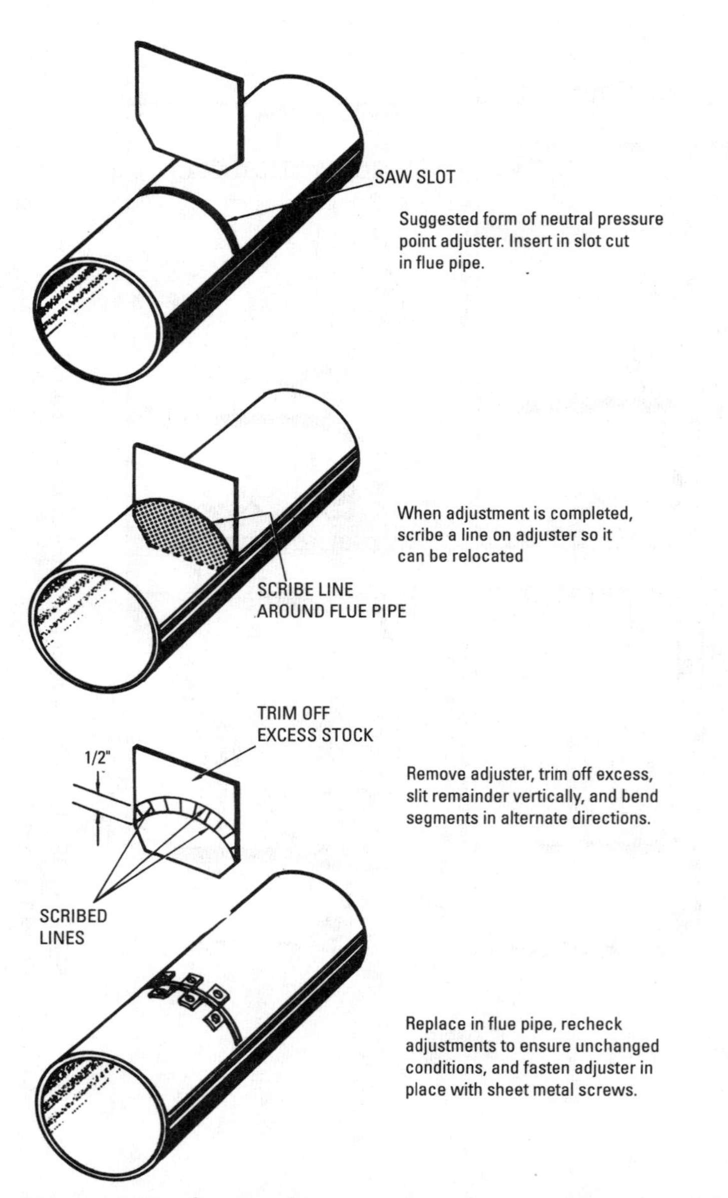

Figure 11-39 Suggested construction of a neutral pressure-point adjuster. *(Courtesy Magic Servant Products Co.)*

adjuster in wide open position until after the burner rating has been established.

Pilot Burner Adjustment

It is sometimes necessary to adjust the pilot flame. On some combination gas controls, a pilot adjustment screw is provided for this purpose. On units using natural gas, the screw should be adjusted until the pilot flame completely envelops the thermocouple. If propane gas is used, the screw should be wide open.

Sometimes a pilot gas regulator is used on a furnace to control pressures in areas where gas pressure variation is great. The pilot valve is placed in wide open position, and pilot flame adjustments are made by adjusting the gas pressure regulator.

Gas Input Adjustment

The gas input of a furnace must not be greater than that specified on the rating plate. However, it is permissible to reduce the gas input to a minimum of 80 percent of the rated input in order to balance the load.

The gas input can be checked by timing the flow through the meter. The procedure for clocking the meter is as follows:

1. Turn off all gas units connected through the meter except the furnace. The main burners of the furnace must be on for the timing check.
2. Time one complete revolution of the hand (or disk) on the meter. The hand should be the one with the smallest cubic footage per revolution (usually ½ or 1 ft^3 per revolution on small meters and 5 ft^3 on large meters).
3. Using the size of the test meter dial (½, 1, 2, or 5 ft^3) and the number of seconds per revolution, determine the cubic feet per hour from the appropriate column in Table 11-1.
4. Determine the actual Btu/h input of the burner by multiplying the cubic feet per hour rate (obtained in Step 3) by the Btu per cubic feet of gas (obtained from the local gas company).

By the way of example, let's assume that you have contacted the local gas company and they have given you a value of 1040 Btu per cubic foot. When you clock the meter, you find that it takes exactly 18 seconds for the hand to make one complete revolution on the ½ ft^3 meter dial. With this information, turn to Table 11-1 and find 18 seconds in the extreme left-hand column. Opposite the value

Table 11-1 Gas Rate in Cubic Feet per Hour

Seconds for One Revolution	Size of Test Meter Dial				Seconds for One Revolution	Size of Test Meter Dial			
	½ ft^3	1 ft^3	2 ft^3	5 ft^3		½ ft^3	1 ft^3	2 ft^3	5 ft^3
	Cubic Feet per Hour					Cubic Feet per Hour			
10	180	360	720	1800	50	36	72	144	360
11	164	327	655	1636	51	35	71	141	353
12	150	300	600	1500	52	35	69	138	346
13	138	277	555	1385	53	34	68	136	340
14	129	257	514	1286	54	33	67	133	333
15	120	240	480	1200	55	33	65	131	327
16	112	225	450	1125	56	32	64	129	321
17	106	212	424	1059	57	32	63	126	316
18	100	200	400	1000	58	31	62	124	310
19	95	189	379	947	59	30	61	122	305
20	90	180	360	900	60	30	60	120	300
21	86	171	343	857	62	29	58	116	290
22	82	164	327	818	64	29	56	112	281
23	78	157	313	783	66	29	54	109	273
24	75	150	300	750	68	28	53	106	265
25	72	144	288	720	70	26	51	103	257
26	69	138	277	692	72	25	50	100	250
27	67	133	267	667	74	24	48	97	243

28	64	129	257	643	76	24	47	95	237
29	62	124	248	621	78	23	46	92	231
30	60	120	240	600	80	22	45	90	225
31	58	116	232	581	82	22	44	88	220
32	56	113	225	562	84	21	43	86	214
33	55	109	218	545	86	21	42	84	209
34	53	106	212	529	88	20	41	82	205
35	51	103	206	514	90	20	40	80	200
36	50	100	200	500	94	19	38	76	192
37	49	97	105	486	98	18	37	74	184
38	47	95	189	474	100	18	36	72	180
39	46	92	185	462	104	17	35	69	173
40	45	90	180	450	108	17	33	67	167
41	44	88	176	440	112	16	32	64	161
42	43	86	172	430	116	15	31	62	155
43	42	84	167	420	120	15	30	60	150
44	41	82	164	410	130	14	28	55	138
45	40	80	160	400	140	13	26	51	129
46	39	78	157	391	150	12	24	48	120
47	38	77	153	383	160	11	22	45	113
48	37	75	150	375	170	11	21	42	106
49	37	73	147	367	180	10	20	40	100

Courtesy Dunham-Bush, Inc.

18 seconds under the ½ ft^3 column heading read 100 ft^3. The rest is simple multiplication:

$$100\ ft^3 \text{ per hour} \times 1040 \text{ Btu per } ft^3 = 104{,}000 \text{ Btu/h}$$

If the Btu/h rate is not within 5 percent of the desired value, change the burner orifices or adjust the pressure regulator. Pressure regulator setting changes are used only for minor adjustments in Btu input. Major adjustments are made by changing the burner orifices.

When making adjustments, remember the following: Btu input may never be less than the minimum or more than the maximum on the rating plate.

There is no meter to clock on furnaces fired with propane. The burner orifices on these units are sized to give the proper rate at 11 in W.C. (water column) manifold pressure.

After the correct Btu input rate has been determined, adjust the burner for the most efficient flame characteristics (see *Combustion Air Adjustment*).

High-Altitude Adjustment

For elevations above 2000 ft, the Btu input should be reduced (derated) 4 percent for each 1000 ft above sea level. This adjustment is required because of the increase in air volume at higher elevations. Air expands at approximately 4 percent per 1000 ft of elevation. At elevations above 2000 ft, the volume of air required to supply enough oxygen for complete combustion exceeds the maximum air-including ability of the burners. Because the primary air volume cannot be increased, the gas flow (input) must be decreased (that is, derated).

Changing Burner Orifices

A gas-fired furnace is supplied with standard burner orifices (spuds) for the gas shown on the rating plate. The orifice size will depend on the calorific value of the gas (Btu per cubic foot) and the manifold pressure (3.5 in W.C. for natural gas and 11 W.C. for LP gas). The burner orifice size partially determines the firing rate by allowing only a predetermined volume of gas to pass. An orifice that is too large will allow too much gas to pass through into the burner. This condition sometimes results in overheating the heat exchanger or restricting the air-inducing ability of the burner.

The gas burner orifices supplied with a furnace are suitable for average calorific values of a gas at the listed manifold pressures. Table 11-2 lists the various orifice capacities for different drill sizes. The procedure for determining the suitability of a particular drill size is as follows:

1. Divide the total Btu input of the furnace by the total number of burner orifices in the unit.
2. From Table 11-2, select the burner orifice size closest to the calculated value (plus or minus 5 percent) for the gas used.

Table 11-2 Orifice Capacity Table

	*Approximate Orifice Capacity (But/h)**	
Drill Size	***Natural Gas @ 3 ½" W.C.***	***Propane Gas @ 11" W.C.***
60	4650	13,000
59	4900	13,750
58	5150	14,450
57	5400	15,150
56	6300	17,600
55	7900	22,000
54	8800	24,750
53	10,400	28,800
52	11,750	33,000
51	13,000	36,700
50	14,250	40,000
49	15,500	43,500
48	16,750	47,200
47	17,900	50,200
46	19,000	53,500
45	19,500	55,000
44	21,500	60,050
43	23,000	
42	25,250	
41	26,750	
40	27,800	
39	28,600	
38	29,900	
37	31,100	
36	33,000	
35	35,000	

*Values determined using the following data: propane at 1.52 specific gravity/2500 Btu/ft^3 and natural gas at 0.62 specific gravity/1000 Btu/ft^3.

Courtesy The Trane Company

This procedure can be illustrated by determining the most suitable drill size for a 150,000-Btu-input, gas-fired furnace. This furnace has four burners, with two orifices per burner, and uses natural gas.

$$\text{capacity of each burner orifice} = \frac{150{,}000}{4 \times 2} = \frac{150{,}000}{8} = 18{,}750$$

In Table 11-2, the figure closest to 18,750 in the left-hand column (natural gas) is 19,000. The drill size opposite this figure is No. 46.

Manifold Pressure Adjustment

A suitable manifold pressure is important for efficient furnace operation. If the gas pressure is too low, it will cause rough ignition, incomplete and inefficient combustion, and incorrect fan control response. An excessively high manifold pressure may cause the burners to overfire the heat exchanger. Overfiring the heat exchanger not only reduces the life of this component, but it also may result in repeated cycling of the burner on the high limit control.

Manifold pressure should be set at 3.5 in W.C. for natural gas. Control manufacturers normally preset the pressure regulator in natural-gas valves at the time of manufacture so that natural-gas units are fired at this 3.5 in W.C. rate. Only small variations in the gas flow should be made by adjusting the pressure regulator. This adjustment should never exceed plus or minus 0.3 in W.C. Major changes in the gas flow should be made by changing the size of the burner orifice (see below).

The manifold gas pressure can be tested by using a U-tube water manometer (Figure 11-40). The manometer is connected to the manifold through an opening covered by a ⅛-in plug cap. The pressure test must be run while the unit is operating. On Dunham-Bush gas-fired furnaces, the test is run through an opening in the gas valve (Figure 11-41). The furnace manufacturer's installation literature should contain instructions for testing the manifold pressure.

Once you have determined the manifold pressure, you may find it necessary to adjust it for better operating characteristics. Manifold pressure can be increased by turning the adjusting screw clockwise, and decreased by turning it counterclockwise.

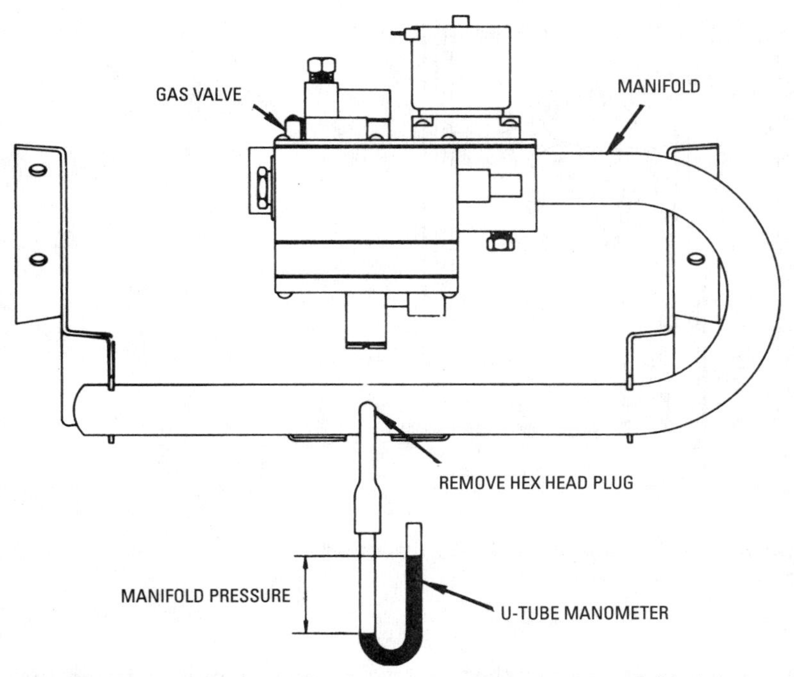

Figure 11-40 Applying a U-tube manometer to gas manifold.
(Courtesy Trane Co.)

> LP gas units are fired at 11 in W.C. manifold pressure. LP gas (propane and butane) is heavier than natural gas and has a higher heating value. As a result, LP gas needs more primary air for combustion. Thus, a higher manifold pressure is required to induce the greater volume of primary air into the burner than is the case in natural gas units

An LP furnace does not have a pressure regulator in the gas valve. Pressure adjustments are made by means of the regulator at the supply tank. These adjustments must be made by the installer and regularly checked by the serviceman for possible malfunctions.

Combustion Air Adjustment

Primary air shutters are provided on all furnaces to enable adjustment of the primary air supply. The purpose of this adjustment is to obtain the most suitable flame characteristics.

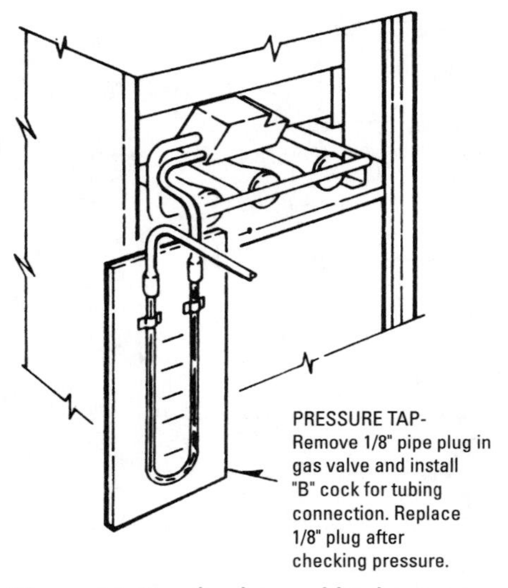

Figure 11-41 Applying a U-tube manometer to gas valve. *(Courtesy Dunham-Bush, Inc.)*

Experience has shown that the most efficient burner flame for natural gas has a soft blue cone without a yellow tip. For propane gas, there should be a little yellow showing in the tips of the flame.

The procedure for adjusting the primary air on a Coleman gas furnace may be summarized as follows:

1. Light the pilot burner.
2. Turn up the setting on the room thermostat until the main burners come on.
3. Allow the main burners 10 minutes to warm up.
4. Loosen locknut on adjusting screw (Figure 11-42).
5. Turn adjusting screw in (clockwise) until the yellow tip appears in the flame.
6. Turn adjusting screw out (counterclockwise) until the yellow tip just disappears.
7. Hold adjusting screw and tighten locknut.
8. Repeat Steps 4 through 7 on each of the other burners.

Figure 11-42 Position of the locknut and adjusting screw on a Coleman gas furnace. *(Courtesy Coleman Co., Inc.)*

On a Fedders gas furnace, a locking screw is located above the primary air opening on each gas burner (Figure 11-43). The primary air adjustment procedure is as follows:

1. Loosen the locking screw at the base of the burner.
2. Adjust the air-shutter opening to a position that gives a slight yellow tip on the end of the flame.
3. Open the air shutter until the yellow tip just disappears.
4. Tighten the locking screw.
5. Repeat Steps 1 through 4 for each of the burners.

A primary air-shutter assembly is provided on all Carrier gas furnaces in order to simplify the adjustment procedure. As shown in Figure 11-44, adjustment of the end burner results in a simultaneous adjustment of all burners.

After the primary air adjustments have been made, check to see that the burners are level. The burner flames should be uniform and centered in the heat exchanger (see *Heat Exchanger* in this chapter).

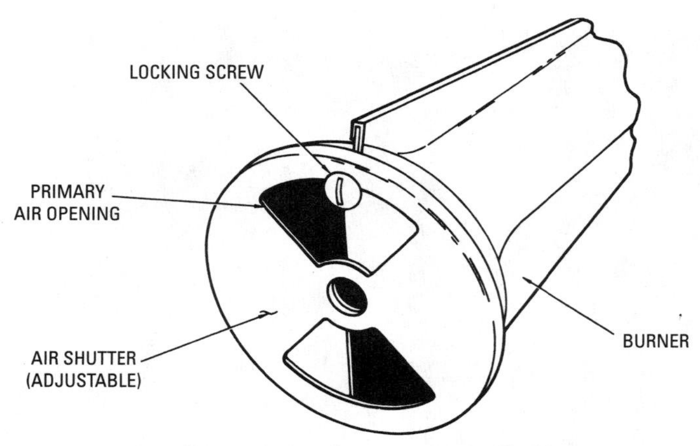

Figure 11-43 Primary air adjustment on a Fedders gas furnace. *(Courtesy Fedders Corp.)*

Before allowing the furnace to continue operating, you must check to see that it has the proper draft. This can be done by passing a match along the draft-hood opening. If the vent is drawing properly, the match flame will be drawn inward (that is, into the draft hood). If the furnace is not receiving proper draft, the products of combustion escaping the draft hood will extinguish the flame. The draft must be corrected before the furnace is operated.

Gas Supply Piping

The gas supply piping (also referred to as the *gas service piping*) must be sized and installed in accordance with the recommendations contained in local codes and ordinances, or, if unavailable, the provisions

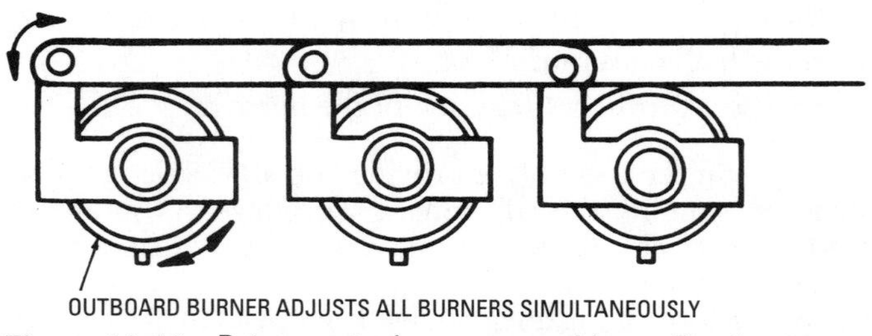

Figure 11-44 Primary air-shutter assembly on Carrier furnace. *(Courtesy Carrier Corp.)*

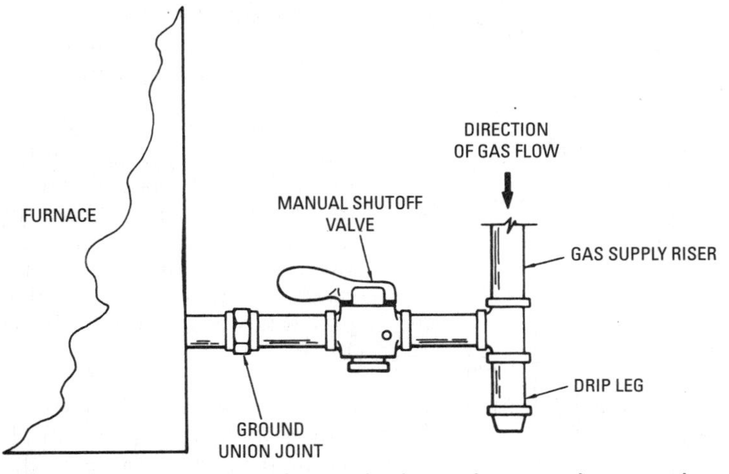

Figure 11-45 Location of manual valve on horizontal gas supply line. *(Courtesy Carrier Corp.)*

of the latest edition of the *National Gas Code* (ANSI Z223.1) and the regulations of the National Fire Protection Association (ANSI/NFPA70). The location of the main shutoff valve will usually be specified by local codes or regulations. Consult the local codes or regulations for recommended sizes of pipe for the required gas volumes. Local codes or regulations always take precedence over national standards when there is a conflict between the two.

Generally, 1-in supply pipe is adequate for furnace inputs up to 125,000 Btu/h. A 1¼-in pipe is recommended for higher inputs. The inlet gas supply pipe size should never be smaller than ½ in. The gas line from the supply should serve only a single unit.

A drip leg (also called a dirt leg or a dirt trap) should be installed at the bottom of the gas supply riser to collect moisture or impurities carried by the gas. A manual main shutoff valve should be installed either on the gas supply riser or on the horizontal pipe between the riser and ground union joint (Figures 11-45 and 11-46). The location of the main shutoff valve will usually be specified by local codes and regulations.

As shown in Figure 11-45, the ground union joint should be installed between the manual main shutoff valve and the gas control valve on the furnace or boiler. The ground union joint provides easy access to the gas controls on the unit for servicing or repair.

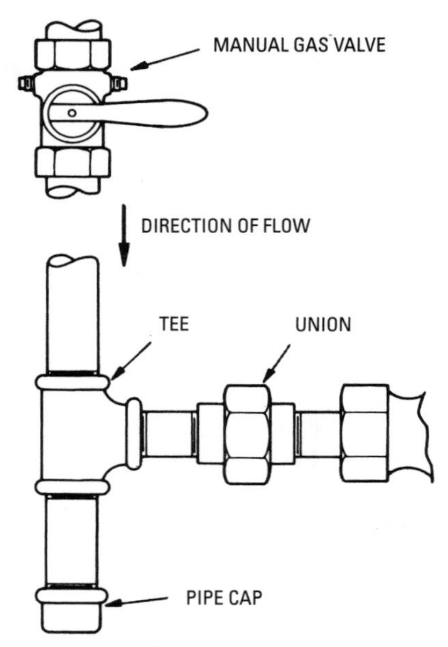

Figure 11-46 Location of union and drip leg for connecting furnace to main gas supply line.

Sizing Gas Piping

The gas supply pipe should be sized per volume of gas used and allowable pressure drop. The pipe must also be of adequate size to prevent undue pressure drop. The diameter of the supply pipe must at least equal that of the manual shutoff valve in the supply riser. Minimum acceptable pipe sizes are listed in Table 11-3.

The volume rate of gas in cubic feet per hour is determined by dividing the Btu/h required by the furnace by the Btu per cubic foot of gas being used. The cubic feet per hour value can then be used to determine pipe size for gas of a specific gravity other than 0.60. The data used for these

Table 11-3 Capacity of Pipe (ft^3/h) (Pressure Drop 0.3—Specific Gravity 0.60)

	Diameter of Pipe (inches)			
Length of Pipe (feet)	*¾*	*1*	*1¼*	*1½*
15	172	345	750	1220
30	120	241	535	850
45	99	199	435	700
60	86	173	380	610
75	77	155	345	545
90	70	141	310	490
105	65	131	285	450
120		120	270	420

Courtesy Janitrol

Table 11-4 Multiplier Table

Specific Gravity	*Multiplier*	*Specific Gravity*	*Multiplier*
0.40	0.813	1.30	1.47
0.50	0.910	1.40	1.53
0.60	1.00	1.50	1.58
0.70	1.08	1.60	1.63
0.80	1.15	1.70	1.68
0.90	1.22	1.80	1.73
1.00	1.29	1.90	1.77
1.10	1.35	2.00	1.83
1.20	1.42		

Courtesy Janitrol

calculations are listed in Tables 11-3 and 11-4. The procedure is as follows:

1. Multiply the cubic feet per hour required by the multiplier (Table 11-4).
2. Find recommended pipe size for length of run in Table 11-3.

Tables 11-3 and 11-4 apply to the average piping installation where four or five fittings are used in the gas piping to the furnace. Existing pipe should be converted to the proper size of pipe where necessary. In no case should pipe less than ¾-in diameter be used. Any extensions to existing piping should conform to Tables 11-3 and 11-4.

Installing Gas Piping

The principal recommendations to be followed when installing gas piping are:

1. A single gas line must be provided from the main gas supply to the furnace.
2. The gas supply pipe must be sized according to the volume of gas used and the allowable pressure drop.
3. The supply riser and drop leg must be installed adjacent to the furnace on either side. Never install it in front of the furnace, where it might block access to the unit.
4. A manual main shutoff valve or plug cock should be installed on the gas supply piping (usually on the riser) outside the unit in accordance with local codes and regulations.

5. Install a tee for the drop leg in the pipe at the same level as the gas supply (inlet) connection on the furnace.
6. Extend the drip leg and cap it.
7. Install a ground union joint between the tee and the furnace controls.
8. Use a pipe compound resistant to LP gas on all threaded pipe connections in an LP gas installation.
9. Test for pressure and leaks. Use a solution of soap or detergent and water to detect any leaks. Never check for leaks with a flame.

Typical Startup Instructions for a Standing-Pilot Gas Furnace

The startup procedure for most standing-pilot gas furnaces is very similar, particularly insofar as use of the thermostat is concerned. The major difference is in the lighting of the pilot burner, and this will depend on the gas valve used.

Caution

Always carefully follow the furnace manufacturer's lighting and operating instructions when attempting to start a furnace. Failure to do so may cause a fire or explosion resulting in possible injury or even death.

The startup procedure for a standing-pilot gas furnace may be summarized as follows:

1. Open all warm air registers.
2. Shut off main gas valve and pilot gas cock, and wait at least 5 minutes to allow gas that may have accumulated in the burner compartment to escape.

Warning

If you smell gas after waiting 5 minutes, immediately call your local gas company (utility) from a telephone outside the house and follow their instructions.

3. Set the room thermostat at the lowest setting.
4. Open the water supply valve if a humidifier is installed in the system.
5. Turn off the line voltage switch if one is provided in the furnace circuit.

6. Light the pilot (see below).
7. Turn electrical power on.
8. Replace all access doors.
9. Light main burners by setting the room thermostat above the room temperature.
10. Set the room thermostat at the desired temperature after the main burners are lighted.

The pilot lighting instructions for a standing-pilot gas furnace should be included with the furnace and prominently displayed. Lighting instructions for the 24-volt Honeywell V8280, V8243, and Robertshaw (Unitrol) 7000 combination gas valves basically summarize the pilot-lighting procedure (Figure 11-47). The procedure is as follows:

1. Depress gas cock dial and turn it counterclockwise to "pilot" position.
2. Keep gas cock dial depressed and light pilot with a match.
3. Keep dial depressed until pilot remains lighted when released.
4. Depress dial and turn counterclockwise to ON position.

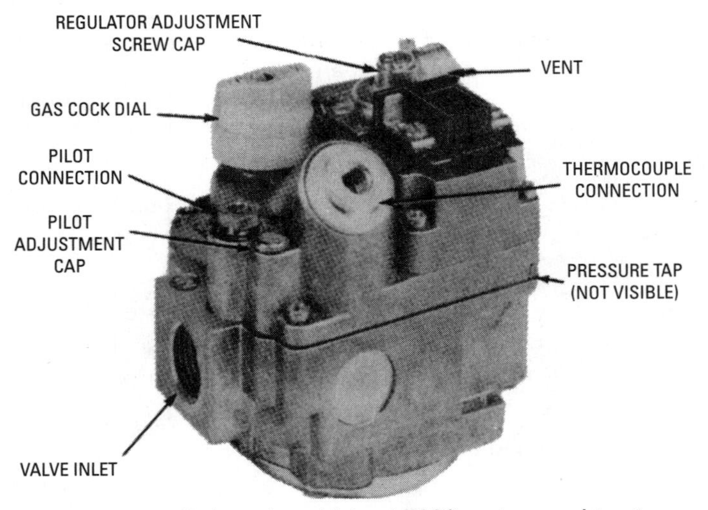

Figure 11-47 Robertshaw Unitrol 7000 series combination gas valve. *(Courtesy Robertshaw Controls Co.)*

Typical Startup Instructions for an Electronic Ignition Furnace

The modern high-efficiency gas furnace uses an electronic ignition system instead of a standing pilot to light the gas burners. The ignition device automatically lights the burners. Do not try to light the gas burners manually in these furnaces.

On gas furnaces equipped with electronic ignition systems, there is a sequence of safety steps that must be followed to light the burner. This information is normally attached to the inside of the burner or blower access door. Follow the procedure according to the manufacturer's guidelines to avoid the risk of fire or explosion. If you have any doubts about the startup procedure, call a qualified technician to start the furnace.

A typical procedure for starting a furnace equipped with an electronic ignition system is offered here simply as a guideline. The steps are as follows:

1. Set the room thermostat to its lowest possible setting.
2. Turn off all electrical power to the furnace.
3. Remove cabinet panel to gain access to the main gas valve.
4. Turn gas valve knob to the OFF position.

Caution

Turn the knob manually. Do not force it, do not use a tool to turn it, and do not attempt to repair it. Doing so may result in a fire or explosion with possible injury or death. If it cannot easily be moved manually, call a qualified HVAC technician for service.

5. Wait 5 minutes for any gas to clear out of the burners and furnace.

Warning

If you smell gas after waiting 5 minutes, immediately call your local gas company (utility) from a telephone outside the house and follow their instructions.

6. If no gas odor has been detected, turn the gas valve knob to the ON position. Do not force the knob.
7. Replace the access panel and restore all electrical power to the furnace.
8. Adjust the thermostat for the desired room temperature setting.

9. Replace the cabinet access panel.
10. Repeat steps 1 through 9 if air needs to be purged from the pilot line.

If the furnace still will not start, shut it off and call a qualified HVAC technician for assistance. The furnace shutoff procedure is as follows:

1. Set the room thermostat to its lowest possible setting.
2. Turn off all electrical power to the furnace.
3. Remove the access panel to the main gas valve.
4. Turn the gas valve knob to the OFF position. Do not force the knob to turn.
5. Replace the access panel.

Blowers and Motors

A forced-warm-air furnace should be equipped with a variable-speed electric motor designed for continuous duty. The motor should be provided with overcurrent protection in accordance with the *National Electrical Code* (ANSI C1-1971).

Blower motors are available as either direct-drive or belt-driven units. On direct-drive units, the blower wheel is mounted directly on the motor shaft and runs at motor speed. The blower wheel of a belt-driven unit is operated by a V-belt connecting a fixed pulley mounted on the blower shaft with a variable-pitch drive pulley mounted on the motor shaft.

The blower motors used in residential furnaces are single-phase electric motors. Depending on the method used for starting, they may be classified as:

1. Shaded pole
2. Permanent split capacitor
3. Split-phase
4. Capacitor-start motors

A direct-drive blower generally uses a shaded pole (SHP) or permanent capacitor (PSC) motor. Belt-driven blower motors are usually split-phase (SPH) types up to the ⅓-hp size. Larger motors are capacitor-start, split-phase motors.

Blower motors are fractional horsepower motors commonly available in the following sizes: 1⁄12, 1⁄10, ⅙, ¼, ⅓, ½, and ¾ hp. A fractional horsepower motor should be protected with temperature- or current-sensitive devices to prevent motor winding temperatures from exceeding those allowed in the specifications.

Provisions should be made for the periodic lubrication of the blower and motor. Use only the type and grade of lubricant specified in the operating instructions. Do not overlubricate. Too much oil can be just as bad as too little. A direct-drive blower should be capable of operating within the range of air temperature at the static pressures for which the furnace is designed and at the voltage specified on the motor nameplate.

Motors using belt drives must be supplied with adjustable pulleys. The only exception is a condenser motor used on a forced-warm-air furnace with a cooling unit. Always provide the means for adjusting a belt-driven motor in an easily accessible location.

Most forced-warm-air furnaces are shipped with the blower motors factory-adjusted to drive the blower wheel at a correct speed for heating. This is a slower speed than the one used for air conditioning, but it is relatively easy to alter the air delivery to meet the requirements of the installation (see *Air Conditioning* in this chapter).

Air Delivery and Blower Adjustment

The furnace blower is used to force air through the heat exchanger in order to remove the heat produced by the burners. This removal of heat by the blower serves a twofold purpose: (1) It distributes the heat through the supply ducts to the various spaces in the structure; and (2) it protects the surfaces of the heat exchanger from overheating.

Most blowers are factory-set to operate with a specific air temperature rise. For example, Janitrol Series 37 gas-fired furnaces are certified by the American Gas Association to operate within a range of 70 to 100°F air temperature heat rise. In other words, the temperature of the air within the supply ducts may be 70 to 100°F higher than the air temperature in the return duct. A temperature rise of 85 to 90°F with approximately 0.1-in external pressure in the area beyond the unit and filter is generally considered adequate for most residential applications.

The furnace air delivery rate should be adjusted at the time of installation to obtain a temperature rise within the range specified on the furnace rating plate. Many furnace manufacturers recommend that the temperature rise be set below the maximum to ensure that temperature rise will not exceed the maximum requirement should the filter become excessively dirty.

Although the blower and motor are factory-set to operate within a specific temperature-rise range, the unit can be field adjusted to deliver more or less air as required. The air delivery rate is adjusted by changing the blower speed. On direct-drive motors, this is accomplished by moving the line lead to the desired terminal. Belt-driven motors are adjusted by changing the effective diameter of the driver pulley (see *Direct-Drive Blower Adjustment* and *Belt-Drive Blower Adjustment*).

Direct-Drive Blower Adjustment

The blower (fan) of a direct-drive unit is operated by a drive shaft connected to the motor. Three common direct-drive motors used with blowers are:

1. Two-speed, direct-drive shaded pole motors
2. Three-speed, direct-drive PSC motors
3. Four-speed, direct-drive PSC motors

These blower motors are shipped from the factory set for low-speed heating operation. If a higher operating speed is desired, the red blower motor wire must be moved from the low-speed red terminal to a higher speed terminal.

Belt-Drive Blower Adjustment

Air-temperature rise can be decreased by increasing the blower speed. A belt-drive unit can be adjusted for a higher speed as follows:

1. Loosen the set screw in the movable outer flange of the motor pulley.
2. Reduce the distance between the fixed flange and the movable flange of the pulley.
3. Tighten the set screw.

The procedure for decreasing the blower speed (and thereby increasing the air-temperature rise) is similar except that in Step 2 above, the distance between the fixed flange and the movable flange of the pulley are increased.

The motor and blower pulley should be aligned to minimize belt wear. The belt should be parallel to the blower scroll. Adjustments can be made by moving the motor or adjusting the pulley on the shaft.

Correct belt tension is also important. Belt slippage occurs when there is too little tension. Too much tension will cause motor overload and bearing wear. Correct belt tension will usually be indicated

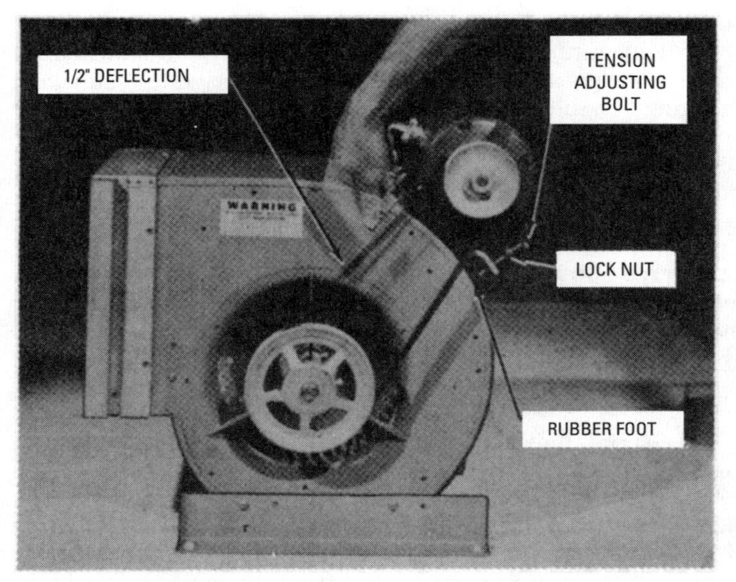

Figure 11-48 Checking belt tension and alignment.

by the amount of depression in the belt midway between the blower and motor pulleys. Usually this will be a 1-in depression at the center of the belt, but it may vary among different furnace manufacturers. Check the furnace manufacturer's specifications and service manual for the preferred amount of belt tension (Figure 11-48).

Air Filters

A forced-warm-air, gas-fired furnace is supplied with either a disposable air filter or a permanent (washable) one. The type of filter will be indicated on a label attached to the filter. Never use a filter with a gravity warm-air furnace, because it will obstruct the airflow.

Proper maintenance of the air filter is very important to the operating efficiency of the furnace. A dirty, clogged air filter reduces the airflow through the furnace and causes air temperatures inside the unit to rise. This increase in air temperature results in a reduced life for the heat exchanger and a lower operating efficiency for the furnace.

If a disposable air filter is used, it should be inspected on a monthly basis. If it is dirty, it should be replaced. This is usually done at the beginning and middle of the heating season. The same schedule should be followed during the summer months when air conditioning is used.

If a permanent filter is provided, it must be removed and cleaned from time to time. Cleaning the filter involves the following procedure:

1. Shake it to remove dust and dirt particles.
2. Vacuum it.
3. Wash it in a solution of soap or detergent and water.
4. Allow time for it to dry.
5. Replace it in the furnace.

Read the label on the filter for any special instructions that may apply to cleaning it. The location of filters in upflow and downflow furnaces is shown in Figures 11-49 and 11-50. Note that, in each case, the air filter is placed directly in the path of the air entering the furnace.

A filter should be installed in the return plenum or duct when air conditioning is added to the system. Because air conditioning normally requires a greater volume of air than heating, the filter should be sized for air conditioning rather than heating.

Access to internally installed filters is generally provided through a service panel on the furnace. Instructions for cleaning or replacing the filters should be placed on the service panel. It would be a good idea to include the dimensions of the replacement filter(s) in the information.

An adequate filter should provide at least 50 in^2 of area per cfm of air circulated. For an air velocity of 1000 cfm, this would mean a filler area of 400 in^2. Filters sized on 50 in^2 for each 100 cfm meet the requirements of air conditioning—a greater volume of air than heating.

Never use a smaller-size filter than the one required in the specifications. To do so would create excessive pressure drop at the filter, which would reduce the operating efficiency of the furnace.

Additional information about air filters can be found in Chapter 13, "Air Cleaners and Filters" in Volume 3.

Air Conditioning

If you intend to install air conditioning equipment at some future date, the duct sizing should include an allowance for it. Air conditioning involves a greater volume of air than heating. Quite often,

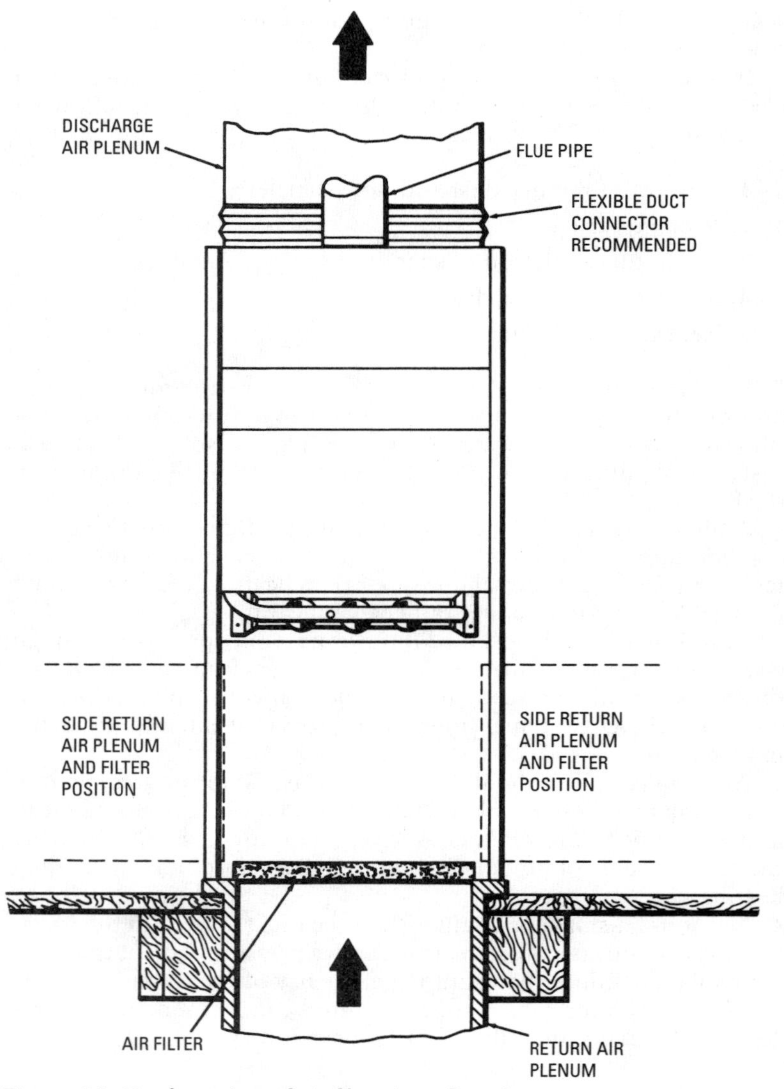

Figure 11-49 Location of air filter in upflow furnace.

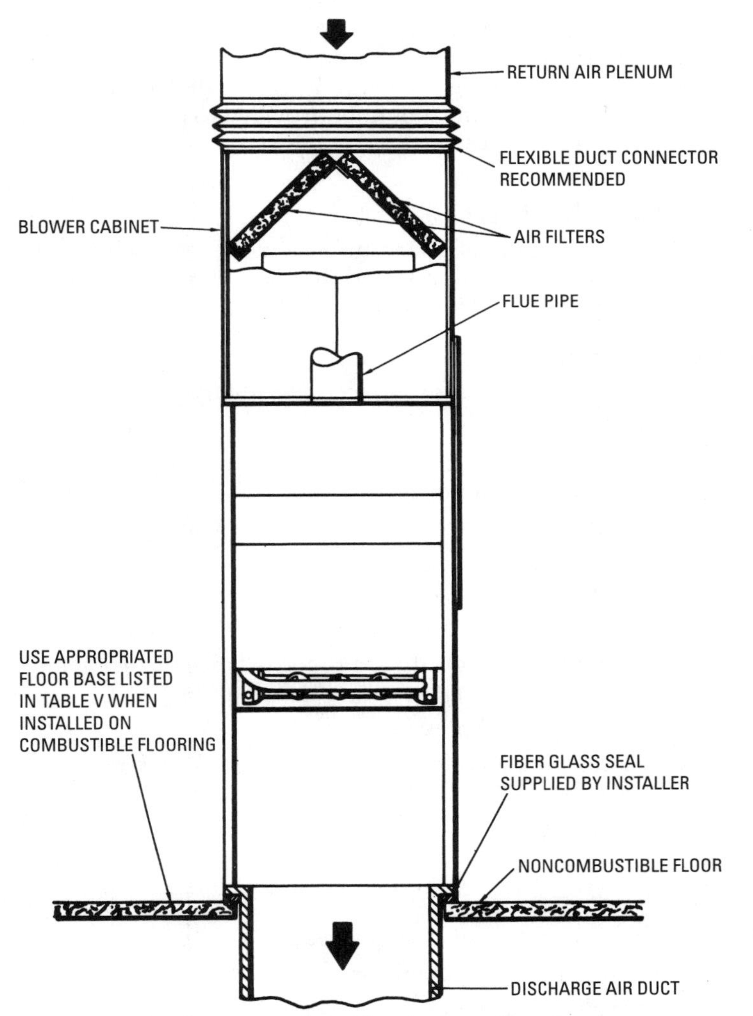

Figure 11-50 Location of air filter in downflow furnace.

the blowers and motors of the furnace will be sized for the addition of cooling with matching evaporator coil and condensing unit. Check the specifications to be sure.

All ductwork located in unconditioned areas (for example, attics, crawl spaces) downstream from the furnace should be insulated. A furnace used in conjunction with a cooling unit should be installed in parallel or on the upstream side of the evaporator coil to avoid condensation on the heating element. In a parallel installation, dampers or comparable means should be provided to prevent chilled air from entering the furnace.

Installation Checklist

A properly installed and maintained furnace will operate efficiently and economically. Figure 11-51 and the following installation checklist is offered as a guide for the installer:

1. Adjust primary air shutter.
2. Provide sufficient space for service accessibility.
3. Check all field wiring.
4. Supply line fuse or circuit breaker must be of proper size and type for furnace protection.
5. Line voltage must meet specifications while furnace is operating.
6. Airflow must be sufficient.
7. Ventilation and combustion air must be sufficient.
8. Ductwork must be checked for proper balance, velocity, and quietness.
9. Measure the manifold pressure.
10. Check all gas piping connections for gas leaks (use soap and water, not a flame).
11. Cycle the burners.
12. Check limit switch.
13. Check fan switch.
14. Adjust blower motor for desired speed.
15. Make sure air filter is properly secured.
16. Make sure all access panels have been secured.
17. Pitch air conditioning equipment condensation lines toward a drain.
18. Check thermostat heat anticipatory setting.
19. Check thermostat for normal operation.
20. Clear and clean the area around the furnace.

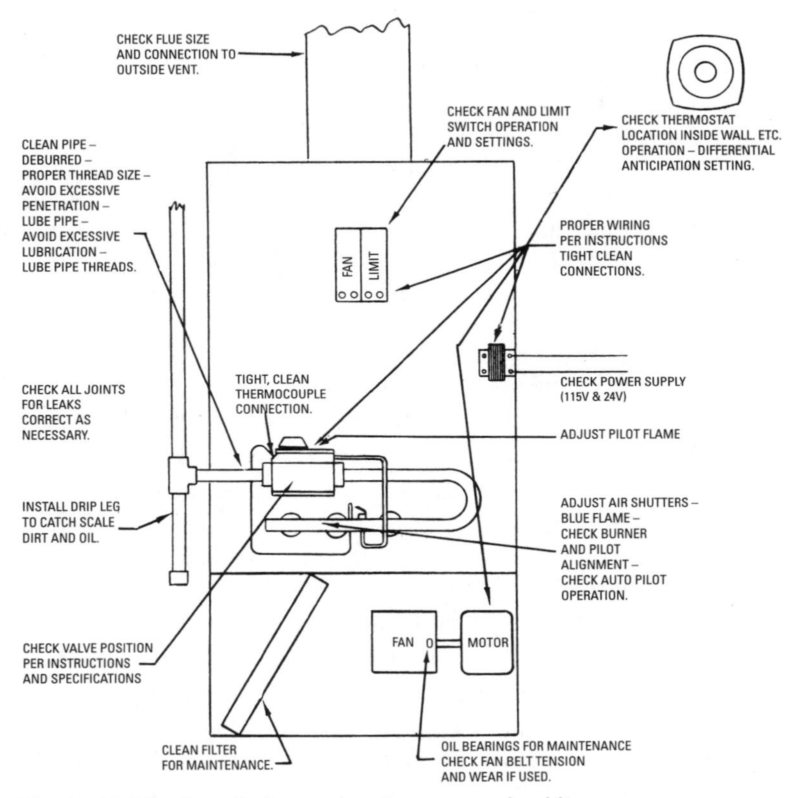

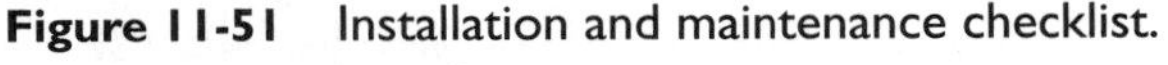
Figure 11-51 Installation and maintenance checklist.

(Courtesy Robertshaw Controls Co.)

Gas Furnace Inspections, Service, and Maintenance Tips

Gas furnaces should be inspected annually, preferably before the beginning of the heating season, by a qualified service technician to ensure continued safe operation. The inspection should include the following:

Caution

Always carefully read the service and maintenance instruction for the furnace first. As a rule, the electrical power is shut off first at the disconnect switch if service or an inspection is to be performed. After switching off the power, turn off the gas valve.

- Inspect the vent pipe for water accumulation, sagging piping, dirt, loose joints, and damage.
- Check the return air duct for a tight connection to the furnace. The duct connection must provide an airtight seal at the furnace and must terminate at its other end outside the room or space where the furnace is located.
- Inspect the furnace wiring for burnt or damaged wires and loose connections.
- Inspect interior and exterior furnace surfaces for dirt or water accumulation.
- Make sure the blower access door is tightly closed.
- Inspect the burners for dirt, rust, or signs of water.
- Make sure the fresh air grilles and louvers are open, clean, and unobstructed.
- Inspect and clean the condensate traps and drain to prevent water accumulation in the furnace.
- Inspect the blower wheel and remove any debris.
- Check the furnace support and base. The base of the furnace must form a tight seal with the support.

If the above items all pass inspection, restore power and start the furnace according to the furnace manufacturer's instructions. It will involve first turning on the gas valve and then turning on electrical power to the furnace. Continue the inspection as follows:

Caution

If you smell gas, call the local gas company immediately from a telephone outside the house and follow their instructions.

- Run the furnace and observe its operation. The furnace should operate smoothly and quietly. Check the vent pipe and return duct to make sure they are not leaking.
- Analyze the combustion gases to make sure they meet the furnace manufacturer's specifications.

Note

Although the hot surface igniter used in a hot surface ignition (HSI) gas furnace may be made of a material (such as silicon carbon) capable of resisting the high temperatures encountered when firing the gas-air mixture, these igniters are fragile and easily damaged if not handled carefully. The damage most commonly occurs during shipment from the furnace manufacturer or supplier, or during installation. The damage is not always visible to the naked eye and a cracked igniter initially can function according to specifications. However, eventually, its operating efficiency will degrade and its service life will be much shorter. Igniter damage too small to see can be discovered by checking for inconsistencies in its glow pattern immediately after installation.

Troubleshooting a Gas Furnace

Table 11-5 contains the most common operating problems associated with residential and small commercial gas furnaces. Each problem is given in the form of a symptom, the possible cause, and a suggested remedy. The table is intended to provide the operator with a quick reference to the cause and correction of a specific problem.

Table 11-5 Troubleshooting Gas Furnaces

Symptom and Possible Cause	*Possible Remedy*
No heat (standing-pilot furnace vented to chimney)	
(a) Thermostat not turned to HEAT.	(a) Turn to HEAT.
(b) Thermostat temperature setting above room temperature.	(b) Lower thermostat heat setting until furnace turns on.
(c) Blown fuse.	(c) Replace fuse. Call electrician if problem reoccurs.
(d) Tripped circuit breaker.	(d) Reset circuit breaker. Call electrician if problem reoccurs.
(e) Dirty filter.	(e) Clean or replace filter.
(f) Faulty gas valve.	(f) Replace gas valve.
(g) Pilot light out.	(g) Light pilot according to the furnace manufacturer's instructions or request a service call from an HVAC technician.

(continued)

Table 11-5 *(continued)*

Symptom and Possible Cause	*Possible Remedy*
(**h**) Emergency power switch off.	(**h**) Turn switch on.
(**i**) Defective thermostat.	(**i**) Replace thermostat.
(**j**) Defective thermocouple.	(**j**) Replace thermocouple.
(**k**) Open plenum switch.	(**k**) Close.
(**l**) Defective plenum switch.	(**l**) Replace switch.
(**m**) Open limit switch (plenum or exhaust).	(**m**) Close limit switch or replace.
(**n**) Blocked flue, vent, or chimney.	(**n**) Locate and remove obstruction.
No heat (high-efficiency furnace vented through PVC pipe)	
(**a**) Thermostat not turned to HEAT.	(**a**) Turn to HEAT.
(**b**) Thermostat temperature setting above room temperature.	(**b**) Lower thermostat heat setting until furnace turns on.
(**c**) Blown fuse.	(**c**) Replace fuse. Call electrician if problem reoccurs.
(**d**) Tripped circuit breaker.	(**d**) Reset circuit breaker. Call electrician if problem reoccurs.
(**e**) Dirty filter.	(**e**) Clean or replace filter.
(**f**) Defective thermostat.	(**f**) Replace thermostat.
(**g**) Defective hot surface igniter.	(**g**) Replace hot surface igniter.
(**h**) Faulty spark igniter.	(**h**) Replace spark igniter.
(**i**) Faulty or tight blower motor.	(**i**) Repair or replace blower motor.
(**j**) Defective fan relay.	(**j**) Replace fan relay.
(**k**) Defective plenum switch.	(**k**) Replace plenum switch.
(**l**) Blocked or restricted PVC vent outlet.	(**l**) Locate and clear blockage.
(**m**) Blocked or restricted PVC intake opening.	(**m**) Locate and clear blockage.
(**n**) Defective gas valve.	(**n**) Replace gas valve.
(**o**) Open limit switch (plenum or exhaust).	(**o**) Close limit switch or replace.
(**p**) Defective pressure switch.	(**p**) Replace pressure switch.
(**q**) Blower door off, open, or not properly closed.	(**q**) Close door firmly and secure.

(continued)

Table 11-5 *(continued)*

Symptom and Possible Cause	*Possible Remedy*
Not enough heat	
(a) Thermostat set too low.	(a) Raise setting.
(b) Lamp or some other heat source too close to thermostat.	(b) Move heat source away from thermostat.
(c) Thermostat improperly located.	(c) Relocate thermostat.
(d) Dirty air filter.	(d) Clean or replace.
(e) Thermostat out of calibration.	(e) Recalibrate or replace.
(f) Limit set too low.	(f) Reset or replace.
(g) Fan speed too low.	(g) Check motor and fan belt and tighten if too loose.
Too much heat	
(a) Thermostat set too high.	(a) Lower setting.
(b) Thermostat out of calibration.	(b) Recalibrate or replace.
(c) Short in wiring.	(c) Locate and correct.
(d) Valve sticks open.	(d) Replace valve.
(e) Thermostat in draft or on cold wall.	(e) Relocate thermostat to sense average temperature.
(f) Bypass open.	(f) Close bypass.
Flame too large	
(a) Pressure regulator set too high.	(a) Reset using manometer.
(b) Defective regulator.	(b) Replace.
(c) Burner orifice too large.	(c) Replace with correct size.
Noisy flame	
(a) Too much primary air.	(a) Adjust air shutters.
(b) Noisy pilot.	(b) Reduce pilot gas.
(c) Burr in orifice.	(c) Remove burr or replace orifice.
Yellow Tip flame	
(a) Too little primary air.	(a) Adjust air shutters.
(b) Clogged burner ports.	(b) Clean ports.
(c) Misaligned orifices.	(c) Realign.
(d) Clogged draft hood.	(d) Clean.
Floating flame	
(a) Blocked venting.	(a) Clean.
(b) Insufficient primary air.	(b) Increase primary air supply.

(continued)

Table 11-5 *(continued)*

Symptom and Possible Cause	*Possible Remedy*
Delayed Ignition	
(a) Improper pilot location.	(a) Reposition pilot.
(b) Pilot flame too small.	(b) Check orifice; clean, increase pilot gas.
(c) Burner ports clogged near.	(c) Clean ports.
(d) Low pressure.	(d) Adjust pressure regulator.
Failure to ignite	
(a) Pilot light off.	(a) Light pilot according to the furnace manufacturer's instructions or request a service call from an HVAC technician.
(b) Main gas supply off.	(b) Open manual valve.
(c) Blown fuse.	(c) Replace fuse. Call an electrician if the problem reoccurs.
(d) Circuit breaker tripped.	(d) Reset circuit breaker. Call an electrician if the problem reoccurs.
(e) Limit switch defective.	(e) Replace.
(f) Poor electrical connections.	(f) Check, clean, and tighten.
(g) Defective gas valve.	(g) Replace.
(h) Defective thermostat.	(h) Replace.
Burner will not turn on	
(a) Pilot flame too large or too small.	(a) Readjust.
(b) Dirt in pilot orifice.	(b) Clean.
(c) Too much draft.	(c) Shield pilot.
(d) Defective automatic.	(d) Replace.
(e) Defective thermocouple.	(e) Replace.
(f) Improper thermocouple.	(f) Properly position thermopile.
(g) Defective wiring.	(g) Check connections; tighten and repair shorts.
(h) Defective thermostat.	(h) Check for switch closure and repair or replace.
(i) Defective automatic valve.	(i) Replace.
Burner will not turn off	
(a) Poor thermostat location.	(a) Relocate.

(continued)

Table 11-5 (continued)

Symptom and Possible Cause	*Possible Remedy*
(b) Defective thermostat.	(b) Check calibration; check switch and contacts; replace.
(c) Limit switch maladjusted.	(c) Replace.
(d) Short circuit.	(d) Check operation at valve; check for short and correct.
(e) Defective or sticking automatic valve.	(e) Clean or replace.
Rapid burner cycling	
(a) Clogged filters.	(a) Clean or replace.
(b) Excessive anticipation.	(b) Adjust thermostat anticipatory for longer cycles.
(c) Limit setting too low.	(c) Readjust or replace limit.
(d) Poor thermostat location.	(d) Relocate.
Rapid fan cycling	
(a) Fan switch differential too low.	(a) Readjust or replace.
(b) Blower speed too high.	(b) Readjust to lower speed.
Blower will not stop	
(a) Manual fan on.	(a) Switch to automatic.
(b) Fan switch defective.	(b) Replace.
(c) Shorts.	(c) Check wiring and correct.
Noisy blower and motor	
(a) Fan blades loose.	(a) Replace or tighten.
(b) Belt tension improper.	(b) Readjust (usually allow 1 in slack).
(c) Pulleys out of alignment.	(c) Realign.
(d) Bearings dry.	(d) Lubricate.
(e) Defective belt.	(e) Replace.
(f) Belt rubbing.	(f) Reposition.
Blower will not run	
(a) Power not on.	(a) Check power switch; check fuses and replace if necessary.
(b) Fan control adjustment.	(b) Readjust or replace.
(c) Loose wiring.	(c) Check and tighten.
(d) Defective motor overload, protector, or motor.	(d) Replace motor.
(e) Defective or tight blower motor.	(e) Repair or replace.

Some Preliminary Troubleshooting Tips:

- Check the thermostat to make sure it is set higher than the actual room temperature. If you have a programmable thermostat, make sure it has fresh batteries.
- Make sure the selector switch is on heat if the system is equipped with central air or if the system is zoned.
- Check the emergency switch (usually a red switch plate at the top of the cellar stairs or on the side of the furnace) to see that it is on.
- If you are familiar with the fuse or circuit breaker panel, see whether the fuse is burned or the breaker tripped. Correct the problem at once. If it repeats, call a serviceman.
- On standing-pilot furnaces, the burner will not light if the pilot has gone out. If you are not familiar with the function of the gas valve or lighting the pilot, call for service.
- Do not disconnect any piping to check for gas supply. An instrument is used to check for pressure. This should be done by a qualified serviceman.
- If the furnace is vented through PVC (white plastic pipe) out the side of the building, examine the ends of the pipe or pipes outside. Blockage of any kind will cause a shutdown. Many high-efficiency, condensing furnaces vented through PVC pipe are self-diagnostic. These furnaces have a steady burning red light. If the light starts blinking, the furnace is indicating there is a problem that requires attention.

Troubleshooting Charts

Some manufacturers include troubleshooting charts in their furnace manuals. A typical gas furnace troubleshooting chart for a Thermo Pride gas furnace (Thermo Products, LLC) is illustrated in Figure 11-52. These troubleshooting charts are organized in a yes/no format, guiding the technician through a list of steps that eventually leads to the specific operating problem and its remedy.

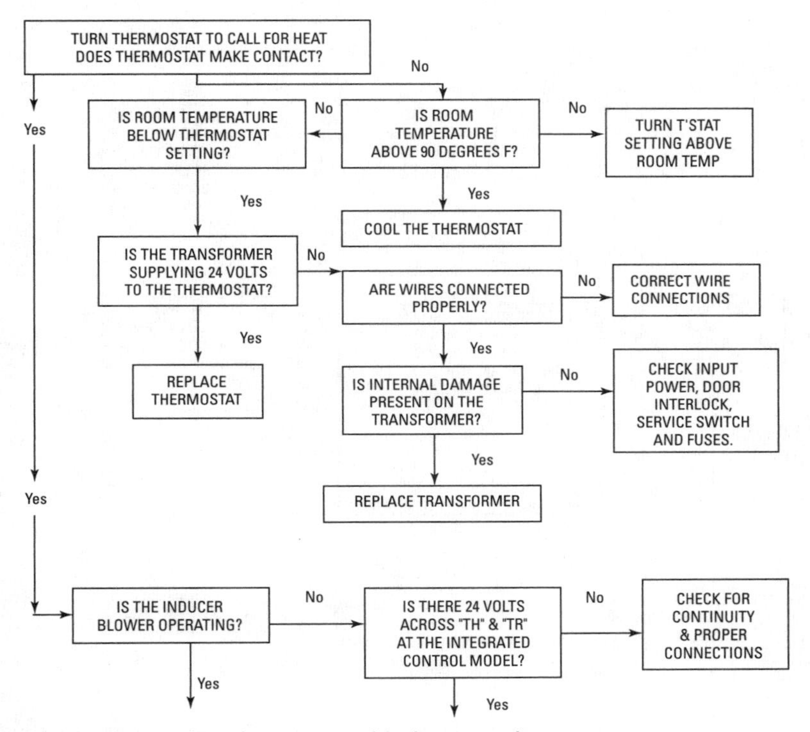

Figure 11-52 Gas furnace troubleshooting chart. *(Courtesy Thermo Pride)*

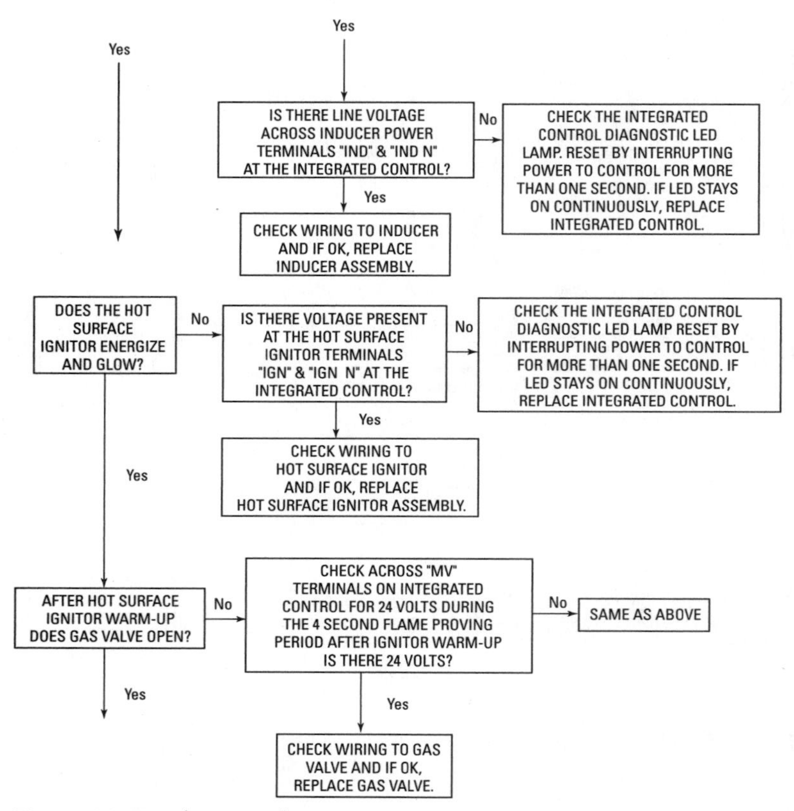

Figure 11-52 *(continued).*

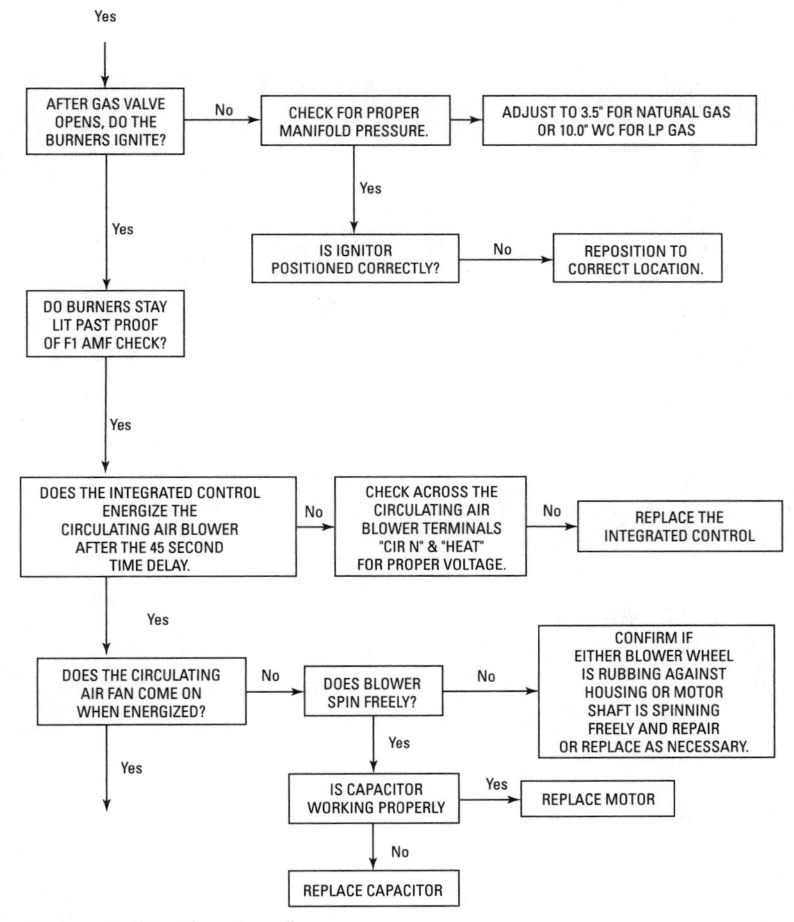

Figure 11-52 *(continued).*

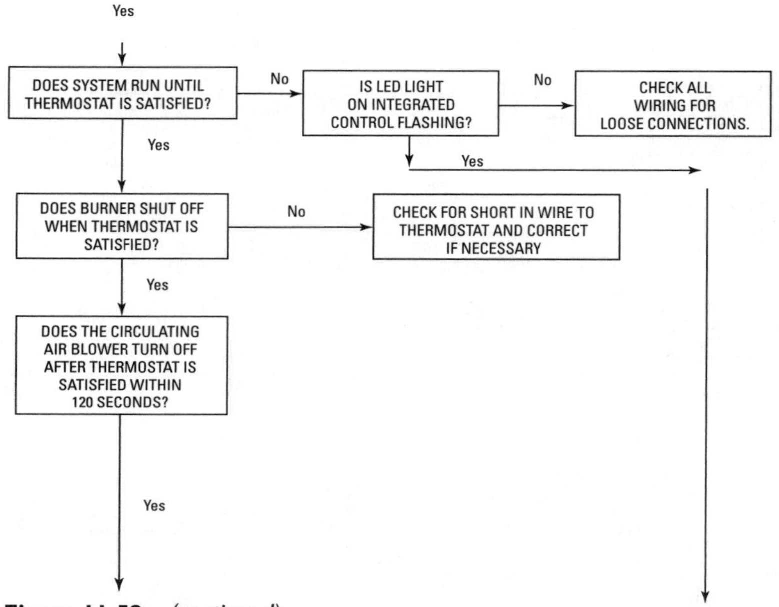

Figure 11-52 *(continued).*

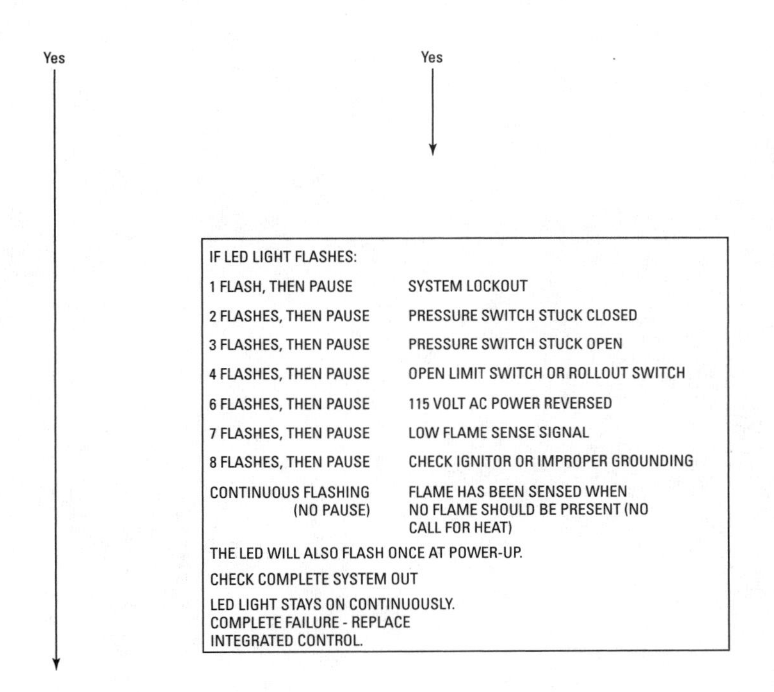

Figure 11-52 *(continued).*

Chapter 12

Oil Furnaces

Oil furnaces are available in upflow, downflow (counterflow), or horizontal-flow models and in a wide range of heating capacities for installations in attics, basements, closet spaces, or at floor level (Figures 12-1, 12-2, and 12-3). The furnace, oil burner, and controls are generally available from the manufacturer as a complete package for residential installations. All internal wiring is done at the factory.

Oil furnaces should be listed by Underwriters Laboratories, Inc. (UL), for construction and operating safety. Furnaces approved by the agency are marked *UL Approved*. Furnaces not made under UL standards should *not* be purchased or installed. Always closely follow the furnace manufacturer's installation, maintenance, and operating instructions. Note also that any local building codes covering furnace installation will take precedence.

Other sources of information concerning standards for oil furnace selection and installation are the publications *Standard for the Installation of Oil Burning Equipment, 2001 Edition* (NFPA No. 31) and *Installation of Warm Air Heating and Air Conditioning Systems*, 2002 Edition (NFPA 90B) from the National Fire Protection Association. See also *American Standard Performance Requirements for Oil-Powered Central Furnaces* (ANSI Z91.1-1972), a publication of the American National Standards Institute.

Conventional Oil Furnace

Most residential oil furnaces still in use today are the noncondensing, conventional-types installed before the mid 1980s. Since the 1980s, furnaces installed in new heating systems, and some replacement furnaces, are the newer mid-efficiency and high-efficiency oil furnaces. A conventional oil furnace with a cast-iron head burner has an AFUE (seasonal efficiency rating) of 60 percent.

The heating cycle of a conventional oil furnace begins when the thermostat calls for heat. This is accomplished in one of two ways: (1) The room temperature falls below the heat setting on the thermostat and the thermostat makes an automatic call for heat, or (2) someone turns the thermostat up manually to a warmer setting. In either case, the call for heat closes an electric circuit to a control relay, and power is sent to both the burner ignition transformer and the fuel pump motor.

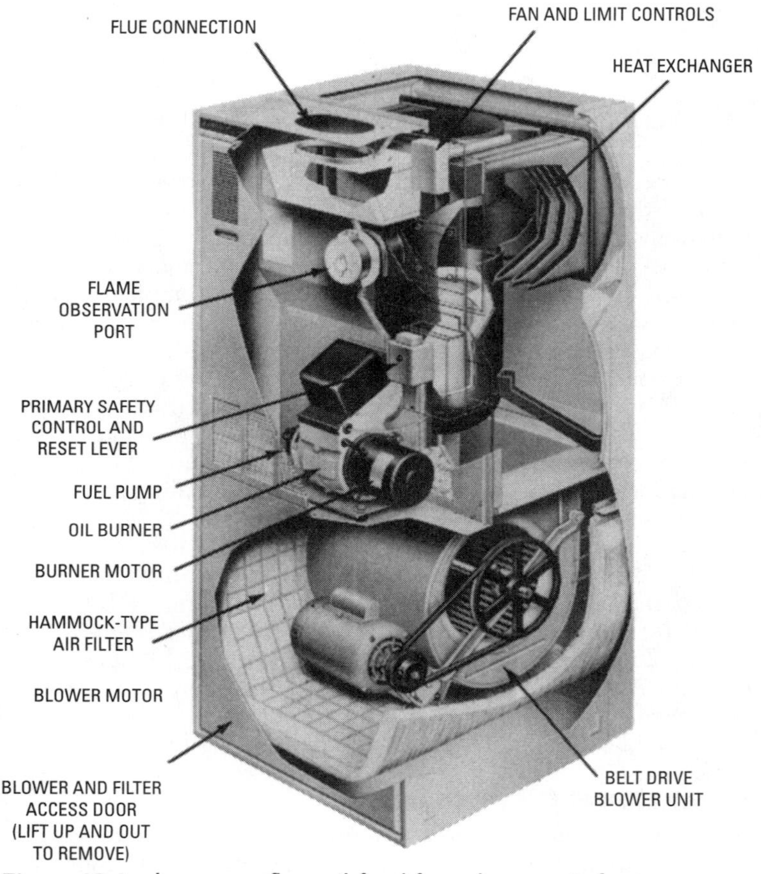

Figure 12-1 Lennox upflow oil-fired forced warm-air furnace.
(Courtesy Lennox Industries Inc.)

The fuel pump motor operates the oil pump, which draws fuel oil from the supply tank and sends it to the nozzle in the gun assembly, where it is combined with the combustion air and atomized. The combustion air is drawn into the combustion chamber by a blower wheel, which is attached to the pump motor shaft. The ignition transformer sends a high-voltage current to the electrodes in the gun assembly to ignite the atomized fuel-oil mixture. The oil furnace heating cycle begins.

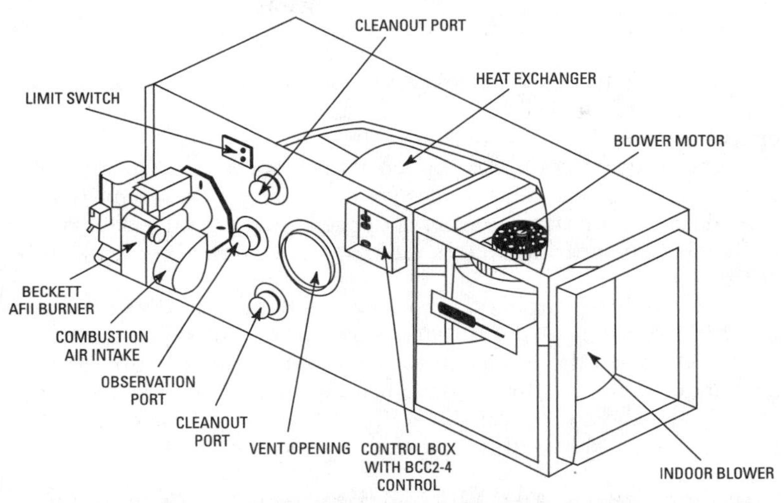

Figure 12-2 Lennox horizontal oil-fired forced warm-air furnace.
(Courtesy Lennox Industries Inc.)

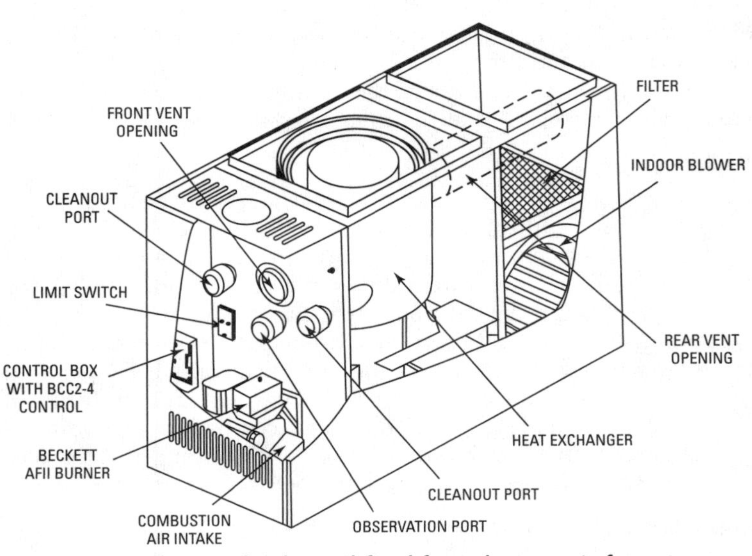

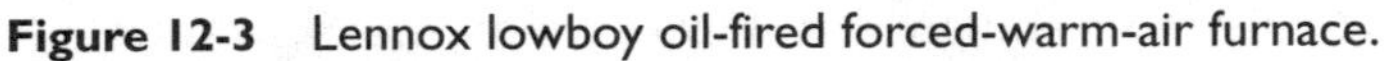
Figure 12-3 Lennox lowboy oil-fired forced-warm-air furnace.
(Courtesy Lennox Industries Inc.)

As soon as the heating cycle begins, a cadmium selenide photocell (called a *cad cell*) is activated as a safety device. The function of the cad cell is to detect the existence of a flame in the combustion chamber. If the cad cell fails to detect a flame within 15 seconds, the circuit is opened and the burner is shut off.

Some oil furnaces are equipped with a stack relay instead of a cad cell. The stack relay is designed to sense heat. If no heat is sensed, the circuit to the burner is opened and the unit is shut off. In both cases, the heating cycle is interrupted until the problem can be corrected.

As in gas-fired furnaces, a fan and limit control is used to regulate the operation of the furnace blower. The fan switch turns the blower on when there is a call for heat and warm air from the heat exchanger is forced through the heating ducts and into the structure. When the temperature reaches the setting on the thermostat, the fan control shuts off the blower to prevent overheating.

Mid-Efficiency and High-Efficiency Oil Furnaces

As was mentioned in Chapter 11 ("Gas Furnaces"), the government and furnace manufacturers have significantly improved the heating efficiency of residential gas furnaces. Similar improvements have been made in oil furnace technology. Two widely used methods, the AFUE measurement and the Energy Star program of the United States Environmental Protection Agency (EPA), are described in the sidebars. A furnace that meets the minimum efficiency requirements of these ratings is referred to as a *high-efficiency oil furnace.*

AFUE Rating

The energy efficiency of an oil furnace is measured by its annual fuel utilization capacity (AFUE). The AFUE ratings for furnaces manufactured today are listed in the furnace manufacturer's literature. Look for the EnerGuide emblem for the efficiency rating of that particular model. The higher the rating, the more efficient the furnace. The government has established a minimum rating for furnaces of 78 percent. Mid-efficiency furnaces have AFUE ratings ranging from 78 to 82 percent. High-efficiency furnaces have AFUE ratings ranging from 88 to 97 percent. The traditional oil furnaces have AFUE ratings of approximately 60 to 65 percent.

Energy Star Certification

The Energy Star Program is an energy performance rating system created by the EPA to identify and certify energy efficient products, including furnaces, boilers, heat pumps, and air conditioners. A broader description of the program is given in Chapter 11, "Gas Furnaces."

Mid-Efficiency (Noncondensing) Oil Furnace

A typical noncondensing, mid-efficiency oil furnace uses 28 to 33 percent less fuel oil than a conventional oil furnace to produce the same amount of heat. Its operation is characterized by significantly lower combustion and dilution air requirements.

The internal components of a typical mid-efficiency oil furnace are illustrated in Figure 12-4. The high-static oil burner fires into the combustion chamber, commonly made of a heat-resistant ceramic material. Air for combustion is drawn into an intake opening on the burner assembly by a small motor where it mixes with the fuel oil and ignites. The hot combustion gases then pass through the heat exchanger and are eventually expelled through an insulated flue pipe to the outdoors. A separate, larger blower (sometimes called the *indoor blower* or *furnace fan)*, forces air across the outer surface of the heat exchanger, extracts heat, and forces the heat out into the rooms and spaces of the house. The two air streams never mix.

Many of these mid-efficiency furnaces do not require a connection to a chimney. They pass the combustion gases though an insulated vent pipe that extends directly through the sidewall of the structure. A barometric damper is not required in the more efficient furnaces. Some use the forced draft of a high-static oil burner to vent the combustion gases. Others use sealed combustion with a high-static oil burner.

The mid-efficiency oil furnace is equipped with a safety shutoff device in case of draft problems. The other controls are similar to those found on conventional oil furnaces.

High-Efficiency (Condensing) Oil Furnace

A high-efficiency oil furnace (sometimes called a *condensing oil furnace*) is equipped with two heat exchangers to extract heat from the combustion gases before they leave the furnace (Figure 12-5). The second heat exchanger, which is commonly made of stainless steel, is used to recover the latent heat in the water vapor produced by combustion. This is accomplished by reducing the temperature of the combustion gases. The temperature of the flue gases exiting the furnace from the second heat exchanger drops to approximately 100 to 120°F. The lowered temperature causes the water vapor trapped in the flue gas to condense in the heat exchanger and

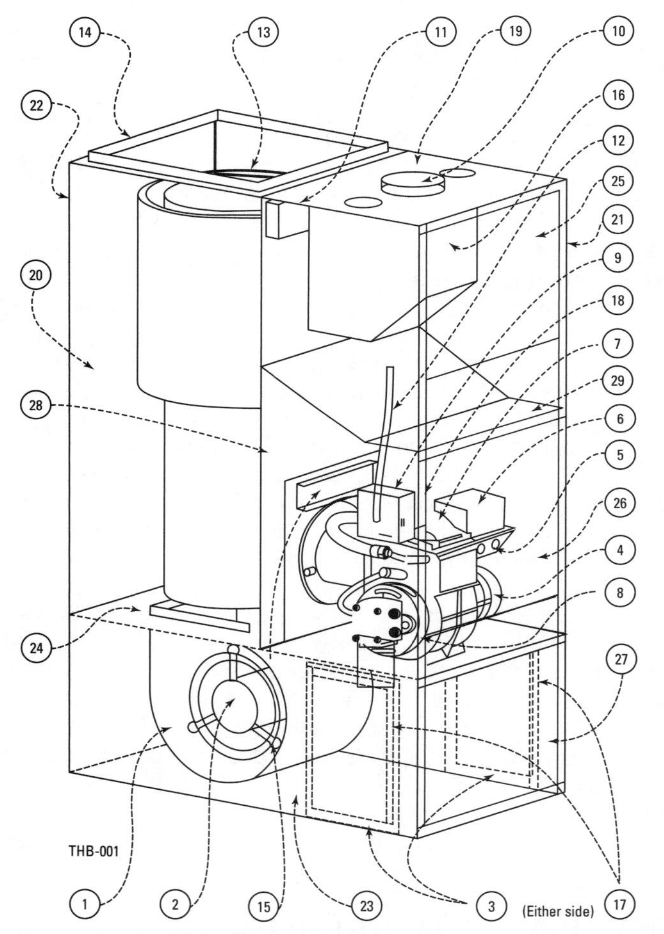

Figure 12-4 Mid-efficiency oil furnace.

1 Blower:
2 Blower motor.........⅓
3 16″ × 25″ Filter (either side)
4 On burner motor
5 Burner junction box
6 Primary safety control
7 Oil burner ignitor
8 Oil burner pump
9 Main junction box
10 Flue connection
11 Fan and limit control
12 Wiring harness
13 Heat exchanger
14 Supply air opening
15 Observation door
16 Flue collector
17 Return air opening (either side)
18 Fen center (direct-drive units only)
19 Top panel, jacket
20 Left panel jacket
21 Right panel, jacket
22 Rear panel, jacket
23 Base assembly
24 Blower pan
25 Top door
26 Burner access panel
27 Blower access door
28 Front divider
29 Spillway baffle

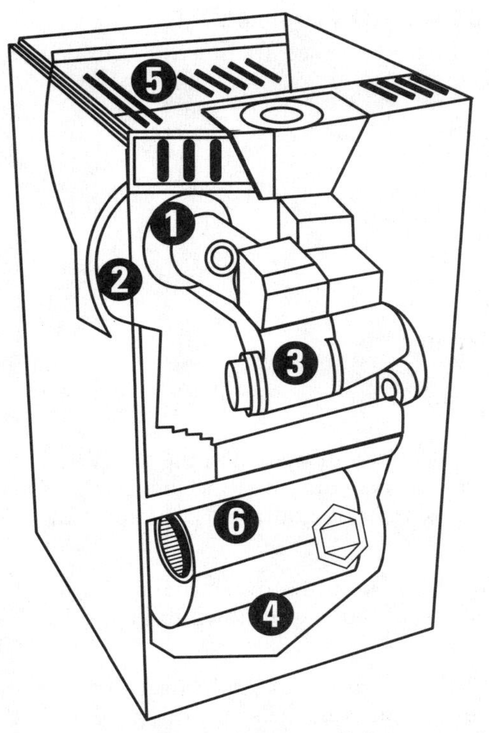

Figure 12-5 High-efficiency oil furnace.

release its latent heat. The recovered latent heat is circulated through the house, the condensate flows out of the bottom of the furnace to a nearby drain, and the cooled combustion gases are exhausted through a plastic vent pipe installed in a sidewall.

The combustion process of an oil furnace produces only about half the water vapor that a gas furnace does. This means that there is significantly less water vapor trapped in the exhaust gases of an oil furnace and, consequently, much less latent heat to recover. The smaller amount of water vapor produced in these furnaces means that the flue gas also has a lower dew point. As a result, the furnace has to work harder to recover less energy. As a result, the high-efficiency oil furnace is not much more energy efficient than the mid-efficiency oil furnace, in contrast to the situation with gas furnaces.

Basic Components of an Oil Furnace

The principal components of an oil furnace are illustrated in Figure 12-6. They are listed as follows:

1. Furnace controls
2. Heat exchanger
3. Burner assembly
4. Fuel pump and motor
5. Blower and motor
6. Combustion blowers
7. Cleanout and observation ports
8. Vent openings
9. Air filter(s)

Each of these components is briefly described in the sections that follow. More detailed information is contained in Chapter 1, "Oil Burners," and Chapter 5, "Gas and Oil Controls" of Volume 2.

Furnace Controls

A number of different controls are used to govern the operation of an oil furnace. Some, but not all of them, are shown in Figure 12-7.

Thermostat

A room thermostat controls the operation of the furnace. The thermostat senses air-temperature changes in the space or spaces being heated, closes an electrical circuit, and starts the ignition process. See Chapter 4, "Thermostats and Humidistats" of Volume 2 for a detailed description of thermostats.

Cad Cell

The cad cell is the principal safety device on an oil furnace. It is located inside the burner assembly in back of the burner access panel. The function of the cad cell is to prove (detect and verify) the burner flame at the start of the heating cycle. In this respect, it provides the same function as the thermocouple in a gas furnace. If the cad cell cannot prove the burner flame, it opens the circuit to the burner, which shuts off the burner motor and the ignition transformer. A detailed description of cad cells can be found in Chapter 5, "Gas and Oil Controls" in Volume 2.

Fan Controls

The *fan and limit control* is a switching and safety device commonly located in a metal box attached to the outside of the furnace

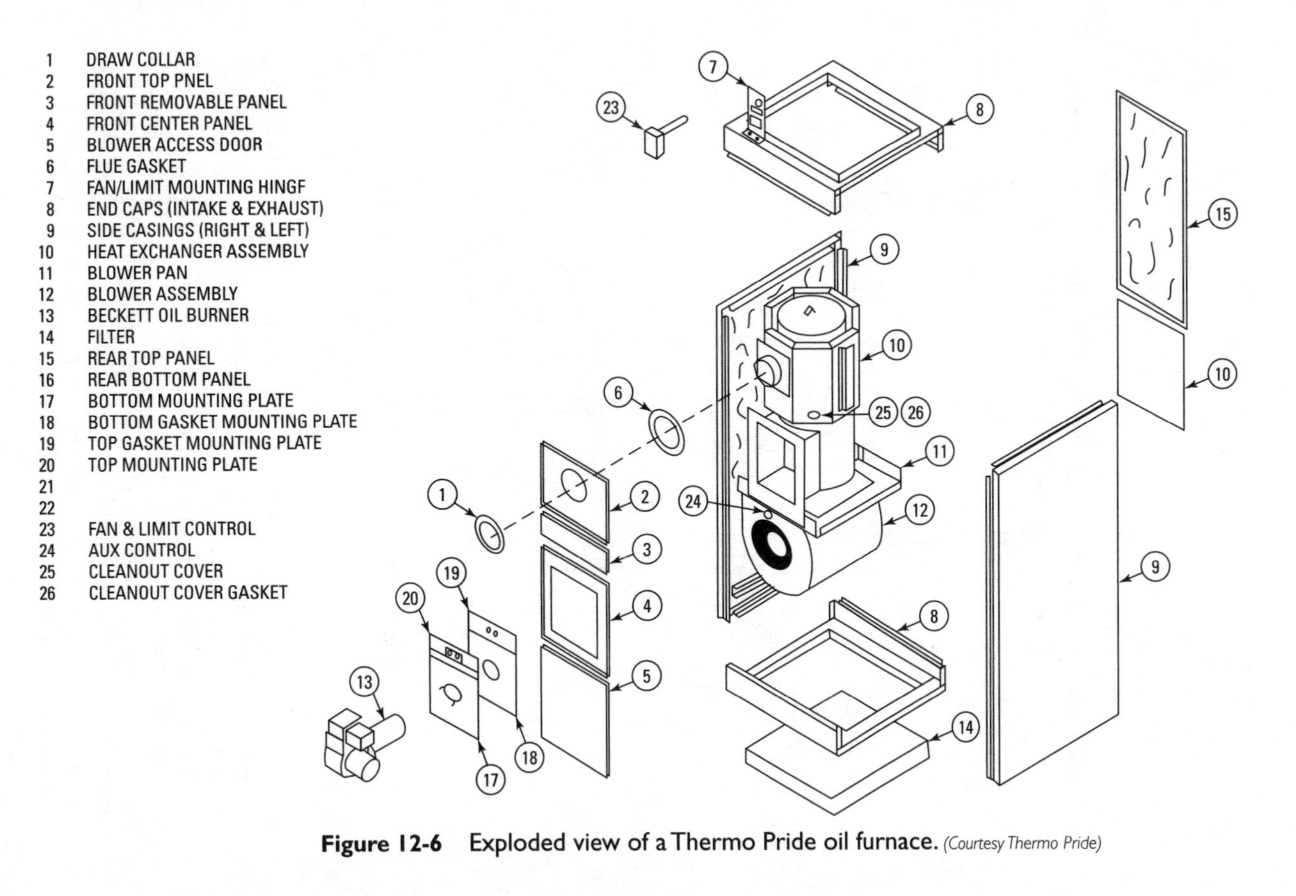

Figure 12-6 Exploded view of a Thermo Pride oil furnace. *(Courtesy Thermo Pride)*

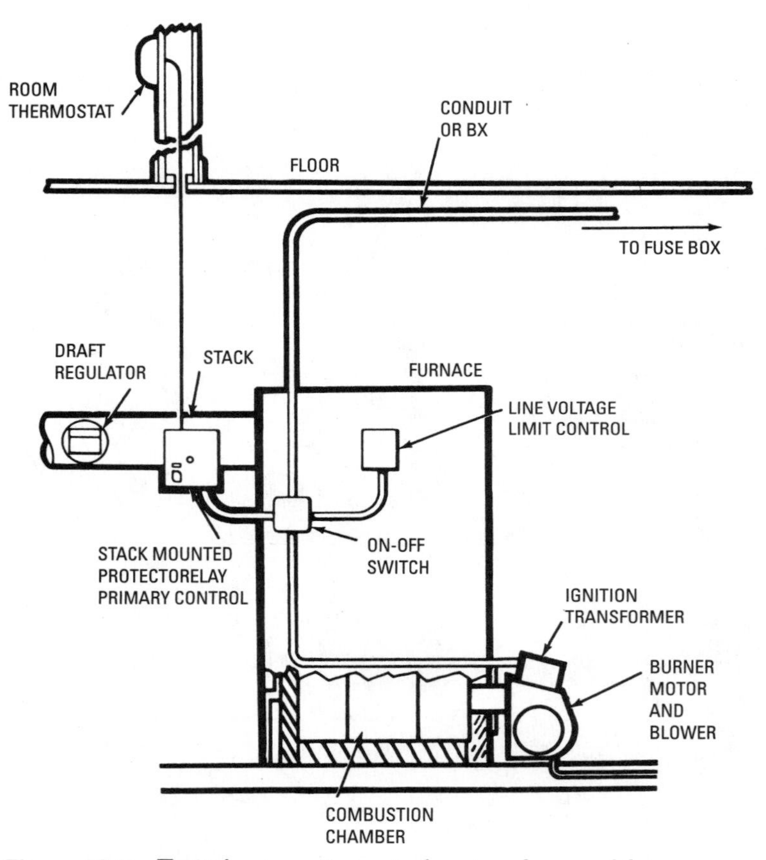

Figure 12-7 Typical automatic control system for an oil furnace.
(Courtesy Honeywell Tradeline Controls)

(Figure 12-8). Its function is to turn the blower on and off within a set temperature range and to shut off the burner if the furnace gets too hot.

The fan and limit control has two temperature settings: an upper one and a lower one. When the temperature in the house reaches the upper setting on the fan control, the fan control circuit opens and turns off the burner. The blower continues to run until the temperature in the heat exchanger reaches the lower temperature setting on the fan control, opens the circuit, and turns off the blower.

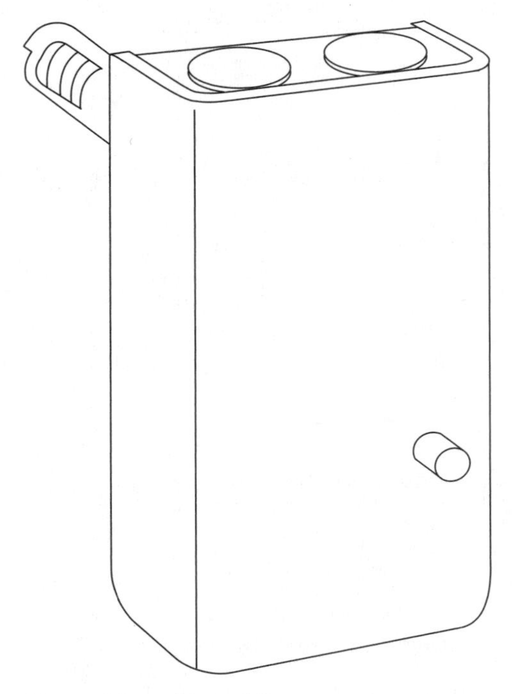

Figure 12-8 Typical fan control.
(Courtesy Honeywell, Inc.)

Figure 12-9 illustrates the internal components of a typical fan control box. The Fan On and Fan Off pointers on the dial are used to establish the temperature range settings. The Safety Limit control is normally set by the manufacturer and should not be adjusted. Always set the Auto/Manual Switch to Auto before reinstalling the fan control cover. A detailed description of fan and limit controls is found in Chapter 6, "Other Automatic Controls" of Volume 2.

Some modern oil furnaces are equipped with electronic fan controls instead of the fan and limit control switches. The control either is a separate box attached to the furnace or is an integral part of the electronic (solid-state) control board.

Delayed-Action Solenoid Valve

A delayed-action solenoid valve fitted to a conventional oil burner will prevent oil furnace spillage (see *Oil Furnace Spillage* in this chapter). The valve is designed to delay the flow of oil to the

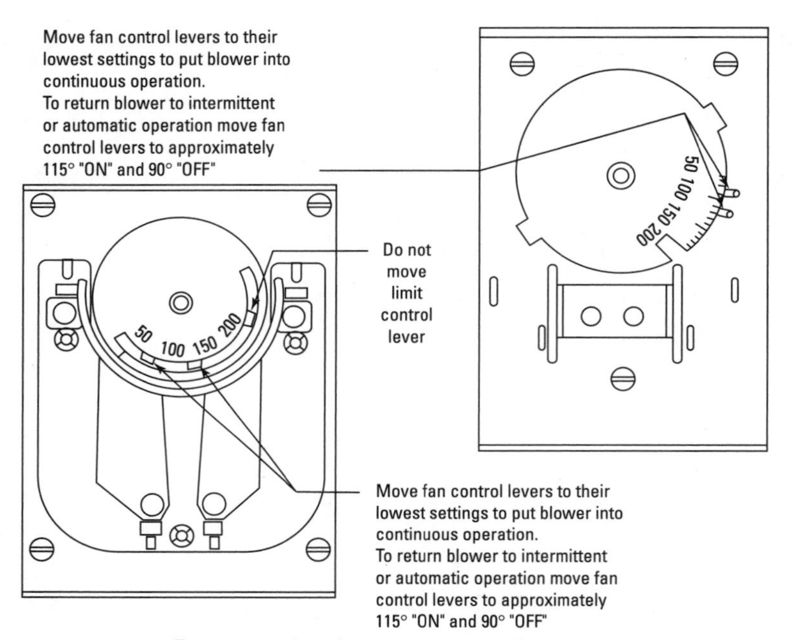

Figure 12-9 Fan control with cover removed. *(Courtesy Honeywell, Inc.)*

combustion chamber a few seconds after the burner starts. The delay allows the burner time to establish a draft for a more complete and efficient combustion of the oil. It also eliminates sooty buildup and backdrafting.

Heat Exchanger

The heat exchanger assembly is the section of the furnace used to transfer the heat of the combustion process to the air being circulated through the ducts of the heating system. The heat exchanger is generally constructed of heavy gauge steel (14 and 16 gauge). The wrap-around, radiator-type heat exchanger is one of the most commonly used designs. It consists of an upper and lower chamber, each with an extension, or pouch, that must be aligned when assembled. The lower part contains the combustion chamber.

The combustion chamber is that portion of the furnace within which the combustion process takes place. It is surrounded by the lower portion of the heat exchanger. The combustion chamber must be made of a material capable of withstanding the high temperatures of the combustion process. Typical combustion chamber materials are stainless steel or some sort of refractory material such as firebrick or a kiln-fired ceramic clay. As shown in Figure 12-10, the nozzle and gun assembly of the oil burner extends through an opening in the heat exchanger to another opening in the combustion chamber.

Round exchangers are the most common type, although square, rectangular, and other shapes do occur. The octagonal heat exchanger illustrated in Figure 12-11 is specific to Thermo Pride oil furnaces.

Some modern oil furnaces are equipped with sealed combustion chambers. Their purpose is to prevent spillage and backdrafts, while at the same time increasing furnace efficiency. Furnaces with sealed combustion chambers sometimes experience startup problems

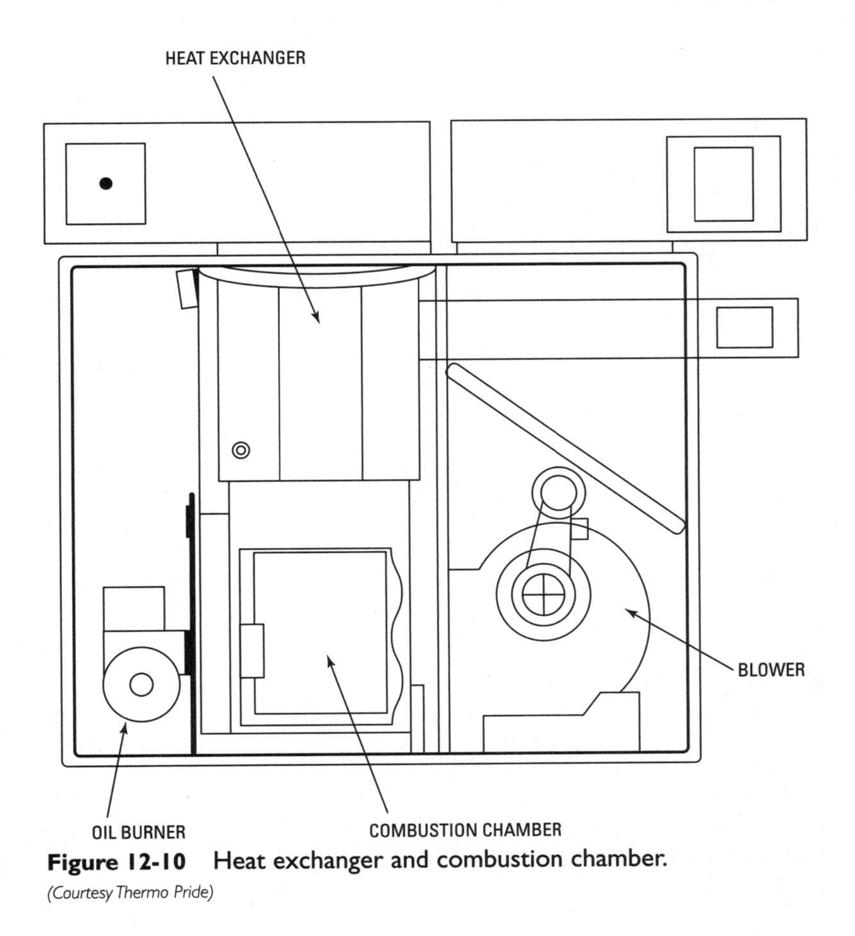

Figure 12-10 Heat exchanger and combustion chamber.
(Courtesy Thermo Pride)

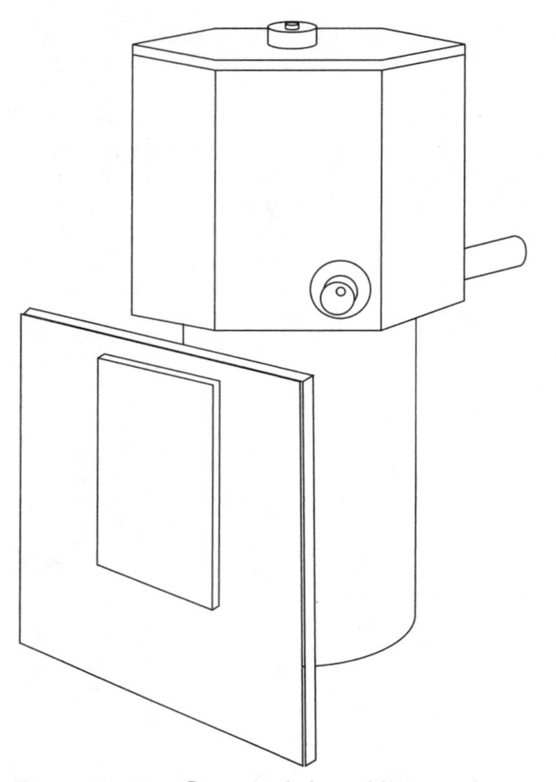

Figure 12-11 Octagonal shaped heat exchanger.
(Courtesy Thermo Pride)

because air delivered to the burners is too cold. The cold combustion air cools the fuel oil, causing the burner ignition problem. It can be solved by warming the air before it reaches the burner.

Oil Burner Assembly

The oil burner assembly consists of the ignition transformer, cad cell, gun assembly, oil nozzle, and electrodes (Figure 12-12). The oil burner is a mechanical device used to combine fuel oil with the proper amount of air for combustion and deliver it to the point of ignition. Oil burners and burner assemblies are described in considerable detail in Chapter 1, "Oil Burners" of Volume 2.

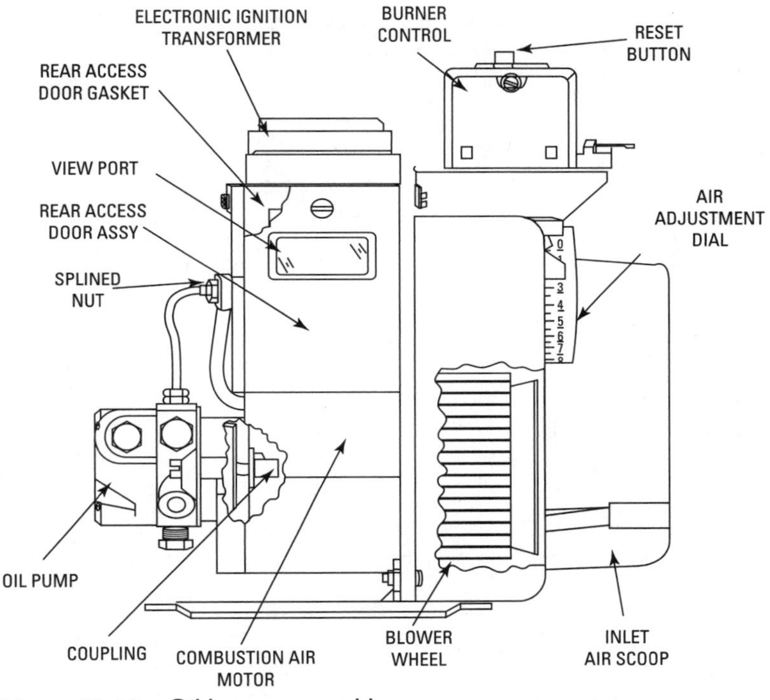

Figure 12-12 **Oil burner assembly.** *(Courtesy Lennox Industries Inc.)*

The *pressure-type oil burner* is the one most commonly used in residential oil furnaces, hot-water, and space heating systems. A *vaporizing oil burner* is sometimes used in small residential furnaces.

One of the first improvements in oil burner technology after the oil crisis in the 1970s was the introduction of more burners with flame-retention heads. These efficient designs were followed by the even more efficient high-static oil burners.

Fuel Pump and Fuel Pump Motor

The fuel pump (Figure 12-13) draws oil from the storage tank and sends it to the gun assembly at pressures between 100 and 140 psi. The fuel pump is connected to the pump motor by the burner coupling. Both the fuel pump and the fuel pump motor are described in Chapter 1, "Oil Burners" in Volume 2.

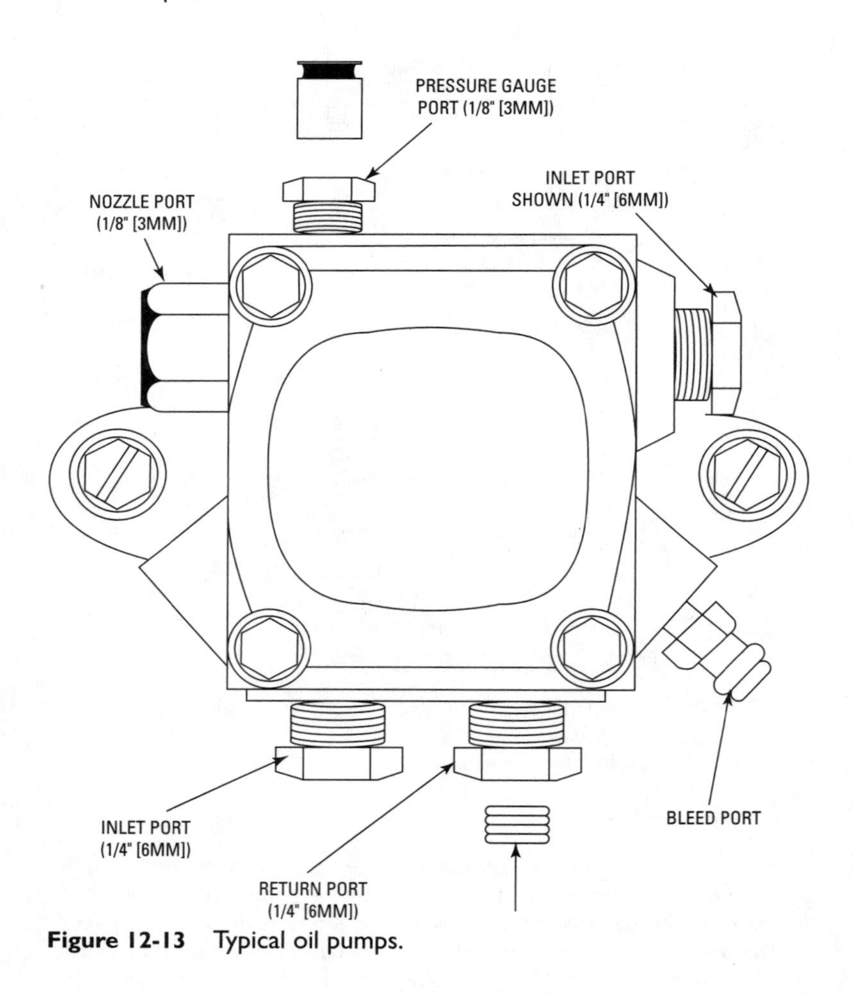

Figure 12-13 Typical oil pumps.

Blower and Motor

The blower assembly consists of the blower wheel housing, the blower wheel, the blower motor, and the blower motor capacitor. The function of the blower is to draw air into the furnace through the return air duct, force the air around the internal surfaces of the heat exchanger to pick up heat, and then deliver the warm air to the interior of the house through the supply ducts. The combustion air required for the combustion process is drawn in through an air inlet scoop on the side of the blower wheel housing. The amount of combustion air is regulated by an air adjustment dial located on the outside of the housing. It can be manually changed by moving the dial setting.

As shown in Figure 12-14, the blower motor is attached to the side of the blower wheel housing. The motor shaft rotates the blower wheel inside the blower wheel housing. The blower motor capacitor, which assists in starting the blower motor, is attached to the outside edge of the blower wheel housing.

Both direct-drive and belt-drive blowers are used with oil-fired furnaces. These blowers and motors are identical to those used in gas-fired furnaces. Read the section *Blowers and Motors* in Chapter 11, "Gas Furnaces" for additional information. The function of the blower is to blow heat through the duct system into the rooms of the structure.

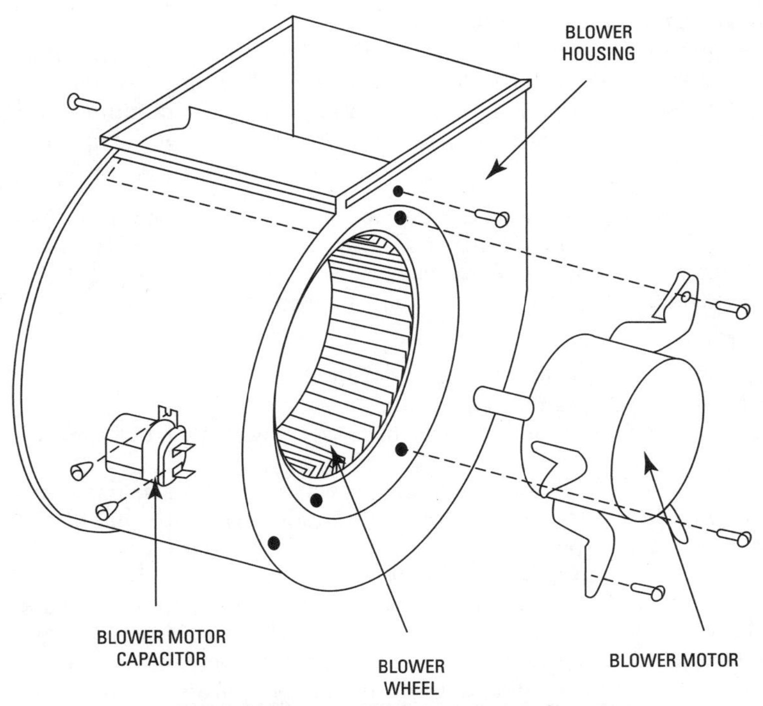

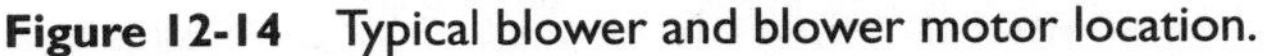
Figure 12-14 Typical blower and blower motor location.

(Courtesy Lennox Industries Inc.)

Oil furnaces equipped with direct-drive blowers operated by commutating motors are capable of adjusting the furnace heat output and blower speed to the heat requirements of the structure. In other words, their operation increases with the demand for heat and decreases when there is none, thereby improving indoor comfort while reducing energy costs. These furnaces are sometimes called *dual-capacity, variable-speed oil furnaces*.

Combustion Air Blowers

High-efficiency oil furnaces are often equipped with small blowers designed to provide a steady supply of combustion air to the burner. These small combustion air blowers operate on not more than 1⁄6 horsepower or less.

Pressure switches are used with combustion air blowers to sense blower speed and whether the flue is blocked. Older conventional oil furnaces used centrifugal switches instead of pressure switches.

Cleanout and Observation Ports

As shown in Figure 12-2, an oil furnace typically is provided with one or two cleanout ports to allow access to the heat exchanger for cleaning (soot removal). The observation port allows monitoring of the inside of the heat exchanger during the operation of the furnace.

Vent Opening

The vent opening connects the furnace to a flue pipe for venting the combustion gases to the outdoors. Many of the modern oil furnaces connect the flue pipe to a sidewall of the house. Traditional oil furnaces vented these gases through a flue pipe connected to the chimney.

Air Filter

A forced-warm-air oil furnace is supplied with either a disposable air filter or a permanent (washable) one. A filter is not used with a gravity warm-air oil furnace, because it will obstruct the airflow.

A disposable air filter should be inspected on a regular basis and replaced when dirty. Always replace the air filter with one of the same size and type of filter media. This information is usually found on a label attached to the filter.

A permanent air filter must also be regularly inspected. When it is dirty, it must be removed and cleaned. The usual method is to vacuum it and then to wash it in a soap or detergent and water solution. Additional information about air filters is found in Chapter 11, "Gas Furnaces," and Chapter 12, "Air Cleaners and Filters" in Volume 3.

Installing an Oil Furnace

An oil furnace must be properly set up and installed by a certified oil heat technician, a representative of the oil furnace manufacturer, or an HVAC professional with equivalent experience.

The size (Btu output capacity) of the oil furnace selected for the heating system is determined by the calculated heat loss for the structure. See Chapter 4, "Sizing Residential Heating and Air Conditioning Systems."

Some Installation Recommendations

The installation of an oil furnace must comply with the local building codes and the manufacturer's installation recommendations. As was mentioned at the beginning of this chapter, additional useful and important information is contained in the National Fire Protection Association's *Installation of Warm Air Heating and Air-Conditioning Systems,* 2002 edition and the *Standard for the Installation of Oil Burning Equipment,* 2001 Edition (NFPA No. 31). The appropriate sections of the *Uniform Mechanical Code*, 2003 edition, should also be consulted.

New furnaces for residential installations are shipped preassembled from the factory with all internal wiring completed. The fuel supply tank must be installed and the fuel line connected to the furnace by the installer. The electrical service from the line-voltage main and the low-voltage thermostat will also have to be connected. Directions for making these connections are included in the furnace manufacturer's installation literature.

Make certain you have familiarized yourself with all local codes and regulations that govern the installation of an oil-fired furnace. Local codes and regulations *always* take precedence over national standards. Read the section *Installation Recommendations* in Chapter 11, "Gas Furnaces" for additional information.

Location and Clearance

An oil furnace should be located as near as possible to the center of the heat-distribution system and chimney. A centralized location for the furnace usually results in the best operating characteristics because long supply ducts and the heat loss associated with them are eliminated. The furnace should be installed as close as possible to the chimney to reduce the horizontal run of flue pipe.

An oil furnace should be located a safe distance from any combustible material. Observe the minimum clearances suggested by the manufacturer for the furnace except where they may come into conflict with local codes and regulations (Table 12-1). Local standards *always* take precedence over the manufacturer's specifications or national standards. Where there are no local standards specifying clearances, use those listed in *Installation of Oil Burning Equipment 1972* (NFPA No. 31-1972). Contents of the 1972 edition have been approved by the American National Standards Institute (ANSI).

Allow access for servicing the furnace and oil burner. Sufficient clearance should also be provided for access to the barometric draft-control assembly.

Wiring

Always follow all national and local codes when wiring a unit. The local code should take precedence when there is a conflict between

Table 12-1 Minimum Clearances Specified for Carrier Oil-Fired Upflow (58HV), Lowboy Upflow (58HL), Downflow (58HC), and Horizontal (58HH) Furnace Models

	Furnace				
	58HV	***58HL***	***58HC***	***58HH***	
From	***Clearance (in)***				
Top of furnace or plenum	2	2	2	6	2[a]
Bottom of furnace	0	0	—	6	2[a]
Front of furnace	15	24	15	24	48[a]
Rear of furnace	1	6	2	6	2[a]
Sides of furnace	1	6	1	6	4[a]
Top of horizontal warm-air duct within 6 ft of furnace	2	—	4	6	3[a]
Flue pipe: horizontally or below pipe	9	—	9	—	—
Vertically	18	18	15	18	9[a]
Any side of supply plenum and warm-air duct within 6 ft of furnace	2	—	6	—	—

[a]*Alcove installation.*

Courtesy Carrier Corp.

the two. If there are no local ordinances and codes, do all electrical wiring in accordance with the *National Electrical Code* (ANSI/NFPA 70-1999).

Closely follow the wiring diagrams provided by the furnace manufacturer and those who manufacture the automatic controls. Failure to do this may result in losing the factory warranty.

A typical wiring diagram for an oil furnace is shown in Figure 12-15. The comments made in Chapter 11, "Gas Furnaces" concerning line-voltage and control-voltage wiring connections also pertain to the wiring of oil furnaces.

Duct Connections

Detailed information concerning the installation of an air-duct system is contained in the following two publications of the National Fire Protection Association:

1. *Installation of Air Conditioning and Ventilating Systems of Other Than Residence Type, 2002 edition* (NFPA No. 90A)
2. *Residence Type Warm Air Heating and Air Conditioning Systems, 2002 edition* (NFPA. No. 90B)

Additional information about duct connections can be found in Chapter 7 of Volume 2, "Ducts and Duct Systems." The specific recommendations concerning furnace duct connections and the air-distribution ducts contained in Chapter 11, "Gas Furnaces" also apply to gas-fired furnaces.

Ventilation and Combustion Air

An oil-fired furnace should be located where a sufficient supply of air is available for combustion, proper venting, and the maintenance of suitable ambient temperature. In buildings of exceptionally tight construction, an outside air supply should be introduced for ventilation and combustion purposes.

If the furnace is located in a confined space such as a furnace room, closet, or utility room, the enclosure should be provided with two permanent openings for the passage of the air supply. One opening should be located approximately 6 in from the top of the enclosure, and the other opening approximately 6 in from the bottom. Each opening should have free area of at least 1 in^2 per 1000 Btu/h (minimum size 100 in^2) of the total input rating, or 1 ft^2 per gallon of oil per hour.

The size of the openings must correspond to the bonnet capacity of the furnace. Furnace manufacturers generally recommend

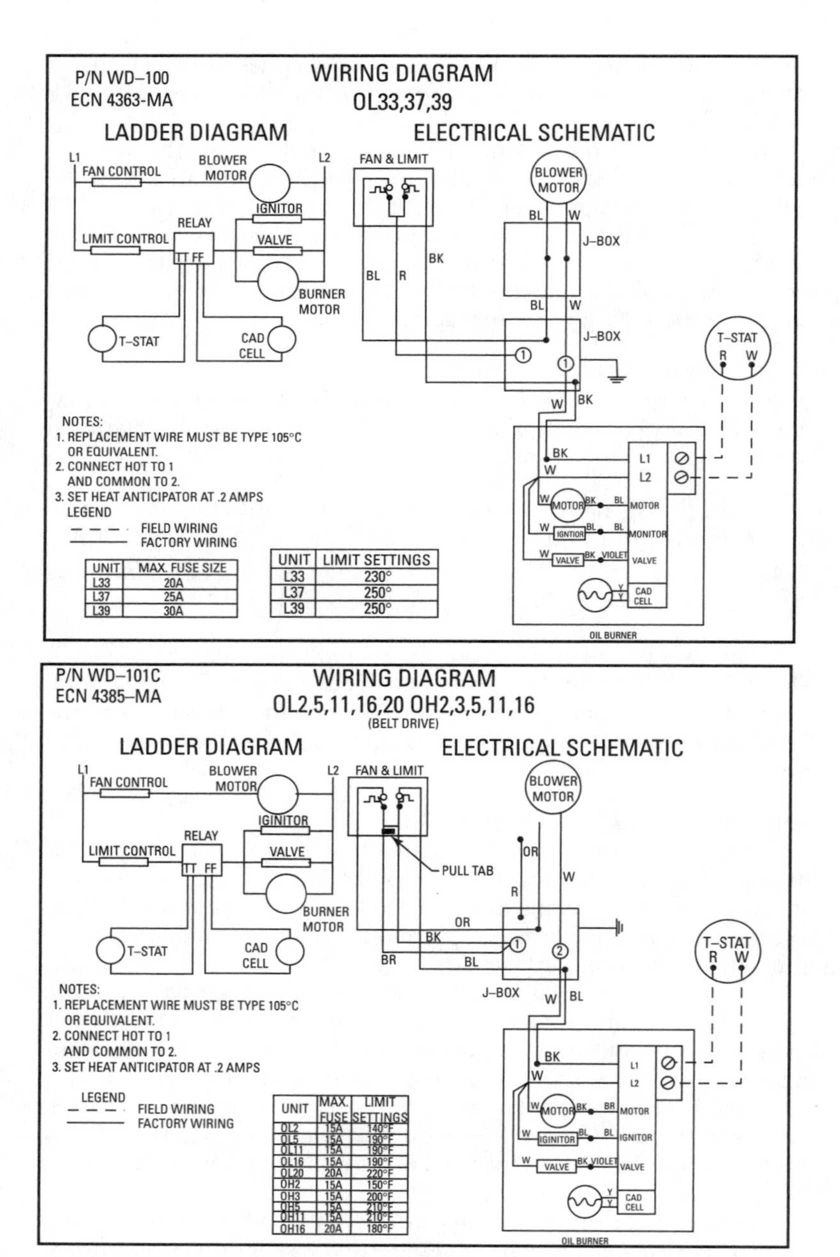

UNIT	MAX. FUSE SIZE
L33	20A
L37	25A
L39	30A

UNIT	LIMIT SETTINGS
L33	230°
L37	250°
L39	250°

UNIT	MAX. FUSE	LIMIT SETTINGS
OL2	15A	140°F
OL5	15A	190°F
OL11	15A	190°F
OL16	15A	190°F
OL20	20A	220°F
OH2	15A	150°F
OH3	15A	200°F
OH5	15A	210°F
OH11	15A	210°F
OH16	20A	180°F

Figure 12-15 Wiring diagrams for various Thermo Pride oil furnace models. *(Courtesy Thermo Pride)*

Table 12-2 Recommended Sizes of Ventilation Openings for Trane Counterflow Oil Furnace

Bonnet Capacity, 1000 Btu/h	*Length, in*	*Height, in*
85	24	12
100	24	12
125	26	13

Courtesy The Trane Co.

suitable sizes for air openings in their equipment specifications (Table 12-2).

Air from an attic or a crawl space can also serve as a ventilation and combustion air supply to a confined furnace. Suitable openings near (or on) the floor and near (or on) the ceiling must be provided for the passage of the air.

When an oil furnace is installed in an unconfined area, such as a full basement, the full amount of air necessary for combustion is generally supplied by infiltration. Only when the construction is exceptionally tight is it necessary to provide openings to the outdoors.

Combustion Draft

An oil-fired furnace requires sufficient draft for the proper burning of oil. The average combustion draft is usually 0.02 in wg over the fire (the so-called *overfire draft*), and between 0.04 and 0.06 in wg draft in the stack (the *flue draft*) for units having inputs up to 2.0 gph. For higher inputs, 0.06 to 0.08 in wg will be the rule. *Always* consult the furnace manufacturer's operating manual first for recommended drafts.

Venting

Oil furnaces use the same methods for venting the products of combustion to the outside air as do gas furnaces, except when the venting method is specifically limited to a particular fuel (for example, Type B gas vents). Additional information about venting methods is contained in Chapter 11, "Gas Furnaces." The type of venting to be used will be recommended by the furnace manufacturer in the installation literature.

Most high-efficiency (condensing) oil furnaces vent their exhaust gases through a single plastic flue pipe extending from the furnace through a nearby sidewall. The flue vent pipe in these installations is not connected to the chimney.

In some condensing furnace installations, two vents are used. One vent consists of the plastic flue pipe running horizontally from the furnace to the sidewall; the other consists of the conventional chimney and flue liner. Both flue vents remain open during the combustion process.

The strong forced draft produced by a high-static oil burner is sufficient to propel combustion gases through the flue in sidewall-venting installations. Some oil furnaces use a small induced fan installed in the flue. The fan vents the combustion gases through a sidewall vent opening.

Flue Pipe

Use 24-gauge galvanized flue pipe (or its equivalent) for the flue connection between the furnace and the chimney. The horizontal run of flue pipe should slope upward at least 1/4 in per running foot. The pitch should not exceed 75 percent of the vertical vent. Locate the unit so that the furnace flue outlet is no more than 10 flue pipe diameters from the chimney connection. If the distance is greater, use the next-size-larger flue pipe.

Insulate the flue pipe when it passes near combustible material. The UL requires that any uninsulated flue pipe be installed with a *minimum* clearance of 6 in from any combustible materials.

Narrow plastic pipe is used to exhaust the flue gases from mid-efficiency and high-efficiency oil furnaces. The trade names of these plastic vent systems are *ULTRAVENT®* (gray plastic flue pipe with gray fittings), *SELVENT®* (black plastic flue pipe with black fittings), and *PLEXVENT®* (black plastic flue pipe with yellow or gray fittings).

Warning

> The plastic venting systems of mid-efficiency and high-efficiency oil furnaces installed after 1988 may crack or separate at the joints and cause carbon monoxide to leak into the house. The entire vent system should be replaced to avoid possible exposure to carbon monoxide.

Barometric Damper

A *barometric damper (also called a draft regulator* or *barometric draft regulator)* is a device designed to maintain a proper draft in an oil furnace by automatically reducing (diluting) the chimney draft (that is, the flue draft) to the desired value (Figure 12-16). Its function is similar to that of the draft hood on a gas furnace. It isolates

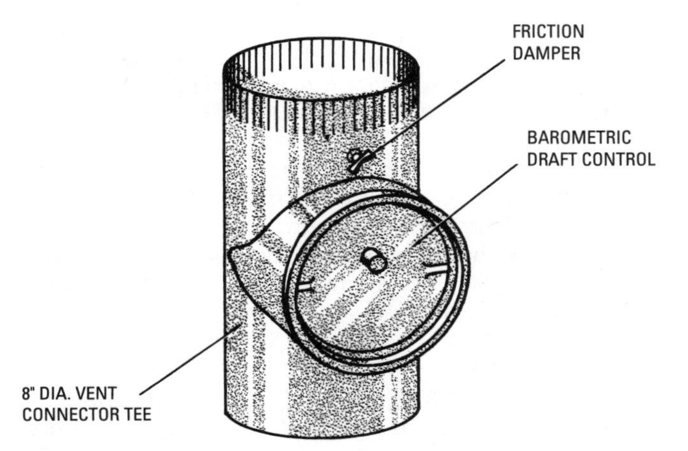

Figure 12-16 Barometric draft-control assembly.
(Courtesy Hydrotherm, Inc.)

the oil burner from changes in pressure at the chimney opening by pulling heated warm air into the exhaust.

Draft regulators are recommended on all installations for each oil-fired appliance connected to a chimney *unless* the appliance is listed for use without one.

Do not use a small draft regulator, because it will not do a satisfactory job. The draft regulator should be installed in the horizontal flue pipe as close as possible to the chimney. When this flue pipe is too short for a satisfactory installation, the draft regulator should be installed in the chimney either above or below the flue pipe entry into the chimney. Some authorities oppose installing the draft regulator in the chimney, but experience has shown that this can be a very satisfactory installation.

Chimneys and Chimney Troubleshooting

The chimney used with an oil-fired furnace should be of sufficient height and large enough in its cross-sectional area to meet the requirements of the furnace. The chimney area should be at least 20 percent greater than the area of the flue outlet. To ensure the adequate removal of flue gases, the height and size of the chimney must be sufficient to create 0.02 to 0.04 in wg draft over the fire for inputs up to 2.0 gph. For higher inputs, 0.06 to 0.08 in wg is recommended. If the stack height is limited, it may be necessary to apply mechanically induced draft to the installation.

Ideally, *only* the furnace should be connected to the chimney. Read the section *Chimneys* in Chapter 11, "Gas Furnaces" for additional information. See also Chapter 3, "Fireplaces, Stoves, and Chimneys" in Volume 3 for additional information about chimneys.

Read the section *Chimney Troubleshooting* in Chapter 11, "Gas Furnaces" for a description of common chimney problems and suggested remedies. The chimneys used with oil-fired furnaces are identical to those used with gas furnaces. Additional information about chimney and flues is covered in Chapter 3, "Fireplaces, Stoves, and Chimneys" of Volume 3.

Installation Checklist

A properly installed furnace will operate efficiently and economically. The following installation checklist is offered as a guide to the installer.

1. Be sure there is at least 0.01-in wg draft over the fire.
2. Check for sufficient combustion and ventilation air.
3. Eliminate downdraft or backdraft.
4. Provide sufficient space for service accessibility.
5. Check all field wiring.
6. Supply-line fuse or circuit breaker must be of proper size and type for furnace.
7. Line voltage must meet specifications while furnace is operating.
8. Ductwork must be checked for proper balance, velocity, and quietness.
9. Check the fuel line for leaks.
10. Cycle the burner.
11. Check the limit switch.
12. Check the fan switch.
13. Make final adjustments to the fire.
14. Adjust the blower motor for desired speed.
15. Make sure air filter is properly secured.
16. Make sure all access panels have been secured.
17. Pitch air conditioning equipment condensate lines toward a drain.
18. Check thermostat heat anticipator setting.

19. Check thermostat for normal operation. Observe at least five ignition cycles before leaving the installation.
20. Clear and clean the area around the furnace.

Fuel Supply Tank and Fuel Line

The fuel oil for an oil furnace is stored in a *fuel supply tank*. The tank is commonly connected directly to the fuel pump in the burner assembly of the furnace.

The fuel supply tank and fuel line must be installed in accordance with the requirements of UL and any local codes and regulations. Local codes and ordinances take precedence over national standards.

Note

> Although an existing tank probably has already met the requirements of the local codes and regulations, it would be prudent to check just to make certain. This is especially true of older, buried fuel supply tanks, because they occasionally leak. A leaking fuel supply tank violates the environmental laws in many areas of the country.

Additional information about fuel supply tanks can be obtained from *Installation of Burning Equipment 1972* (NFPA No. 31), a publication of the National Fire Protection Association (NFPA). Use only an approved tank and line for the installation.

A one-pipe system is recommended for an oil-tank installation when the following conditions are present:

1. Tank is installed in the basement or similar location and/or below the pump inlet port.
2. Vacuum at pump does not exceed 2 in.

A two-pipe (return line) system is recommended for a tank installation when the following conditions exist:

1. Tank is located outside and/or below the oil-pump inlet port.
2. Vacuum at pump does not exceed 10 in.

Use a two-stage pump for installations having especially long lines or high lifts *if* the vacuum at the inlet port does not exceed 20 in.

A shutoff valve should be installed in the suction line close to the oil burner. This valve enables the operator to disconnect the line without draining it. Do *not* place a shutoff valve in the return line.

Closing a shutoff valve on the return line while the oil burner is operating could damage the pump. All joints in the fuel line should be sealed with a suitable oil-piping compound.

Note

> A leak in the oil line will cause a loss of pump prime and a failure of the burner to light within the required 15 seconds of the ignition cycle.

Fuel Tank Location

The location of the oil supply tank is subject to local regulations. These must be consulted before a *new* tank is installed.

The supply tank can be located inside or outside the building, above the level of the oil burner or below it. Furthermore, outside tanks can be located underground or above it. It is recommended that larger supply tanks be located outside and underground.

Figures 12-17 through 12-20 illustrate four ways to locate oil supply tanks. When installing a fuel tank, the following suggestions should prove helpful:

1. The filler pipe should be a *minimum* of 2 inches in diameter; the vent pipe 1¼ inches.
2. Use wrought-iron pipe with malleable-iron fittings for both the filler and vent pipes.

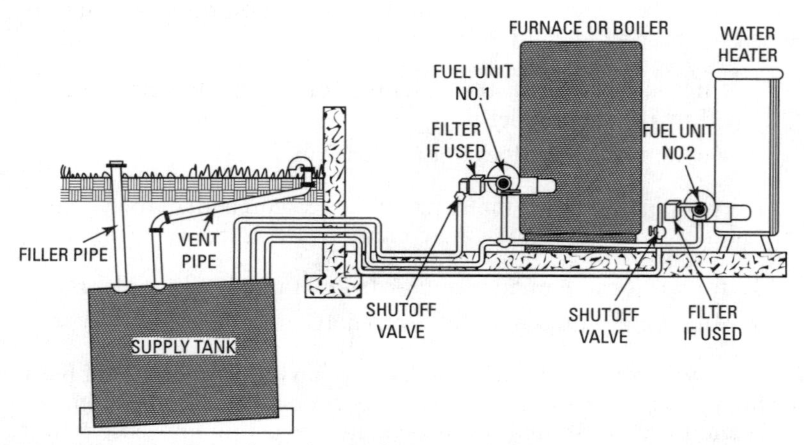

Figure 12-17 Outside tank installation with the tank located below the level of the oil burner. *(Courtesy Sundstrand Hydraulics)*

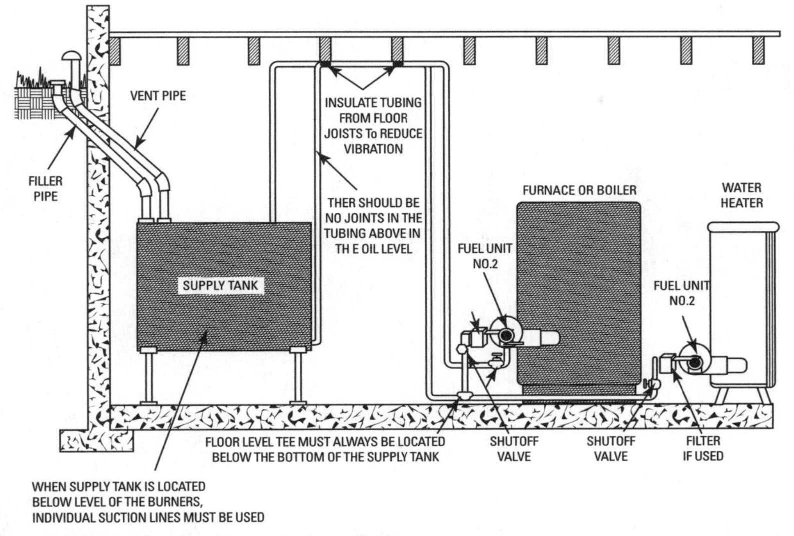

Figure 12-18 Inside tank installation. *(Courtesy Sundstrand Hydraulics)*

3. Coat only the *male* thread of the pipe with a pipe compound suitable for use with oil burning equipment.
4. Oil supply lines between the oil supply tank and the oil burner should be made of copper tubing (diameters will vary depending

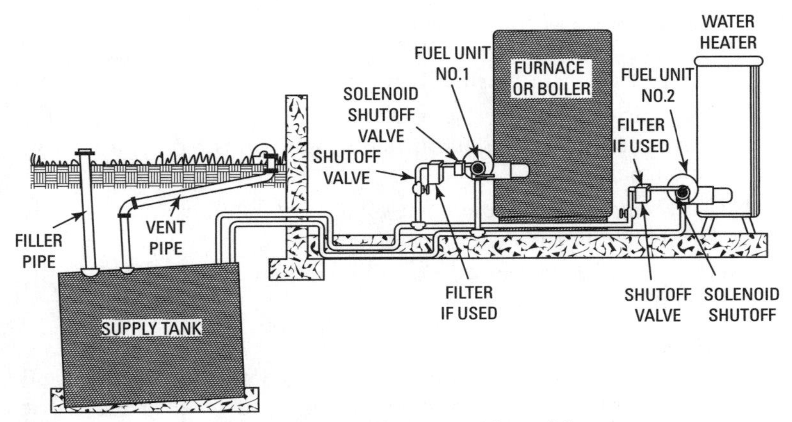

Figure 12-19 Outside tank installation involving lift.
(Courtesy Sundstrand Hydraulics)

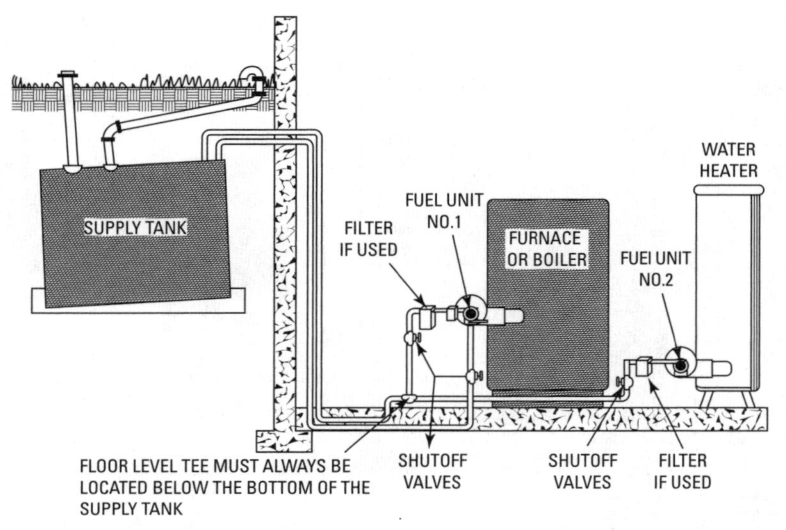

Figure 12-20 Outside tank installation with the tank located above the level of the burner. *(Courtesy Sundstrand Hydraulics)*

on local regulations, pipe length, and the specifications of the oil burner being used).

5. Use a floor-level tee if the oil supply lines run overhead.
6. The oil burners for the water heater and furnace (or boiler) may be connected to a common feed line in conventional gravity feed installations (Figure 12-20).
7. No return line is necessary when the supply tank is installed above the level of the oil burner and the oil is fed by gravity to the burner.
8. Use a single-line system if the oil is gravity fed from the supply tank to the burner.
9. Use a two-line system if the oil tank is buried and below the level of the burner.

Filler Pipe

The *filler pipe* is the filling connection located on the top of the oil supply tank and terminating above the ground level at least 2 ft from the outside building wall. This is commonly a 2-in-diameter pipe of a design and material specified by the local authorities.

Generally, the specifications will mandate the use of a corrosion-resistant material.

It is important that the filler pipe be connected to one opening and the vent pipe to a separate opening on the tank. No cross-connection of vent pipe is permitted with the filler pipe or the return line from the oil burner. Some authorities demand double-swing connections at the oil supply tank.

The termination point of the filler point at ground level should be equipped with a watertight metal cap. The termination point should be of such design that oil spillage is minimized when the oil hose is disconnected.

Tank Vent Pipe

The vent pipe is attached to the oil supply tank at a point separate from that of the filler pipe, and it should terminate at a point above the ground at least 2 ft from the outside wall of the building. The vent pipe must be equipped with an approved vent hood or weatherproof cap to prevent water and other contaminants from entering the pipe and gaining access to the oil supply tank.

Oil Filter

A cartridge-type oil filter should be installed on the fuel line inside the building and as close to the fuel tank as possible. One filter is adequate on oil burners operating at more than 1.00 gph.

Change the filter cartridge at least once a year. The filter body should be thoroughly cleaned before a new cartridge is installed.

Note

> A cartridge-type oil filter in the oil line will prevent sludge in the oil clogging the fuel pump strainer or the oil burner nozzle. Both of these problems are a common cause of system failure.

Blowers and Motors

Both direct-drive and belt-drive blowers are used with oil-fired furnaces. These blowers and motors are identical to those used in gas-fired furnaces. Read the section *Blowers and Motors* in Chapter 11, "Gas Furnaces" for additional information.

Waste Oil Furnaces

Waste oil furnaces have been used for years in garages and other commercial establishments where a sufficient supply of waste oil is produced. Instead of paying a third party to haul away the waste oil, it can be cleanly burned in a waste oil furnace. Burning the waste oil on the premises not only provides heat for the structure (at no cost) but also eliminates the expense of having the oil hauled away. A waste oil burner and furnace are designed to burn the waste oil without emitting smoke and with very little odor.

Waste oil is a surprisingly efficient heating fuel. Tests have shown that it contains from 183,000 to 240,000 Btu of energy per gallon. That is more than twice the energy potential of a gallon of natural gas or propane.

Note

> Waste oil furnaces are not UL approved for use inside a residential structure. On the other hand, they can be installed in a separate structure with the heat being forced into the residence through ducts.

A typical waste oil furnace is illustrated in Figure 12-21. It consists of a multi-fuel burner, pumps, tank strainer, one-way check valve, a filter, a vacuum gauge, and a low-volt wall thermostat. Some waste oil furnaces are also provided with barometric dampers, pumps, and air compressors.

Waste oil furnaces are relatively free of operating problems. Those problems that do occur are commonly traced to incorrect installation procedures. They include (1) using fuel pipes that are too small in diameter, (2) contaminants in the fuel, and (3) improper chimney installation or parts.

Periodic maintenance includes cleaning the reusable filter by rinsing it in a solvent bath and vacuuming the ash and dust residue from inside the dual cleanout doors at the end of each day. Both procedures will maximize the operating efficiency of the waste oil furnace.

Air Conditioning

A furnace used in conjunction with a cooling unit should be installed in parallel or on the upstream side of the evaporator coil to avoid condensation on the heating element. In a parallel installation, dampers or comparable means should be provided to prevent chilled air from entering the furnace.

Duct sizing should include an allowance for air conditioning, even though it may not be initially installed with the furnace. Air conditioning involves a greater volume of air than heating.

All ductwork located in unconditioned areas (attics, crawl spaces, and so on) *downstream* from the furnace should be insulated to prevent unwanted heat gain.

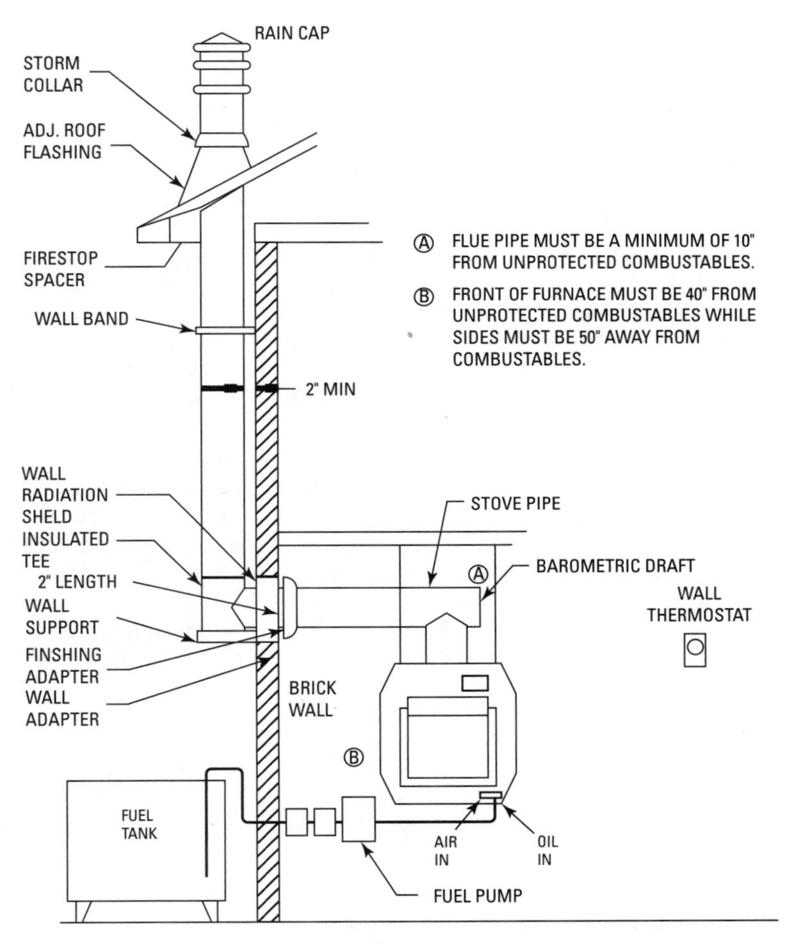

Figure 12-21 KAGI waste oil furnace. *(Courtesy Kagi)*

Starting the Burner

The procedure for starting an oil burner may be summarized as follows:

1. Open all warm-air registers.
2. Check to be sure all return air grilles are unobstructed.

3. Open the valve on the oil supply line.
4. Reset the burner primary relay.
5. Set the thermostat above the room temperature.
6. Turn on the electric supply to the unit by setting the main electrical switch to the *on* position.
7. Change the room thermostat setting to the desired temperature.

The oil burner should start after the electric power has been switched on (Step 6). There is no pilot to light as is the case with gas-fired appliances. The spark for ignition is provided automatically on demand from the room thermostat. Allow the burner to operate at least 10 minutes before making any final adjustments. Whenever possible, use instruments to adjust the flame.

Oil Furnace Spillage

In oil-fired furnaces, the inefficient incomplete combustion of the oil will cause sulfur dioxide (SO_2) gas to spill from the furnace. Spillage commonly occurs during the few seconds during the initial startup of a conventional oil furnace, which is not a problem, but it can occur more often, and well beyond the initial startup period, in a poorly maintained furnace. Sulfur dioxide gas in large quantities can be harmful to the lungs. It has a strong, unpleasant odor. The problem can be eliminated by:

1. Fitting a delayed-action solenoid valve to the gas burner (see *Delayed-Action Solenoid Valve* in this chapter) to increase the efficiency of the combustion process.
2. Install a well-balanced barometric damper in the flue pipe.
3. Seal all flue pipe joints against leaks.

Oil Furnace Inspection, Service, and Maintenance Tips

Oil furnaces should be inspected annually, preferably before the beginning of the heating season, by a qualified service technician to ensure continued safe operation. The inspection should include the following:

Caution

Always carefully read the service and maintenance instruction for the furnace first. As a rule, the electrical power is shut off first at the disconnect switch if service or an inspection is to be performed. After the power is switched off, the gas valve is then turned off.

1. Inspect the vent pipe for water accumulation, sagging piping, dirt, loose joints, and damage.
2. Check the return air duct for a tight connection to the furnace. The duct connection must provide an airtight seal at the furnace and must terminate at its other end outside the room.
3. Inspect the furnace wiring for burnt or damaged wires and loose connections.
4. Inspect interior and exterior furnace surfaces for dirt or water accumulation.
5. Make sure the blower access door is tightly closed.
6. Inspect the burners for dirt, rust, or signs of water.
7. Make sure the fresh air grilles and louvers are open, clean, and unobstructed.
8. Inspect and clean the condensate traps and drain to prevent water accumulation in the furnace (condensing furnaces).
9. Inspect the blower wheel and remove any debris.
10. Check the furnace support and base. The base of the furnace must form a tight seal with the support.
11. Run the furnace and observe its operation. The furnace should operate smoothly and quietly.
12. Check the vent pipe and return duct to make sure they are not leaking.
13. Analyze the combustion gases to make sure they meet the furnace manufacturer's specifications.
14. Clean or change the air filter once a month or as needed, if less than a month.
15. Clean and adjust oil burner before the start of the heating season.
16. Inspect oil supply tank for water and possible contaminants daily or at the beginning of the heating season.
17. Drain oil supply tank at the end of the heating season and check for cracks or other damage.
18. Clean heat exchanger and combustion chamber at the beginning of the heating season and whenever there appears to be a

significant accumulation of soot and dirt. Remove soot and dirt with a long-handled brush and a vacuum hose.

19. Inspect the inside and outside of the chimney for damage.
20. Inspect and clean the chimney before the beginning of the heating season.
21. Clean the furnace flue pipe, barometric damper, and chimney base.
22. Clean the fan/blower blades.
23. Clean or replace the air filter.
24. Fill the small oil cups of a belt-driven blower motor once or twice during the heating season. Direct-drive blowers with internal motors do not require periodic oiling.
25. Check the oil pressure in the burner.
26. Check all burner fittings for leaks.
27. Clean the oil filter bowl and replace, if required.
28. Clean the burner.
29. Check the condition of the burner nozzle and replace if dirty. Dirty oil burner nozzles cannot be cleaned and returned to service.
30. Check the operation of the cad cell flame detector.
31. Check the operation of the thermostat and fan controls.

Oil Furnace Adjustments

After inspecting the heating system and making any necessary service and repairs, the oil furnace should be adjusted for maximum operating efficiency. This is accomplished by taking four measurements through a small (½ inch diameter) hole in the flue pipe very close to the furnace. First, run the furnace for at least 15 minutes to establish a steady and uniform flue pressure. Then, take the following measurements:

1. Check the smoke contact of the flue gases.
2. Check the draft pressure of the flue gases.
3. Check the temperature of the flue gases.
4. Check the carbon monoxide or oxygen content of the flue gases.

Adjustments to the oil furnace and the burner will be made on the basis of the four measurements listed above. Adjustment must be made in accordance with the furnace, burner, or control manufacturer's procedures listed in its service, maintenance, and repair manuals.

Troubleshooting Oil Furnaces

Table 12-3 contains the most common operating problems associated with oil-fired furnaces. Each problem is given in the form of a symptom, the possible cause, and a suggested remedy. This list is intended to provide the operator with a quick reference to the cause and correction of a specific problem.

Table 12-3 Troubleshooting an Oil-Fired Furnace

Symptom and Possible Cause	*Possible Remedy*
Change in size of fire	
(a) Dirty nozzle.	(a) Clean or replace.
(b) Low pressure.	(b) Adjust at pump.
(c) Plugged strainer.	(c) Clean.
(d) Cold oil.	(d) Adjust pressure.
No oil flow	
(a) Oil level below intake line in the supply tank.	(a) Fill tank.
(b) Clogged strainer.	(b) Remove and clean strainer.
(c) Clogged nozzle.	(c) Clean or replace.
(d) Air leak in intake line.	(d) Tighten fittings and plugs; check valves.
(e) Restricted intake line (high vacuum reading).	(e) Replace kinked tubing; check valves, filters.
(f) Two-pipe system air bound.	(f) Check bypass plug.
(g) Single-pipe system air bound.	(g) Loosen gauge port and drain oil until foam is gone.
(h) Slipping or broken coupling.	(h) Tighten or replace coupling.
(i) Frozen pump shaft.	(i) Replace.
Oil spray but no ignition	
(a) Dirty electrodes.	(a) Clean or replace.
(b) Improper spacing.	(b) Reset.
(c) Cracked porcelain.	(c) Replace.
(d) Dead transformer.	(d) Replace.
(e) Loose connection.	(e) Tighten.
(f) Faulty relay.	(f) Replace.

(continued)

Table 12-3 *(continued)*

Symptom and Possible Cause	*Possible Remedy*
Burner motor does not start	
(a) Defective thermostat.	(a) Replace.
(b) Fuse burned out.	(b) Replace.
(c) Limit control open.	(c) Check setting and correct.
(d) Contact dirty or open on primary relay.	(d) Clean or replace relay.
(e) Relay transformer burned out.	(e) Replace relay.
(f) Motor stuck or burned out or overload protector out.	(f) Replace if burned out.
(g) Primary relay off on safety.	(g) Push reset button.
Noisy operation	
(a) Bad coupling alignment.	(a) Loosen fuel unit or motor.
(b) Loose coupling.	(b) Tighten setscrews.
(c) Air in oil line.	(c) Bleed oil line; look for leaks.
(d) Pump noise.	(d) Continued running sometimes works in gears. If not, replace.
(e) Hum vibration.	(e) Isolate pipes from structural members.
(f) Combustion noise.	(f) Adjust noise.
(g) Furnace too small.	(g) Check heat loss to be sure furnace properly sized.
(h) Burner noisy.	(h) Check mounting and position; adjust air.
(i) Blower noisy.	(i) Oil bearing. Tighten shaft collars; adjust belt tension; align and tighten pulleys; position rubber isolators.
Burner will not run continuously	
(a) Lockout timing too short.	(a) Replace primary control.
(b) Poor flame due to too much air; too little oil.	(b) Check nozzle, air adjustment, oil pressure, and size of nozzle.
(c) Water or air in oil.	(c) Look for leak in supply.
(d) Control wired wrong.	(d) Check and rewire.
Pulsation	
(a) Air adjustment.	(a) Readjust air.
(b) Pressure over fire.	(b) Correct draft to 0.02 in W.C. negative.

(continued)

Table 12-3 *(continued)*

Symptom and Possible Cause	*Possible Remedy*
(c) Dirty or improperly set electrodes.	(c) Clean and reset; wire primary control for continuous ignition.
(d) Too much oil impingement.	(d) Check nozzle and pump pressure; check nozzle size and angle and position of drawer assembly.
Short cycling of fan	
(a) Fan control setting.	(a) Set lower turn on (115°F).
(b) Input too low.	(b) Check burner input.
(c) Temperature rise too low due to excessive speed of blower.	(c) Slow blower down and check ventilation.
Short cycling on limit control	
(a) Limit setting low.	(a) Reset to maximum.
(b) Input too high.	(b) Check burner input.
(c) Temperature rise too high due to blower running too slow.	(c) Increase blower speed.
(d) Temperature rise too high due to restricted returns or outlets.	(d) Open dampers or add additional outlets or returns.
(e) Fan control setting too high.	(e) Reset lower (115°F).
(f) Control out of position.	(f) Place cad cell in proper position.
High fuel consumption	
(a) Input too high.	(a) Check burner input.
(b) Flue loss too great.	(b) Measure CO^2 and flue-gas temperature; if loss is more than 25%, reset air, check input, and speed up blower. Check static pressure in return and outlet plenum and correct to recommended values.
Not heating	
(a) Low input.	(a) Check nozzles and input.
(b) Insufficient air circulating.	(b) Speed up blower. Check size and location of ducts and outlets. Set fan control and blower for continuous air circulation.

Troubleshooting Charts

Some manufacturers include troubleshooting charts in their furnace manuals. A typical oil furnace troubleshooting chart for a *Thermo Pride* oil furnace is illustrated in Figure 12-22. These troubleshooting charts are organized in a yes/no format, guiding the technician through a list of steps that eventually lead to the specific operating problem and its remedy.

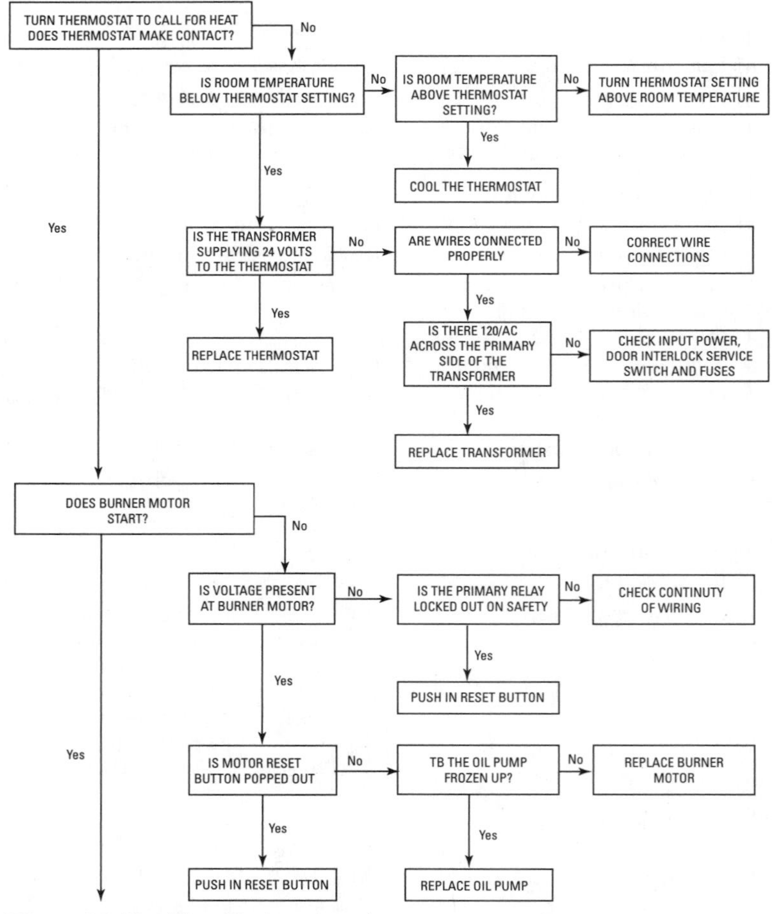

Figure 12-22 Troubleshooting chart. *(Courtesy Thermo Pride)*

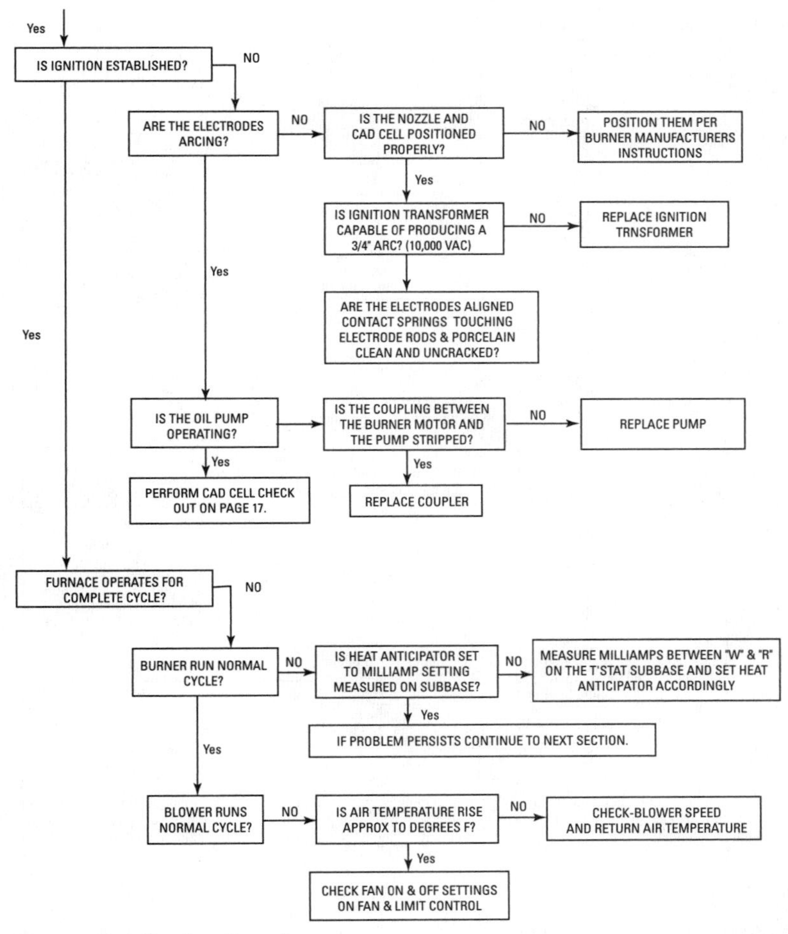

Figure 12-22 *(continued).*

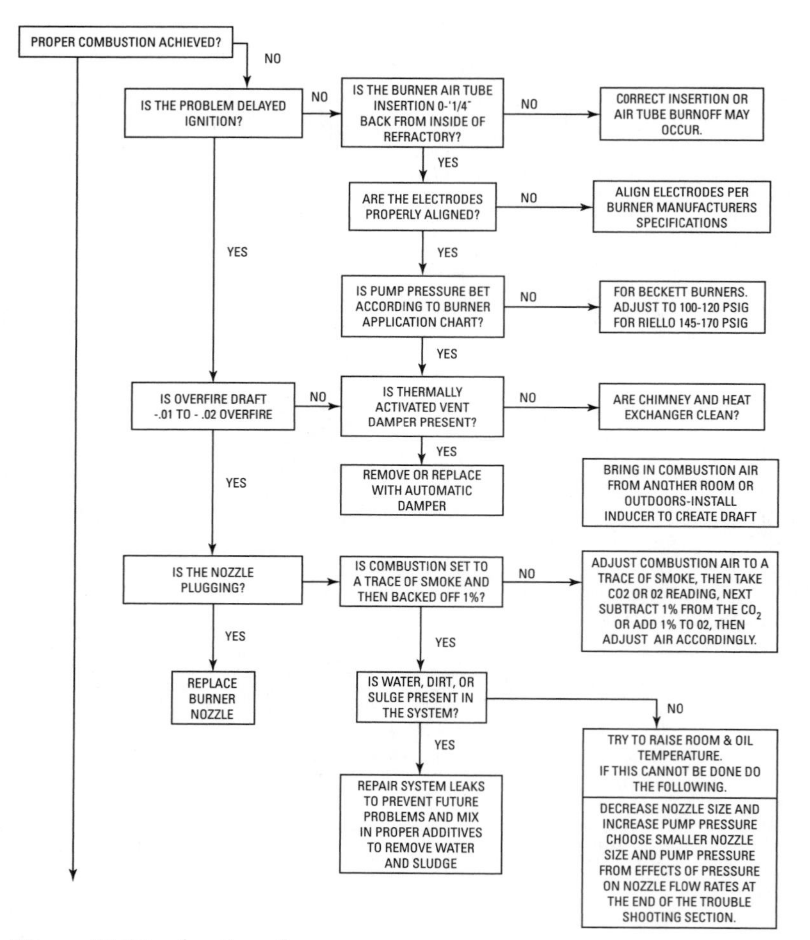

Figure 12-22 *(continued).*

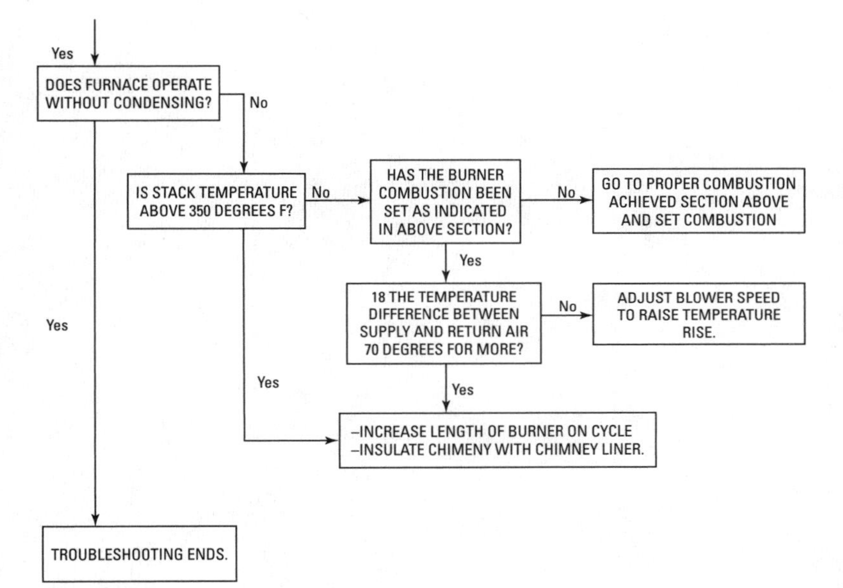

Figure 12-22 *(continued).*

Chapter 13

Coal Furnaces, Wood Furnaces, and Multi-Fuel Furnaces

Coal and wood furnaces burn solid fuels. The former burns coal or coke; the latter, various types of wood. Solid-fuel forced-warm-air furnaces are rated in accordance with formulas provided by the former National Warm Air Heating and Air Conditioning Association (now the Air Conditioning Contractors of America). Check the latest edition of the publication *Commercial Standard for Solid-Fuel-Burning Forced Air Furnaces*. Some solid-fuel furnaces are designed to burn coal in combination with one or more other fuels, such as gas, oil, or wood (see *Multi-Fuel Furnaces* in this chapter).

The operating principle of coal, wood, or multi-fuel furnaces is relatively simple. The combustion process takes place in a sealed firebox located inside the furnace cabinet. A blower forces the heated air over an exchanger and then through the ducts to the living areas inside the house or building. Automatic controls are used to control its safe operation.

Coal Furnaces

Coal furnaces are either hand fired or fired with a coal stoker. A stoker is an automatic coal feeding device that carries the coal from the storage bin to the furnace as needed.

Early coal furnace models were gravity-feed types (Figure 13-1). The heat rose from the furnace through registers in the floor. No fan or blower was used to force the heated air through ducts. Later furnace models were equipped with blowers so that they could be used in forced-warm-air heating systems (Figure 13-2). The compact design of a modern coal furnace is illustrated in Figure 13-3.

Sizing Requirements

The first step in planning a heating system is to calculate the maximum heat loss and gain for the structure. This should be done in accordance with procedures described in the manuals of the Air Conditioning Contractors of America (e.g., *Manual J, Residential Load Calculation*) or by a comparable method (see Chapter 4 in this volume). This is very important, because the data will be used

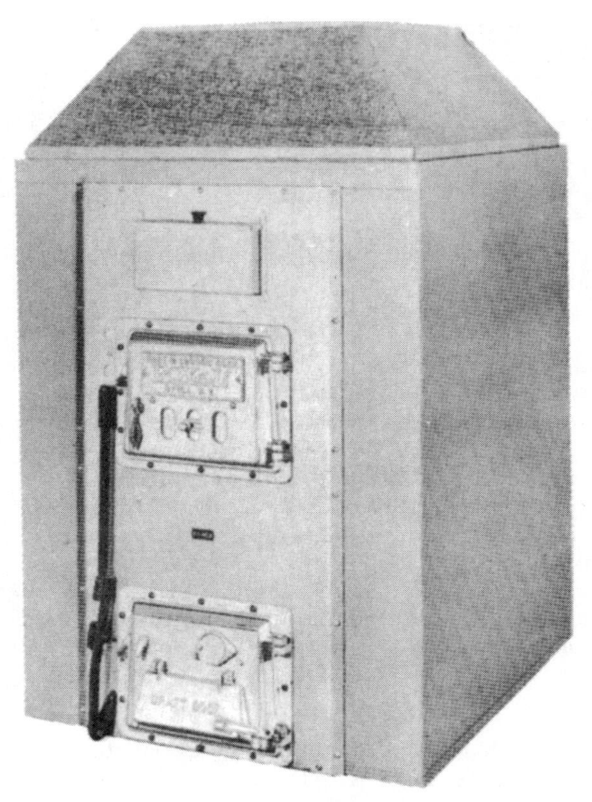

Figure 13-1 Oneida coal-fired gravity warm-air furnace. *(Courtesy Oneida Heater Co., Inc.)*

to determine the size (capacity) of the furnace selected for the installation.

Location and Clearance

A coal furnace should be located a safe distance from any combustible materials. Consult the local codes and regulations for the required clearances.

Never obstruct the front of the furnace. Access must be provided to the fire and ashpit doors in order to operate the furnace.

A centralized location for the furnace usually results in the best operating characteristics, because long supply ducts and the heat loss associated with them are eliminated.

Figure 13-2 Oneida coal-fired forced-warm-air furnace.
(Courtesy Oneida Heater Co., Inc.)

Installation Recommendations

Make certain you have familiarized yourself with all local codes and regulations that govern the installation of coal furnaces and coal stokers. Local codes and regulations take precedence over national standards. See *Installation, Operating and Maintenance Instructions for Coal, Wood, and Multi-Fuel Furnaces* in this chapter for additional information.

Duct Connections

The furnace supply plenum contains air at very high temperatures. Because of these high temperatures, make certain there are safe clearances between the ductwork and any combustibles (including wall and ceiling materials). Check the furnace manufacturer's recommendations.

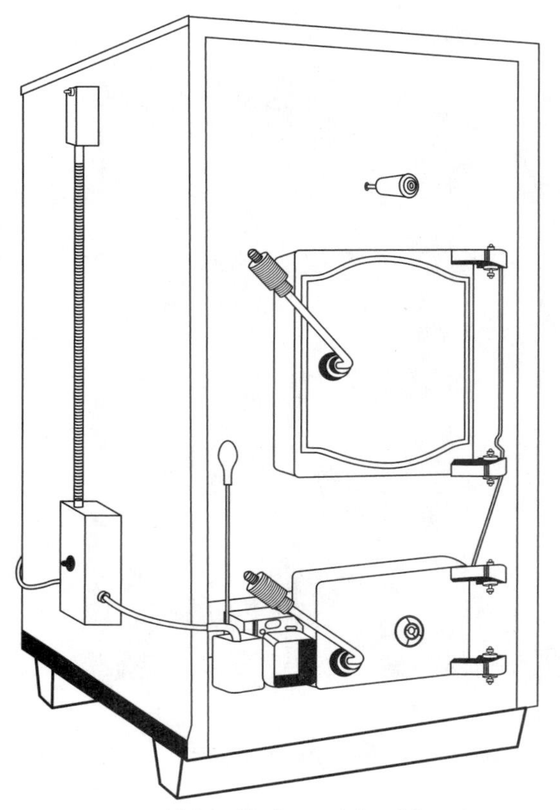

Figure 13-3 Yukon/Eagle coal-fired furnace.
(Courtesy Yukon/Eagle)

Note

The same fire prevention precautions should also be taken for combustible materials near air conditioning and evaporator coils.

Detailed information concerning the installation of an air-duct system is contained in the following two publications of the National Fire Protection Association:

1. *Installation of Air Conditioning and Ventilating Systems of Other Than Residence Type* (NFPA No. 90A).

2. *Residence Type Warm Air Heating and Air Conditioning Systems* (NFPA No. 90B).

Additional information about duct connections can be found in Chapter 7 of Volume 2 ("Ducts and Duct Systems"). The comments in Chapter 11, "Gas Furnaces" about furnace duct connections and air distribution ducts also apply to ducts used with coal-fired forced-warm-air furnaces.

Electrical Wiring

A wiring diagram specifying the electrical connections between the various controls should be provided by the furnace and stoker manufacturers.

Caution

All electrical wiring *must* be done in accordance with the *National Electrical Code* and the local authorized code.

Warning

Turn off the electric power at the circuit breakers (or fuse box) before making any line voltage connections.

As shown in Figure 13-4, a room thermostat operating through the stoker time relay starts and stops the stoker in response to temperature conditions. The time relay operates the stoker to keep the fire alive during periods when heat may not be required by the room thermostat. A stoker time relay should include a device to shut down the stoker immediately after a shutdown call from the thermostat. This action eliminates fuel waste by preventing the overshooting of room temperature.

A high-limit control is used to protect the system against excessive temperatures. This control takes the form of an air switch in a warm-air heating system.

A snap switch should be installed in the wiring between the stoker time relay and the stoker motor to open the stoker motor circuit when the firebox or ashpit is being cleaned (Figure 13-4).

A wiring diagram for a combination wood-oil furnace is shown in Figure 13-5. Note that the wood controls and oil controls are wired as separate systems. A similar method of wiring the controls as two separate systems is used for a combination wood-gas furnace

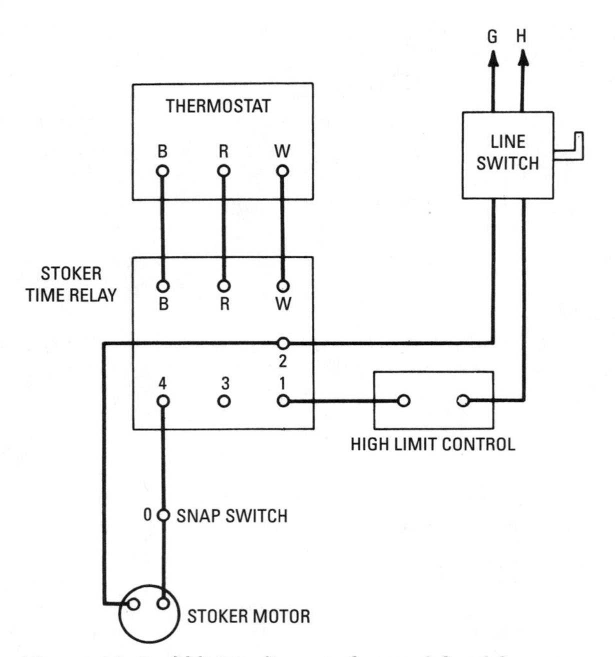

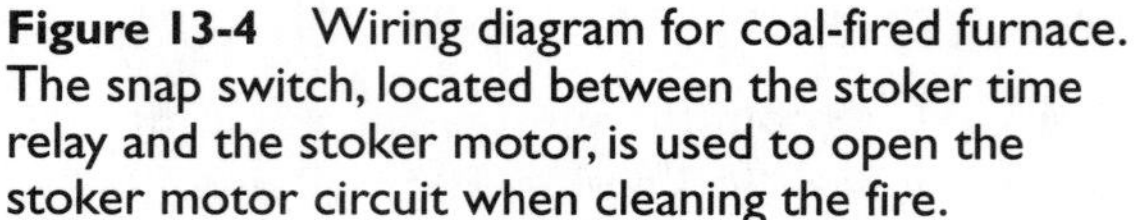
Figure 13-4 Wiring diagram for coal-fired furnace. The snap switch, located between the stoker time relay and the stoker motor, is used to open the stoker motor circuit when cleaning the fire.

(Figure 13-6). This is a common practice on furnaces designed to burn more than one fuel.

The wiring diagrams shown in Figures 13-7 and 13-8 are for gas and oil heating systems with air conditioning.

Ventilation and Combustion Air

The total draft requirement of a coal-fired furnace is greater than those of furnaces that burn gas or oil because the chimney draft must overcome the resistance of the fuel bed.

Both a primary and secondary air supply are necessary for the combustion of solid fuels. The primary air passes through the fuel bed, generally on an upward path from the ashpit. The secondary air is usually admitted through slots in the furnace fire

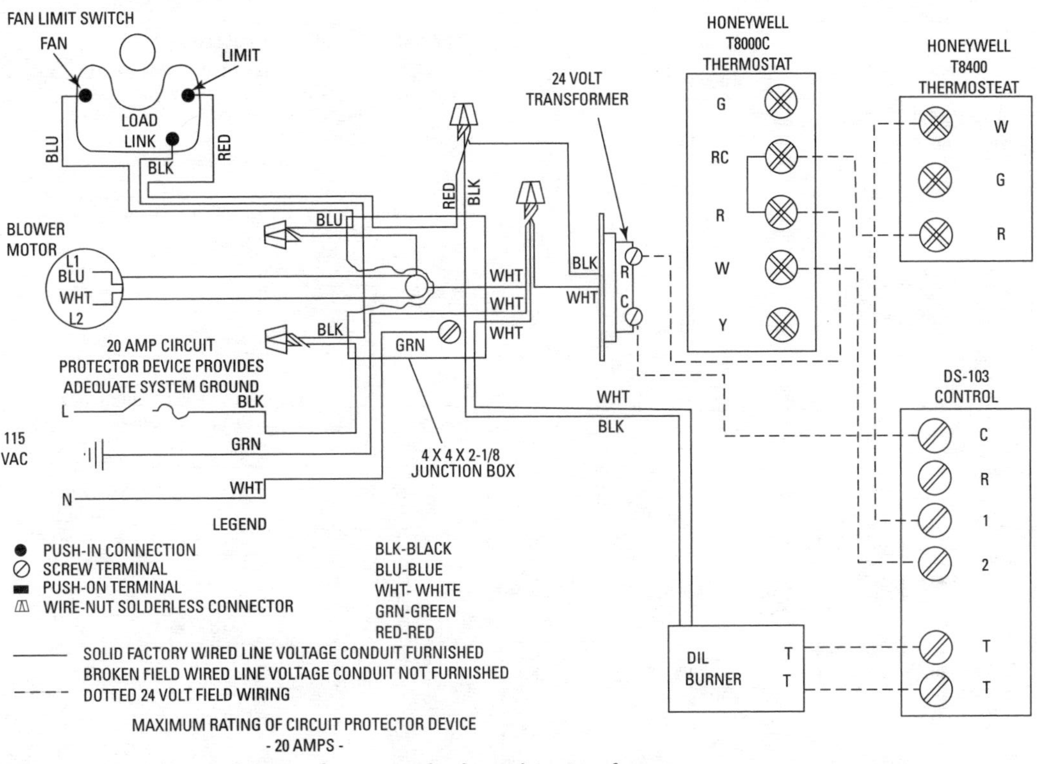

Figure 13-5 Wiring diagram for a wood-oil combination furnace. *(Courtesy Yukon/Eagle)*

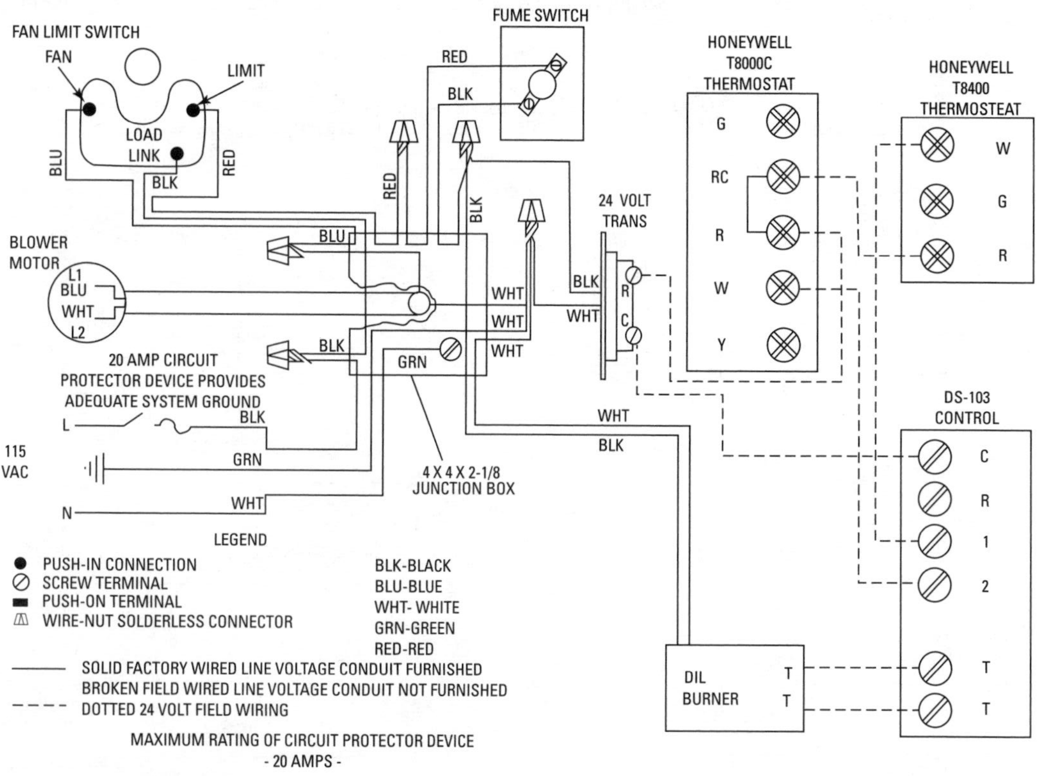

Figure 13-6 Wiring diagram for a wood-gas combination furnace. *(Courtesy Yukon/Eagle)*

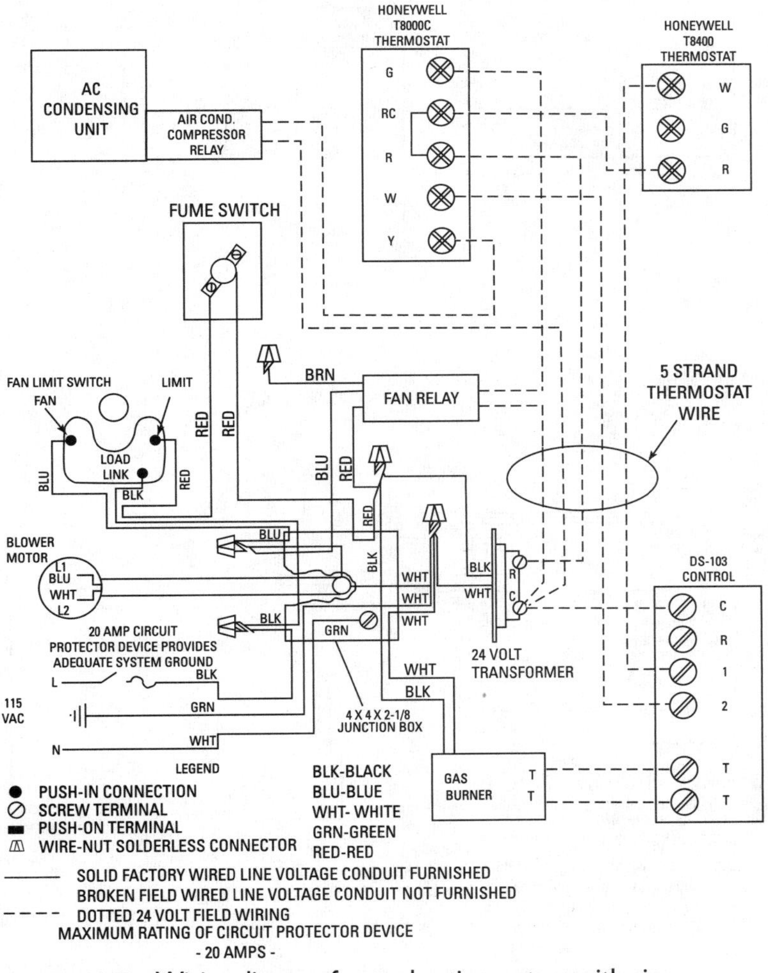

Figure 13-7 Wiring diagram for gas heating system with air conditioning. *(Courtesy Yukon/Eagle)*

door and passes *over* the fire to complete the combustion process (Figure 13-9).

On residential furnaces, the secondary air slots in the fire door should be kept *closed*. There is usually enough air leakage to

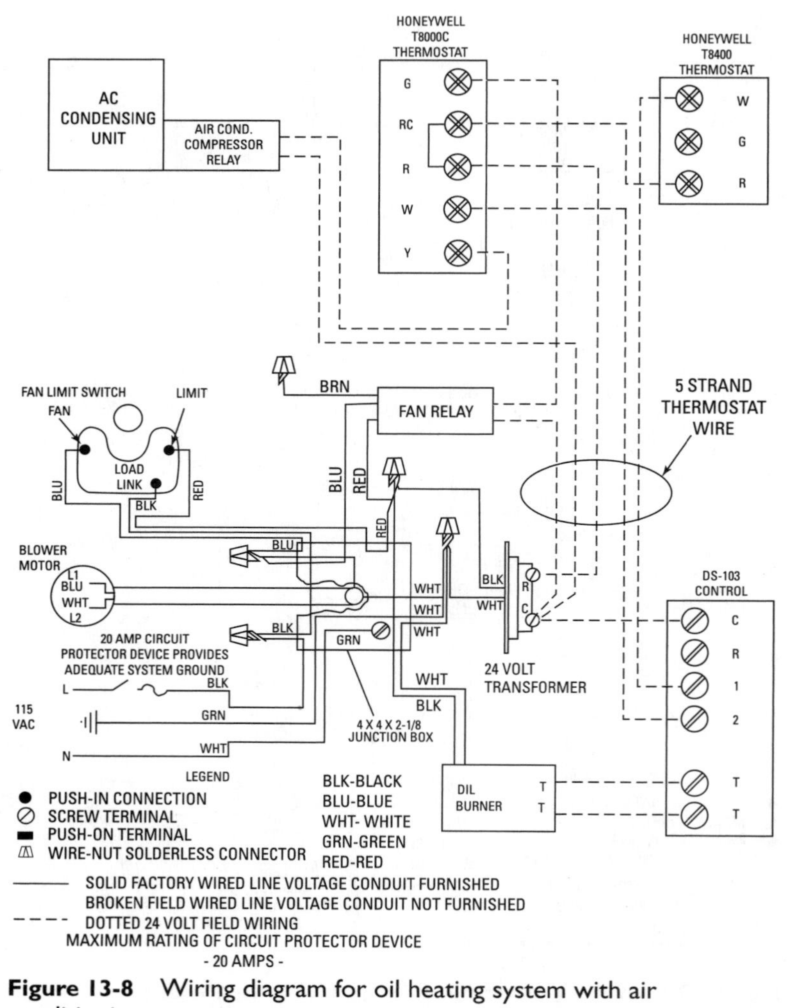

Figure 13-8 Wiring diagram for oil heating system with air conditioning. *(Courtesy Yukon/Eagle)*

admit a suitable amount of secondary air without having to open the fire door. An excessive amount of secondary air (particularly after the fire has taken hold) will actually reduce the efficiency of the furnace.

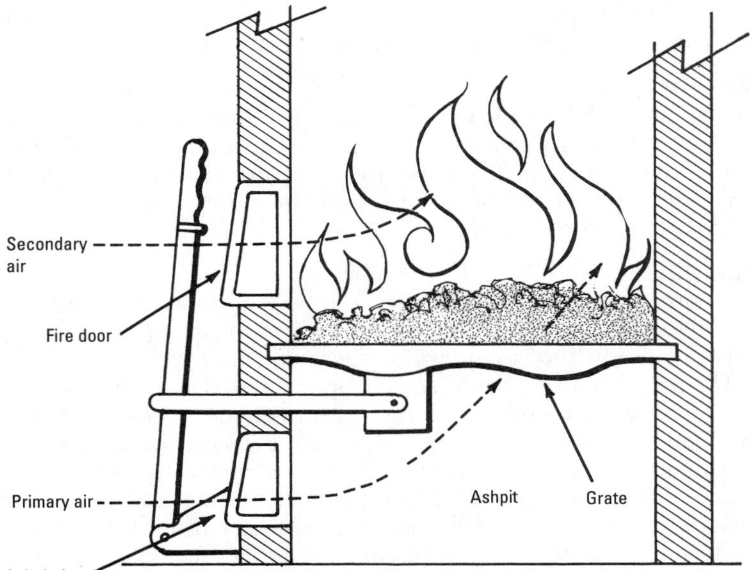

Figure 13-9 Entry points for primary and secondary air.

The draft will determine the rate at which the fuel is burned. The draft in furnaces using coal or coke will depend upon the following factors:

1. Type of fuel (anthracite or bituminous, coke, etc.)
2. Size of fuel
3. Fuel-bed thickness
4. Ash and clinker accumulation
5. Flue resistance to the flow of gas
6. Soot accumulation in flue
7. Grate area

The rate at which the fuel is burned will depend upon the heating load demand of the specific location. The burning rate can be varied by regulating the draft. One of the most effective methods of doing this is by placing a damper in the flue outlet. This method proportionally reduces both primary and secondary air.

An ashpit damper is required for low combustion rates. A cold-air check damper is necessary when there is excessive chimney draft.

Venting

Provisions must be made for venting the products of combustion to the outside in order to avoid contamination of the air in the living or working spaces of the structure. Masonry chimneys and low-heat Type A prefabricated chimneys are the most common methods of venting coal-fired furnaces.

Flue Pipe

Coal furnaces require flue passages—larger than those used—for gas and oil furnaces. This is due to the far greater volume of smoke, soot, and other products of combustion associated with burning solid fuels.

The flue pipe extending vertically up through the chimney should have a constant cross-sectional area throughout its length (Figure 13-10). The chimney flue (i.e., the passage between the furnace and the chimney) must have a cross-sectional area no less than that of the furnace flue collar. A diameter 1 inch larger is recommended. Consult the local building code for flue sizing recommendations if more than one furnace or appliance is to be connected to a single flue.

The chimney thimble, which connects the smoke pipe to the chimney, should be made of heat resistant fire clay. Its inside diameter should be roughly equal to the outside diameter of the smoke pipe to ensure an airtight fit at the chimney. Do not allow the chimney thimble to extend beyond the chimney flue lining.

Chimneys and Chimney Troubleshooting

A chimney used with a coal furnace should be of sufficient height and area to meet the requirements of the furnace. The top of the chimney should be at least 2 ft higher than the highest portion of the roof within 10 ft horizontally. There should also be a minimum of 3 feet between the top of the chimney and the point at which it passes through the roof (see Figure 13-10). For best results, only the furnace should be connected to the chimney. Read the section *Chimneys* in Chapter 11, "Gas Furnaces."

A chimney connected to a coal or wood burning furnace must have a cleanout door at the bottom. This door allows ashes and other residue to be removed from the bottom of the chimney. The cleanout door must have a tight fit. It must never be left open during the combustion process.

The chimneys used with coal furnaces are basically the same as those used with gas and oil units. Read the section *Chimney Troubleshooting* in Chapter 11, "Gas Furnaces" for a description of common chimney problems and suggested remedies.

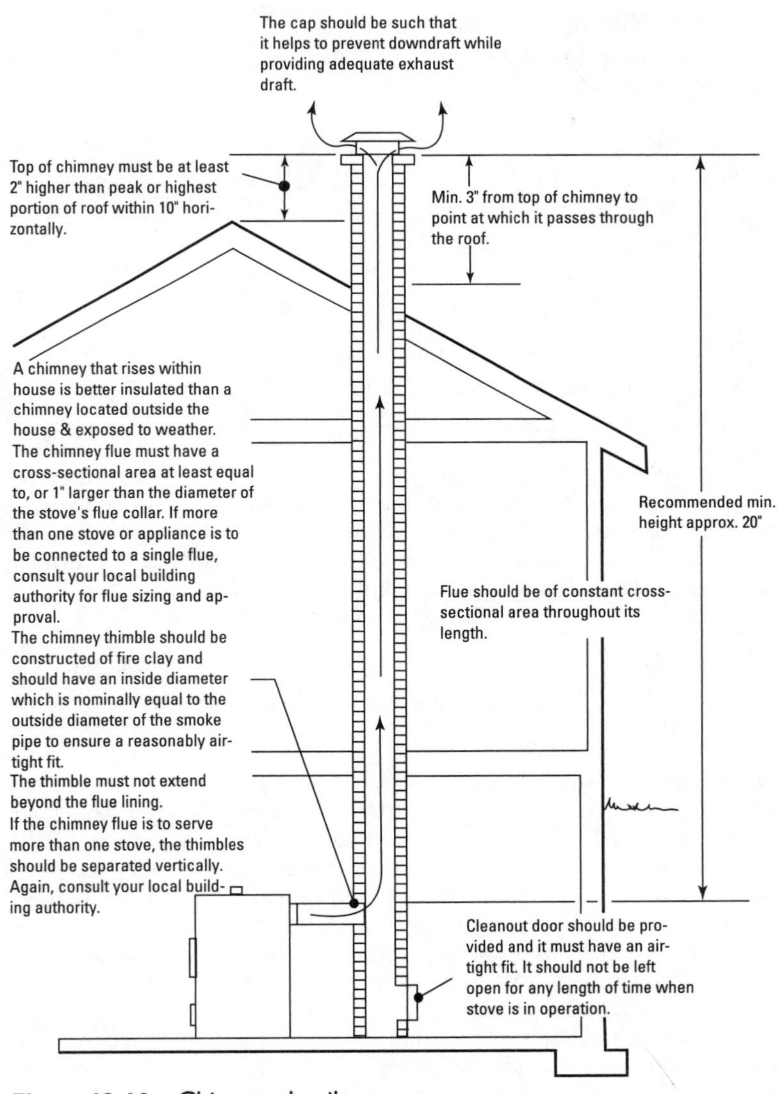

Figure 13-10 Chimney details. *(Courtesy Yukon/Eagle)*

Furnace Components

The basic components of a coal furnace used in a forced-warm-air heating system are:

1. Cabinet or jacket
2. Firebox
3. Grate
4. Heat exchanger
5. Blower and motor
6. Access door or doors for stoking and cleaning
7. Small blower to fan fire when more heat is required
8. Coal stoker (for automatic feeding)
9. Automatic controls

Each of these components is described in the sections that follow. Most of these components are also found in wood and multi-fuel furnaces. Additional information is contained in Chapter 10, "Furnace Fundamentals" and the various chapters in which furnace controls and fuel-burning equipment are described.

Note

Solid- and multi-fuel furnaces may also include backup devices such as a gas burner, an oil burner, or electric heating elements.

Automatic Controls

The automatic controls of a coal furnace equipped with a stoker are shown in Figure 13-11. This control system is very similar to the one used for an oil burner except for the automatic timer included with the stoker relay and transformer. During the heating season, the fire of a coal furnace must burn continually. The timer is a device designed to operate the coal stoker for a few minutes every hour or half-hour in order to keep the fire alive during those periods when little heat is required.

A stack thermostat or similar device is recommended for use with stokers in areas subject to electric power failures. An electric power failure will shut down the stoker. If the shutdown period is long enough, the fire will die for lack of fuel. When the electricity comes on again, the stoker will feed coal to a cold firepot. A stack thermostat or light-sensitive electronic device will monitor the stack heat or fire and prevent the stoker operation when the fire is out.

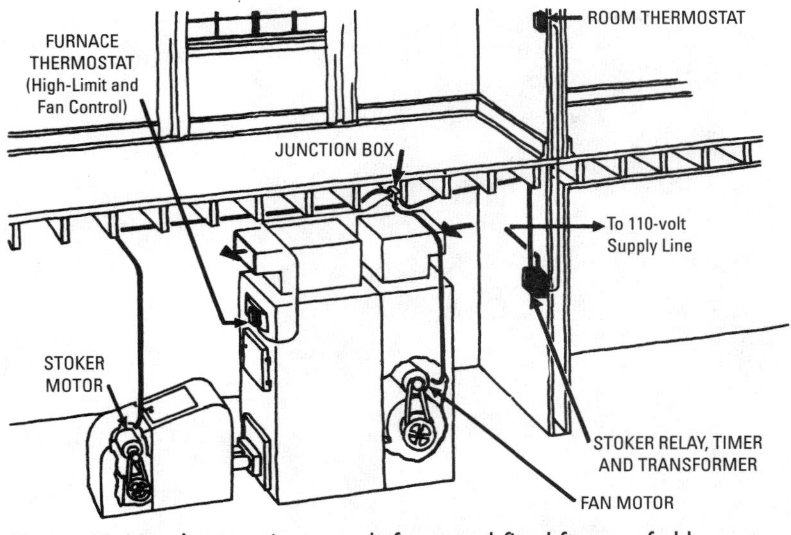

Figure 13-11 Automatic controls for a coal-fired furnace fed by a stoker. *(Courtesy U.S. Department of Agriculture)*

A blower fan control is generally included with most forced-warm-air furnaces. Many manufacturers will also provide a hand-operated draft control with a coal-fired furnace. Electrically operated dampers for draft control are generally available, but at extra cost.

The operation of a furnace is controlled by the room thermostat. In heating systems in which a coal-fired furnace is used, the thermostat opens or shuts dampers to increase or decrease the supply of air to the fire. When the supply of air is increased, the fire becomes hotter, and the amount of heat generated by the furnace is increased. Decreasing the supply of air to the fire has the opposite effect. Older coal-fired furnaces were not controlled by a room thermostat. Hand-operated dampers were used instead.

Furnace Grate

An essential part of any coal-fired furnace is the metal grate on which the fuel is burned (Figure 13-12). The grate should be designed to allow sufficient primary air to pass upward through the fuel bed for the combustion process. The ashes will drop to the ashpit below through the same openings when the grate is shaken.

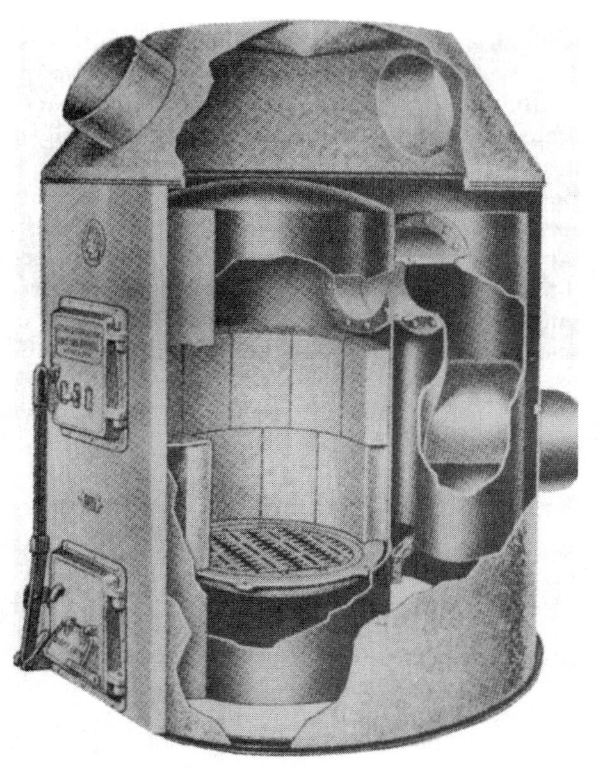

Figure 13-12 Cutaway view of a coal-fired furnace showing grate and combustion chamber details.
(Courtesy Oneida Heater Co., Inc.)

There should be a metal heat transfer surface (heat exchanger) of sufficient size above the fire to transfer the heat from the combustion process to the water or air in the heating system.

Coal Stoker

A coal stoker is a device used to feed coal to a coal-fired furnace or boiler automatically. See Chapter 3 of Volume 2, "Coal Firing Methods" for additional information about coal stokers.

Hand-Firing Methods

The method used to hand-fire a furnace will depend largely upon the type of solid fuel used in it. Coke and the various types of coals each have their own special hand-firing methods. Some of these firing methods are described in Chapter 3 of Volume 2, "Coal Firing Methods."

Blower and Motor Assembly

The blowers used on coal furnaces are the same centrifugal types found on gas, oil, and electric forced-warm-air furnaces. The blowers are usually installed on either side of the furnace or at the back. The latter arrangement is especially recommended for wood-burning and multi-fuel furnaces. A detailed description of blowers and blower motors is included in Chapter 11, "Gas Furnaces."

Blower Adjustment

Instructions for making blower adjustments can usually be obtained from the furnace manufacturer. A brief description of methods used to make blower adjustments is included in Chapter 11, "Gas Furnaces."

Air Filter

The number, size, and type of air filter will be recommended by the furnace manufacturer in the specifications. See also Chapter 12, "Air Cleaners and Filters" in Volume 3.

System Accessory Devices

A coal furnace used in a central forced-warm-air heating system may include some or all of the following accessories:

1. Electronic air filter
2. Humidifier
3. Dehumidifier
4. Electronic air cleaner
5. High performance media filter
6. Central air evaporator coil
7. Domestic hot-water coil
8. Condensate pump

Each of these accessories is covered in Volumes 2 and 3. They may also be used with wood and multi-fuel furnaces.

One method of adding summer air conditioning to a coal-fired heating installation is to install a separate and independent air conditioning system. This is expensive, because it includes the equipment and separate ducts, but it avoids many complications.

If you are considering the idea of adding air conditioning at some future date, you should select a solid-fuel furnace capable of meeting your cooling needs. For example, Oneida All-Fuel and Two-in-One furnaces are equipped with blowers large enough to handle up to 4 tons of air conditioning and electronic air cleaning.

Wood Furnaces

A wood furnace is very similar in design to a coal furnace except that it burns wood. The components are identical (sealed firebox, exchanger, blower, etc.), and it can be combined with the same types of accessories (electronic air cleaner, humidifier, etc.) Figure 13-13 shows a wood furnace.

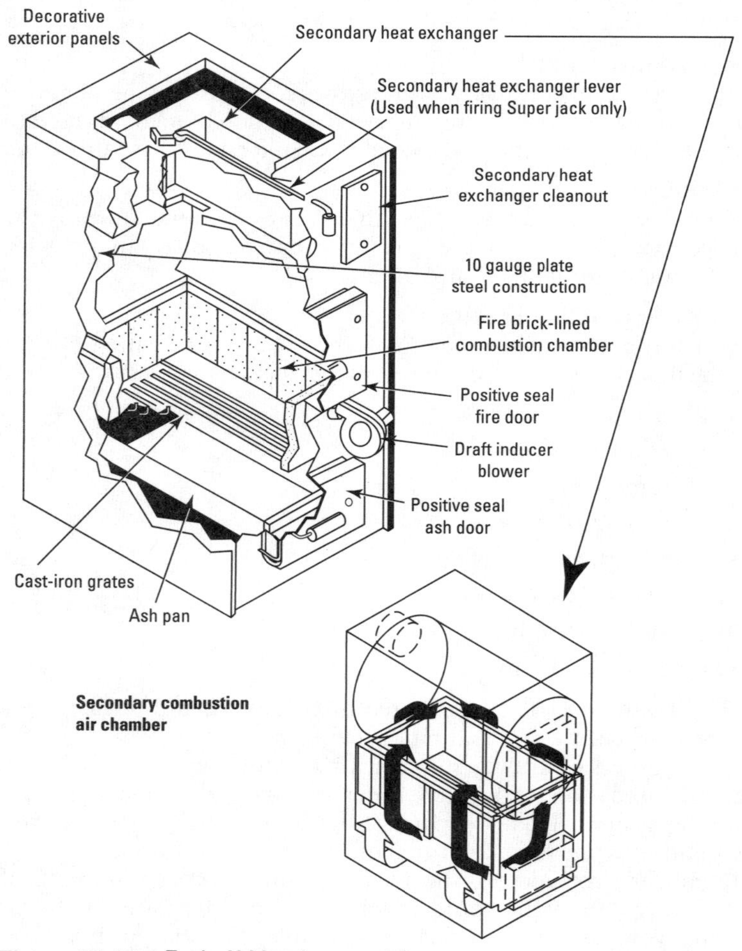

Figure 13-13 Eagle I Wood or coal furnace. *(Courtesy Yukon/Eagle)*

Wood Add-On Furnaces

Wood add-on furnaces are designed to be added to existing oil furnaces as a backup heat source (Figure 13-14). Their construction details are identical to those of a standalone wood furnace.

Caution

Make sure the existing furnace can accept an add-on unit. Not all furnaces can do this.

There are two types of add-on installations: independent and parallel. The parallel installation uses the existing blower in addition to the blower on the wood burning furnace (Figure 13-15).

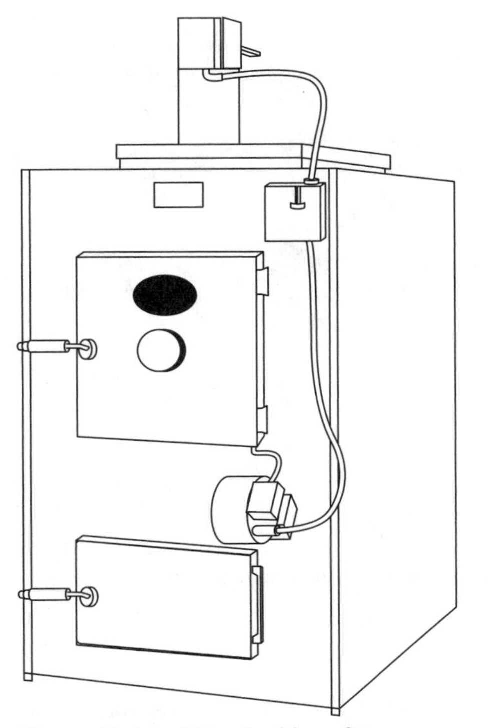

Figure 13-14 Wood add-on furnace.

(Courtesy Yukon/Eagle)

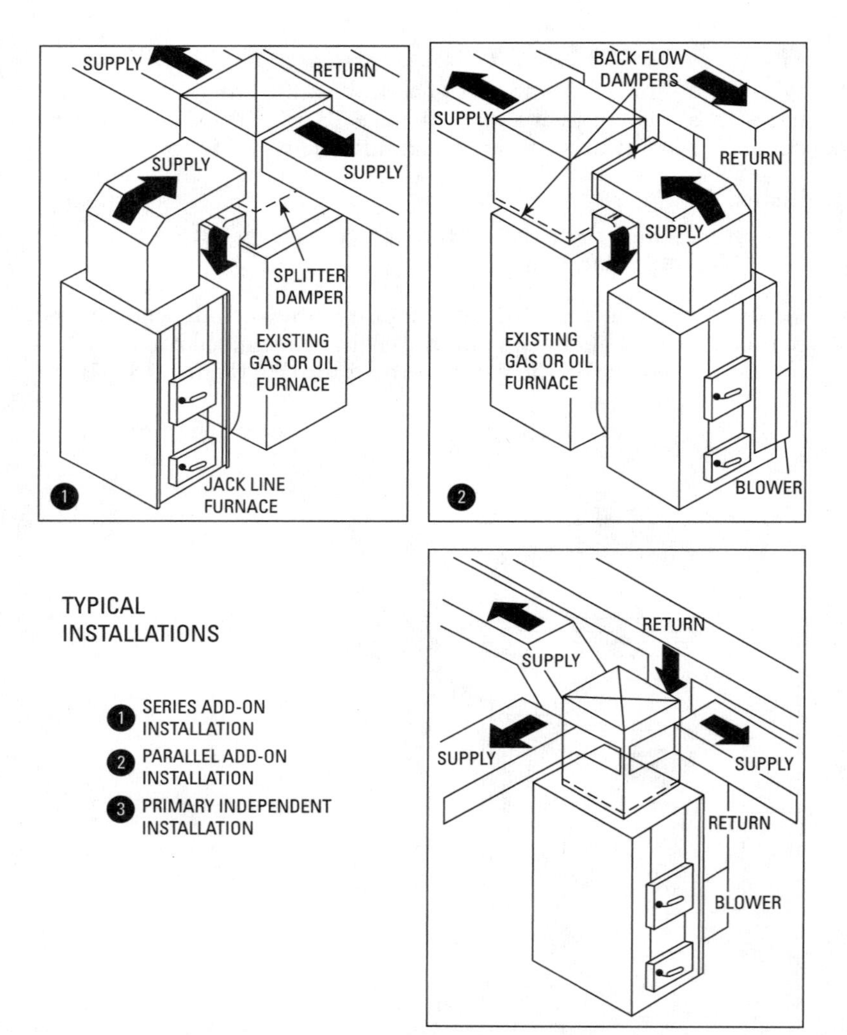

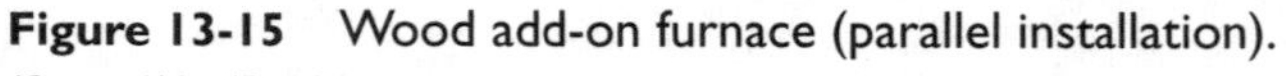

Figure 13-15 Wood add-on furnace (parallel installation).
(Courtesy Yukon/Eagle)

Multi-Fuel Furnaces

Furnaces designed to burn more than one fuel are called *multi-fuel furnaces* or *combination furnaces*. These furnaces are designed to burn oil or gas in one combustion chamber, wood or coal in the other, and automatically switch between the fuels when the situation warrants (Figures 13-16 and 13-17).

The solid-fuel combustion chamber of these furnaces burns either coal or wood and does so much more efficiently than standard coal or wood stoves. The other chamber uses an oil burner, gas burner, or electric resistance heater to generate heat in the same manner as in a single-fuel furnace. Either combustion chamber is capable of heating an entire house.

The operation of the furnace shown in Figure 13-16 is controlled by a Honeywell two-stage thermostat. The thermostat is equipped with indicator lights to indicate which fuel is being burned.

Wood or coal can be used as the primary fuel with oil, gas, or electricity as the automatic backup; or oil, gas, or electricity can be used as the primary fuel with wood or coal serving as the backup.

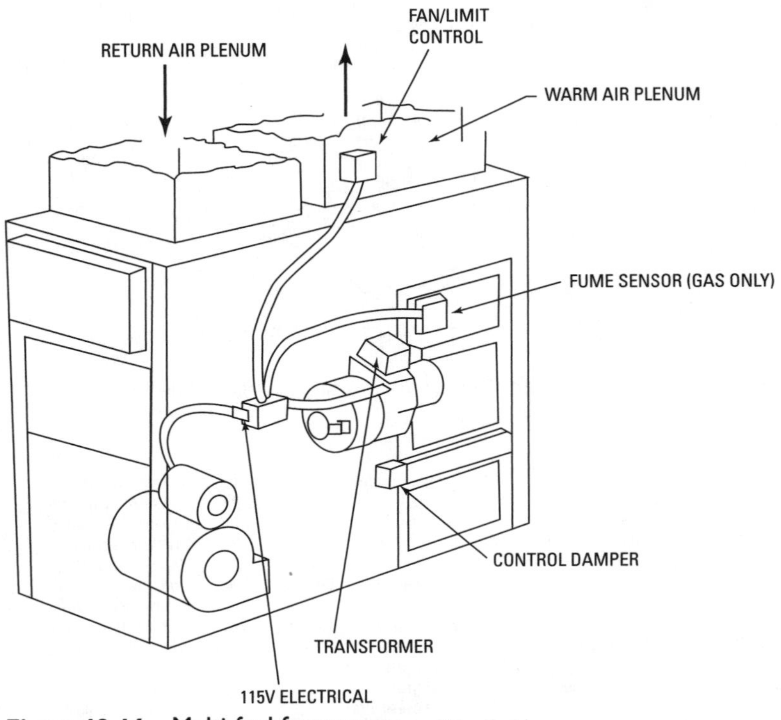

Figure 13-16 Multi-fuel furnace. *(Courtesy Yukon/Eagle)*

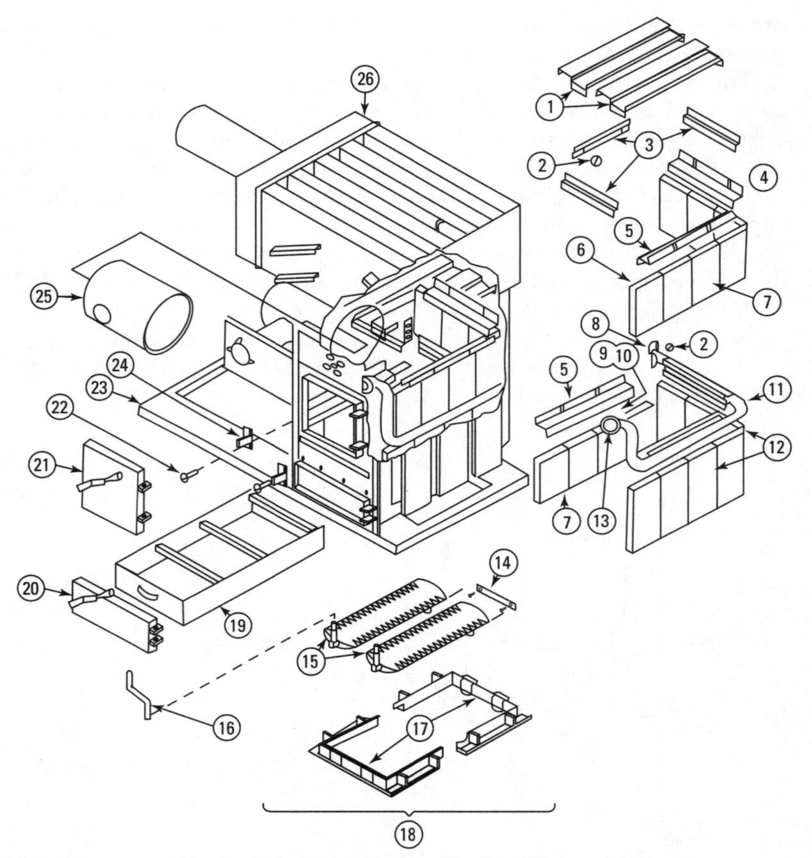

Figure 13-17 Exploded view of Yukon/Eagle oil/gas solid fuel combination furnace. *(Courtesy Yukon/Eagle)*

1 Smoke baffle.
2 Hex nut (½ inch)
3 Baffle bracket
4 Top brick retainer (rear)
5 Top brick retainer (side)
6 Fire brick (9 × 4½ × 2-inch)
7 Fire brick (9 × 6 × 2-inch)
8 Tube support bracker
9 Fender washers (1/16-inch I.D. × 9-inch O.D.)
10 Thread cutting screw
11 Secondary air tube weldment
12 Fire brick (12 × 6 × 2-inch)
13 Secondary air tube gasket
14 Connecting rod assembly
15 Coal grate
16 Coal grate handle
17 Coal grate frame
18 Coal grate shaker assembly
19 Ash pan weldment
20 Ash door assembly
21 Fire door assembly
22 Machine screw
23 Base weldment
24. Door latch
25 Refractory pot liner
26 Combination chamber weldment

Installation, Operating, and Maintenance Instructions for Coal, Wood, and Multi-Fuel Furnaces

Make certain you have familiarized yourself with all local codes and regulations governing the installation, operation, and maintenance of coal furnaces and coal stokers, wood furnaces, and multi-fuel furnaces. Local codes and regulations take precedence over national standards.

Assembly and Installation Recommendations

Coal, wood, and multi-fuel furnaces are shipped disassembled with complete assembly instructions. The installer must assemble the furnace in accordance with these instructions. If the furnace is fired by a coal stoker or contains some automatic controls, the installer must connect the electrical service from the line voltage main. The low-voltage thermostat must also be connected.

Caution

> NEVER attempt to assemble and install a furnace unless you have the necessary qualifications. Furnace assembly is a task best left to those with the necessary training and experience. It is not a do-it-yourself project. An improperly assembled and installed furnace is a fire hazard and could lead to serious injuries or even death.

Note

> Always check the local building codes first before installing a furnace. The installation must comply with the local code requirements.

NEVER install a wood furnace so that the supply air from the wood furnace feeds into the return air of the conventional furnace. Doing so may result in the following problems:

1. Shortened blower motor life caused by high air temperatures from the wood furnace.
2. High temperatures in the return air ducts caused by reverse air flow during a power failure.
3. High temperatures causing the blower to trip off on a thermal overload.
4. Toxic fumes pulled from the heat exchanger and forced into the rooms.

Locate the furnace as close to the chimney or flue as possible and near the center of the heat distribution system (i.e., the ductwork) or next to a primary furnace. Be sure to maintain the minimum clearances specified in the local building code. The NFPA (National Fire Protection Agency) Standard No. 90B also lists and describes the minimum standard clearances from combustible surfaces. Recommended clearances are listed in Table 13-1.

Reduced clearances are permitted if the area between the furnace and combustible ducts are protected according to NFPA Bulletin 90B. In any case, the furnace must be so located that it receives sufficient combustion air.

Caution

> Shielding a wall with sheet metal or masonry will not protect it. These materials are excellent heat conductors, and any combustible wall material behind them will still be a fire hazard.

Maintenance Instructions

Always follow a regular service and maintenance schedule to ensure safe and efficient operation of the furnace. Keep a service and maintenance list next to the furnace and check off the date of each inspection with comments on what was done.

1. Make certain the duct connections to the furnace are tight and secure.
2. Follow a regular service and maintenance schedule of the furnace and chimney for safe and secure operation.

Table 13-1 Recommended Minimum Clearances

Furnace clearances	
Above top of plenum	18"
From the front of the furnace	48"
From sides and back of furnace	18"
From flue pipe	18"
From existing furnace	6"
Clearances from horizontal warm air duct	
Within 3 feet of plenum	18"
Within 3 to 6 feet of plenum	6"
Beyond 6 feet of plenum	1"
All heat runs	1"

Courtesy Yukon/Eagle

3. Frequently inspect and clean the heat exchanger, smoke pipe, and chimney of soot and/or creosote.
4. Periodically check the furnace cabinet (jacket) for cracks.
5. Make certain all access doors fit tightly.
6. Inspect and clean the air filter monthly. Change the air filter at least twice a year.
7. Regularly inspect and clean the air cleaner grids on furnaces equipped with electronic air cleaners.
8. Periodically clean the humidifier (if so equipped).
9. Oil the blower motor twice a year (blowers with oil caps).

Wood and Coal Furnace Maintenance

1. Remove ash and other debris left over from the combustion process on a daily basis during the heating season.

Note

Never spread ash on your garden or lawn. It contains toxic substances that will kill plants or grass.

Caution

Never allow ashes to build up to the grate level in the furnace. Doing so will shorten the service life of the grate.

2. Keep the heating surfaces and flues clean. Soot will reduce heating efficiency. A dirty furnace requires much more fuel to produce the required heat than a clean one does.
3. Hard clinkers lodged between the grate bars should be removed with a poker or slice bar.

Multi-Fuel Furnace Maintenance

Multi-fuel furnaces, or combination furnaces as they are also called, burn coal or wood with oil, gas, or electricity. Because they burn solid fuels (wood or coal), the maintenance instructions for coal and wood furnaces will also apply to the solid-fuel burning sections of a multi-fuel furnace.

1. Inspect and perform any necessary maintenance on a gas or oil burner at least twice a year, preferably at the end and beginning of the heating system. Gas burner and oil burner service and maintenance is covered in the second volume of the *Heating, Ventilating, and Air Conditioning Library.*

Caution

Maintaining, adjusting, and repairing oil and gas burners requires special instruments and training. These tasks are best left to a trained HVAC technician or someone with equivalent experience.

2. Remove and clean the oil burner strainer at the end of the heating system.
3. Fill the oil storage tank at the end of the heating season to prevent water vapor from collecting.
4. Lubricate the oil burner every 3 months.
5. Have a qualified HVAC technician check and service the oil burner at the beginning of the heating season.
6. Clean, inspect, and service the furnace, smoke pipe, and chimney at the beginning of the heating system.
7. Check the furnace and burner electrical system before the beginning of the heating system (primary relay, limit control, thermostat, electrodes, electrode terminals, and transformer terminals).
8. Clean the solid fuel combustion chamber according to the procedures for cleaning a coal or wood furnace.

Wood and Coal Furnace Operation

Any solid fuel has approximately 12,000 Btu per pound if its moisture content is zero. This holds true for coal as well as any of the woods burned in a furnace. The Btu per cord of air dried wood are listed in Table 13-2. Ignition temperatures of coal and wood are listed in Table 13-3.

Wood Furnaces

The wood burning furnace shown in Fig. 13-13 is designed to allow the primary air under the grate to create the initial burning. As the wood burns, gases containing as much as 40 percent of the energy of the wood rise to the top of the flame. The secondary air system (i.e., the round tubes between the firebrick) draws room air into the tubes and provides oxygen to the firebox to burn these gases before they can escape up the chimney flue. The procedure for starting and maintaining a wood fire in a furnace may be outlined as follows:

1. Open the manual draft spinner three or four turns.
2. Make sure the smoke pipe damper is open.
3. Place several pieces of crumpled paper in the center of the furnace firebox.

Table 13-2 Btu Per Cord of Air Dried Wood

Type	*Pound Weight Per Cord*	*Btus Per Cord of Air Dried Wood*	*Equivalent Value #2 Fuel Oil Gallons*
White Pine	1800	17,000,000	120
Aspen	1900	17,500,000	125
Spruce	2100	18,000,000	130
Ash	2900	22,500,000	160
Tamarack	2500	24,000,000	170
Soft Maple	2500	24,000,000	170
Yellow Birch	3000	26,000,000	185
Red Oak	3250	27,000,000	195
Hard Maple	3000	29,000,000	200
Hickory	3600	30,500,000	215

Courtesy Yukon/Eagle

4. Place a few handfuls of kindling wood in a criss-cross pattern across the paper, and then add a few pieces of firewood.

Caution

Never start a coal or wood fire with gasoline, kerosene, thinners, or any other type of volatile substance.

5. Ignite the paper and close the door to the firebox. Resist the desire to open the door immediately after igniting the paper to see how the fire is doing. Opening the door at this point may cause a flame flash out. It will take a few minutes for the fire to establish itself.

Table 13-3 Ignition Temperatures of Coal and Wood

Type	*Ignition Temperature*
Paper	350°F
Wood	435°F
Western lignite coal	630°F
Low volatile bituminous coal	765°F
High volatile bituminous coal	870°F
Anthracite coal	925°F

Courtesy Yukon/Eagle

6. Start adding larger pieces of wood when you see some red-hot burning embers. Build the fire slowly. Smaller pieces of wood burn cleaner because they have more surface area exposed to the fire. Use wood a maximum of 4 to 6 inches thick and about 2 to 4 inches narrower than the width of the furnace firebox.
7. Once the fire is established, adjust the draft according to your needs. The draft control on the furnace controls the burning time.

Caution

Always open the door to the firebox slowly. Rapidly opening the door could cause flames to leap out of the firebox.

Coal Furnaces

Burning coal is more difficult than burning wood. It requires patience and a regular procedure. If you are not careful, a coal fire can extinguish itself in a relatively short time. Once a coal fire starts to go out, it is next to impossible to reverse it. If the coal fire goes out, remove all the coal from the furnace and then start the process all over again.

Coal comes in various sizes and types. As a rule, anthracite coal is recommended, because it burns with little smoke when burned properly. Select a coal size that will not fall through the air spaces in the furnace grate. These air spaces are ½ inch wide, so the coal size must be larger than that. Coal is available in three sizes: pea size, nut size, and stove size. Nut-size coal (1³⁄₁₆ to 1⅝ inch) is recommended for coal furnaces.

How to Start a Coal Fire

1. Place a small amount of crumpled paper and sticks of kindling wood on the ash-covered grates of the furnace.
2. Ignite the paper and wait until the wood starts to burn well.
3. Cover the burning paper and wood with a thin layer of evenly distributed coal. The pieces of coal must touch each other to ignite.
4. Soon after the first layer of coal is ignited, gradually add more coal until the fire bed is built up to about 6 inches deep. As you add the fresh coal always leave some of the burning coal uncovered. Be careful not to smother the fire when adding fresh coal.
5. Draw the top red coals toward the front of the firebox and pile fresh coals toward the back.

Note

Protect the furnace grate from direct contact with the fire by a 1- to 2-inch thick layer of ash. This will help prevent the cast-iron grates from overheating and coal from falling through the grate openings.

How to Maintain a Coal Fire

1. Shake the grates once or twice a day to prevent excessive ash accumulation on the grates.
2. Shake the grates with a few short strokes and stop as soon as you see red coals appearing in the ash pan.

Note

Shaking the grate too much may extinguish the fire; shaking it not enough may restrict the amount of air reaching the fire.

3. Never poke or break up a coal fire. Poking the fire tends to bring ash to the surface of the coal bed, where it fuses into lumps or clinkers. Clinkers interfere with efficient burning of the coal.

How to Bank a Coal Fire

Coal fires are reduced (banked) during the night, when less heat is required. If you are using anthracite coal, pile the coal higher toward the back of the firebox and allow it to slope forward toward the firebox door. Leave some red or burning coal uncovered in the front of the firebox.

If you are using bituminous coal, shake the fire and add coal, forming the center cone. Allow enough time for the volatiles to burn off before closing the door to the firebox.

Multi-Fuel Furnace Operation

Begin by loading the coal or wood into the furnace firebox. The automatic controls of the furnace will then take over, and the oil burner or gas burner will ignite the coal or wood when heat is required.

Note

Electric heating is not automatic (i.e., thermostat controlled) and must be manually started.

Set the two-stage thermostat. The thermostat controls both combustion chambers at separate temperature settings. The oil or gas burner will automatically start and provide supplemental, backup heat if the wood or coal fire cannot maintain the temperatures set on the thermostat.

The wood or coal fire is automatically controlled by the wood or coal thermostat setting. After the wood or coal is ignited and the furnace begins delivering the heat called for by the thermostat, the oil or gas burner automatically shuts off. The thermostat automatically maintains the heat levels inside the structure by controlling the amount of combustion air to the wood or coal fire.

As the wood or coal fire dies down, the temperature inside the structure declines until it reaches the thermostat temperature setting selected for the oil, gas, or electric heat. When this temperature setting is reached, the oil burner, gas burner, or electric heaters automatically start, and the backup combustion chamber then functions as a conventional furnace.

Troubleshooting Coal, Wood, and Multi-Fuel Furnaces

The troubleshooting sections that follow list the most common problems associated with coal, wood, and multi-fuel furnaces. Each problem is given in the form of a symptom, the possible cause or causes, and some suggested remedies (Tables 13-4, 13-5, and 13-6).

Table 13-4 General Heating Problems

Symptom and Possible Causes	*Suggested Remedies*
Inadequate fire	
(a) Insufficient draft	(a) Clean ashpit and remove obstruction to primary air supply; adjust damper control.
(b) Dirty furnace	(b) Clean furnace.
(c) Dirty flue	(c) Clean flue
(d) Poor-quality fuel	(d) Replace with better-quality fuel.
(e) Grate clogged with slate, clinkers, or other combustion residue	(e) Dislodge with poker, or dump grate and rebuild fire.
Excessive coal in firebox	
(a) Stoker feeding too much coal	(a) Reduce coal feed rate on stoker.
(b) Insufficient air	(b) Open manual draft.

(continued)

Table 13-4 (continued)

Symptom and Possible Causes	*Suggested Remedies*
(c) Stoker wind box full of siftings	(c) Clean out wind box.
(d) Accumulations of clinkers in fire	(d) Clean fire.
Excessive fuel use	
(a) Dirty furnace	(a) Clean furnace.
(b) Dirty flue	(b) Clean flue.
(c) Incorrect draft	(c) Adjust dampers for correct rate of combustion.
Not enough heat or no heat	
(a) Thermostat set too low	(a) Raise thermostat setting.
(b) Thermostat in wrong location	(b) Relocate thermostat.
(c) Thermostat out of calibration	(c) Recalibrate or replace thermostat.
(d) Lamp or other heat source too near thermostat	(d) Remove heat source. Relocate thermostat if heat source cannot be moved.
(e) Dirty air filter	(e) Clean or replace air filter.
(f) Limit set too low	(f) Reset or replace thermostat.
(g) Fan speed too low	(g) Check for loose fan belts and tighten if too loose; check motor and repair or replace.
(h) Oil or gas burner is not firing properly	(h) Contact your local serviceman.
(i) No oil in storage tank	(i) Check oil tank gauge. If empty, fill tank and start the burner.
(j) Valve in oil line is closed	(j) Open valve in oil line and start burner.
(k) No power to furnace	(k) Check fuse or circuit breaker. Replace blown fuse, or reset tripped circuit breaker. Check the shutoff switch to make sure it is in the ON position.

Table 13-5 Blower/Fan Troubleshooting

Symptom and Possible Causes	*Suggested Remedies*
Blower will not run	
(a) No power	(a) Check blower/fan power switch; check fuses, circuit breakers. Make corrections to re-establish power.
(b) Fan control adjustment too high	(b) Readjust or replace.
(c) Loose wiring	(c) Tighten connections.
(d) Defective wiring	(d) Replace wiring.
(e) Defective motor overload, protector, or motor	(e) Replace motor.
Blower will not stop	
(a) Fan set on manual	(a) Switch to automatic.
(b) Defective fan switch	(b) Replace fan switch.
(c) Short in wiring	(c) Check wiring for damage or loose connection and correct.
Rapid fan cycling	
(a) Fan switch differential too low	(a) Readjust or replace fan switch
(b) Blower speed too high	(b) Readjust to lower speed
Noisy blower motor	
(a) Loose fan blades	(a) Tighten or replace fan blade assembly.
(b) Incorrect belt tension	(b) Check specifications and readjust to allow the required slack (commonly about 1 inch).
(c) Pulleys out of alignment	(c) Realign pulleys.
(d) Dry pulley bearings	(d) Lubricate bearings.
(e) Defective belt	(e) Replace belt.
(f) Belt rubbing another surface	(f) Adjust belt tension.

Table 13-6 Coal Stoker Troubleshooting

Symptom and possible causes	*Suggested remedies*
Coal stoker stops	
(a) No power	(a) Check main power switch, fuses/circuit breakers, and correct to restore power.
(b) Obstruction in feed screw	(b) Remove obstruction.
(c) Dirty fire	(c) Clean fire.
Stoker motor fails to start	
(a) No power	(a) Check main power switch, fuses/circuit breakers, and correct to restore power.
(b) Overload condition	(b) Push reset button on transmission; push reset buton on stoker.
(c) Blown fuses	(c) Replace fuses.
(d) Limit control contacts open	(d) Allow furnace to cool off.
Stoker operates continuously	
(a) Controls out of adjustment	(a) Readjust controls or call local sales representative for adjustment.
(b) Dirty fire	(b) Clean fire.
(c) Fire out	(c) Rebuild fire.
(d) Dirty furnace	(d) Clean furnace.

Caution

Gas odors around a multi-fuel (gas/coal or gas/wood) furnace may indicate a safety problem. If you smell gas, (1) immediately open some windows; (2) extinguish any open flame; (3) do not touch any electrical switches; and (4) immediately call the local gas company for advice.

Chapter 14

Electric Furnaces

Electric heating is the only heat produced almost as fast as the thermostat calls for it. It is nearly instantaneous because there are no heat exchangers to warm up. The heating elements start producing heat the moment the thermostat calls for it. Unlike electric baseboard heating systems, electric furnaces are designed to provide the same heating/cooling advantages as gas or oil furnaces.

Because there is no flame with electric heat, there is no need to vent smoke or flue gases to the outside. Furthermore, there is no chimney loss with electric heat. It is 100 percent efficient, compared with the 60 to 95 percent efficiency of furnaces using other fuels.

Electric furnaces are available in upflow, downflow, or horizontal-flow models and in a wide range of sizes. For example, Carrier electric furnaces are available in 15 standard models from 5 to 35 kW in 5-kW increments. Other manufacturers offer a similar range of models (Figures 14-1 and 14-2).

An electric furnace should be listed by Underwriters Laboratories, Inc. (UL) for construction and operating safety. Furnaces approved by the agency are marked *UL Approved.*

Electric furnaces should be installed in accordance with local codes and regulations, the *National Electrical Code,* and recommendations made by the National Fire Protection Association.

Electrical Power Supply

Contact the local power company and make certain adequate electrical service is available for the furnace load *plus* all other appliances that will be on the line.

Check the *National Electrical Code* and the local code requirements. All wiring (including sizing) *must* comply with the requirements of these codes. When there is any conflict, the local codes and regulations take precedence. The manufacturer's requirements are also important. For example, Janitrol discourages the use of aluminum wire, although it is acceptable to the *National Electrical Code*. Use of aluminum wire in this case could jeopardize the furnace warranty.

Be sure to consult the manufacturer's wiring diagrams for electrical power requirements. Read the instructions carefully, and be sure you understand them thoroughly before you begin work.

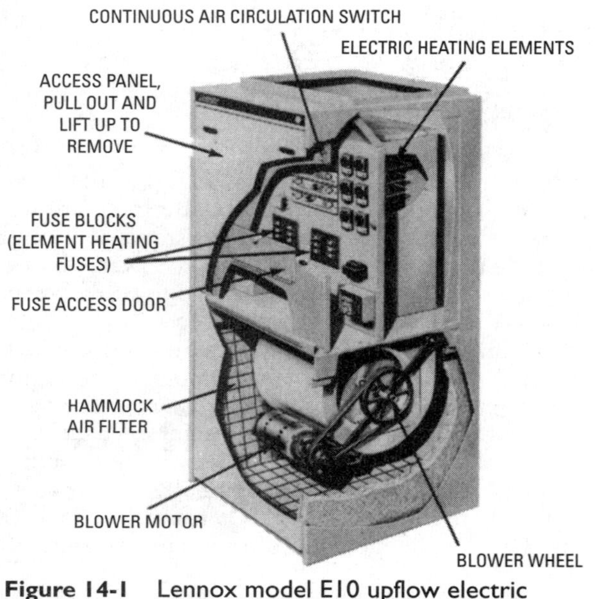

Figure 14-1 Lennox model E10 upflow electric furnace. *(Courtesy Lennox Air Industries Inc.)*

Planning Suggestions

If an electric furnace is being planned for a new structure, the maximum heat loss for each heated space must be calculated in accordance with procedures described in the manuals of the National Warm Air Heating and Air Conditioning Association or by a comparable method. This is very important, because these data will be used to determine the size (capacity) of the furnace selected for the installation.

Do not consider an electric heating system unless the structure is insulated properly. This insulating will be more extensive than that used with other types of heating systems. For example, ceilings should have a *minimum* of 6 in of blanket or loose-fill insulation, and cavities between studs in exterior walls should be filled with insulation completely. A description of the insulation requirements for a structure in which an electric heating system is used is included in Chapter 9, "Electric Heating Systems."

If the heating or heating/cooling installation is to be approved by either the Federal Housing Authority (FHA) or the Veterans' Administration (VA), heat loss and heat gain calculations should be

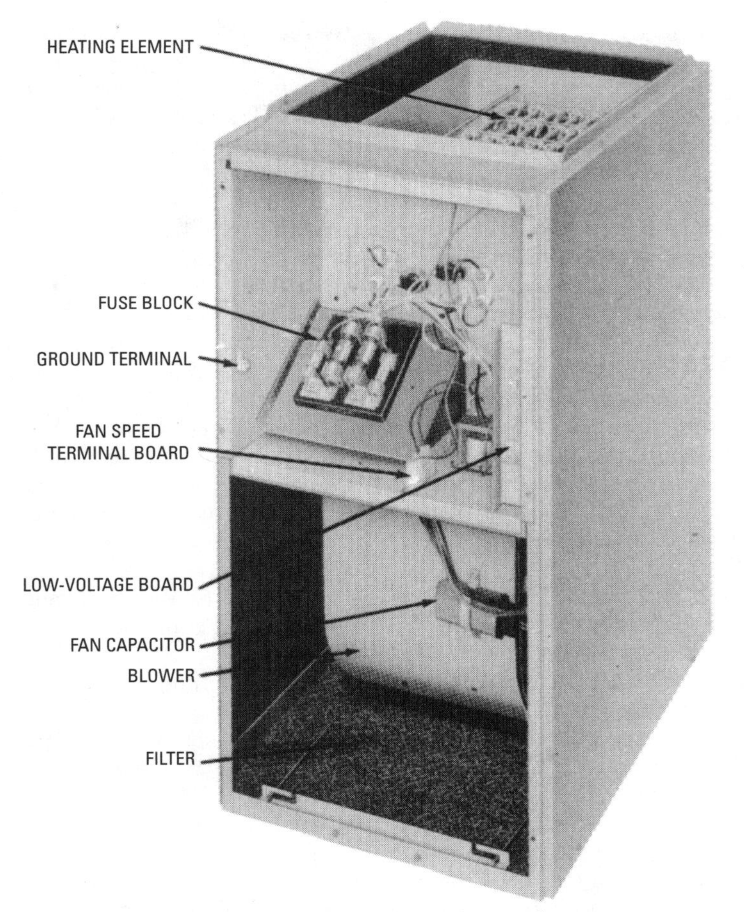

Figure 14-2 Carrier forced-warm-air electric furnace. *(Courtesy Carrier Corp.)*

made in accordance with the procedures described in the Air Conditioning Contractors of America's *Manual J*.

Location and Clearance

An electric forced-warm-air furnace should be located as near as possible to the center of the heat distribution system. Centralizing the furnace eliminates the need for one or more exceptionally long supply ducts. Long supply ducts are uneconomical because they are subject to a certain amount of heat loss. The number of elbows should be kept to a minimum for the same reason.

Electric furnaces are not vented, because electric heat is not produced by the combustion process. No flue gases or other toxic products of the combustion process occur with electric heat. As a result, a chimney and flue pipe are not required, and it is not necessary to consider these factors when locating the furnace.

A clearance of 24 to 30 inches in front of the furnace access panel should be provided for servicing and repairs. There is no minimum clearance requirement for ductwork and combustible materials. Electric furnaces may be installed with zero clearance between the cabinet and combustible materials, because the heat from the furnace is not produced by a flame.

Installation Recommendations

New electric furnaces for residential installation are shipped from the factory with all internal wiring completed. These furnaces are also generally shipped as a preassembled unit. In order to install the new furnace, the electric service from the line voltage main and the low-voltage thermostat must be connected. Directions for making these connections are found in the furnace manufacturer's installation instructions.

Familiarize yourself with all local codes and regulations that govern the installation of an electric furnace. Local codes and regulations take precedence over national standards.

Check the insulation of the structure to determine whether it is properly insulated for electric heat. The insulation should be installed in accordance with recommendations in *All Weather Comfort Standard of Electrically Heated and Air Conditioned Homes* (Electric Heating Association).

The furnace should be mounted on a level surface. If the unit is not level, it may develop serious vibrations. An insulating material can be placed under the furnace in most installations to reduce sound vibrations when the unit is operating. A noncombustible base is recommended for counterflow models.

Duct Connections

Detailed information concerning the installation of an air-duct system is contained in the following two publications of the National Fire Protection Association:

1. *Installation of Air Conditioning and Ventilating Systems of Other Than Residence Type* (NFPA No. 90A).
2. *Residence Type Warm Air Heating and Air Conditioning Systems* (NFPA No. 90B).

Additional information about duct connections can be found in Chapter 7, "Ducts and Duct Systems" of Volume 2. The comments made in Chapter 11, "Gas Furnaces" about furnace duct connections and air distribution ducts apply for the most part to ducts used with electric forced-warm-air furnaces.

Basic Components

An electric forced-warm-air furnace will generally consist of the following basic components (Figure 14-3):

1. Automatic controls
2. Heating elements
3. Safety controls

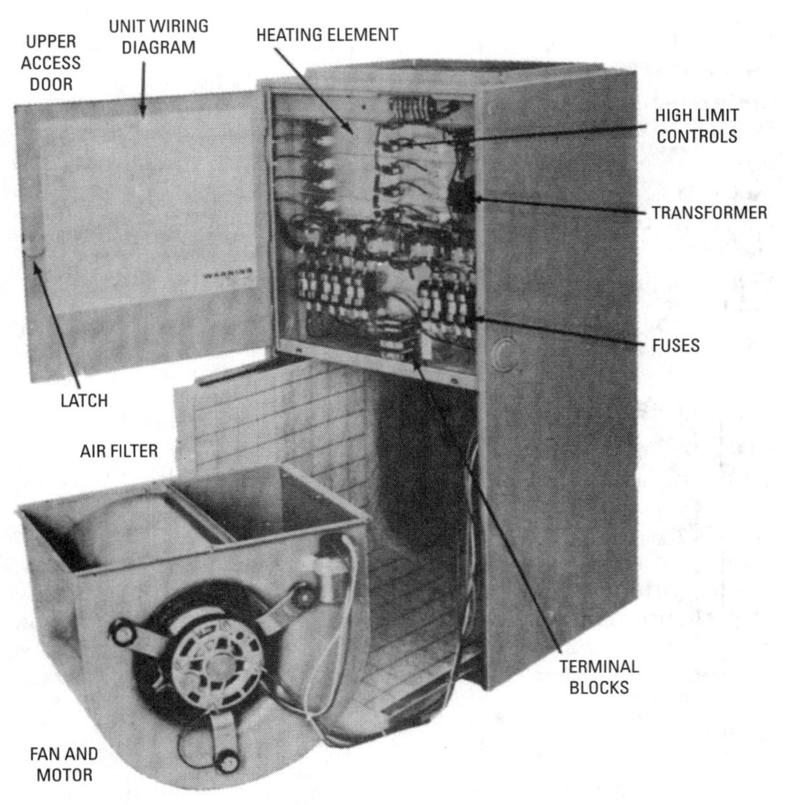

Figure 14-3 Principal components of an electrical furnace. *(Courtesy Trane Co.)*

4. Blowers and motors
5. Air filter(s)

Some electric furnaces also include an electronic air cleaner, an air conditioning evaporator coil, a humidifier, or some combination of these accessories. An electric heating/cooling system may also include a condensate pump if the dehumidifying process produces excessive amounts of water. In a zoned electric heating system, a zone control panel and motor-actuated dampers will be attached to either the furnace or the ducts.

Each of these components is described in the sections that follow. Additional information is contained in Chapter 10, "Furnace Fundamentals" and the various chapters in which furnace controls are described.

Automatic Controls

The automatic controls used in an electric heating system are designed to ensure its safe and efficient operation. Detailed descriptions of these controls are found in Chapter 4, "Thermostats and Humidistats" of Volume 2. This section is primarily concerned with outlining the operating principles of the automatic controls used with an electric furnace. These controls include:

1. Room thermostat
2. Thermostat heat anticipator
3. Timing sequences

In a central heating system, the wall-mounted room thermostat is the control that governs the normal operation of the furnace. The operating principle is simple. The temperature selector on the thermostat is set for the desired temperature. When the temperature in the room falls below this setting, the thermostat will call for heat and cause the first heating circuit in the furnace to be turned on. There is generally a delay of about 15 seconds before the furnace blower starts. This prevents the blower from circulating cool air in the winter. After about 30 seconds, the second heating circuit is turned on. The other circuits are turned on one by one in timed sequence.

Note

Use a heat pump thermostat or a conventional thermostat containing an electric setting to operate an electric furnace.

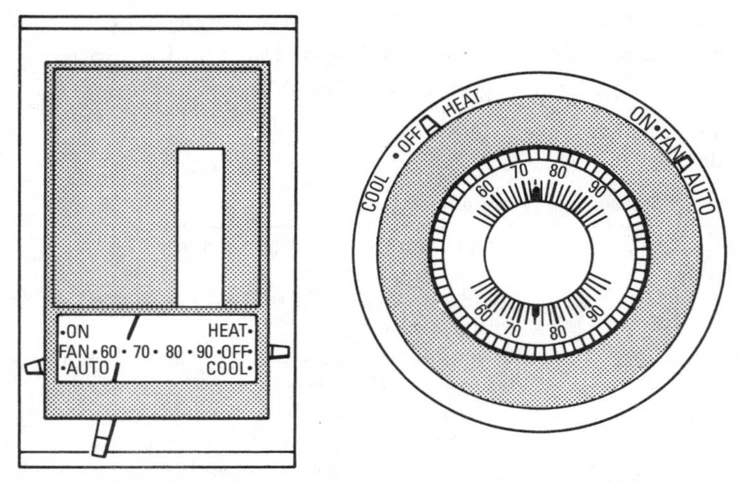

Figure 14-4 Typical room thermostats used in central electric heating systems. *(Courtesy Lennox Industries Inc.)*

When the temperature reaches the required level, the thermostat opens. After a short time, the first heating circuit is shut off. The others are shut off one by one in timed sequence. The blower will continue to operate until the air temperature in the furnace drops below a specified temperature.

A typical room thermostat will have a fan switch, a system switch, and a temperature selector (Figure 14-4). The temperature selector (a dial or lever device) on the thermostat is used to select the desired temperature. The actual operation of the heating system is governed by the positions of the fan and system switches. The switch positions and their functions are listed in Tables 14-1 and 14-2.

Most room thermostats contain a *heat anticipator.* This is a device designed to assist the thermostat in controlling closer to the desired temperature range (Figure 14-5).

When timing sequences are used (see below), the current flowing through the first time-delay sequencer (relay) must also flow through the heat anticipator. In order to obtain satisfactory operation, the heat-anticipator setting must be equal to the current draw of the sequencer.

The furnace manufacturer will generally recommend the setting for the heat-anticipator adjustment for each size unit. For example,

Table 14-1 Thermostat Operation in Single-Stage Heating/Two-Stage Cooling; Two-Stage Heating/Single-Stage Cooling; Two-Stage Heating/Two-Stage Cooling

Thermostat Switch Setting		
Fan	***System***	***Function***
Auto	Off	System completely shut down. Blower only, continuous operation.
On	Off	Provides air circulation when no cooling or heating is desired.
Auto	Cool	Blower and cooling system cycle on and off as thermostat demands.
On	Cool	Blower runs continuously; cooling system cycles on and off as thermostat demands.
Auto	Heat	Blower and furnace cycle on and off as thermostat demands.
On	Heat	Blower runs continuously; furnace cycles on and off as thermostat demands.
Auto	Auto	Unit cycles for either cooling or heating as per demand of thermostat.
On	Auto	Blower runs continuously; cooling and heating cycle as per demand of thermostat.

Courtesy Fedders Corporation

the setting recommended for a Trane Model EUADH 07 electric furnace is 0.45. This thermostat adjustment will vary depending upon the type of time-delay sequencer used, the furnace manufacturer, and the size of the furnace. This may be illustrated by the recommended heat-anticipator settings given by Coleman for its 10-kW, 15-kW, and 20-kW furnace models (Table 14-3). All Coleman 25-kW models require a heat-anticipator setting of 0.60.

After you have adjusted the heat anticipator to the suggested setting, operate the furnace several hours and observe the results. If there is *insufficient* heat, it may be caused by short furnace cycles. This can be corrected by moving the heat-anticipator pointer to a slightly higher setting. If there is too much heat, then long furnace cycles are overheating the structure. This can be corrected by moving the heat-anticipator pointer to a slightly lower setting. After

Table 14-2 Thermostat Operation in Single-Stage Heating/Cooling

Thermostat Switch Setting		
Fan	***System***	***Function***
Auto	Off	System completely shut down. Blower only, continuous operation.
On	Off	Provides air circulation when no cooling or heating is desired.
Auto	Cool	Blower and cooling system cycle on and off as thermostat demands.
On	Cool	Blower runs continuously; cooling system cycles on and off as thermostat demands.
Auto	Heat	Blower and furnace cycle on and off as thermostat demands.
Heat	On	Blower runs continuously; furnace cycles on and off as thermostat demands.

Courtesy Fedders Corporation

making these thermostat adjustments, allow the furnace to operate several hours to determine whether further adjustment is required. Additional information about the thermostat heat anticipator is contained in Chapter 11, "Gas Furnaces."

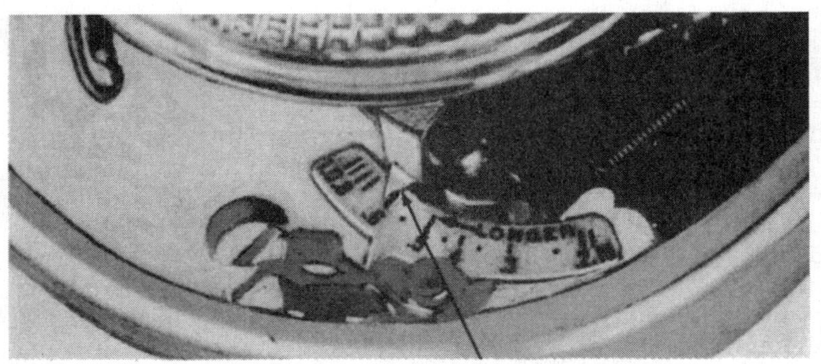

Figure 14-5 Thermostat heat anticipator. *(Courtesy Trane Co.)*

Table 14-3 Recommended Heat-Anticipator Settings for Coleman 10-, 15-, and 20-kW Electric Furnaces

Sequencer	*10 kW*	*15 kW*	*20 kW*
White-Rogers	0.15	0.15	0.10
Klixon	0.15	0.30	
Honeywell	0.40	0.40	

Courtesy The Coleman Company, Inc.

Heating Elements

A heating element must operate at a temperature well above its surroundings to deliver heat at a useful high rate. If it is a radiant heating element, it may operate red-hot or nearly white-hot over a relatively long period of time.

Wires with a high resistance, such as iron, chromium, nickel, manganese, and their alloys, are commonly used for heating elements. The heat output of a wire can be varied by changing its material composition or by changing its size or diameter. A smaller-diameter wire will have a higher resistance than a larger-diameter wire.

The heating elements in an electric furnace are resistance coils made from high-temperature chrome nickel heat-generating wire. Most manufacturers design their furnaces so that the entire heating element assembly can be removed for easy maintenance or repair (Figure 14-6).

Open elements are used in noncentral heating units, such as radiant or convective heaters, where it is desirable for the element to operate at a relatively high temperature. These elements reach operating temperatures very rapidly when energized. The exact operating temperature depends on the material used in the element and the type of heat desired. A radiant heater, for example, would probably operate at higher temperatures than would a wall-mounted convective heater.

Another type of element used in noncentral heating is the encapsulated, or completely enclosed, element. The simplest form of encapsulated element is ceiling cable, which has a layer of plastic insulation over it that can withstand the heat. Ceiling cable is designed to operate at very low temperatures.

A more complicated enclosed element is that used in baseboard convectors. An outer sheath of ceramic or metal protects the resistance wire from damage, corrosion, and deterioration. The

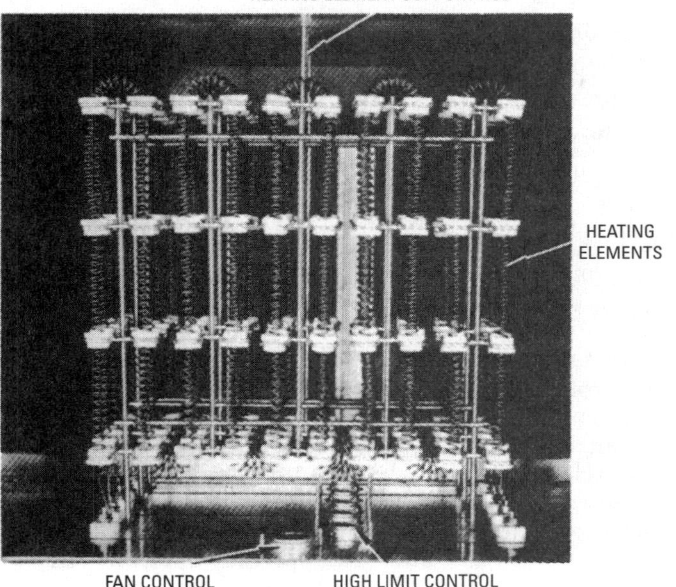

Figure 14-6 Heating elements and controls. *(Courtesy Trane Co.)*

heat-dissipating fins add surface area to increase the rate of heat transfer to the surrounding medium (air or water).

Timing Sequences

If all the heating elements in an electric furnace turned on simultaneously, there would be a momentarily excessive demand on the power supply, resulting in a temporary interruption in the electric service. This problem of power drains and surges is eliminated by timing the heating elements so that they start one at a time in predetermined increments. Sequencers or relays are used for this purpose. In a typcal application the sequencer has a small electric heater powered by the 24-volt control voltage from the thermostat. The control voltage activates a thermo-disk and a timing sequence. The timing sequence is set to a predetermined number of seconds for turning on and off the heating elements. The on-off sequence is also staggered so that the heating elements do not turn on and off all at the same time. The first squencer or relay commonly turns on both the first heating element and the furnace fan.

Safety Controls

Most electric furnaces are equipped with a variety of different safety controls to protect the appliance against current overloading or excessive operating temperatures. These safety controls are:

1. Temperature limit controls
2. Secondary high-limit control
3. Furnace fuses
4. Circuit breakers
5. Control voltage transformer
6. Thermal overload protector

Temperature Limit Controls

In the line control diagram of a Trane EUADH 07A model electric furnace shown in Figure 14-7, each heating element is shown with a high-limit control device located between the heating element and the time-delay relay. These high-limit control devices are designed to limit the outlet air temperature on Trane electric furnaces to 200°F. If the temperature of the outlet air should exceed 200°F, the high-limit control will open and interrupt the supply of electric power to the heating element.

Secondary High-Limit Control

Secondary high-limit protection is provided by a fusible link in each electric heating element (Figure 14-8). This device is designed to shut off the current when temperatures in the furnace become excessive. It functions as a backup system in case of limit switch failure.

Furnace Fuses

Furnace fuses are used to provide protection against possible overload conditions and to ensure correct, safe field wiring. Each heating element circuit is protected by two branch fuses as shown in Figure 14-9. These are sized to limit the current draw of each heater and are designed to open on a short-circuit or an overloaded-circuit condition. For overload protection, the blower motor and relay are also safeguarded with a separate fuse.

Circuit Breakers

Some electric furnaces are equipped with circuit breakers and a terminal board. The wiring diagram for the Coleman 25-kW electric furnace illustrated in Figure 14-10 indicates that both the terminal board and the circuit breakers are located in the power supply feed line to the furnace.

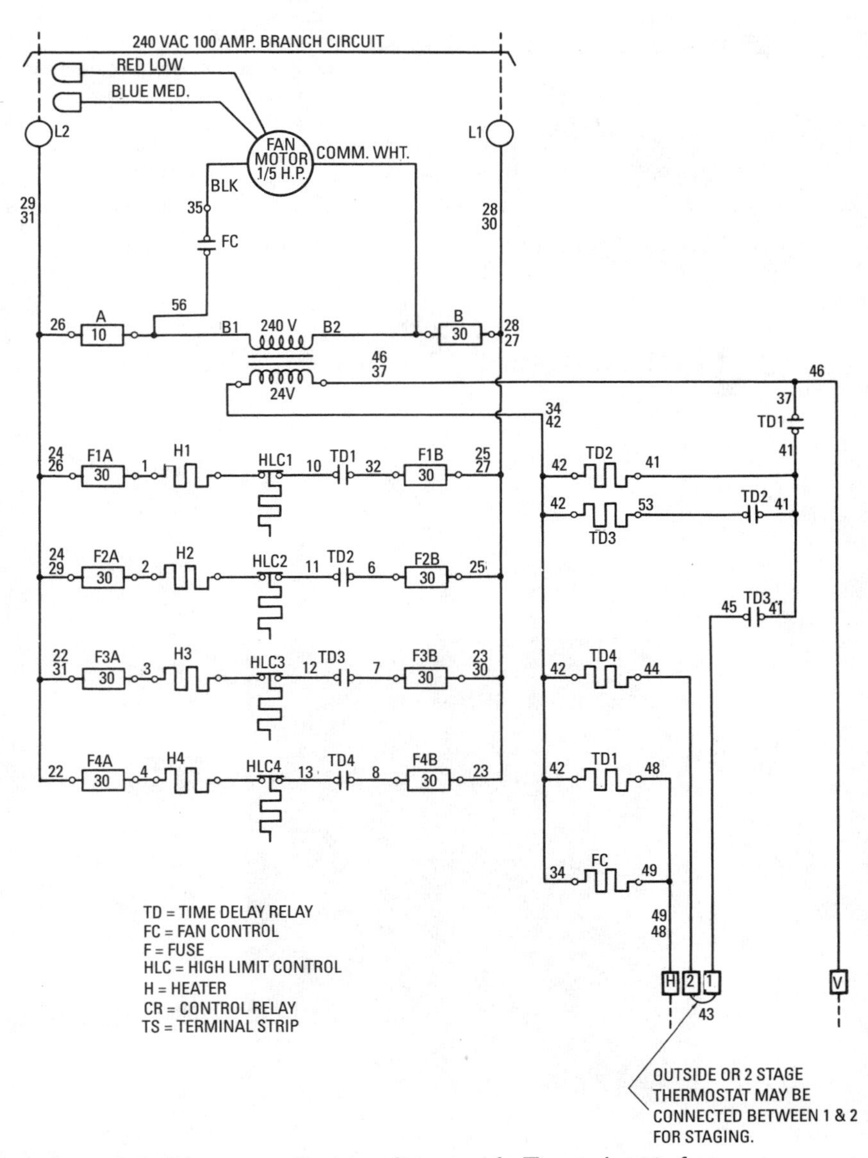

Figure 14-7 Line-control wiring diagram of a Trane electric furnace.

(Courtesy Trane Co.)

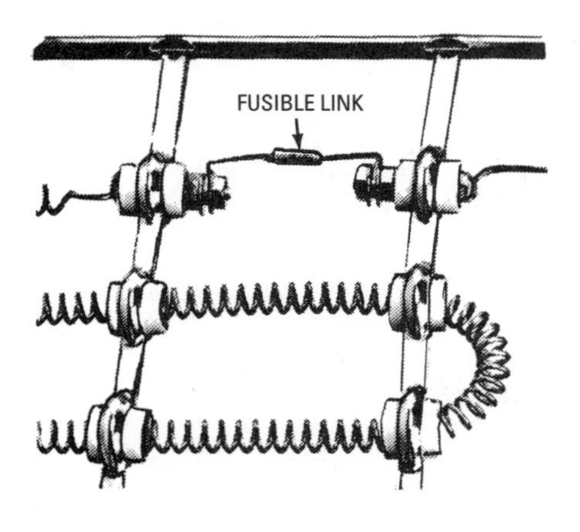

Figure 14-8 Fusible link connection in electric heating element. *(Courtesy International Heating and Air Conditioning Corp.)*

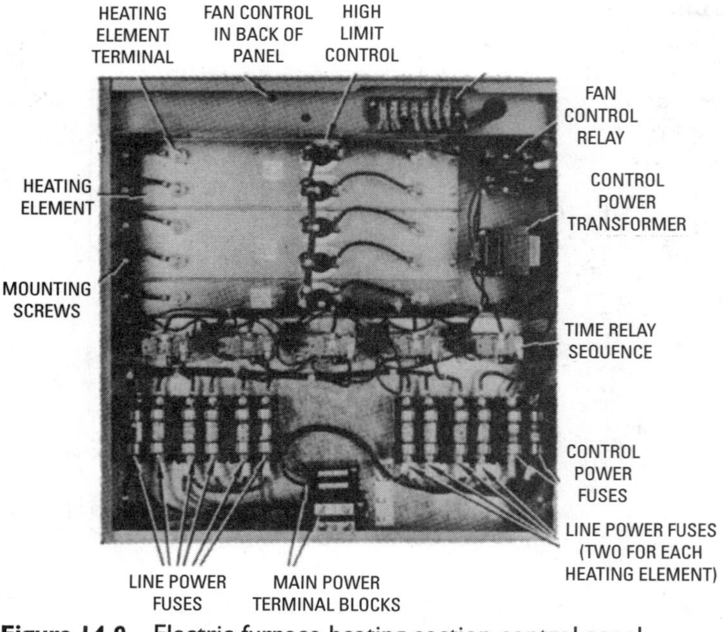

Figure 14-9 Electric furnace heating-section control panel showing location of line power fuses. *(Courtesy Trane Co.)*

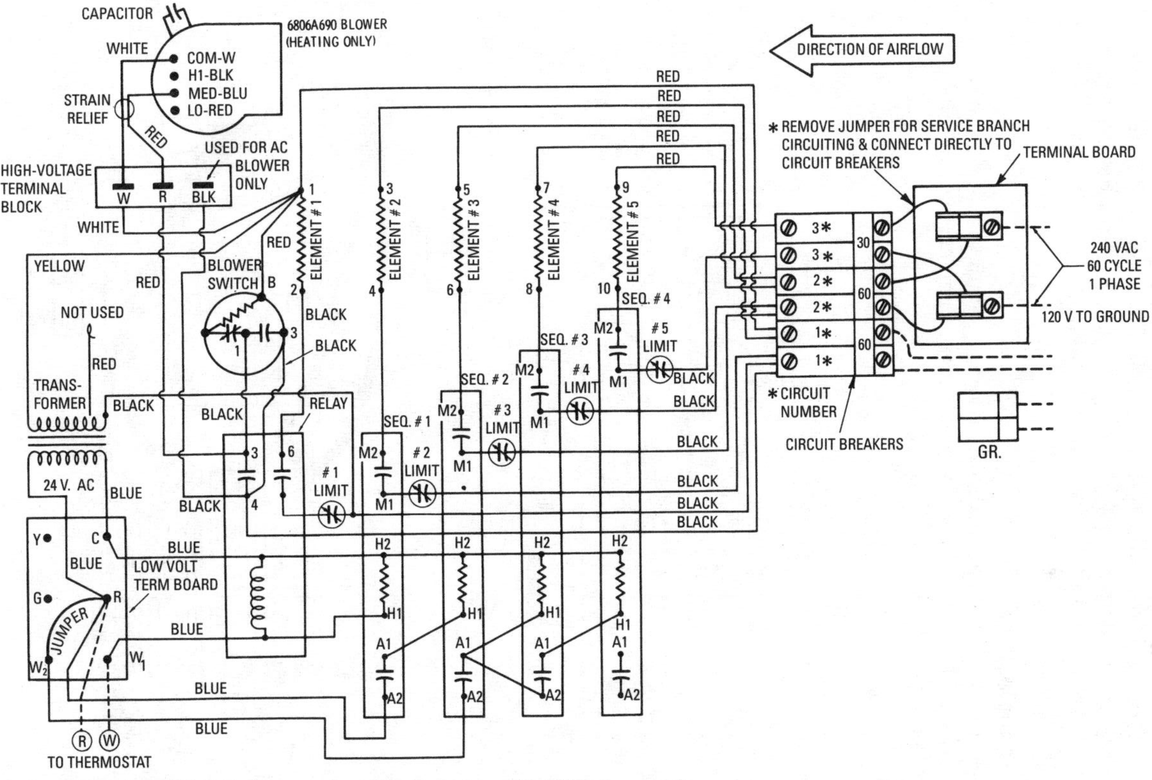

Figure 14-10 Wiring diagram for the Coleman 25-kW electric furnace. *(Courtesy Coleman Co., Inc.)*

When it is desirable to run branch circuits to the circuit breakers (bypassing the terminal board), the jumper wires between the circuit breakers and the wiring are connected as shown in Figures 14-11 and 14-12.

Transformer

A control voltage transformer is used to limit the amount of output current. Limiting the amount of output current permits the use of open control wiring.

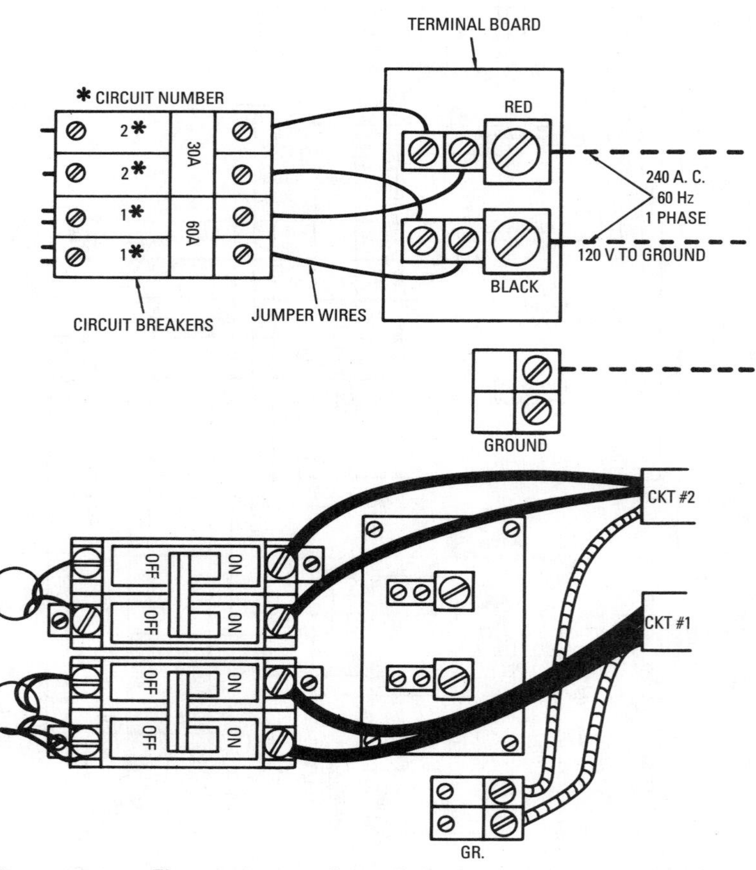

Figure 14-11 Electrical wiring for a Coleman 15- or 20-kW electric furnace with circuit breakers. *(Courtesy Coleman Co., Inc.)*

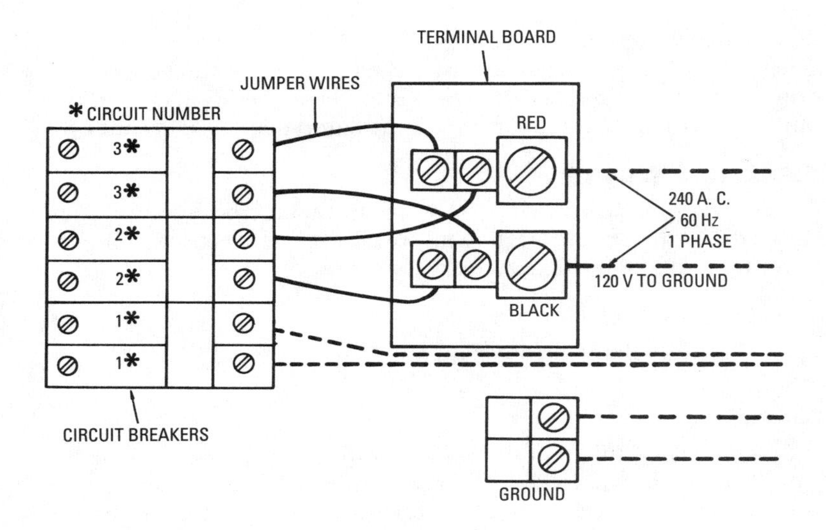

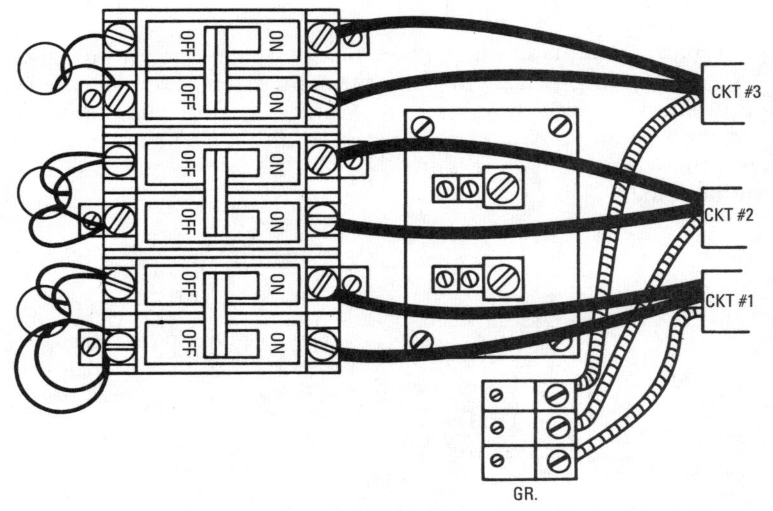

Figure 14-12 Electric wiring for a Coleman 25-kW electric furnace with circuit breaker. *(Courtesy Coleman Co., Inc.)*

Thermal Overload Protector

The fan motor is protected against locked rotor or overheated conditions by thermal overload devices. When these conditions occur, the fan motor circuit is automatically opened, and the motor is shut off.

Electrical Wiring

All internal furnace wiring is done at the factory before it is shipped. At the site, the following two types of electrical connections are required to field wire the unit:

1. Line voltage field wiring
2. Control voltage field wiring

Caution

DEADLY HIGH-VOLTAGE conditions exist inside the cabinets of electric furnaces. Only a trained HVAC technician or someone with equivalent experience should attempt to service or repair electrical components.

A typical example of line voltage field wiring is shown in Figure 14-13. Line voltage wiring involves the connection of the furnace to the building power supply. Line voltage wiring runs directly from the building power panel to a fused disconnect switch. From there, the wiring runs to terminals L_1 and L_2 on the power-supply terminal block.

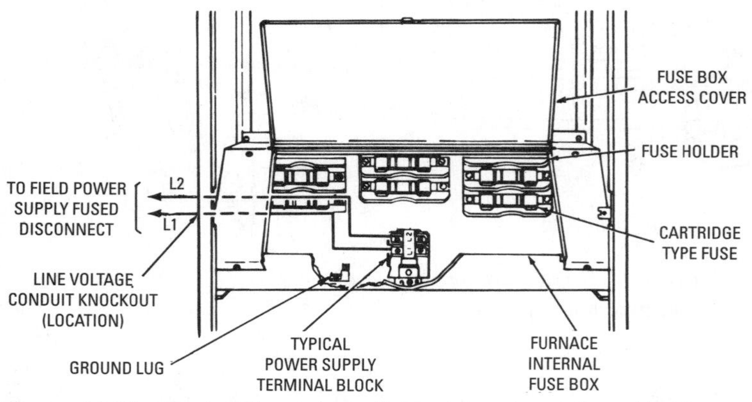

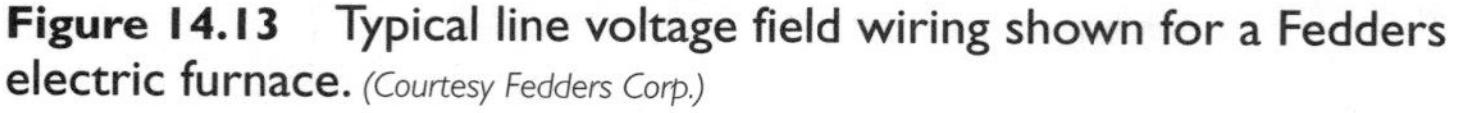

Figure 14.13 Typical line voltage field wiring shown for a Fedders electric furnace. *(Courtesy Fedders Corp.)*

The unit must be properly grounded either by attaching a grounded conduit for the supply conductors (knockouts in the side panels of Fedders electric furnaces are provided for this purpose) or by connecting a separate wire from the furnace ground lug to a suitable ground.

The external control voltage circuitry consists of the wiring between the thermostat and the low-voltage terminal block located in the control voltage section of the furnace. Instructions for control voltage wiring are generally shipped with the thermostat.

Some typical examples of control voltage field wiring used with Fedders electric furnaces are shown in Figures 14-14 and 14-15. Control voltage field wiring connections used with Coleman electric furnaces are shown in Figure 14-16.

The *National Electrical Code* requires that furnaces larger than 10 kW be supplied with branch circuit fusing. Power connections on units of this size are usually made to lugs on the fuse blocks.

Blowers and Motors

The blowers and motors used with electric furnaces are identical to those used in gas-fired furnaces. Read the section *Blowers and Motors* in Chapter 11, "Gas Furnaces" for additional information.

Air Delivery and Blower Adjustment

It is sometimes necessary to adjust the blower speed to produce a temperature rise through the furnace that falls within the limits stamped on the furnace nameplate. Blower adjustment procedures are described in full detail in Chapter 11, "Gas Furnaces."

Air Filter

The air filters used in electric forced-warm-air furnaces are either permanent types that can be removed and cleaned on a periodic basis or replaceable, throwaway filters.

More detailed information concerning furnace air filters is contained in Chapter 12, "Air Cleaners and Filters" of Volume 3. See also the comments about air filters in the section *Maintenance and Operating Instructions* in this chapter.

Air Conditioning

A furnace should be installed in parallel or on the upstream side of the cooling unit to avoid condensation in the heating section. Parallel installation will require dampers or some other means to prevent cool air from entering the furnace.

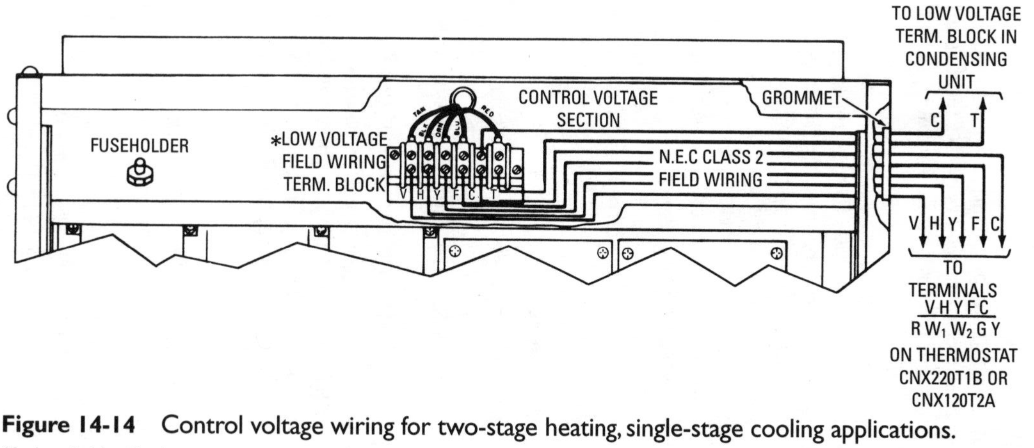

Figure 14-14 Control voltage wiring for two-stage heating, single-stage cooling applications.
(Courtesy Fedders Corp.)

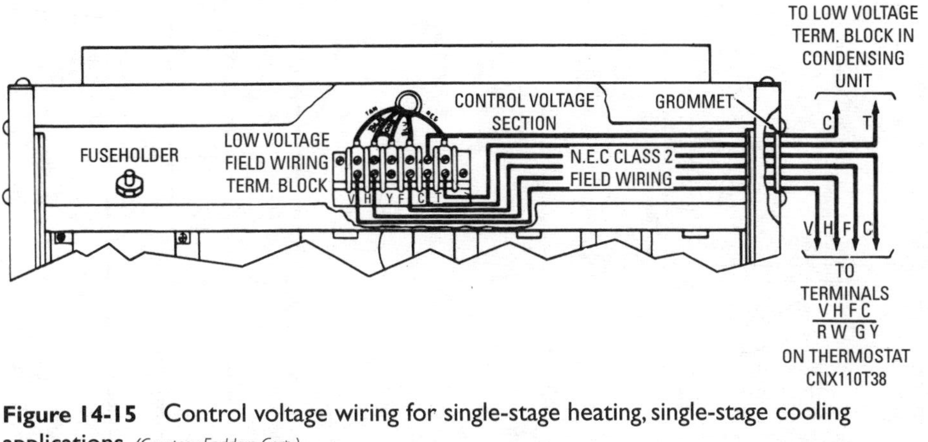

Figure 14-15 Control voltage wiring for single-stage heating, single-stage cooling applications. *(Courtesy Fedders Corp.)*

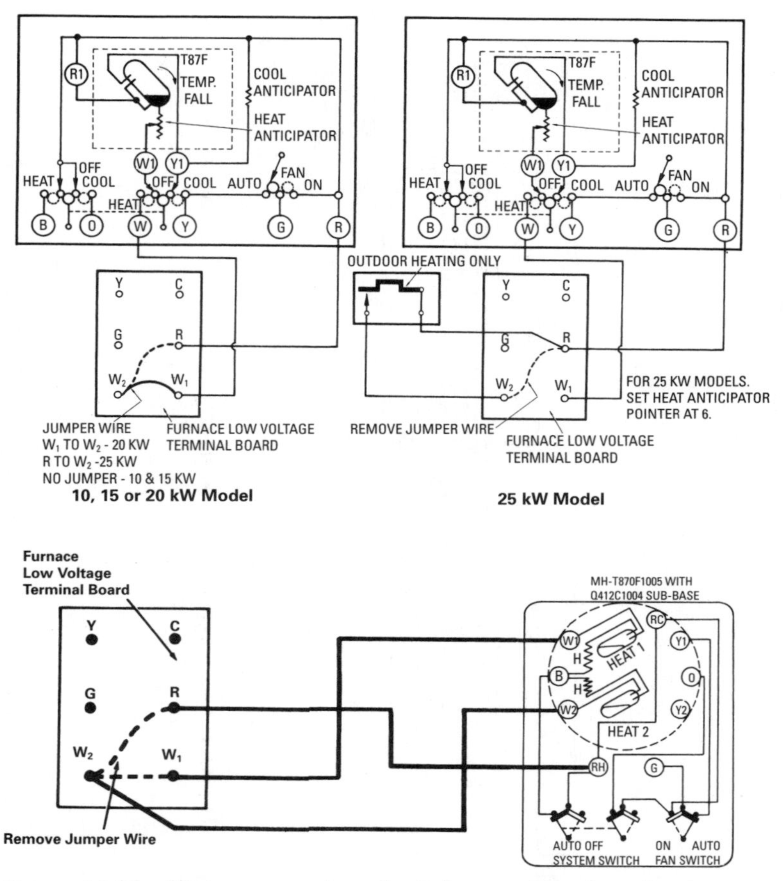

Figure 14-16 Three connections for indoor and outdoor heating (only) thermostats for a Coleman model 6806 electric furnace.

(Courtesy Coleman Co., Inc.)

The air conditioning component of a typical electric heating and cooling system generally consists of an outdoor condensing unit, indoor coils, and a cabinet to house the cooling coils. Most furnace manufacturers provide detailed instructions for adding air conditioning to the heating unit. The important thing to remember is to size the ducts for the larger volume of air used in air conditioning.

Additional information about air conditioning can be found in Chapters 8, 9, and 10 in Volume 3.

Maintenance and Operating Instructions

Maintenance and operating instructions will normally be provided for the furnace by the manufacturer. If no instructions are available, try contacting a field representative or writing the company for a duplicate copy of the owner's manual.

Caution

> *Always* shut off the electrical power supply to the furnace before attempting to service it. This is very important because DEADLY HIGH-VOLTAGE CONDITIONS EXIST. Be sure to open all furnace fused-disconnect switches before servicing.

Furnace Air Filters, Electronic Air Cleaners, and Humidifers

The air cleaner is one of the most important parts of an electric forced-air furnace. A clogged one will cause the furnace to run longer and harder to deliver the desired amount of heat. The result will be higher energy use and higher energy bills.

Air filters should be—inspected on a periodic basis. Clean or replace the filters on a monthly basis during the heating season. Do the same every 3 or 4 months the rest of the year. If the system has an electronic air cleaner, periodically inspect and wash the grids. If the system is equipped with a humidifier, inspect and clean it on a periodic basis.

Caution

> Steam-generating-type humidifers are line voltage powered. They must be shut off before servicing.

Heating Elements and Heating Control Wiring

Inspect the heating element and heating control wiring to make certain connections are tight and clean. Check for burned or frayed wires. Check for any breaks or cracks in the wire insulation. These can cause shorting and are a potential fire hazard. All terminal block wiring connections should also be tight and clean.

Caution

> Do not attempt to service the controls inside the furnace cabinet unless you are a qualified HVAC technician. DEADLY HIGH-VOLTAGE CONDITIONS EXIST, which can result in serious injury or death.

Check the manufacturer's maintenance and/or troubleshooting manual for instructions on testing the elements to see which ones are drawing current. The method (and equipment) will varying among different furnace manufacturers. For example, General Electric/Trane requires the use of either an ohmmeter or clamp on ammeter.

Check the furnace wiring diagram to make certain the fuses are of the correct type and amperage.

Blowers/Fans

Some blower/fan motors are permanently lubricated and will not require further attention. Others have ports for oiling and require periodic lubrication. See *Blowers and Motors* in Chapter 11, "Gas Furnaces."

Periodically check the blower/fan belts (if used) for belt tension and adjust if necessary. If the belts are frayed or in anyway damaged, replace them. Brush the blower/fan blades and the entire enclosure area to keep dust from being blown through the ducts into your rooms. Blower/fan maintenance should be done every time you clean or replace the air filter. You should also inspect and clean the room registers at the end of the ducts at this time.

Ducts

Check the ducts for loose connections or damage, and correct as necessary. Seal all duct seams and joints in the ductwork, and seal the connection between the furnace plenum and the ducts. Doing so will increase the amount of air (and heat or cool air) delivered through the ducts to the rooms.

Insulate ductwork in basements and crawl spaces to reduce heat loss and lower your energy use and costs.

Thermostats

A combined heating and cooling system is designed to operate year-round without being shut down. The only changeover required are the room thermostat settings and periodic maintenance.

Check the thermostat setting against the actual temperature in a room. The actual temperature can be measured with a thermometer placed in the center of the room. You may find that the actual room temperature is several degrees lower than the thermostat setting. This is a common characteristic of electric heating systems. You will probably have to set the thermostat a bit higher to get the desired results. Just do a little experimenting. On the other hand, the thermostat may be allowing the room temperature to rise above the setting. This may indicate a faulty thermostat. If you suspect a faulty thermostat, have it serviced by an HVAC technician.

Periodically check the batteries in a programmable thermostat. Your heating/cooling problem may be nothing more than an inexpensive battery replacement. Lower the setting on a programmable thermostat when you are sleeping, away at work, or on a vacation. This will result in lower energy costs.

Furnaces without Air Conditioners

When an electric furnace is installed *without* an air conditioning unit, the furnace should be shut down at the end of the heating season. The procedure for doing this will be found in the furnace owner's manual. It is a simple operation that generally consists of opening the main fused-disconnect switch (or switches) in the power-supply lines serving the furnace.

Troubleshooting an Electric Furnace

Any appliance may sometimes fail to operate efficiently because of a malfunction somewhere in the equipment. The problems most commonly associated with electric-fired furnaces are listed in Table 14-4.

Caution

> DEADLY HIGH-VOLTAGE CONDITIONS exist within the cabinet (case) of an electric furnace. Do not attempt any troubleshooting if it might involve contact with electric controls and components unless you are a trained HVAC technician or have equivalent experience with electric furnaces.

Table 14-4 Troubleshooting Electric Furnaces

Symptom and Possible Causes	*Suggested Remedies*
Unit fails to operate	
(a) Defective furnace transformer.	(a) Replace.
(b) Blown or defective transformer fuse.	(b) Replace fuse.
(c) Defective thermostat.	(c) Replace thermostat.
(d) Improperly set room.	(d) Change thermostat to proper setting.
(e) Open fused-disconnect switch.	(e) Correct problem and close furnace circuit.
(f) Blown or faulty furnace fuse.	(f) Replace fuse.

(continued)

Table 14-4 *(continued)*

Symptom and Possible Causes	*Suggested Remedies*
Fan operates with low or no heat	
(a) Blown or faulty heater element fuse.	(a) Replace fuse.
(b) Defective time-delay sequencer or relay. *Note:* Some electric furnaces use relays instead of sequencers (e.g., General Electric, Trane, Rheem, Ruud).	(b) Replace sequencer or relay. *Note:* Sequencers and relays are not interchangeable.
Heats without fan operation	
(a) Faulty fan-control relay.	(a) Replace.
(b) Defective fan motor.	(b) Repair or replace.
(c) Faulty fan motor wiring or loose connections.	(c) Repair or replace wiring.
(d) Defective fan motor run capacitor.	(d) Repair or replace.
Individual heater fails to operate	
(a) Blown or faulty heater circuit fuse.	(a) Replace fuse.
(b) Defective high-limit control.	(b) Replace.
(c) Defective time-delay sequencer or relay.	(c) Replace.
(d) Faulty heater element.	(d) Replace.
(e) Defective heat pump thermostat.	(e) Check thermostat operation.
(f) Wrong heat pump thermostat.	(f) Replace with heat pump thermostat, or replace with conventional thermostat having an electric setting (the electric setting will activate the fan), or add a relay or double sequencer.
Fan operates on heating, not on cooling	
(a) Defective cooling-cycle control relay.	(a) Replace.
(b) Improperly connected or faulty room thermostat.	(b) Make proper connections to thermostat.
(c) Defective or improper fan motor connections.	(c) Repair or replace.

(continued)

Table 14-4 *(continued)*

Symptom and Possible Causes	*Suggested Remedies*
Furnace will not operate	
(a) No power to furnace.	(a) Check power source (fuses or circuit breakers in main box).
(b) Blown fuses.	(b) Wrong size fuses. Replace with correct size fuses.
(c) Dirty or loose connections on fuses.	(c) Clean and/or tighten connections.
(d) Heaters (elements) shorting to cabinet case.	(d) Correct.
(e) Circuit breaker incorrect size for amp load.	(e) Test and replace as necessary.
Low heat or no heat	
(a) No power to furnace.	(a) Check power source (fuses or circuit breakers in main box).
(b) Incorrect thermostat setting.	(b) Set thermostat to desired temperature setting.
(c) Low air flow in and out of system.	(c) Check to verify air flow matches system specifications. If not, check for a blocked air filter or duct and correct.
Low humidity levels	
(a) Incorrect humidifier setting	(a) Reset humidifier to recommended setting.
(b) Incorrect humidistat setting.	(b) Reset humidistat to recommended setting.
(c) Incorrect thermostat fan switch setting. Switch set on manual instead of automatic.	(c) Change thermostat fan switch setting from manual to automatic.
(d) Incorrect ventilation switch setting on furnace. Switch set on continuous (low) instead of automatic.	(d) Change ventilation switch setting from continuous (low) to automatic.
High humidity levels	
(a) Additional means required to reduce excessive humidity.	(a) Install humidistat on furnace.
(b) Incorrect humidistat setting.	(b) Change humidistat setting.

(continued)

Table 14-4 *(continued)*

Symptom and Possible Causes	*Suggested Remedies*
(c) Incorrect thermostat fan switch setting. Switch set in automatic instead of manual.	(c) Change thermostat fan switch setting from automatic to manual.
(d) Incorrect ventilation switch setting on furnace. Switch set on automatic instead of continuous (low).	(d) Change ventilation switch setting from automatic to continuous (low).

Chapter 15

Steam and Hot-Water Space Heating Boilers

Boilers are used to supply steam or hot water for heating, processing, or power purposes. This chapter is primarily concerned with a description of the low-pressure steam and hot-water (hydronic) space heating boilers used in the heating systems of residences and small buildings (Figures 15-1 and 15-2).

The basic construction of both low-pressure steam and hot-water space heating boilers fired by fossil fuels consists of (1) an insulated steel jacket enclosing a lower chamber in which the combustion process takes place; and (2) an upper chamber containing cast-iron sections or steel tubes in which water is heated or converted to steam for circulation through the pipes of the heating system.

Steam and Hot Water Boiler Similarities and Differences

Steam and hot-water (hydronic) space heating boilers are very similar physically, but there are some important differences.

- Steam boilers operate only about three-fourths full of water, whereas hot-water boilers operate completely filled with water.
- Steam boilers in residential steam heating systems operate at 2 psi pressure or slightly more, whereas residential hot-water boilers operate at approximately six times that pressure.
- Steam boilers are equipped with a low-water cutoff device to protect the appliance from burning out if it should run out of water. Only large hot-water space heating boilers with a capacity exceeding 400,000 Btu/h are presently required by code to be equipped with low-water cutoffs. (*Note:* Many HVAC contractors who install the smaller residential hot-water boilers strongly recommend the addition of a low-water cutoff device to these appliances to prevent burnout if the boiler loses its water.)

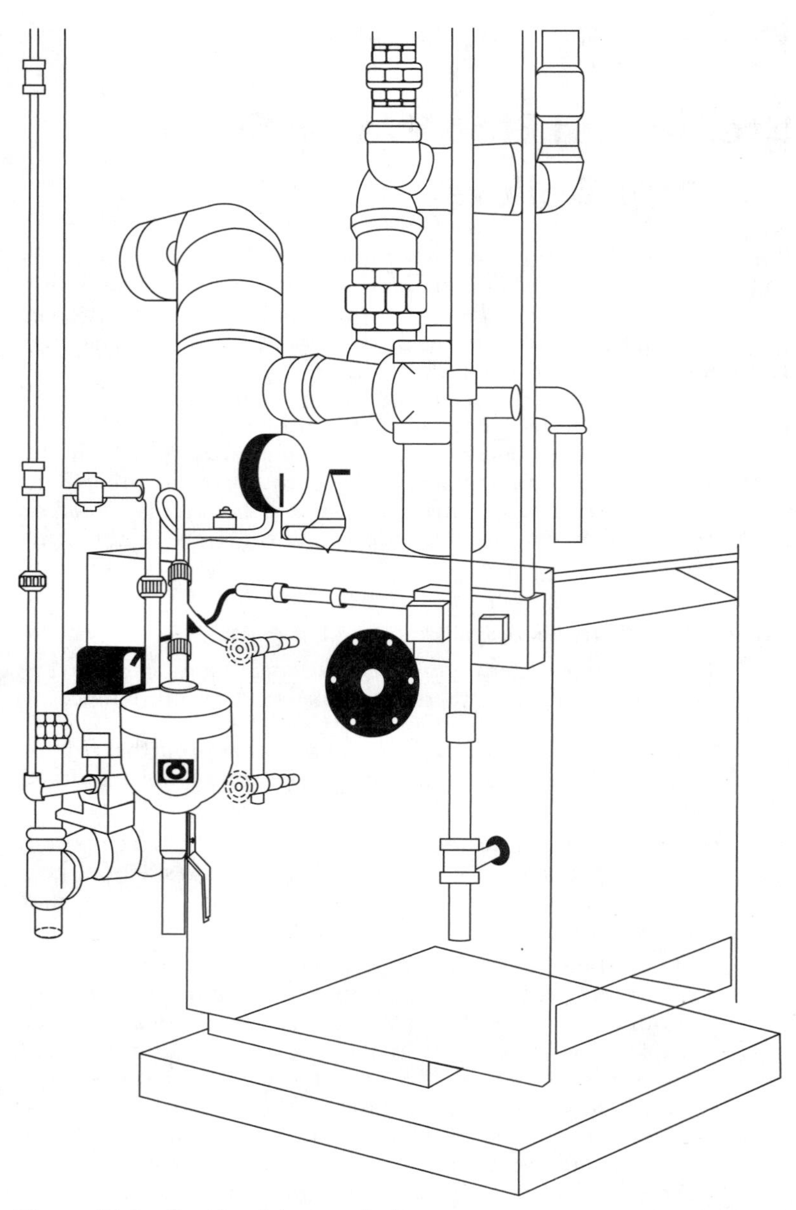

Figure 15-1 Residential steam boiler. *(Courtesy ITT McDonnell & Miller)*

- Steam boilers require makeup feed to replace water lost through evaporation and the production of steam during normal operation. Hot-water boilers can operate with little or no need for makeup water under the same normal operating conditions.

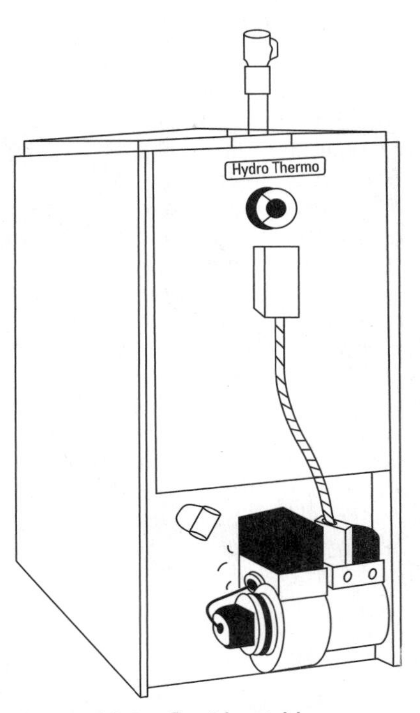

Figure 15-2 Residential hot water (hydronic) boiler. *(Courtesy Hydrotherm, Inc.)*

The design and construction of the lower chamber depends upon the type of fuel used to fire the boiler. It serves as a combustion chamber for coal-fired and oil-fired boilers and as a compartment for housing the gas burner assembly on gas-fired boilers. These gas burner assemblies are commonly designed for easy removal so that they can be periodically cleaned or serviced.

Oil burners are externally mounted with the burner nozzle extending into the combustion chamber. This is also true of gas conversion burners. Gas burner assemblies, on the other hand, are located inside the lower chamber of the boiler.

The cast-iron sections or steel tubes in the upper chamber of the boiler contain water that circulates through the pipes in the heating system in the form of either steam or hot water. The heat from the combustion process in the lower chamber of the boiler is transferred through the metal surface of the cast-iron sections or steel tubes to the water contained in them, causing a rise in temperature. The amount of water contained in these passages is one of the ways in which steam boilers and hot-water space heating boilers are distinguished from one another. In hot-water space heating boilers these passages are completely filled with water; whereas in low-pressure steam boilers only the lower two-thirds are filled. In steam boilers the water is heated very rapidly, causing steam to form in the upper one-third. The steam, under pressure, rises through the supply pipes connected to the top section of the boiler.

A boiler jacket contains a number of different openings for pipe connections and the mounting of accessories. The number and type of openings on a specific boiler jacket depends upon the type of boiler (i.e., steam or hot water). Among the different openings to be found on a boiler jacket are the flue connection, water feed (supply) connection, inspection and cleanout tapping, blowdown tapping, relief valve tapping, control tapping, drain tapping, expansion tank tapping, and return tapping. There are also gas and oil burner connections. Figure 15-3 illustrates the arrangement of control tappings in a Weil-McLain oil-fired boiler.

Most (but not all) of the controls on low-pressure steam and hot-water space heating boilers fired by the *same* fuel are similar in design and function, but there are exceptions. For example, a few boiler controls and fittings are designed to be specifically used on steam boilers; others are found only on hot-water space heating boilers. The various boiler controls and fittings are described in the appropriate sections of this chapter.

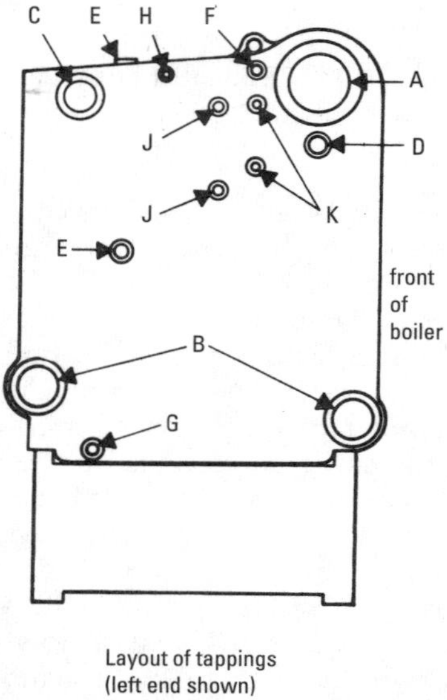

Location on Boiler	Size (Inches)	Steam Boilers	Water Boilers
A	6	Supply	Supply
B	4	Return	Return
C	3	Safety valve	Relief valve
D	1¼	Blow-off	Not required
E	1	Water feeder	Not required
F	¾	Pressure limit	Temperature limit
G	¾	Drain	Drain
H	½	Pressure gauge	Temperature and altitude gauge
J	½	Gauge glass and low water cut-off	Not required
K ■	⅜	Try-cocks	Not required

■ Tappings available on special order only

Layout of tappings (left end shown)

Figure 15-3 Boiler control tapping locations. *(Courtesy Weil-McLain)*

Boiler Rating Method

The construction of low-pressure steel and cast-iron heating boilers is governed by the requirements of the ASME Boiler and Pressure Vessel Code. This is a nationally recognized code used by boiler manufacturers, and any boiler used in a heating installation should clearly display the ASME stamp. State and local codes are usually patterned after the ASME Code.

The location of the identification symbols used by the ASME is specified by the code and determined by the type of boiler. For example, on a water-tube boiler, it appears on a head of the steam-outlet drum near and above the manhole opening. On vertical fire-tube boilers, the stamp bearing the identification symbol should appear on the shell above the fire door and handhole opening. Other types of boilers (e.g., Scotch marine and superheaters) have their own specified location for the identification symbol stamp.

The ASME Boiler and Pressure Vessel Code applies only to boiler construction, specifically to maximum allowable working pressures, not to its heating capacity. A number of different methods are used to rate the heating or operating capacity of a boiler. The boiler manufacturers have developed their own ratings, but these are generally used along with rating methods available from several professional and trade associations.

The Steel Boilers Institute no longer exists, but its SBI rating is still found on many existing steel boilers. The I=B=R (or IBR) logo was created by the now defunct Institute of Boiler and Radiator Manufacturers to indicate the gross output(s) at 100 percent firing rate for most sectional cast-iron boilers. The I=B=R rating logo is now used by the Hydronics Institute Division of the Gas Appliance Manufacturers Association (GAMA).

The Mechanical Contractors Association of America has devised a method for rating boilers not covered by either the SBI or I=B=R codes. Finally, gas-fired boilers are rated in accordance with methods developed by the American Gas Association.

Other rating logos appearing on boilers and in their installation and operation manuals are the Underwriters Laboratories, Inc. (UL) and the Underwriters' Laboratories of Canada logos.

In terms of its heating capacity, the rating of a boiler can be expressed in square feet of equivalent direct radiation (EDR) or thousands of Btu/h. Sometimes a boiler horsepower rating is also given, but this has proven to be misleading.

For steam boilers, 1 square foot of equivalent direct radiation (EDR) is equal to the emission of 240 Btu/h. For a water boiler, 1 square foot of EDR is considered equal to the emission of 150 Btu/h.

A boiler horsepower (bhp) is the evaporation of 34.5 lb of water into dry steam from and at 212°F. For rating purposes, 1 bhp is considered as the heat equivalent of 140 ft^2 of steam radiation per hour. In some cases bhp ratings are obtained by dividing steam SBI ratings by 140.

A boiler is rated according to its operating or heating capacity, but this rating will vary in accordance with the *type* of load used as the basis for the rating. The three types of connected loads used to determine the rating of a boiler are:

1. Net load
2. Design load
3. Gross load

Net load refers to the actual connected load of the heat-emitting units in the steam or hot-water heating system. *Design load* includes the net-load rating plus an allowance for piping heat loss. Finally, *gross load* will equal the net load and the piping heat loss, plus an additional allowance for the pickup load.

Boiler Heating Surface

The boiler *heating surface* (expressed in square feet) is that portion of the surface of the heat transfer apparatus in contact with the fluid being heated on one side and the gas or refractory being cooled on the other side. The *direct* or *radiant* heating surface is the surface against which the fire strikes. The surface that comes in contact with the hot gases is called the *indirect* or *convection* surface.

The *heating capacity* of any boiler is influenced by the amount and arrangement of the heating surface and the temperatures on either side. The arrangement of the heating surface refers to the ratio of the diameter of each passage to its length, as well as its contour (straight or curved), cross-sectional shape, number of passes, and other design variables.

Boiler Efficiency

The *boiler efficiency* is the ratio of the heat output to the caloric value of the fuel. Boiler efficiency is determined by various factors including the type of fuel used, the method of firing, and the control settings. For example, oil- and gas-fired boilers have boiler efficiencies ranging from 70 to 80 percent. A hand-fired boiler in which anthracite coal is used will have a boiler efficiency of 60 to 75 percent.

Boiler Energy Efficiency

Two government programs have been created within the last 20 years to rate the energy efficiency of different heating appliances such as furnaces, boilers, water heaters, and heat pumps. These two programs are (1) the annual fuel utilization capacity (AFUE) program and (2) the Energy Star Certification program.

- **Annual Fuel Utilization Capacity (AFUE).** The energy efficiency of an oil-, gas-, or coal-fired boiler is measured by its annual fuel utilization capacity (AFUE). The AFUE ratings for boilers manufactured today are listed in the boiler manufacturer's literature. Look for the EnerGuide emblem for the efficiency rating of that particular model. The higher the rating, the more efficient the boiler. The government has established a minimum rating for boilers of 78 percent. Mid-efficiency boilers have AFUE ratings ranging from 78 to 82 percent. High-efficiency (condensing) boilers have AFUE ratings ranging from 88 to 97 percent. Conventional (noncondensing) steam and hot-water space heating boilers have AFUE ratings of approximately 60 to 65 percent.
- **Energy Star Certification.** Energy Star is an energy performance rating system created in 1992 by the U.S. Environmental Protection Agency (EPA) to identify and certify certain energy-efficient appliances. The goal is to give special recognition to companies who manufacture products that help reduce greenhouse gas emissions. This voluntary labeling program was expanded by 1995 to include furnaces, boilers, heat pumps, and other HVAC equipment. Both the Energy Star label and an AFUE rating are used to identify an energy-efficient appliance.

Types of Boilers

The boilers used in low-pressure steam and hot-water space heating systems can be classified in a number of different ways. Some of the criteria used in classifying them are:

1. Construction material
2. Construction design
3. Boiler position
4. Number of passes of the hot gases

5. Length of travel of the hot gases
6. Type of heating surface
7. Type of fuel used

Most boilers are constructed of either cast iron or steel. *A few* are constructed from nonferrous materials such as aluminum. Cast-iron boilers generally display a greater resistance to the corrosive effects of water than steel ones do, but the degree of corrosion in steel boilers can be significantly reduced by chemically treating the water.

The heating core of many boilers is formed by joining together a series of cast-iron sections either horizontally (so-called pancake construction) or vertically (Figure 15-4). In the horizontal cast-iron section design, the heating surface of each cast-iron section is exposed at right angles to the rising flue gases (Figure 15-5). The water travels in a zigzag path from section to section in a manner similar to the flow of water in a steel tube boiler (Figure 15-6).

Steel boilers may be classified with respect to the relative position of water and hot gases in the tubular heating surface. In fire-tube boilers, for example, the hot gases pass within the boiler tubes while the water being heated circulates around them. In water-tube boilers, the reverse is true. Flexible steel tubes are used in some boilers for the circulation of the water around the heat rising from the fire (Figure 15-7).

A hot-water (hydronic) copper-fin tube operates on a different principle from the cast-iron and steel boilers. It is designed to transfer heat almost instantly to the water (Figure 15-8). Water flows across the boiler heat exchanger, picks up heat, and then moves through the pipes to the heat convectors, radiators, or panels.

Note

> If the water stops flowing while the burner is still running, heat will build up until the water flashes into steam and damages the boiler. This condition is similar to dry firing in cast-iron and steel boilers. It can be avoided by installing a flow switch in the path of the water. The switch turns off the burner when the water stops flowing.

Boilers can also be classified according to the number of passes made by the hot gases (e.g., one pass, two passes, three passes) The *length* of travel of the hot gases is another method used for classifying boilers. The efficiency of a boiler heating surface depends, in

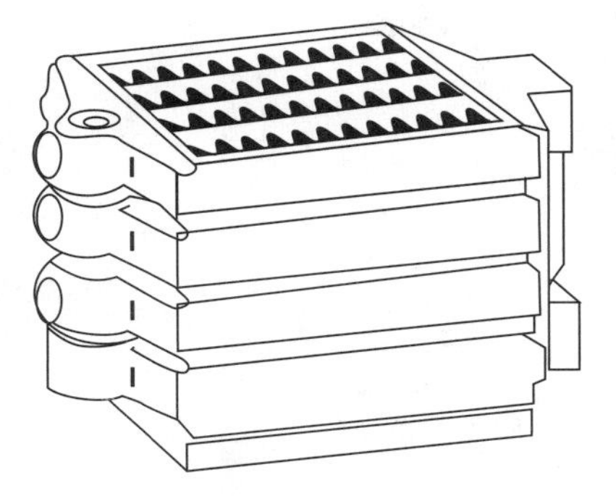

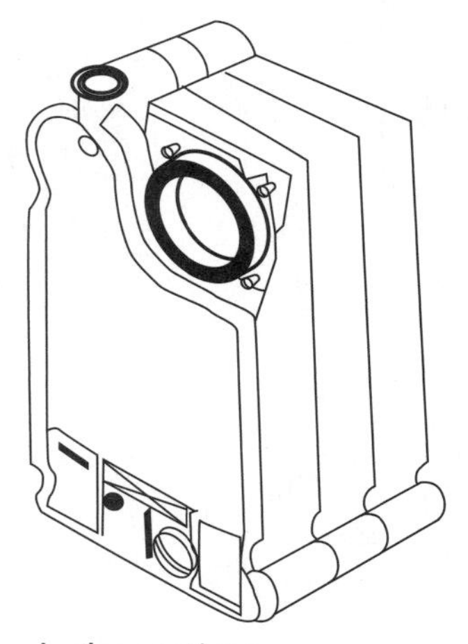

Figure 15-4 Cast-iron boiler sections.

part, upon the ratio of the cross-sectional area of the passage to its length.

Among the various fuels used to fire boilers are oil, gas (natural and propane), coal, and coke. Conversion kits for converting a boiler from one gas to another are available from some manufacturers. Changing from coal (or coke) to oil or gas can be accomplished by using conversion chambers and making certain other modifications. See Chapter 16, "Boiler and Furnace Conversions."

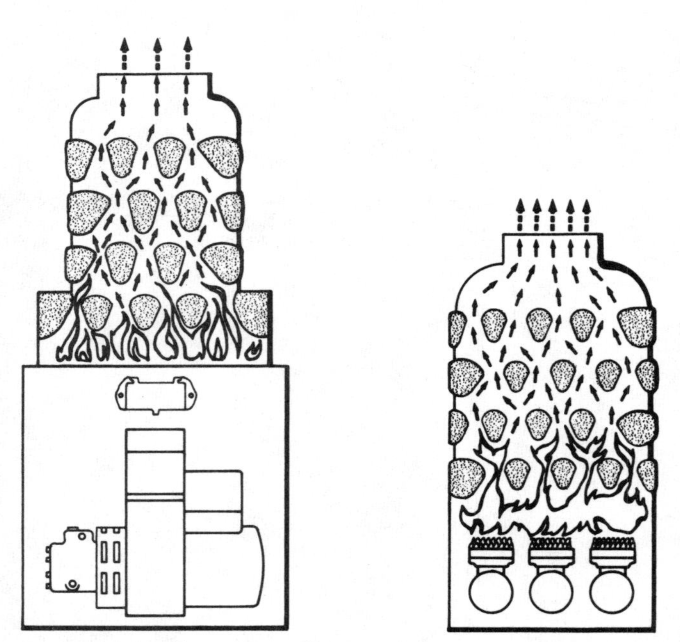

Figure 15-5 Direction of heat travel. *(Courtesy Hydrotherm, Inc.)*

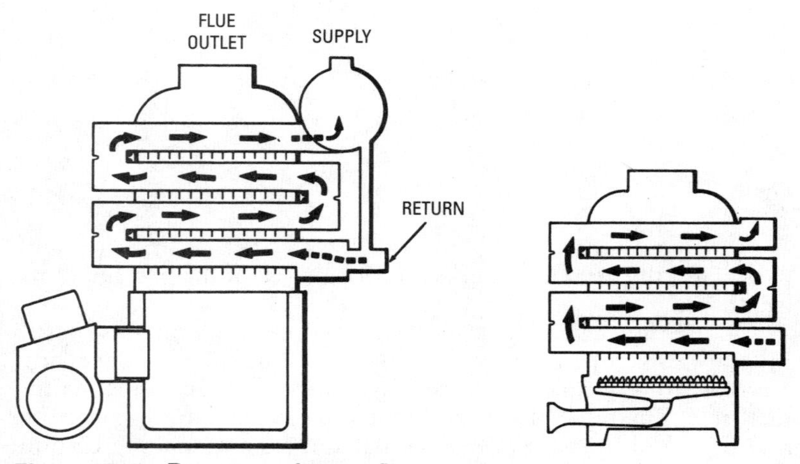

Figure 15-6 Direction of water flow. *(Courtesy Hydrotherm, Inc.)*

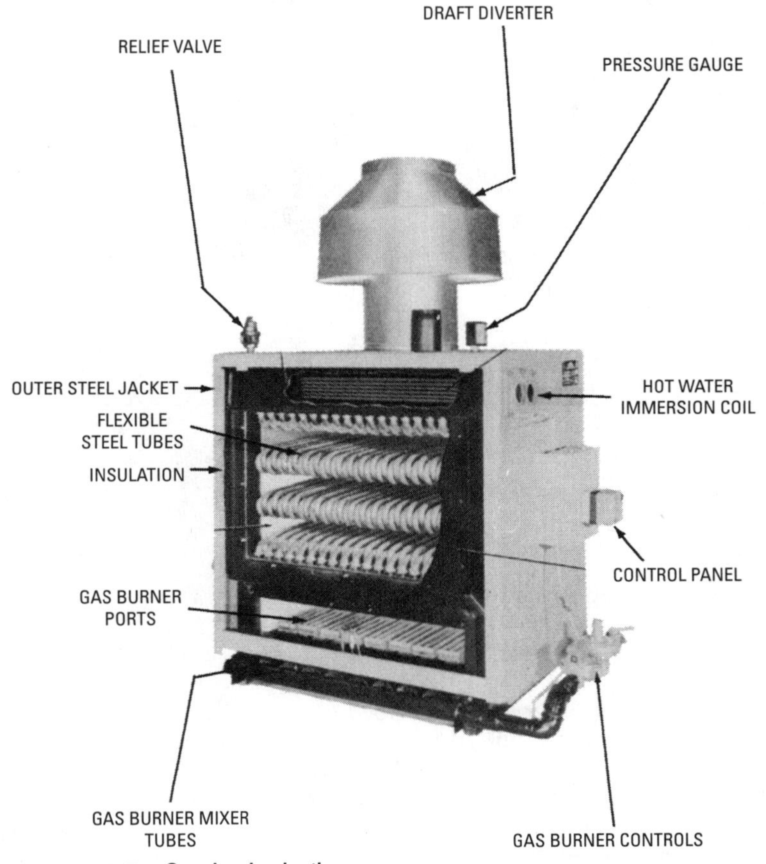

Figure 15-7 Steel-tube boiler. *(Courtesy Bryan Steam Corp.)*

Electricity can also be used to fire boilers. One advantage in using electric-fired boilers is that the draft provisions required by boilers using combustible fuels is not necessary.

Note

Unlike the boilers fired by fossil fuels (oil, gas, coal, etc.), electric boilers do not have an AFUE efficiency rating. They operate at almost 100 percent efficiency.

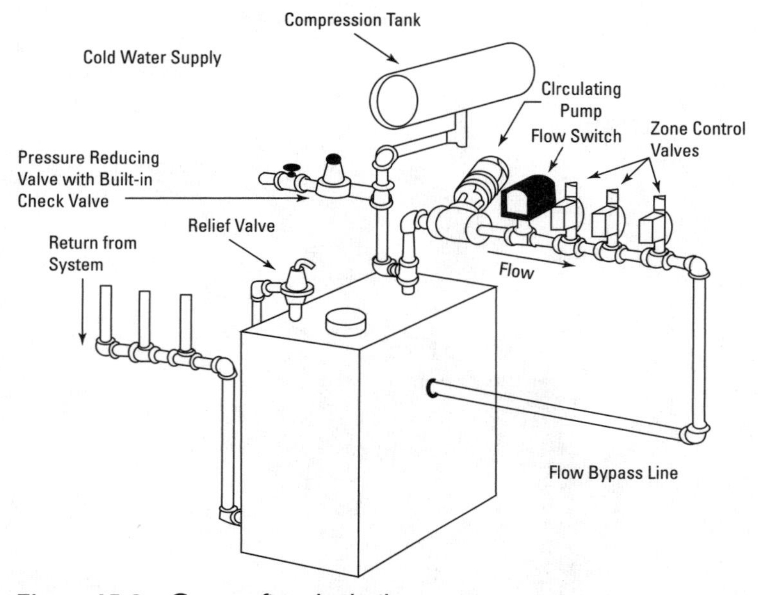

Figure 15-8 Copper fin tube boiler. *(Courtesy ITT McDonnell & Miller)*

The classification criteria described above are selective and limited to the more common types in use. Considering the multiplicity of boiler types and designs available, it is extremely difficult to establish a classification system suitable for all of them.

Gas-Fired Boilers

Gas-fired boilers generally consist of several closely placed cast-iron sections or steel tubes with a series of gas burners (i.e., a gas burner assembly) placed beneath them. The flue gases pass upward between the sections or tubes to the flue collector.

Draft losses are kept low in these boilers because the pressure at which the gas is supplied is generally sufficient to draw in the amount of air necessary for combustion. The fact that there is no fuel-bed resistance, as there is with coal-fired boilers, also contributes to low draft loss.

The draft in gas-fired boilers is generally nullified by the diverter; consequently, the resistance offered by the boiler passages is not an important variable. When there is a problem of excessive draft, it can be controlled by installing a sheet-metal baffle in the flue connection at the boiler [see *Control Excessive Draft (Gas-Fired Boilers)* in this chapter].

Most of the controls and accessories used to operate gas-fired boilers are described in considerable detail in Chapter 2, "Gas Burners"; Chapter 5, "Gas and Oil Controls"; and Chapter 6, "Other Automatic Controls" of Volume 2. The exact placement of these controls and accessories on the steel boiler jacket may differ slightly from one manufacturer to another, but not significantly. The major difference will be in the types of controls and accessories used to govern the temperature, pressure, and of the heat-conveying medium; and this is determined by whether it is a steam or hot-water space heating boiler (see *Steam Boiler Valves, Controls, and Accessories* and *Hot-Water Boiler Valves, Controls, and Accessories* in this chapter). An exploded view of a conventional gas-fired hot-water (hydronic) boiler is shown in Figure 15-9.

OilFired Boilers

An oil-fired boiler contains a heat transfer surface consisting of either cast-iron sections or steel tubes and a special combustion chamber shaped to meet the requirements of an oil burner.

The oil burner is a device designed to mix fuel oil with air under controlled conditions and to deliver the mixture to the combustion chamber for burning. The burner is mounted outside the chamber. The oil burners used in residential heating boilers are usually high-pressure atomizing burners, although other types are used on occasion.

It is possible to convert a coal-fired boiler to oil by redesigning the combustion chamber. However, boilers specifically designed to use oil as a fuel have proven to be more efficient and economical than coal-fired boilers converted to oil.

Read Chapter 1, "Oil Burners"; Chapter 5, "Gas and Oil Controls"; and Chapter 6, "Other Automatic Controls" in Volume 2 for a description of the controls and accessories used to operate oil-fired boilers. Also read the appropriate sections in this chapter (*Steam Boiler Valves, Controls, and Accessories* and *Hot-Water Boiler Valves, Controls, and Accessories*). An exploded view of a conventional oil-fired boiler is is shown in (Figure 15-10).

Coal-Fired Boilers

A typical coal-fired boiler contains a grate for the fuel bed, located beneath metal heat transfer surfaces such as cast-iron sections (Figure 15-11). The mixture of hot gases resulting from combustion

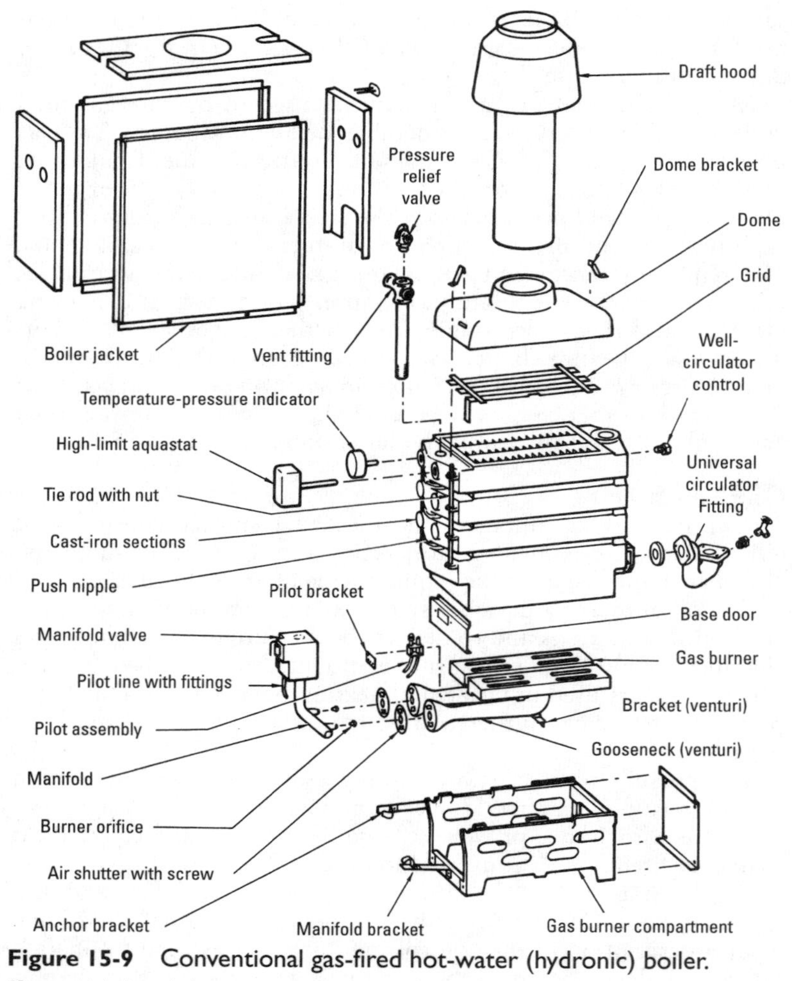

Figure 15-9 Conventional gas-fired hot-water (hydronic) boiler.
(Courtesy Hydrotherm, Inc.)

rises through the passages in the sections, transferring its heat to the water contained in them.

The controls and accessories used to operate a coal-fired boiler are described in Chapter 3, "Coal Firing Methods" of Volume 2. The controls and accessories used to govern the temperature, pressure, and flow rate of the heat-conveying medium are described in

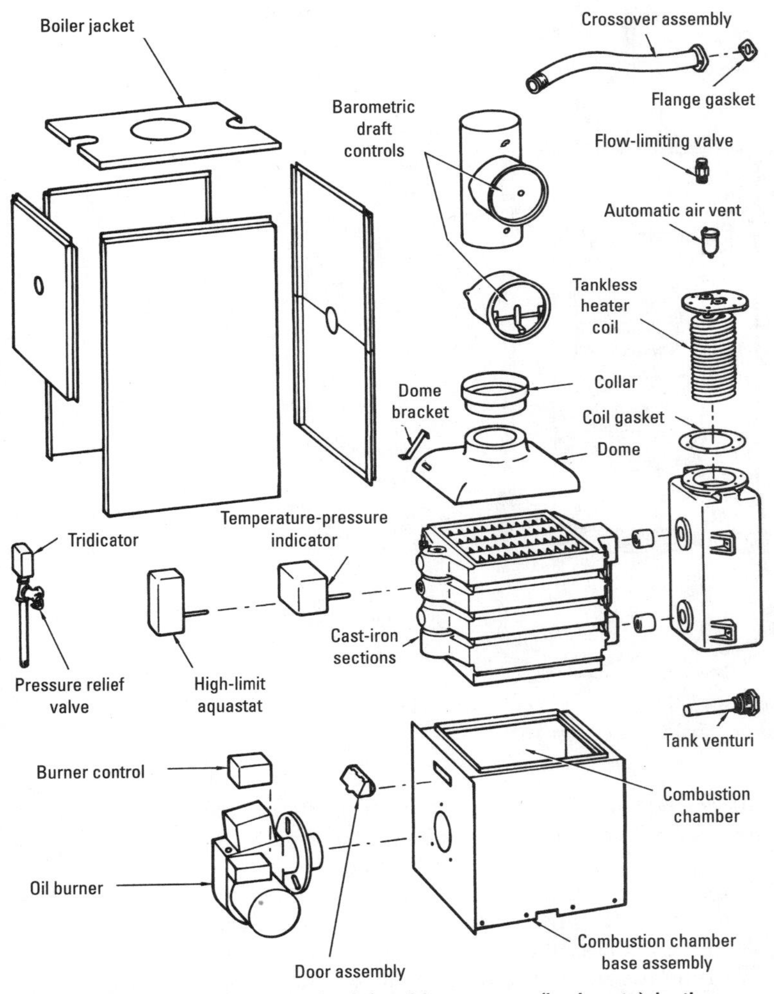

Figure 15-10 Conventional oil-fired hot water (hydronic) boiler.
(Courtesy Hydrotherm, Inc.)

this chapter (see *Steam Boiler Valves, Controls, and Accessories* and *Hot-Water Boiler Valves, Controls, and Accessories*).

A coal-fired boiler requires more draft than gas- or oil-fired boilers because of the resistance of the fuel bed. Special care must be given to the design of the chamber surrounding the grate to ensure

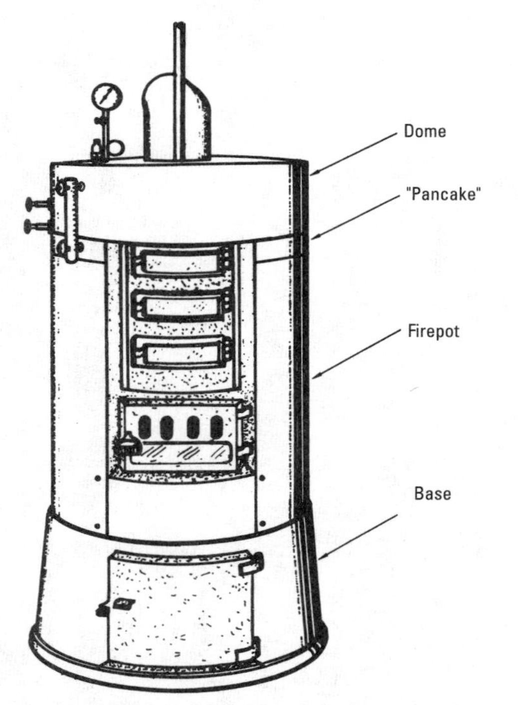

Figure 15-11 Older model of a vertical, round coal-fired boiler.

sufficient volume for a proper mixture of fuel and air. Provisions must also be made for introducing sufficient air through the grate (and fuel bed) for the combustion process.

Before the introduction of oil burners, coal was the fuel in general use, and it is still largely used, especially for heating plants already built and in cases where the owner cannot afford the expense of converting to oil. Moreover, some cast-iron boilers designed to burn coal are not suited to burn oil, which adds to the expense of a conversion job. Finally, the increasing worldwide shortage of oil will find it more and more necessary to find applications for low-polluting uses of coal as a heat source. An important use for coal will be steam and hot-water heating systems equipped with devices to reduce the polluting effect of burning coal.

Coal-burning boilers may be hand-fired or fitted for automatic stoker operation. In many automatic stokers, fuel is carried from the hopper through a feed tube by means of a rotating worm. Intermittent action of the worm agitates the fuel bed, prevents arching of the coal in the retort, and ensures that the incoming air reaches every part of the fire at all times. An auxiliary air connection between the feed tube and the windbox prevents gas accumulating in the tube and eliminates any tendency to "smoke back" through the hopper.

An underfeed stoker is one in which the fuel is fed upward from underneath. The action of a screw or worm carries the fuel back through a retort, from which it passes upward as the fuel is consumed, the ash being finally deposited on dead plates on either side of the retort, from which it can be removed.

Only part of the fuel being burned is actually burned in the fuel bed. Under the influence of the high temperature created in the fuel bed and lack of sufficient air, unburned gases are released above the retort and tuyeres; and unless these gases are mixed with air and burned inside the combustion chamber, they will leave the boiler unburned, carrying away a large percentage of the heat energy value originally contained in the coal.

Nearer the outside of the fuel bed, the fuel burns less violently, and much more air is passed through the fuel bed than is necessary. It is this excess air that must be mixed with the unburned gases issuing from the central point of the fuel bed if high combustion efficiency is to be realized.

Coke can also be used to fire boilers, but it requires different handling. In order to be completely consumed, coke needs a greater volume of air per pound of fuel than coal and therefore requires a stronger draft, which is increased by the fact that it can burn economically only in a thick bed.

Because less coke is burned per hour per square foot of grate than coal, a larger grate and a deep firepot are required to accommodate the thick bed of coke.

The quick-flaming combustion that characterizes coal is not produced by coke because the latter fuel contains very little hydrogen; however, a coke fire is more even and regular.

Electric Boilers

Compact wall-mounted electric boilers are used in residential hot-water (hydronic) heating systems (Figure 15-12). Heat is generated by electric heating elements immersed in water housed in a waterproof cast-iron shell. Although small in size, these boilers are capable of generating as high as 90,000 Btu/h, enough heat for the average eight-room house.

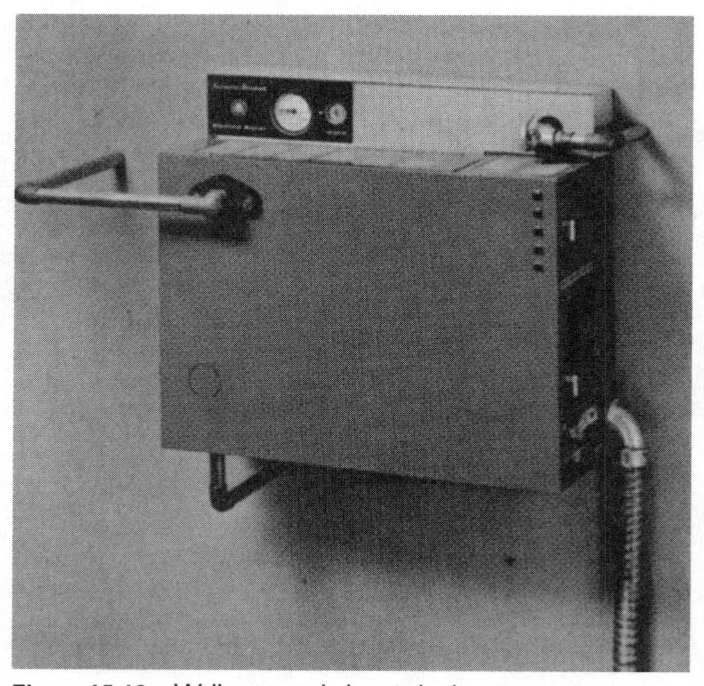

Figure 15-12 Wall-mounted electric boiler. *(Courtesy American Standard)*

The basic components of a conventional electric boiler are shown in Figure 15-13. An automatic air vent located above the cast-iron boiler shell is used for bleeding off trapped air from the water system. Safety devices include a water pressure-relief valve for reducing system pressure, an adjustable limit control that allows selection of maximum boiler water temperature for system design, and preset pressure controls to guarantee safe operation within the prescribed pressure range.

The limit control is an immersion device that shuts off the boiler if the temperature exceeds a predetermined setting. The preset pressure controls consist of a high-pressure switch and a preset safe-fill switch. The high-pressure switch is designed to deenergize the boiler if the pressure reaches 28 psi. This switch is reinforced by the relief valve, which is preset to relieve system pressure at 30 psi. The preset safe-fill switch prevents the electric heating elements from being energized unless pressure in the system is 4 psi. This

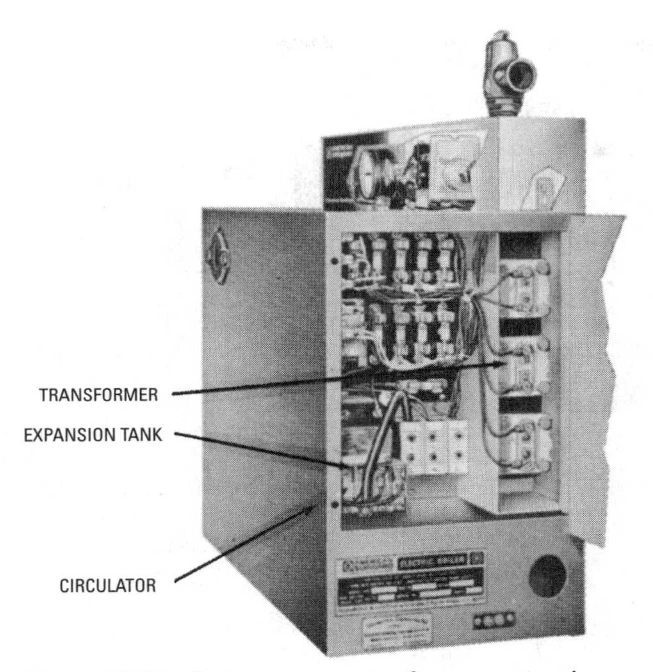

Figure 15-13 Basic components of a conventional electric boiler. *(Courtesy American Standard)*

prevents any chance of the heating elements burning out should an attempt be made to operate the boiler dry.

The circulator, cast-iron boiler shell, and expansion are all enclosed within the outer steel jacket of the boiler. The electric heating elements are mounted inside the ends of the boiler casting. A drain valve mounted below the boiler casting is designed for the connection of standard hose fittings.

All controls are prewired and mounted in place by the manufacturer. The only electrical connections that need to be made during the installation of the boiler are those to the main power supply and room thermostat.

The sequencing relay switch provides for an incremental loading on the electric service line. This reduces line voltage fluctuation and prevents power surge during energizing.

Minimum installation clearances for the boiler just described are shown in Figure 15-14. Two air openings of 108 in^2 (6" × 18") each are required for closet installation.

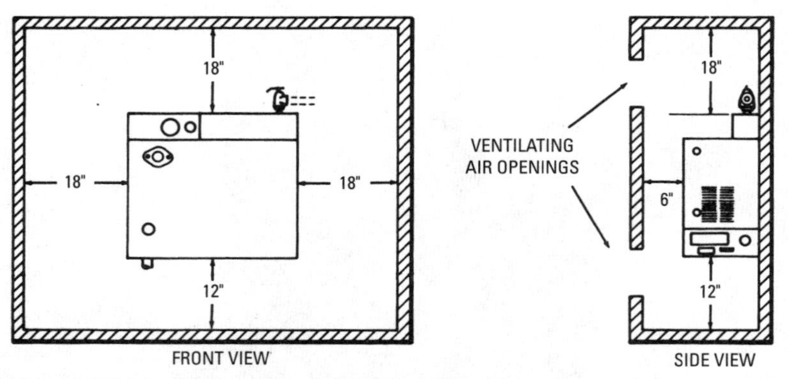

Figure 15-14 Minimum installation clearances.

A typical piping arrangement for a series-loop, hot-water heating system using an electric boiler of this capacity is shown in Figure 15-15. An outline of the procedure used for filling this system is as follows:

1. Close all zone valves except the one for the zone to be purged.
2. Open the boiler drain valve.
3. Open the fill valve.
4. Close the purge valve.
5. Vent air from the boiler by manually opening the relief valve.

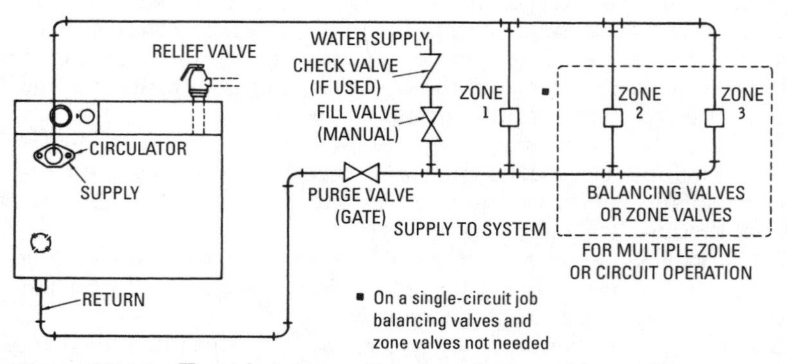

Figure 15-15 Typical piping arrangement for a multizoned hot-water heating system.

6. Open the valve for the second zone to be purged and close the first.
7. After the second zone is purged, open the third zone valve and close the second.
8. Repeat the procedures described in Steps 6 and 7 for as many zones as are in the system.
9. When all zones have been purged, close the boiler drain cock.
10. When water appears at the relief valve, release the lever and allow the valve to close.
11. Continue filling until pressure gauge reads approximately 12 psi. Close fill valve.
12. Open purge valves.

Where multiple circuits are used without zone valves, balancing valves should be installed for balancing. These can be shut off to purge each circuit individually. For a single-loop system, no additional vents, valves, drains, or other accessories are required.

Smaller electric-fired boilers having outputs ranging from approximately 6000 to 20,000 Btu/h are available for individual apartment heating or for residential zoned heating through the use of several units. The basic components of one of these smaller boilers is shown in Figure 15-16. The piping arrangement for a series-loop heating

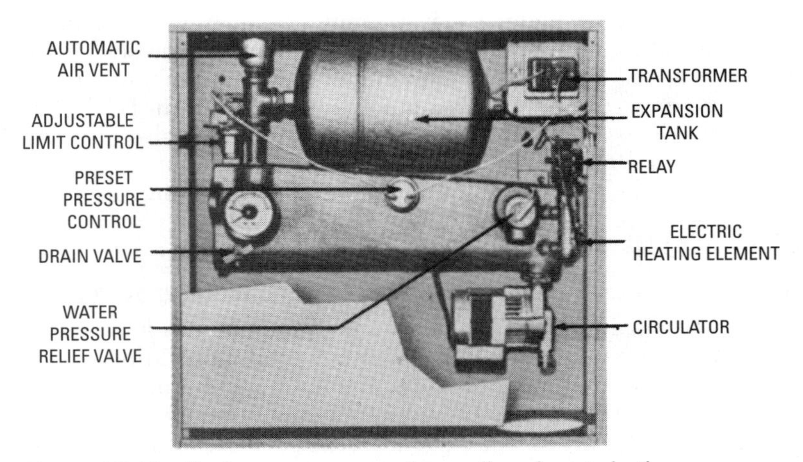

Figure 15-16 Basic components of a smaller electric boiler.

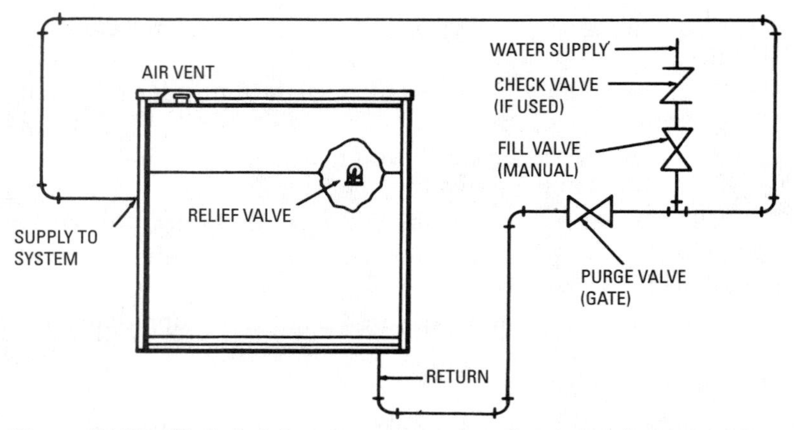

Figure 15-17 Typical piping arrangement for a hot-water heating system.

system using one of these boilers is illustrated in Figure 15-17. The procedure for filling the system is similar to the one just outlined except for the elimination of Steps 6 to 9.

High-Efficiency Boilers

The development of the high-efficiency (condensing) boiler came about as a direct reaction to the oil crisis in the 1970s. Higher heating oil costs and the public's desire for greater fuel efficiency resulted in the development of boilers with a much higher efficiency than the conventional ones. Using the ASFUE ratings, the conventional boilers had a fuel efficiency in the 60 percent range, whereas the new high-efficiency boilers have a fuel efficiency rating of 85 percent or higher.

A high-efficiency boiler vents its combustion gases in PVC pipe through a sidewall instead of the chimney. The boiler also requires an induced-draft fan (power vent) and an outside source air-intake for combustion air. Typical high-efficiency (condensing) oil-fired and gas-fired hot water (hydronic) boilers are illustrated in Figures 15-18 and 15-19.

Steam Boiler Valves, Controls, and Accessories

The boilers used in steam heating systems are fitted with a variety of devices designed to ensure the safe and proper operation of the boiler. These boiler valves, controls, and accessories can be divided by function into two basic categories: (1) indicating or measuring devices, and (2) controlling devices. Sometimes both functions are combined in a single unit.

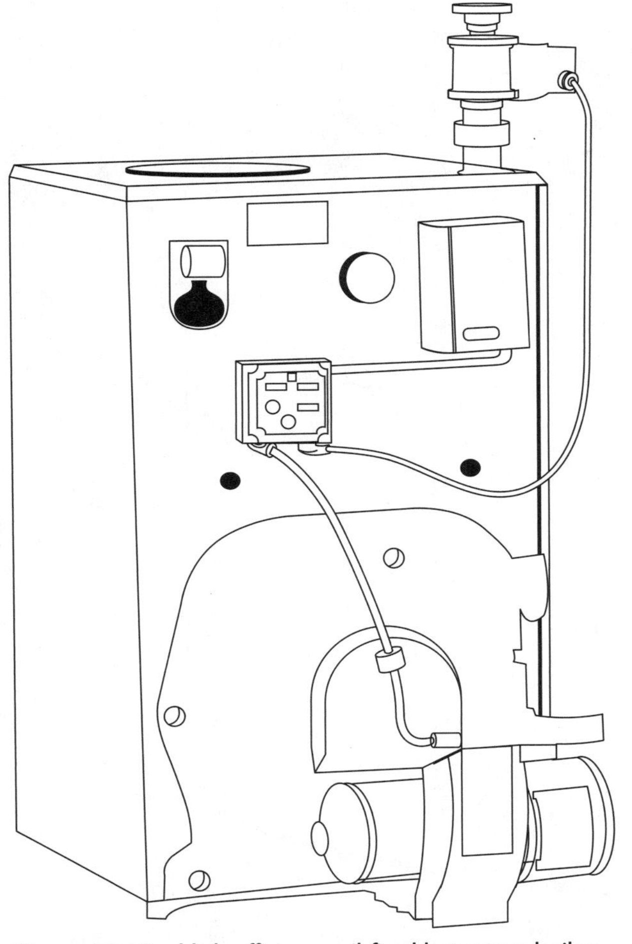

Figure 15-18 High-efficiency oil-fired hot water boiler.
(Courtesy Hydrotherm, Inc.)

Steam boilers operate under high pressures and temperatures. To avoid serious injury from scalding steam or water, never begin any service or repairs shutoff before taking the following precautions:

- Wait for the boiler to cool down to 80°F (27°C) or more.
- Wait for the boiler pressure to drop to 0 psi (0 bar).

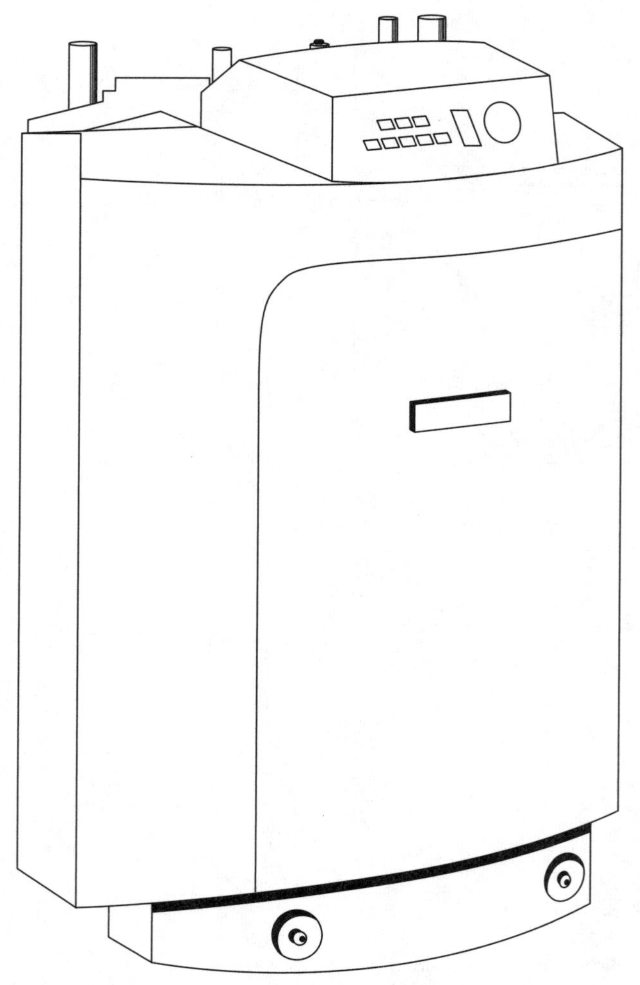

Figure 15-19 High-efficiency gas-fired hot water boiler. *(Courtesy Weil-McLain)*

- Disconnect the electrical power before making any electrical connections.
- Connect a temporary drain pipe to the control opening to prevent exposure to steam discharge.

Principal Steam Boiler Valves, Controls, and Accessories

A steam boiler should be equipped with the following valves, controls, and accessories:

- Water level gauge: A gauge that measures the water level inside the boiler.
- Low-water cutoff: A device that automatically switches off the burner if the boiler water level becomes too low for safe operation.
- Pressure gauge: A gauge that measures the operating pressure inside the boiler.
- Safety relief valve: A valve that discharges excess steam when boiler pressure exceeds the maximum pressure limit on the valve.
- High-pressure limit switch: A switch that shuts off the burner when the boiler pressure exceeds a preset level.
- Condensate pump: A small pump used to return condensate to the boiler.

Indicating devices include water gauges, pressure gauges, and similar types of devices that provide information about the operating conditions in the boiler. They are used to indicate temperatures, pressures, or water levels that fall outside the design limits of the boiler. Controlling devices include boiler equipment designed to cause changes in these conditions. For example, pressure relief valves are used to relieve excess pressure in the boiler.

Steam Boiler Low-Water Cutoffs

Steam boilers must be provided with a water-level control device to shut off the automatic fuel-burning equipment when the water level in the boiler drops to a level too low for safe operation. This water-level control device is referred to as a *low-water cutoff*. The two types of low-water cutoff used on steam boilers are the float type and the probe type.

Note

According to Section 4 of the American Society of Mechanical Engineers (ASME) Boiler and Pressure Vessel Code, *all* residential steam boilers, regardless of their size or location, must have a low-water cutoff device installed.

A float low-water cutoff device consists of a cutoff switch operating in conjunction with a float located in the boiler water or in a float chamber installed next to the boiler. The float is connected through a linkage to a switch that operates a feed water valve. As the water level falls, the float drops with it until it reaches a point at which the feed water switch is actuated. If the water level continues to fall, a second switch connected to the float by a linkage is actuated, and the automatic fuel-burning equipment is shut off.

A probe low-water cutoff device depends on the flow of a low electrical current to control the operation of the automatic fuel-burning equipment. The electrical current flows from the probe through the water to keep the relay energized. When the water level falls, there will be a point at which the probe loses direct contact with the water. As a result, the contact is broken, and the flow of the electrical current is stopped. This, in turn, causes the relay to be energized, which results in shutting off the fuel-burning equipment. A probe low-water device *cannot* be used in the direct operation of feed water valves.

There are variety of ways to attach a low-water cutoff to a steam boiler (Figure 15-20). Some are designed to be attached to the glass gauge on the outside of the boiler. In this type of installation, the elevation of the low-water cutoff is already determined by the location of the glass gauge tappings on the side of the boiler.

Note

> If there is a horizontal cast line on the outside of a gauge-mounted low-water cutoff, make certain it is located above the minimum safe water level specified by the boiler manufacturer.

Another type of low-water cutoff installation is to attach it directly to the side of the boiler. As shown in Figure 15-20, it is not connected to the water gauge. Its elevation is determined by the location of the pipe tappings on the boiler. The body of the control must be located at or above the manufacturer's minimum safe water level.

On some boilers, the low-water cutoff is combined with a water feeder to add water to the boiler when the water level falls below the safe operating limit. An example of one of these combined units is shown in Figure 15-21. They are available with either single or dual switch assemblies. The single double-throw switch assembly provides a combination feeder and burner cutoff switch, with an extra terminal for line voltage with single-pole, double-throw service

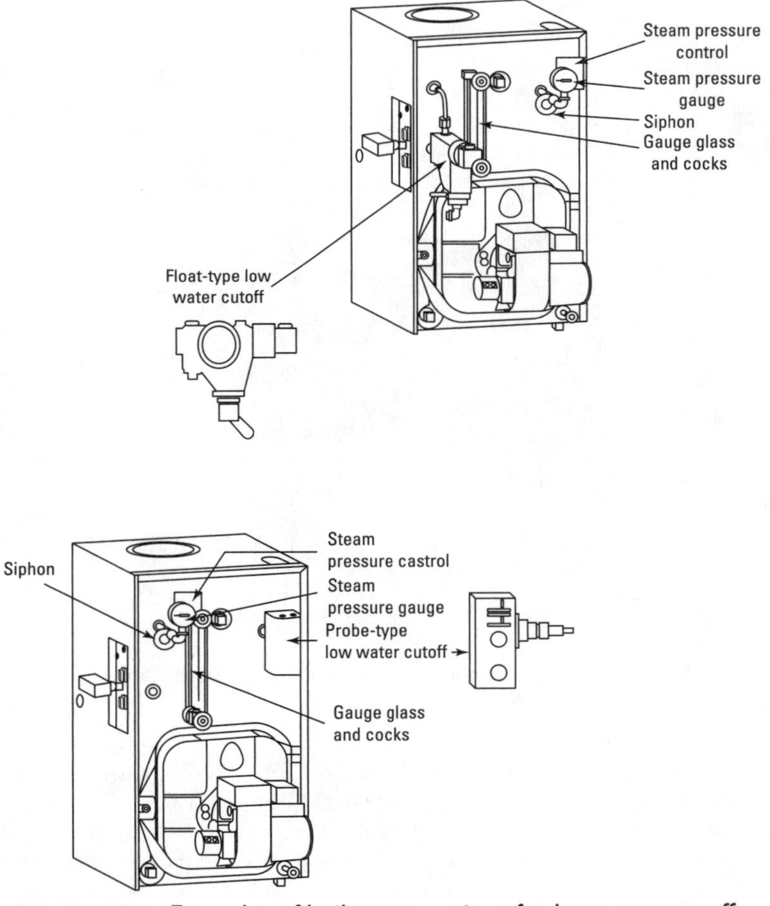

Figure 15-20 Examples of boiler connections for low-water cutoffs.

(Figure 15-22). The dual switch assembly is used for line voltage burner service and for independent low- (or high-) voltage alarm, feed valve, or pump starter service. When an emergency condition occurs, the switch interrupts the current to the burner and shuts it off. When the emergency has passed, the water feeder takes over and feeds makeup water when needed for normal operation. On units equipped with dual switch assemblies, the alarm, feed valve, or pump starter switch actuates just before the burner switch cuts the

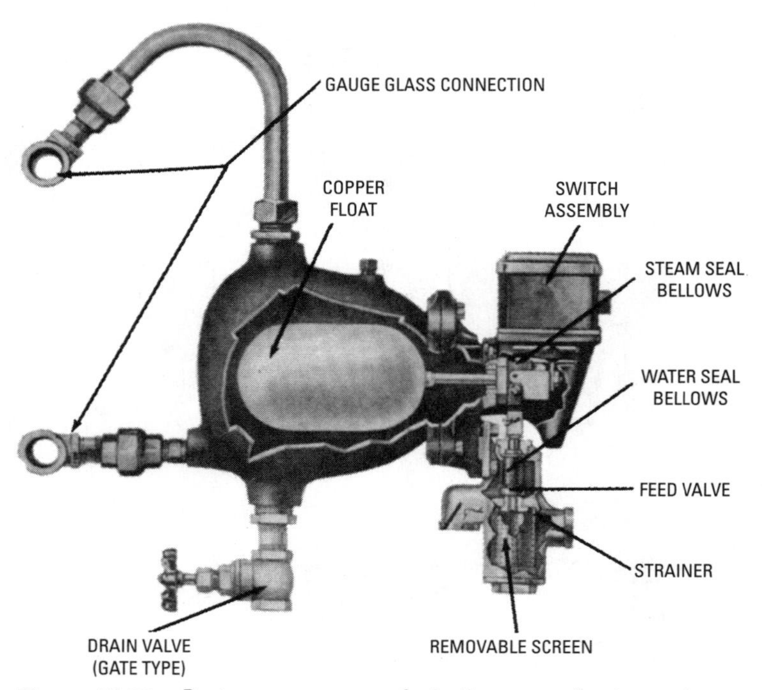

Figure 15-21 Basic components of a boiler water feeder and cutoff. *(Courtesy Watts Regulator Co.)*

firing. A typical installation of a Watts 60 LWD boiler water feeder and cutoff on a steam boiler is illustrated in Figure 15-23.

A low-water cutoff *without* a boiler water feeder is usually adequate for providing automatic low-water safety protection for most small boilers (Figure 15-24). These units can be installed on any boiler having gauge glass connections.

Direct installation (built-in) low-water cutoffs are available for steam boilers with limited space in the boiler jacket (Figure 15-25). Each unit contains a switch assembly, a float and bellows, and a threaded barrel casting that fits into a 2½-in tapping in the side of the boiler.

Installing a Low-Water Cutoff

The Watts No. 89 low-water cutoff illustrated in Figure 15-24 is suitable for low-pressure steam heating boilers with a maximum operating steam pressure of 15 psi. Although installation is relatively simple,

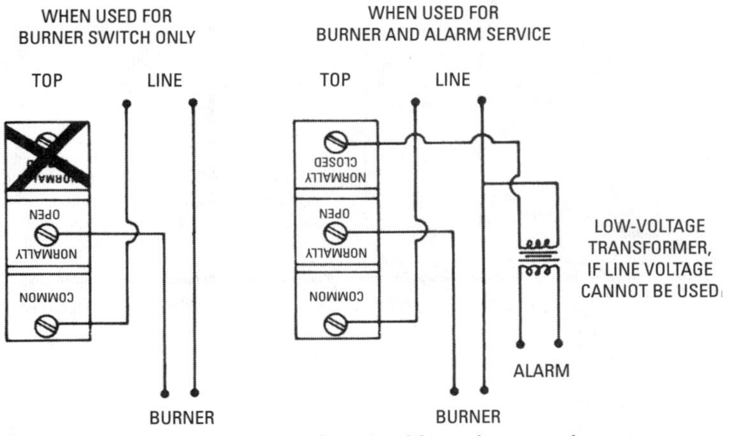

Figure 15-22 A single-switch assembly and wiring diagram.
(Courtesy Watts Regulator Co.)

the instructions should be carefully followed. This is a boiler safety device, and it must operate properly or the boiler can be damaged.

Before installing the Watts low-water cutoff, the gauge glass and cocks must be removed from the boiler. When you have done this, proceed as follows:

1. Install the ½-in tee (with long and short nipples) on the float chamber by inserting the short nipple in the float chamber tapping in the end opposite the switch. Screw the tee tight with the long nipple pointing in the correct direction, depending on the location of the boiler (Figure 15-26).

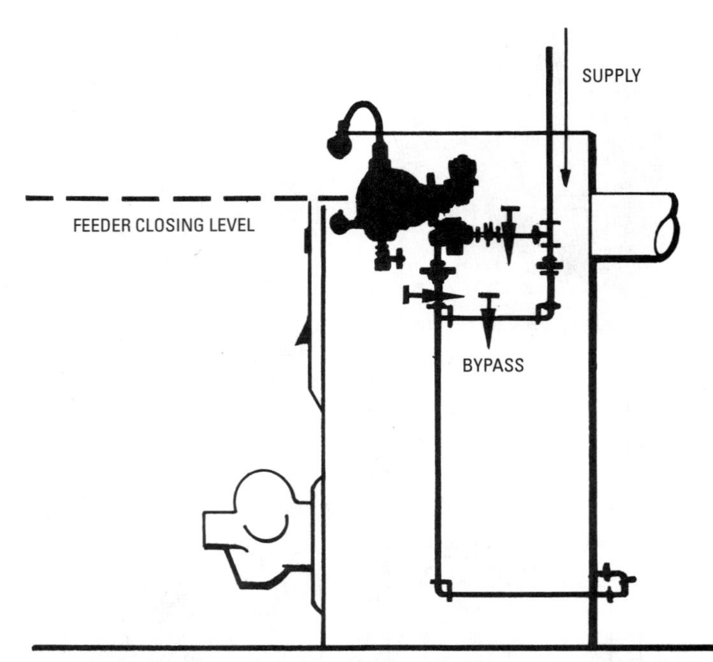

Figure 15-23 Typical installation of a boiler water feeder and cutoff.
(Courtesy Watts Regulator Co.)

2. Insert the long nipple into the lower glass gauge tapping on the boiler.
3. Swing the entire float chamber until the nipple is made up tight. Line up the cutoff so that the top of the switch box is level (Figure 15-27).
4. Screw the nipple in the tee carrying the tubing connector the upper gauge glass tapping and pull up tight. Install a compression coupling in the top float chamber tapping (Figure 15-28).
5. Hold the tube bend in position and mark the tube on a level with the top of the top of the hex on the compression coupling. Cut off the tube at this mark (Figure 15-29).
6. Slide the ring and nut over the end of the connector tube.
7. Slide the end of the tube into the compression coupling in the float chamber, and tighten both couplings.

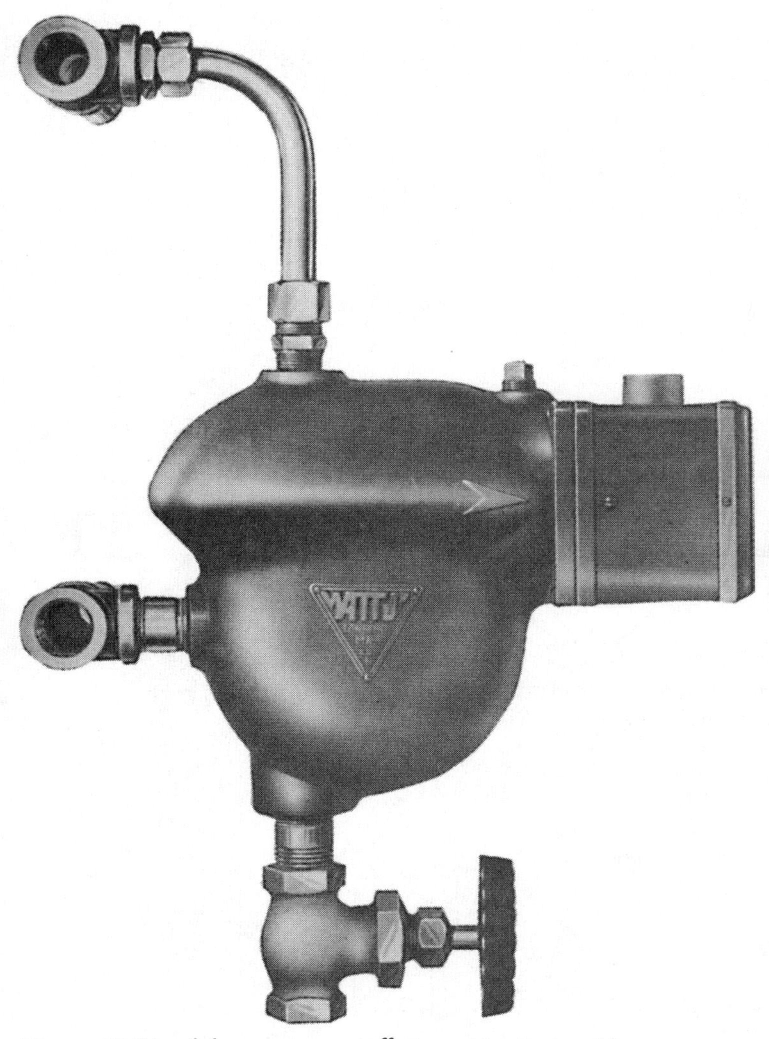

Figure 15-24 A low-water cutoff. *(Courtesy Watts Regulator Co.)*

8. Install a drain valve in the bottom float chamber tapping (Figure 15-30).
9. Reinstall the gauge cocks in the end of the tees and replace the gauge glass.

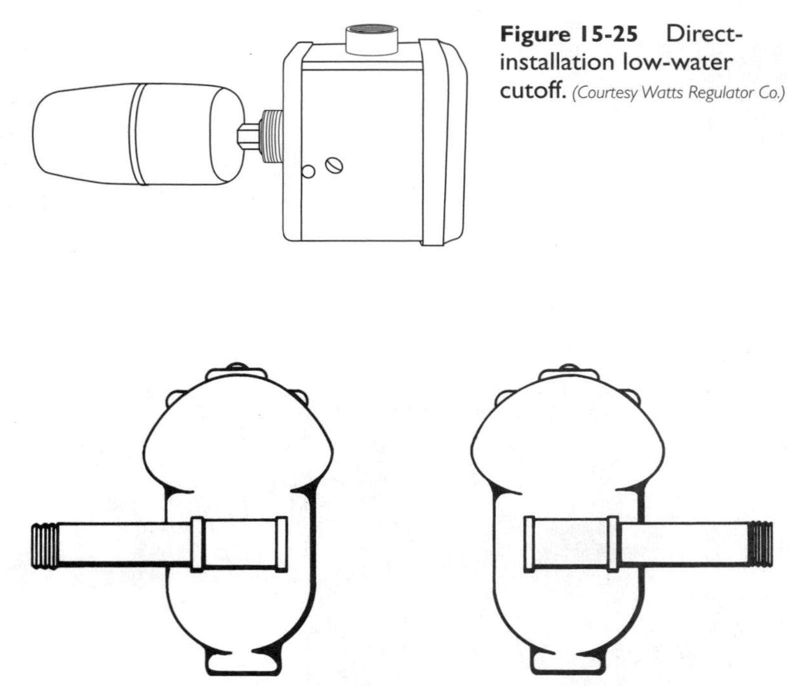

Figure 15-25 Direct-installation low-water cutoff. *(Courtesy Watts Regulator Co.)*

Figure 15-26 Installing the short nipple in the float chamber tapping. *(Courtesy Watts Regulator Co.)*

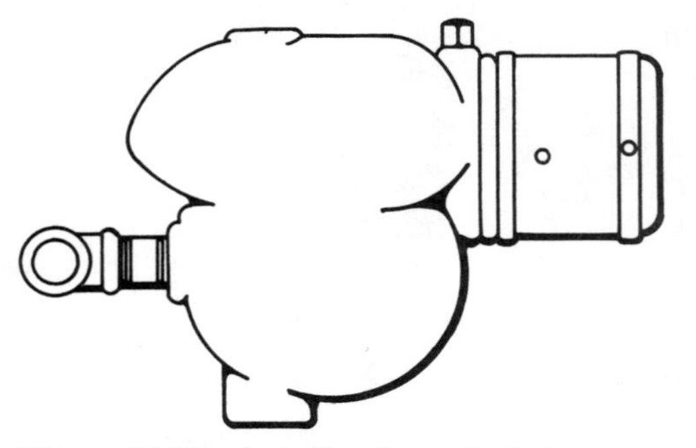

Figure 15-27 Installing long nipple into lower-gauge glass tapping in boiler. *(Courtesy Watts Regulator Co.)*

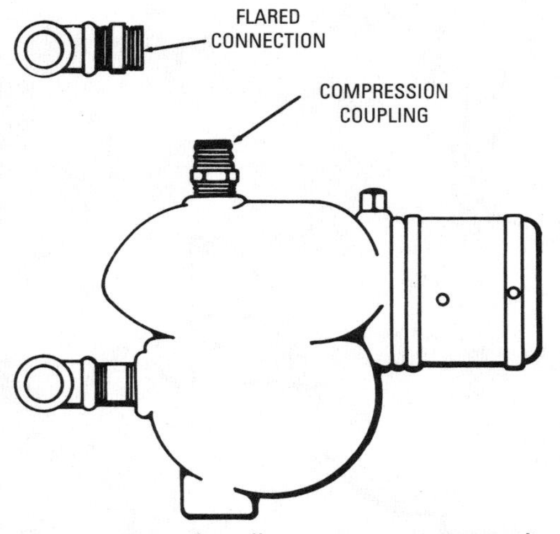

Figure 15-28 Installing compression coupling.
(Courtesy Watts Regulator Co.)

Note

The drain valve on the bottom of the low-water cutoff should be opened once a month (or oftener) during boiler operation to flush out sediments from the float chamber.

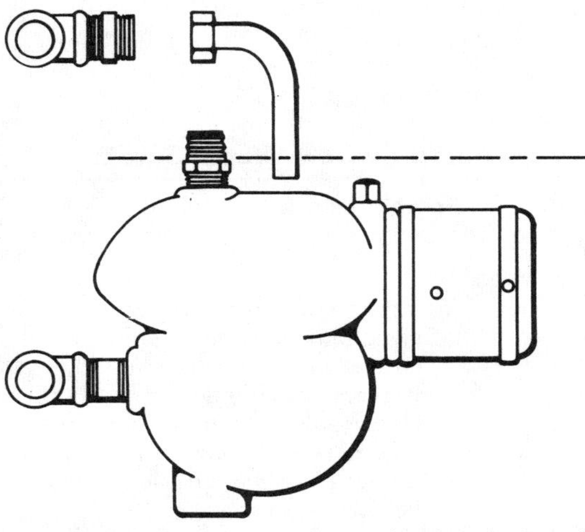

Figure 15-29 Cutting tube on level with top of hex. *(Courtesy Watts Regulator Co.)*

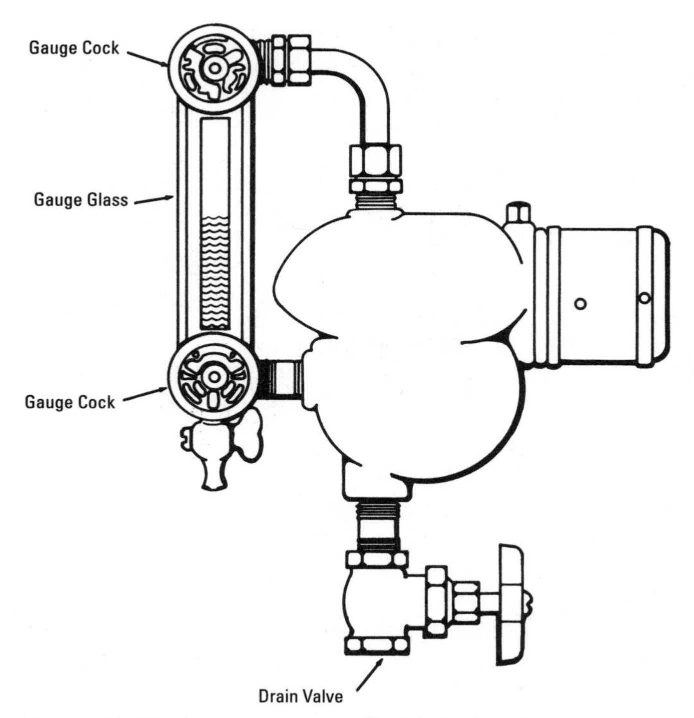

Figure 15-30 Low-water cutoff with drain valve. *(Courtesy Watts Regulator Co.)*

Table 15-1 Troubleshooting a Low-Water Cutoff

Symptom and Possible Cause	*Possible Remedy*
Burner Fails to Shut off When Boiler Reaches Low-Water Stage	
(a) Switch contacts are fused together.	(a) Remove switch and operate switch manually to determine correct switch opoeration. Replace switch if defective. Check electrical load to low-water cutoff control to make sure it is within ratings of switch. Overload may be burning out switch.

(continued)

Table 15-1 *(continued)*

Symptom and Possible Cause	*Possible Remedy*
(**b**) Float chamber is filled with mud, scale, or sediments.	(**b**) Clean float chamber; Check switch terminals 1 and 2 to see if they are open when water level is below the low-water cutoff control. If not, remove the switch and manually operate the terminals to determine if they will open and close freely. If not, replace switch.
Electric Water Feeder in Low-Water Cutoff Will Not Shut Off	
(**a**) Switch contacts are fused together.	(**a**) Remove switch and operate switch manually to determine correct switch operation. Replace switch if defective.
(**b**) Bellows in float chamber covered with mud, scale, or sediment build up.	(**b**) Open float chamber and clean or replace the bellows. Check switch terminals 3 and 4 to see if they are open when water level is above the low-water cutoff control. If they, remove the switch and manually operate the terminals to determine if they will open and close freely. If not, replace switch.

Fusible Plugs

A fusible plug is used on some boilers as a protection against dangerous low-water conditions. Examples of fusible plugs are shown in Figure 15-31. They are generally made of bronze and filled with pure tin. When the temperature (and pressure) in the boiler builds up to about 450°F (the approximate melting point of pure tin), the tin core melts and relieves the pressure within the boiler.

Fusible plugs must comply with ASME standards and are available in several sizes for use on boilers having a steam pressure of less than 250 psi.

Pressure Relief Valves

A steam boiler is equipped with a pressure relief valve (or valves), which opens and releases excess steam at or below the maximum allowable working pressure of the boiler (Figures 15-32 and 15-33). These are pop safety-relief valves designed to comply

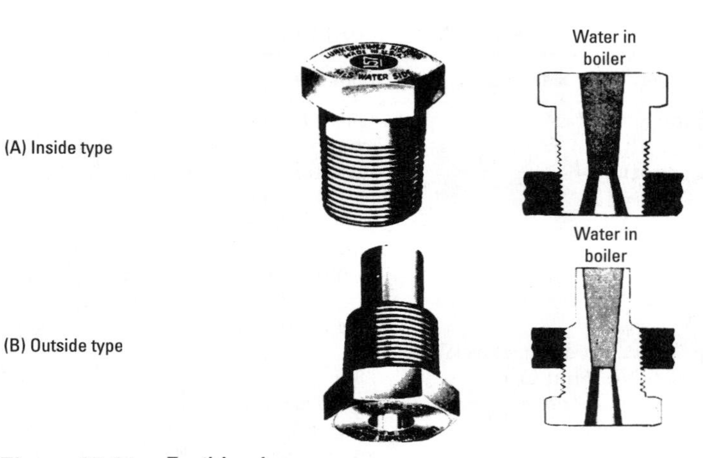

Figure 15-31 Fusible plugs. *(Courtesy Lunkenheimer Co.)*

with the requirements of the ASME Boiler and Pressure Vessel Code. They are also sometimes called *safety valves, safety relief valves,* or simply *relief valves.*

Pop safety valves exhaust the steam through holes drilled around the top of the spring housing. They function by "popping" wide open at the set pressure, remaining in that position until pressure has dropped the predetermined amount (commonly known as *blowback* or *blowdown*), and then snapping shut instantly. Lift levers and drain holes in the discharge side of the valve are required by ASME codes.

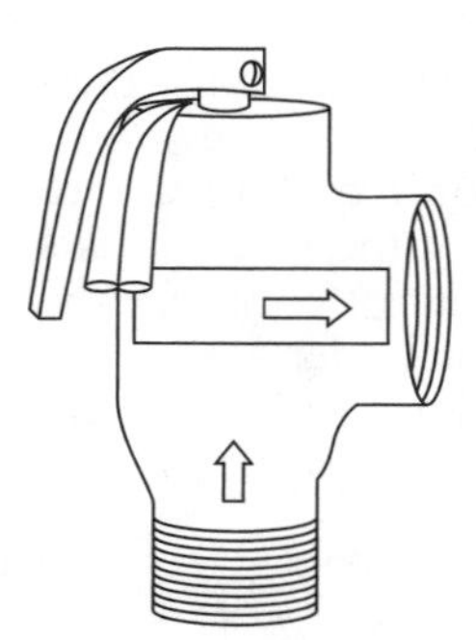

Figure 15-32 Pressure relief valve. *(Courtesy Watts Regulator Company)*

On the low-pressure boilers used in residential steam heating systems, the pressure relief valve is commonly preset to open and release steam when a maximum pressure of 15 psi is reached in the boiler. The valve closes when the steam pressure once again falls below 15 psi.

Warning

NEVER use a pressure relief valve with a pressure relief rating higher than the maximum working pressure of the boiler. If the boiler exceeds its maximum (safe) working pressure, it could rupture and explode at a pressure less than the rating of the pressure relief valve.

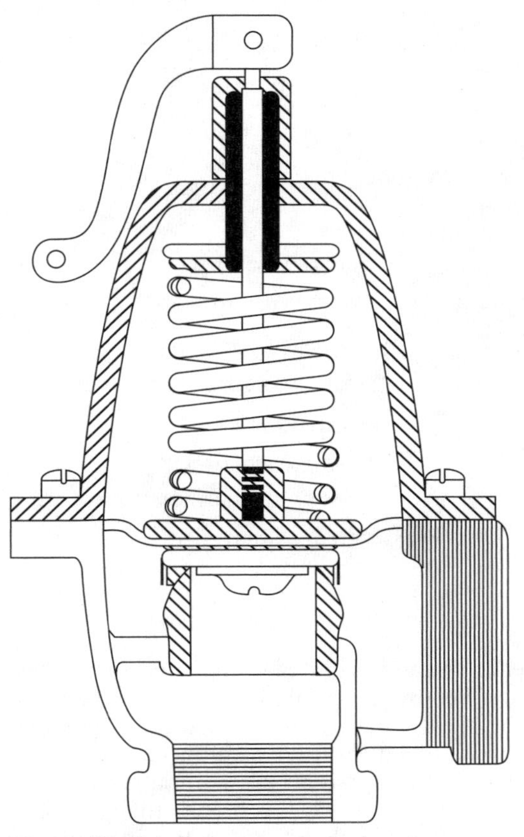

Figure 15-33 Pressure relief valve.
(Courtesy Cash Acme)

Pressure Controllers

In many steam heating systems, a line-voltage pressure controller is used to provide operating control, automatic limit protection, or a manual reset limit for protection in pressure systems of up to 300 psi. One such pressure controller is the Honeywell L404 Pressuretrol® Controller shown in Figure 15-34. It is available in a variety of different models.

Some models of the L404 are designed to provide on-off and proportional control of steam boilers fired by proportional-type burners. These can be field adjusted to operate either in unison (burner starts

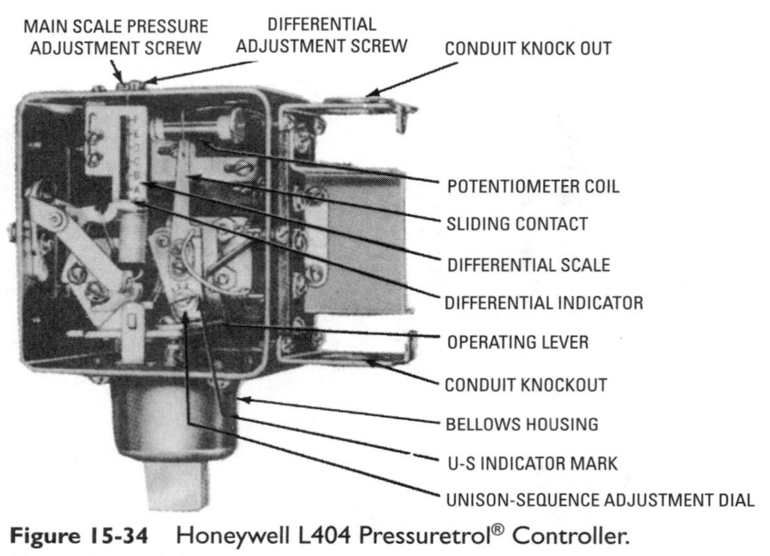

Figure 15-34 Honeywell L404 Pressuretrol® Controller.

(Courtesy Honeywell, Inc.)

at high fire) or in sequence (burner starts at some firing rate other than high fire). A common bellows assembly located in the bellows housing actuates the stop-start switch (on-off operation) and the wiper of the 185-ohm potentiometer (proportional operation).

The L404 operates strictly as a high-limit pressure safety control on steam heating boilers (see Figure 15-35). It breaks the electrical circuit on pressure rise. A variation of this controller is used for suspension-type unit heaters.

Direct control for a proportional motor operating an automatic burner can be obtained by using a pressure controller with two potentiometers operating in unison. This design makes it possible to control two motors simultaneously. It is also provided with an adjustable throttling range.

Pressure controllers are also available for vapor or vacuum systems. A pressure controller with a bellows-operated mercury switch is manufactured by Honeywell for use on vapor heating systems with pressures of up to 4 psi (see Figure 15-36). It can be used as a boiler high-limit control with cut-in and cutout settings in the vacuum range. In such installations, the heating system must include a vacuum pump and a siphon loop.

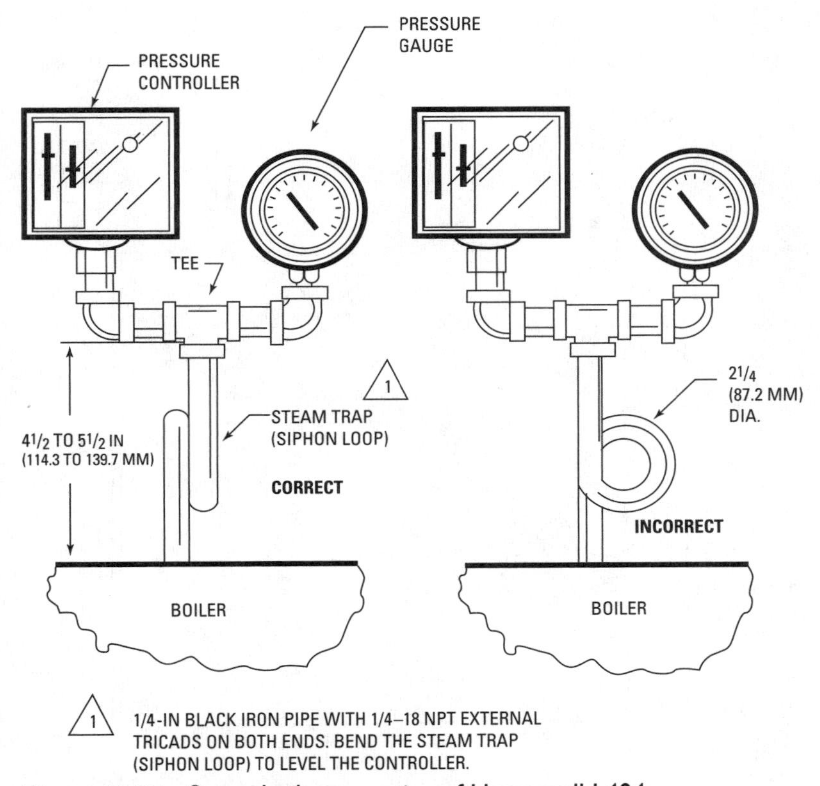

Figure 15-35 Steam boiler mounting of Honeywell L404 Pressuretrol® Controller. *(Courtesy Honeywell, Inc.)*

Vacuum Relief Valve

A vacuum-relief valve (Figure 15-37) is an effective means of providing protection against the buildup of excessive vacuum conditions in steam heating and steam processing systems. It is also used in combination with temperature-pressure relief valves on water heaters.

Vacuum conditions can often occur after the supply line has been shut off. Steam condenses, and a vacuum can be created that not only affects system operation but also can cause damage to equipment. The vacuum relief valve automatically admits air to the system in order to break up the vacuum. These valves operate

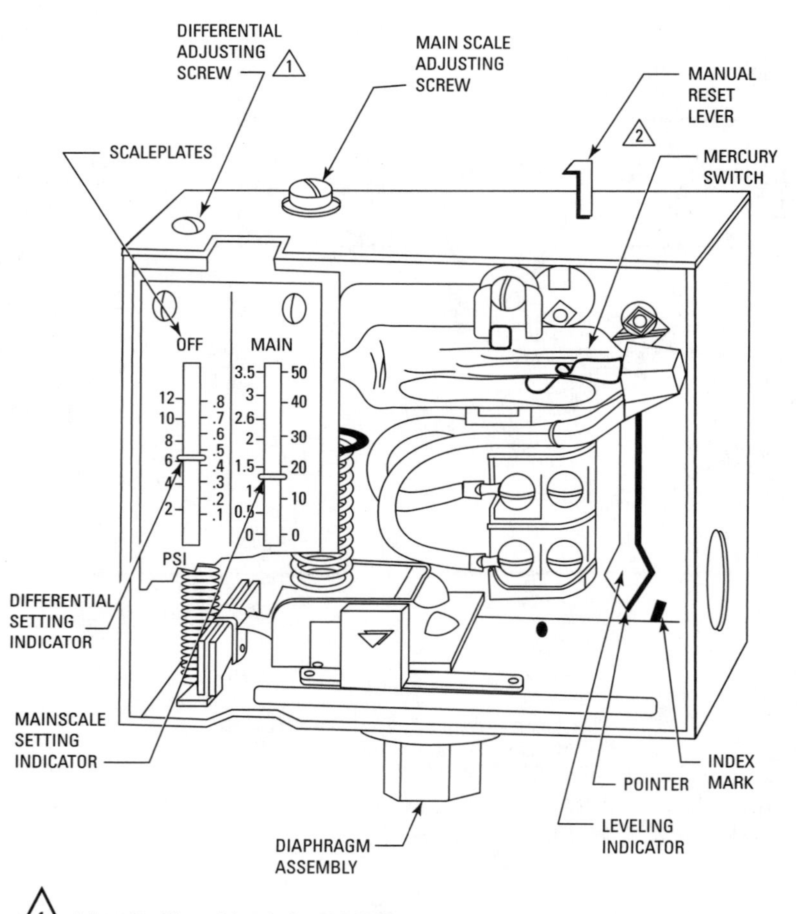

1 Adjustable differential on the L404A,B,F; L404L with a 5 to 150 psi. 0 34 to 10,3 kg/cm² [34 to 134 kPa] operating range and L604A models only.

2 Trip-free manual reset Lever on the L404C,D and L604L models only.

Figure 15-36 Setting a Honeywell Pressuretrol® Controller.
(Courtesy Honeywell, Inc.)

under conditions up to a maximum temperature of 250°F. The valve disk should be constructed from a material capable of withstanding high temperatures, particularly when the valve is used in a steam processing system.

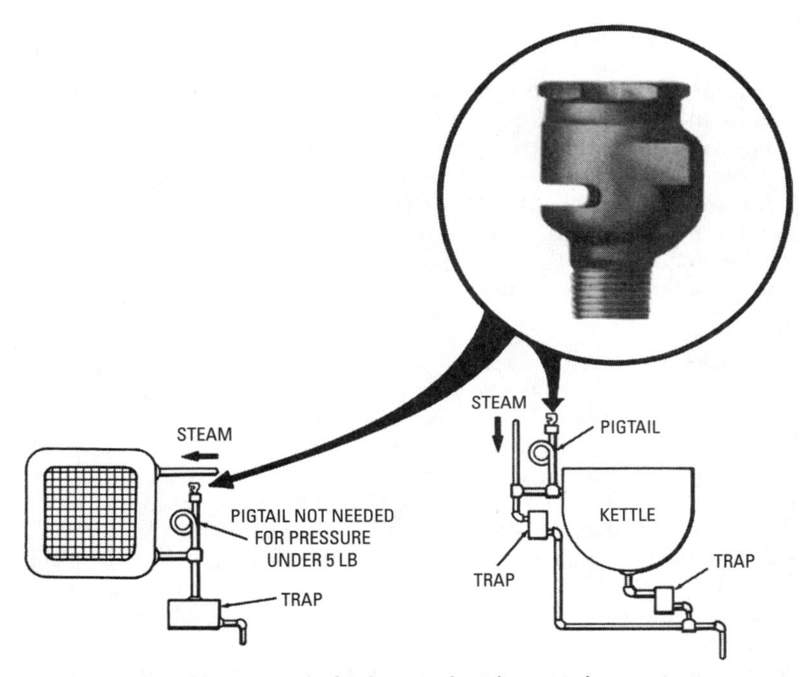

Figure 15-37 Vacuum relief valve used with a unit heater in a steam heating system. *(Courtesy Watts Regulator Co.)*

Steam Boiler Aquastat

Aquastats have been used on some steam boilers to control temperature limits. It is similar in function to the pressure control on a steam boiler or the fan and limit control on a forced-warm-air furnace. Aquastats are used much more often on hot-water boilers to control temperature limits or to operate the circulator (pump). See *Hot-Water Boiler Aquastats* in this chapter.

Blowdown Valve

Sediments and contaminants in the water will settle out over a period of time and accumulate at a low point in the bottom of the boiler. These can be removed through a blowoff valve installed in a line connected to the lowest part of the boiler. The blowoff valve is opened periodically, and the accumulated sediments are drained off.

Steam boilers operating at 15 psig or less require a blowdown valve the full size of the boiler connection. Steam boilers operating at pressures greater than 15 psig require at least two such blowdown valves. One or both of the valves must be the slow opening type.

Each water column and float type low-water cutoff must also be equipped with a blowdown valve.

Foam sometimes forms on the surface of the boiler water. This is usually indicated by drops of water appearing with the steam. This condition is caused by the presence of oil, dissolved salts, or similar organic matter in the water. One method of eliminating this problem is by draining off part of the water in the boiler and adding an equal amount of fresh, clean water. Another method involves blowing the foam from the water with a specially connected pipe or hose. The boiler should have a blowdown tapping for this purpose.

Try Cocks

Try cocks are small valves installed on steam boilers at the safe high-water level and at the safe low-water level (Figure 15-38). When the boiler water column is inoperable, try cocks function as a backup system to determine the water level. The water level is determined by opening *slightly* first one try cock and then the other. The water level is indicated by the color of the plume escaping from the try-cock orifice. A steam plume is characteristically colorless and will indicate that the water level is too low (i.e., below the level of the try-cock orifice). Water, on the other hand, is characterized by a white plume. A false water-level reading will be obtained if the try cock is opened too wide when the level of the water is only slightly below the level of the try-cock orifice. The violent agitation of the water caused by a wide-open try cock will result in some of

Figure 15-38 Try cocks. *(Courtesy Ernst Gage Co.)*

the water escaping, giving the false impression that the level of the water in boiler is at or above the try-cock fitting.

On large steam boilers a third try cock is often installed between the other two. Sometimes a try cock on these larger boilers is at a level too high to reach. When this is the case, chain-operated, lever-type try cocks are installed.

Steam Boiler Injectors

A steam boiler injector is a device used on some boilers to create accelerated steam circulation in a steam heating system. The essential parts of a boiler injector are shown in Figure 15-39.

This device operates on the induction principle. In operation, steam from the boiler, entering the steam nozzle, passes through it, through the space between the steam nozzle and combining tube, and then out through the overflow. This produces a vacuum, which draws in the water through the water inlet. The incoming cold water condenses the steam traversing the combining tube, and the water jet thus formed is at first driven out through the overflow. As the velocity of the water jet increases, sufficient momentum is obtained to overcome the boiler pressure, with the result that the water enters the delivery tube and passes the main check valve into the boiler.

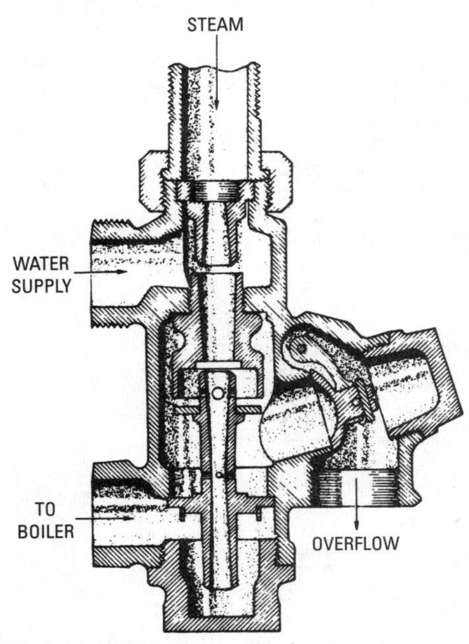

Figure 15-39 Steam boiler injector.

The induction principle of the steam injector is the same as the operating principle of hydraulic boosters used in forced-hot-water heating systems, with the exception that fast-flowing hot water is used instead of steam to obtain the inductive action to accomplish accelerated circulation.

Water Gauges

A water gauge is used to check the level of water in the boiler *visually* (Figure 15-40). If the water level in the boiler is high enough, the glass tube will be approximately ⅔ to ¾ full. Check the boiler specifications, because the safe operating level will vary among manufacturers. If no water is showing in the tube, the boiler must be turned off and refilled to the proper level. Do *not* add the water until the boiler has had time to cool off (Figure 15-41).

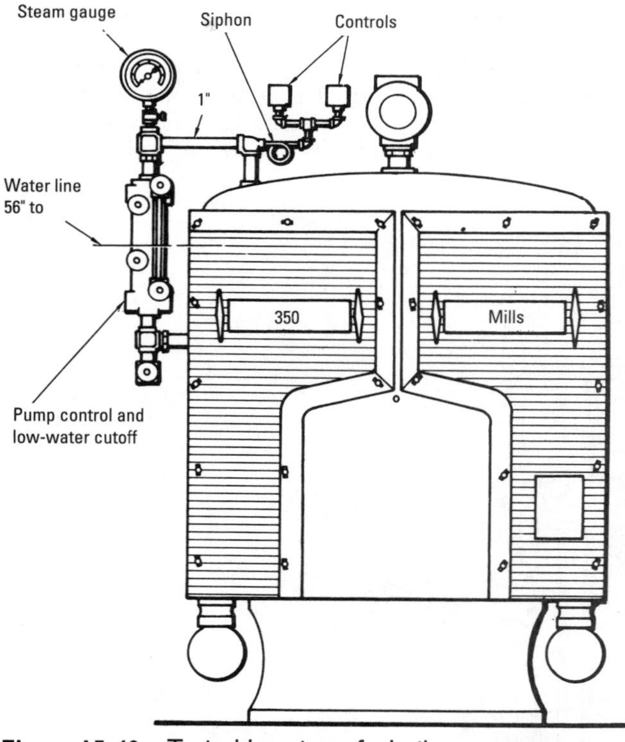

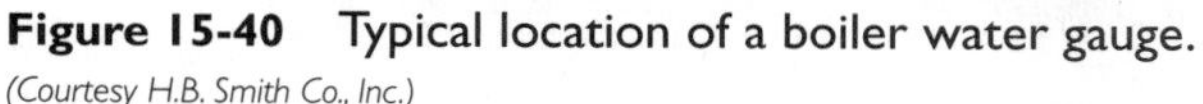

Figure 15-40 Typical location of a boiler water gauge.
(Courtesy H.B. Smith Co., Inc.)

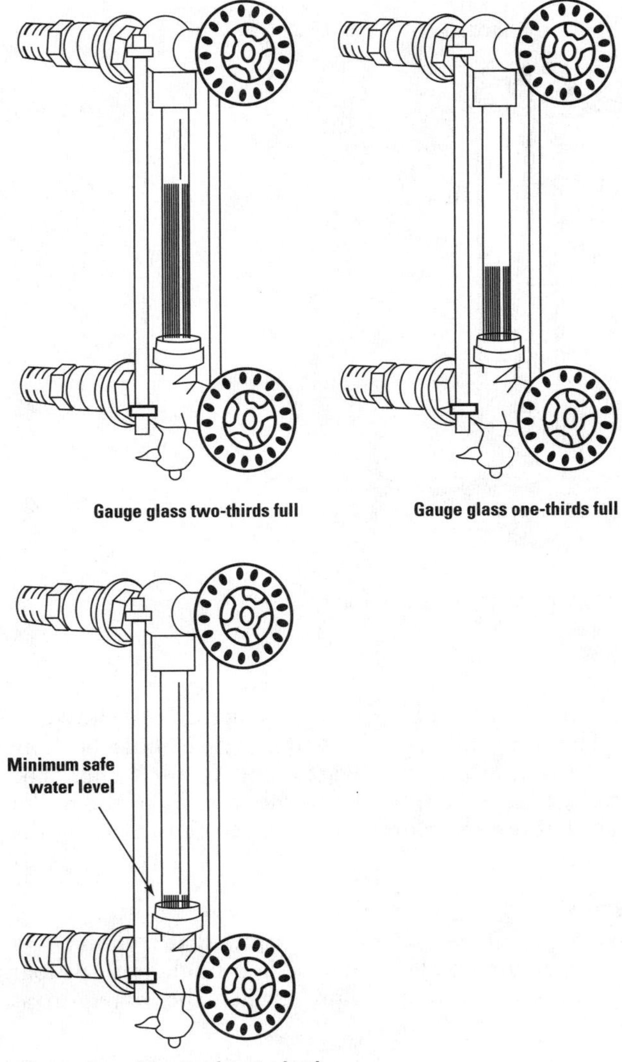

Figure 15-41 Water gauge levels. *(Courtesy ITT McDonnell & Miller)*

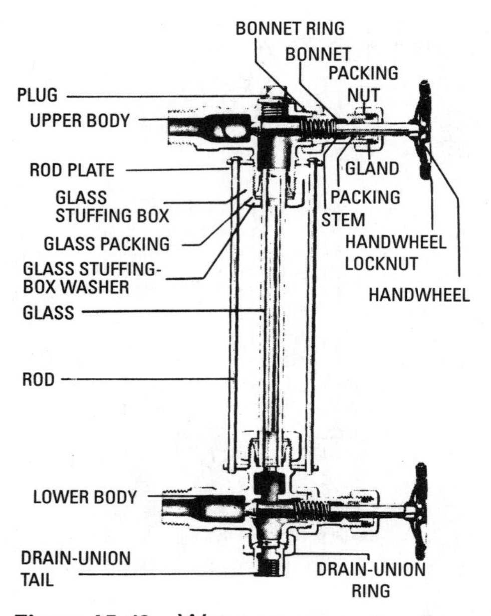

Figure 15-42 Water gauge construction details. *(Courtesy Lunkenheimer Co.)*

The construction details of a typical water gauge are shown in Figure 15-42. The glass tube of the water gauge must be long enough to cover the safe range of water level in the boiler. The ends of the water gauge are connected to the interior of the boiler by fittings located above the safe high-water level and below the safe low-water level.

Water Columns

A water column (Figure 15-43) is a boiler fitting that combines try cocks, a water gauge, and an alarm whistle in a single unit. A float in the column activates an alarm whistle when the water drops below a safe level.

Steam Gauges

The difference between the pressure found on the inside of the boiler and the pressure on the outside of it is indicated by a steam gauge (Figure 15-44). This is a gauge pressure reading and should not be

Figure 15-43 Water column.
(Courtesy Ernst Gauge Co.)

confused with absolute pressure. A steam gauge is used to measure the steam pressure at the top of the boiler. Generally, a reading of 12 psi (pounds per square-inch) indicates a dangerous buildup of pressure in low-pressure steam heating boilers. The boiler should be shut down before the pressure exceeds this level.

If the steam gauge is operating properly, the needle (or pointer) will move with each change of pressure inside the boiler. Shut off the steam, and the gauge needle should drop to zero; turn on the steam, and the needle should rise to the correct reading. It is very important that the steam gauge be regularly checked to ensure that it is operating properly.

The steam gauges used on boilers operate on the bent-tube principle, that is, the tendency of a bent or curved tube to assume a straight position when pressure is applied. As shown in Figure 15-45, one end of the curved tube is attached in a fixed position to a pigtail (connector tube), which, in turn, is attached to the boiler. The gauge needle is mounted on a rack-and-pinion gear attached to the *free* end of the curve tube. The pressure in the curved tube causes its free end to move slightly in its effort to assume a straight position. This slight movement is multiplied by the rack-and-pinion gear, causing the needle to rotate and indicate the steam pressure.

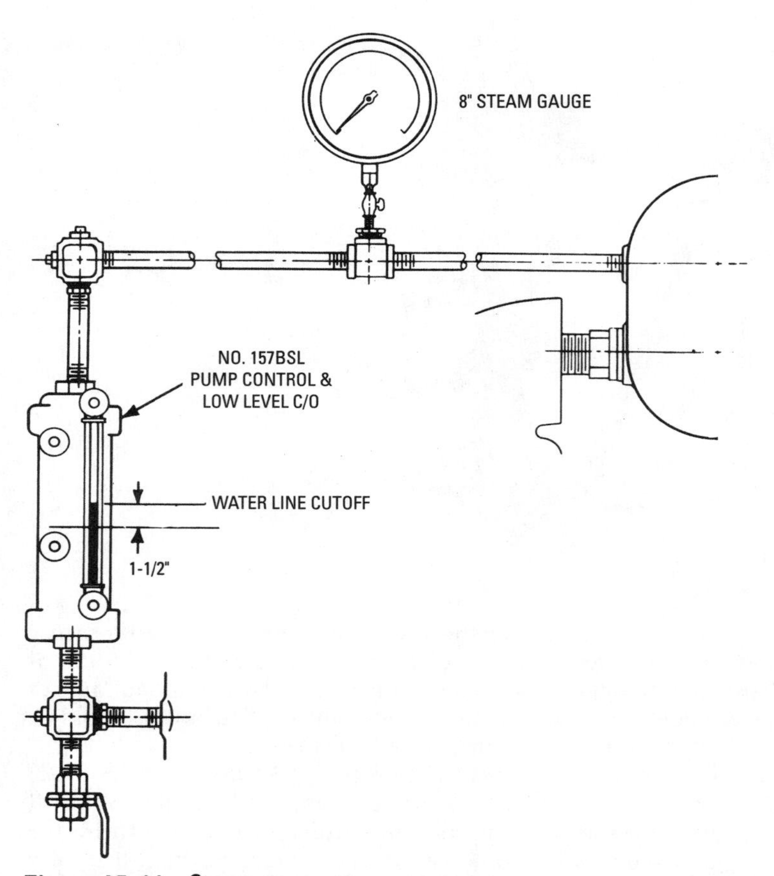

Figure 15-44 Steam gauge. *(Courtesy H.B. Smith Co., Inc.)*

Steam Gauge Pigtails

A steam gauge pigtail (Figure 15-46) is a length of tubing with one or more loops in it used to connect the steam gauge to the boiler. It functions as a protective device by preventing live boiler steam from coming into contact with working parts of the gauge. The steam is denied passage by condensate, which forms in the loop (or loops) of the pigtail. It is recommended that the steam gauge pigtail be filled with water *before* it is attached to the boiler. It should be attached to the boiler in a position where heat and vibration will be at a minimum.

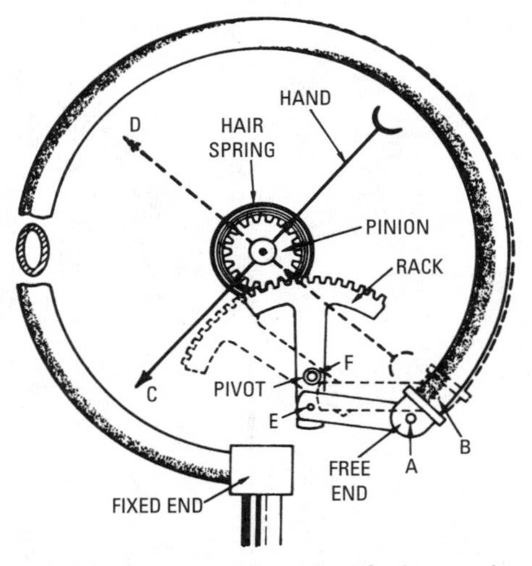

Figure 15-45 Mechanism of a bent-tube steam gauge.

Hartford Return Connection

A Hartford return connection or steam loop (Figure 15-47) is a piping arrangement used in a steam heating system to return condensation to the boiler. This is done to prevent excessive water loss in the boiler when a leak occurs in the return line. The Hartford return connection is described in Chapter 8, "Steam Heating Systems."

Hot-Water Boiler Valves, Controls, and Accessories

The boilers used in hot-water heating systems are also fitted with a variety of valves, controls, and other devices designed to ensure the safe and proper operation. Some are similar to those used on steam boilers (e.g., pressure relief valves and low-water cutoffs). They also share the same measuring and controlling functions described for steam boiler valves, controls, and accessories. Figures 15-48 and 15-49 illustrate their arrangement on a hot-water space heating boiler.

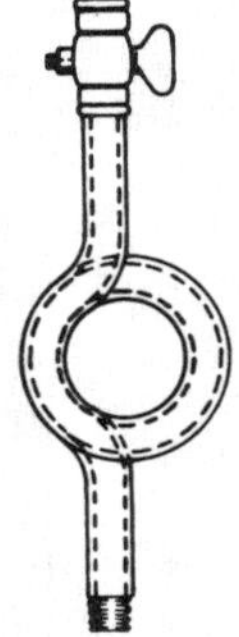

Figure 15-46 Steam gauge pigtail.

Hot water (hydronic) boilers operate under high pressures and temperatures. To avoid serious injury,

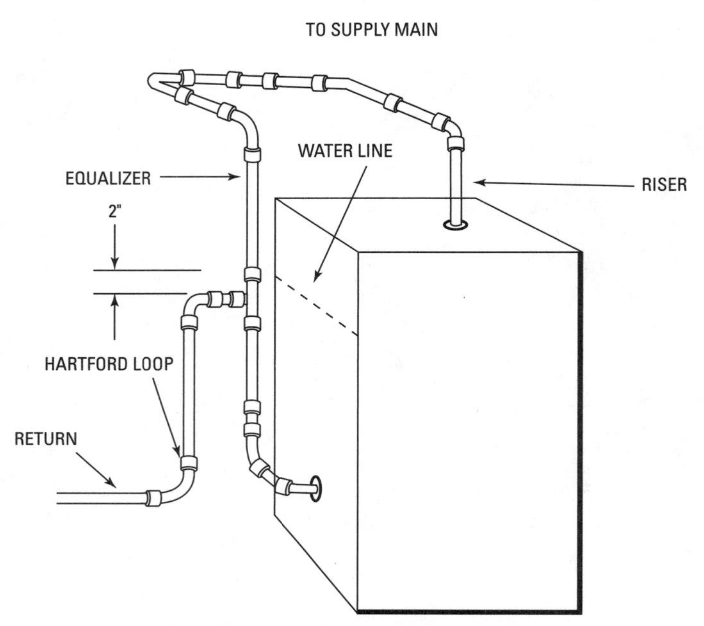

Figure 15-47 The Hartford return connection.

never begin any service or repairs shutoff before taking the following precautions:

- Wait for the boiler to cool down to 80°F (27°C) or more.
- Wait for the boiler pressure to drop to 0 psi (0 bar).
- Disconnect the electrical power before making any electrical connections.

Principal Hot-Water Boiler Valves, Controls, and Accessories

A hot-water boiler should be equipped with the following valves, controls, and accessories:

- Pressure relief valve: A safety valve used to relieve boiler pressure when it exceeds the maximum safe level.
- Low-water cutoff: A device that automatically switches off the burner if the boiler water level becomes too low for safe operation.
- High-pressure limit switch: A safety device used to turn off the burner when boiler pressure exceeds its preset maximum safe operating level.

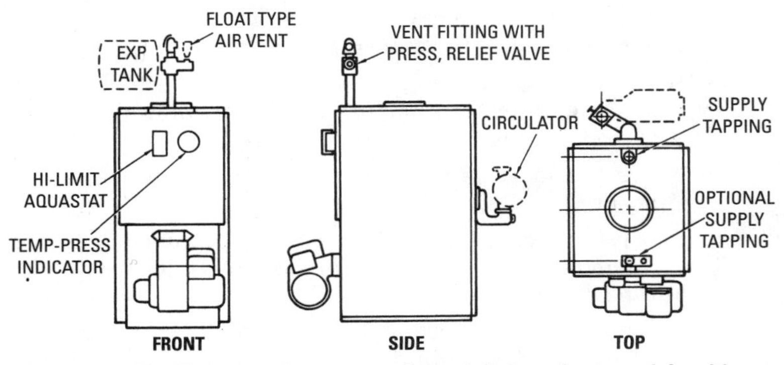

Figure 15-48 Pipings, valves, controls, and fittings for an oil-fired hot-water (hydronic) boiler. *(Courtesy Hydrotherm, Inc.)*

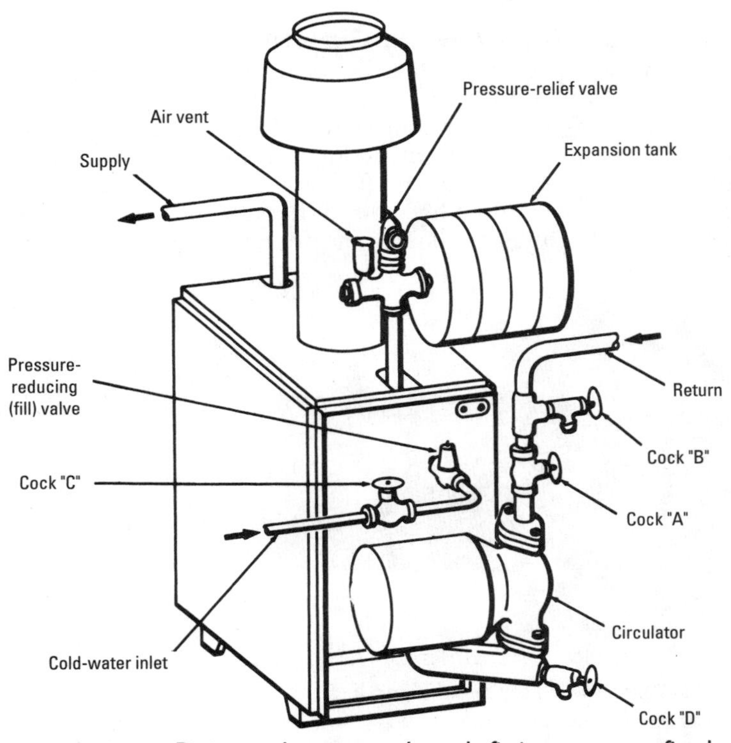

Figure 15-49 Pipings, valves, controls, ands fittings on a gas-fired hot-water (hydronic) boiler. *(Courtesy Hydrotherm, Inc.)*

- Aquastat: An automatic switching device used to control temperature limits or to operate the circulator (pump).
- Water pressure-reducing valve: A valve used to keep the boiler and system automatically filled with water at the desired operating pressure.
- Air vent: A device that releases air from the system to the atmosphere.
- Expansion tank: A tank used to accommodate the expanded volume of heated water in a hot-water (hydronic) heating system.
- Air separator: A device used to separate trapped air bubbles and pockets from the water before it circulates through the hot-water heating system.
- Circulator: A pump used to move the water through the hot-water heating system.

Hot-Water Boiler Low-Water Cutoffs

A low-water cutoff device, such as the one shown in Figure 15-50, can also be used in a hot-water heating system to provide protection against a low-water-level condition in the boiler resulting from runaway firing caused by malfunctioning controls or a break in the return piping. The low-water cutoff device should be installed in the piping so that the raised line cast on the float chamber body is on a level with the top of the boiler. The drain valve located directly

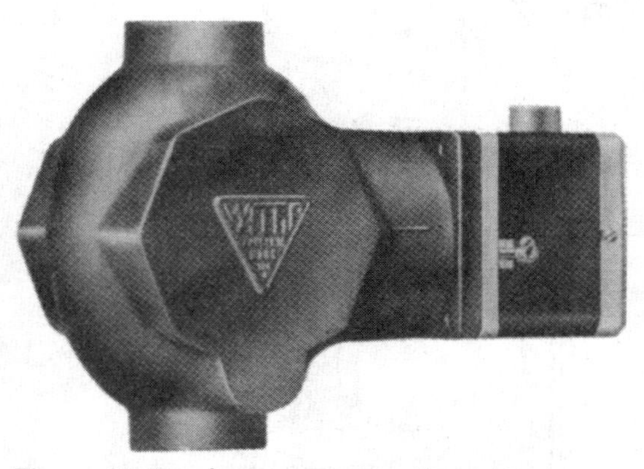

Figure 15-50 Low-water cutoff used in a forced-hot-water heating system.
(Courtesy Watts Regulator Co.)

below the low-water cutoff should be opened periodically to flush mud and sediment from the float chamber. It is recommended that this be done at least once a month.

Note

> The ASME Boiler and Pressure Vessel Code requires that hot-water space heating boilers be equipped with a low-water cut-off device *only* if the boiler input is greater than 400,000 Btu/h. However, on coil-type boilers, where the flow of water is required to prevent overheating, a flow switch or similar device can be used instead of the low-water cutoff switch to shut off burner operation.

Some hot-water (hydronic) space heating systems have piping for the radiators, convectors, or tankless water heaters installed below the minimum safe water level of the boiler. A low-water cut-off should be installed in these systems for low-water protection.

The probe-type low-water cutoff illustrated in Figure 15-51 has a green "power on" LED and a "low-water condition" red LED for

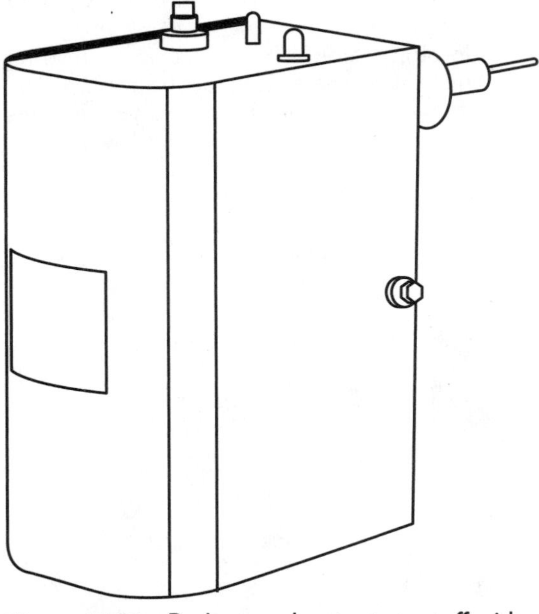

Figure 15-51 Probe-type low-water cutoff with LED indicator. *(Courtesy ITT McDonnell & Miller)*

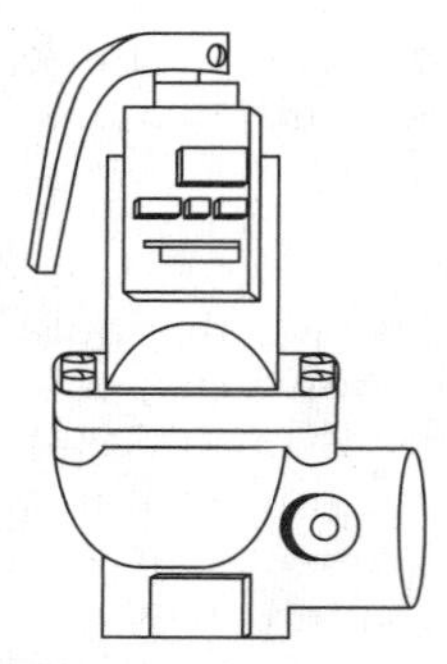

Figure 15-52 Pressure relief valve. *(Courtesy Watts Regulator Company)*

immediate recognition of the water level status of the boiler. This probe-type low-water cutoff can be installed in either the boiler tapping or a supply riser. Two models are available: one for a 24 volt burner circuit, the other for a 120-volt circuit.

Hot-Water Boiler Pressure Relief Valves

ASME rated pressure relief valves are used on hot-water space heating boilers to relieve pressure created by two different conditions: (1) water thermal conditions and (2) steam pressure conditions (Figures 15-52 and 15-53). These are exclusively safety devices and normally should not be used as regulating valves or other units intended to regulate or control the flow pressure. Relief valves are intended to prevent personal injury and damage to property. These valves start to open at the set pressure and require a certain percentage of overpressure to open fully. As the pressure drops, they start to close and shut at approximately the set pressure.

As thermal conditions develop inside a hot-water space heating boiler, pressures may be built up to the setting of the relief valve. When the pressure reaches the safety limit setting, the valve functions as a *water* relief valve and discharges the small amount of water that is expanded in the system.

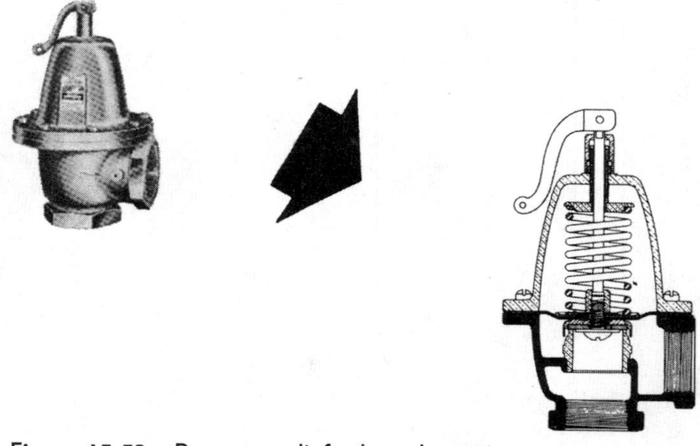

Figure 15-53 Pressure relief valve schematic. *(Courtesy A.W. Cash Valve Mfg. Corp.)*

If *both* water and steam are present in a hot-water space heating boiler, the firing controls probably are malfunctioning or have completely broken down. This may result in runaway firing of the boiler, which could cause the boiler water to reach steam-forming temperatures, creating a steam pressure condition. Under these circumstances, the relief valve functions as a steam safety-relief valve. The steam is discharged at a rate or faster than the boiler can generate it, thus restoring system pressure to a safe level. Although these valves are steam rated and have an emergency Btu steam discharge capacity if runaway firing conditions occur, they should *not* be continuous steam service.

In order to be completely effective, the safety relief valves used on hot-water space heating boilers must be designed to discharge the excessive water pressure created by thermal expansion and also the excessive steam pressure resulting from runaway emergency temperature conditions.

High-Pressure Limit Switch

The high-pressure limit switch is a safety device used to shut down the operation of the gas or oil burner when the pressure boiler pressure exceeds a preset level (commonly 5 to 8 psi). The high-pressure switch is connected to the boiler by a pigtail pipe.

Figure 15-54 Aquastat with immersed heat-sensitive elements.

Hot-Water Boiler Aquastats

An aquastat is a device commonly used on hot-water space heating boilers and some steam boilers to control temperature limit or to operate the circulator (hot-water heating system pump). It is similar in function to the pressure control on a steam boiler or the fan and limit control on a forced-warm-air furnace.

Basically, an aquastat is an automatic switching device consisting of a metal- or liquid-filled heat-sensitive element designed to detect temperature drop or rise of the boiler water. Aquastats can be strapped to the hot-water supply riser or mounted so that the heat-sensitive element is immersed in a boiler well (Figure 15-54).

The type of aquastat used in a heating installation generally depends upon whether it is designed to control temperature limit or to switch on the circulator. If the former is the case, the aquastat will close on temperature drop and open on temperature rise. The aquastat will close on temperature rise if it is used to operate the circulator.

An example of a strap-on type aquastat is the ITT General Controls L-53 Strap-On Hot Water Control illustrated in Figure 15-55. This type of aquastat responds to water temperature changes as conducted through the supply-riser pipe wall to the temperature-responsive base of the device. The enclosed switching is supplied in three basic types, depending upon their operating principles: direct-action (Figure 15-56), reverse-action (Figure 15-57), and double-action (Figure 15-58) models.

A direct-action (normally closed, N.C.) aquastat used as a high-limit control must always be located on the supply riser where it will be subjected to the maximum boiler water temperature. Its location on the riser will therefore have to be as close to the boiler as possible but ahead of any line shutoff valves.

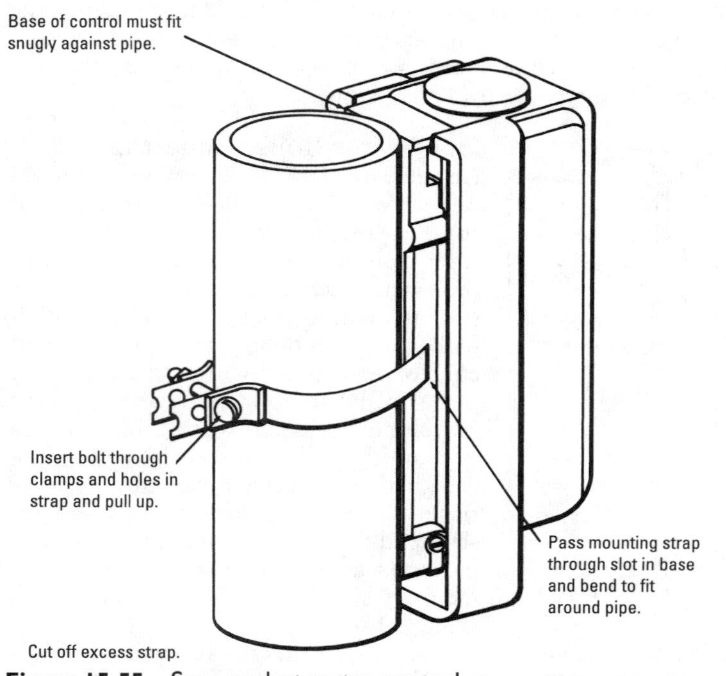

Figure 15-55 Snap-on hot-water control. *(Courtesy ITT General Controls)*

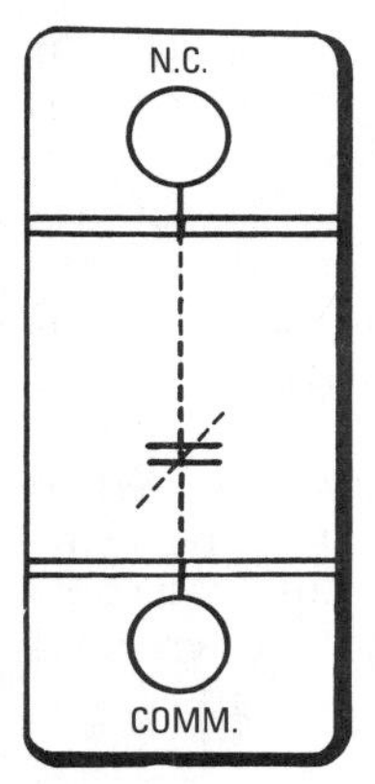

Figure 15-56 Direct-action switch operation. *(Courtesy ITT General Controls)*

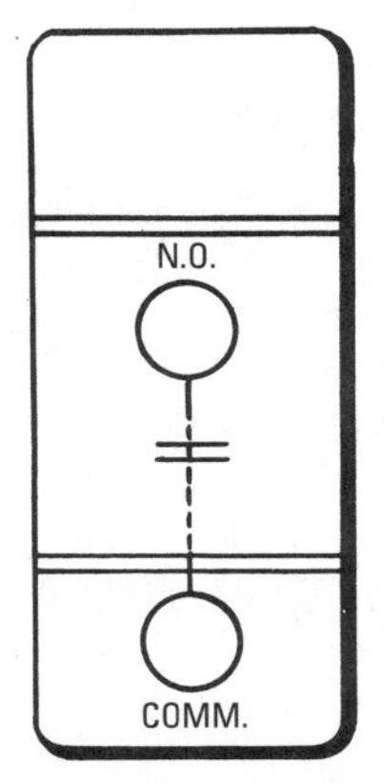

Figure 15-57 Reverse-action switch operation. *(Courtesy ITT General Controls)*

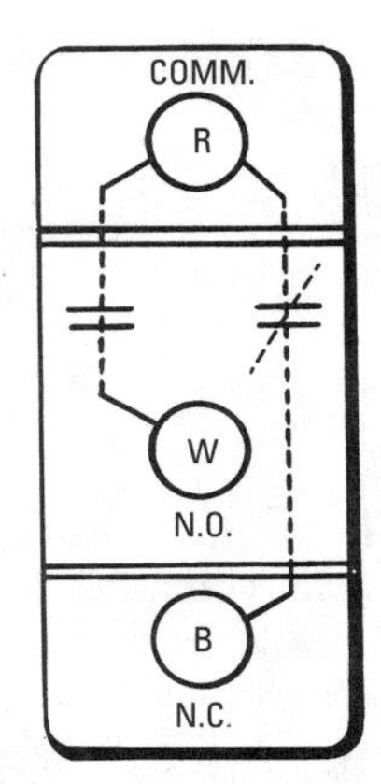

Figure 15-58 Double-action switch operation. *(Courtesy ITT General Controls)*

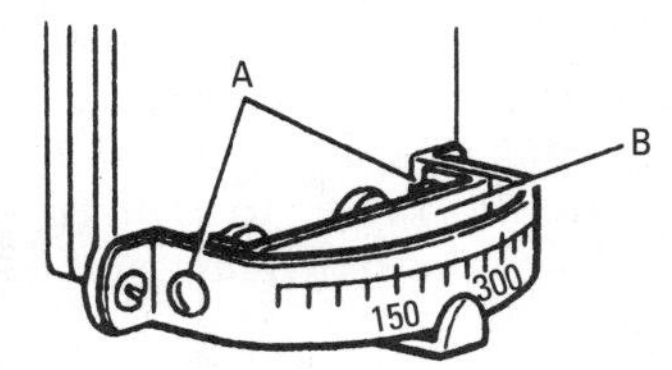

Figure 15-59 Indicator can be set for a temperature higher than 200°F by removing the screws (A) and holding stop bracket to scale plate and removing the bracket (B). *(Courtesy ITT General Controls)*

Direct-action aquastats will open the circuit on temperature rise. Setting the adjustable scale pointer to a position on the scale will result in breaking the circuit (Figure 15-59). When the aquastat is used as a high-limit control, the cutout setting should be as low as possible and still ensure proper heating in cold weather. An initial cutout setting of 170°F is recommended for a gravity hot-water heating system. An initial setting of 200°F is suggested for a forced-hot-water (hydronic) heating system. The mechanical differential of the control illustrated in Figure 15-59 is nonadjustable and is approximately 15°F.

A reverse-action (normally open, N.O.) aquastat should be mounted ahead of any valves or traps *on the return line* when it is used on a unit heater installation. It should be mounted on the largest riser from the boiler if it is used to prevent circulator operation when boiler water temperature is low. A reverse-action aquastat closes the circuit on temperature rise.

A double-action (single-pole double-throw, SPDT) aquastat is used to start circulator operation with a single switch actuation and to function as an operating control to maintain boiler water temperature. This type of aquastat should be mounted on the largest riser ahead of any valve but at a point where it will be subjected to maximum boiler water temperatures. The double-action aquastat illustrated in Figure 15-58 opens *R* to *B* contacts and closes *R* to *W* contacts on temperature rise at scale setting.

A strap-on aquastat can be mounted in any position. When mounting these aquastats, make absolutely certain that the pipe surface is clean and free of rust and corrosion. All rough and high spots should be filed smooth. Nothing should be allowed to interfere with the operation of the temperature responsive base of the control.

Pressure-Reducing Valves

Most hot-water heating systems are equipped with a water pressure-reducing valve designed to feed water into the boiler automatically when the pressure in the system drops below the valve setting. When the pressure returns to the minimum pressure setting, the valve automatically closes. Thus, the function of a water pressure-reducing valve is to keep the system automatically filled with water at the desired operating pressure (Figure 15-60). These valves are also referred to as *pressure-reducing boiler feed valves, water pressure-reducing fill valves,* or *feed water pressure regulators*. Regardless of what they are called, their function remains the same: to deliver water to the boiler at the required pressure. To perform this function,

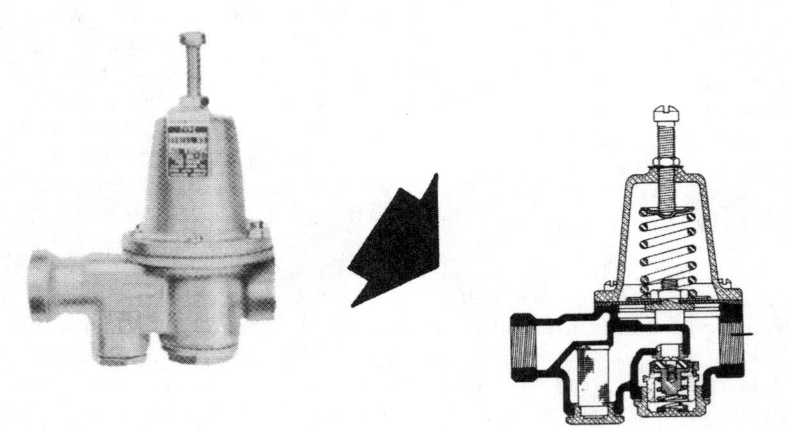

Figure 15-60 Water pressure-reducing valve. *(Courtesy A. W. Cash Valve Mfg. Co.)*

they are *always* installed in the cold-water supply line. If they are part of a combination pressure-reducing and pressure relief valve (dual control), the pressure relief part of the combination valve is always located between the boiler and the pressure-reducing part.

A manually operated feed valve was used on older hot-water heating systems to provide the same function as the automatic water-pressure-reducing valve. On these older systems, the manually operated feed valve should be *partly* opened and water added to the boiler when the movable (white or red) arrow on the altitude gauge drops below the setting of the stationary (black) arrow (see *Altitude Gauges* in this chapter).

Some pressure-reducing valves are equipped with a built-in check valve in the supply inlet to prevent a backflow of contaminated boiler water into the domestic water supply. This backflow drainage of boiler water generally occurs when the supply pressure falls below the feeder valve setting. Not only can this drainage cause possible damage to the boiler, it can also contaminate the domestic water supply. Some pressure-reducing valves are also equipped with integral strainers to trap foreign matter that may clog the valve. The strainer screen should be removed for cleaning at the beginning of each heating season. This can be done by removing the bottom plug from the strainer and withdrawing the screen. Clean and carefully replace screen and plug.

The principal components of a water-pressure-reducing valve are shown in Figure 15-61. This particular valve is constructed with an integral strainer. It is also possible to use separate units in combination, as Figure 15-62 illustrates. The two units are joined by a short threaded connection. A typical installation in which a water-pressure-reducing valve is used is shown in Figure 15-63.

As mentioned previously, a water-pressure-reducing valve should be installed in the water supply line to the boiler. It should also be installed at a level *above* the boiler and in a horizontal position. Before installing the valve, flush out the supply pipe to clear it of chips, scale, dirt, and other foreign matter that could interfere with its operation. Install a shutoff valve *ahead* of the regulator, and then connect the supply line to the inlet (usually marked "in" on the valve casting).

To fill the system, open the shutoff valve located ahead of the pressure-reducing valve. Water will flow into the system until it is full and under pressure. The shutoff valve must always be kept open when the system is in operation.

Water-pressure-reducing valves are usually set by the manufacturer to feed water to the boiler at approximately 15 lb of pressure.

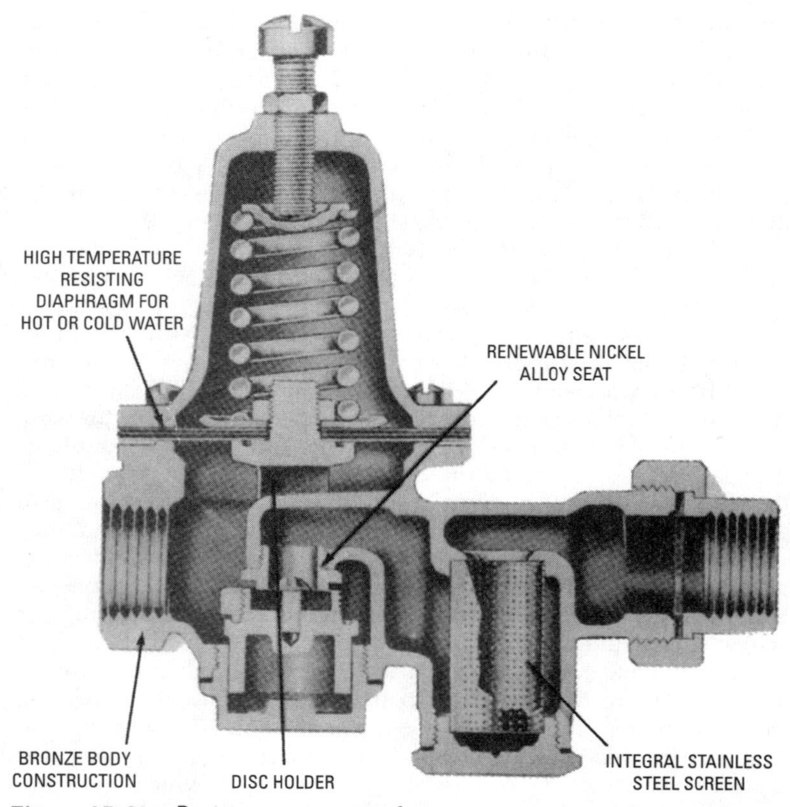

Figure 15-61 Basic components of a water pressure-reducing valve with integral stainless steel strainer. *(Courtesy Watts Regulator Co.)*

This pressure setting is sufficient for residences and houses up to a three-story building in size. For higher buildings in which the pressure may not be sufficient to lift the water to the highest radiator, it may be necessary to reset the water-pressure-reducing valve for higher pressure. To do this, calculate the number of feet from the regulator to the top of the highest radiator. Multiply this by 0.43 and add 3 lb. This will give the pressure needed to raise the water to the highest radiator and keep it under pressure. Loosen the locknut on the valve, and turn the adjustment screw clockwise slowly until the gauge indicates the pressure calculated. When the desired pressure is obtained, tighten the locknut.

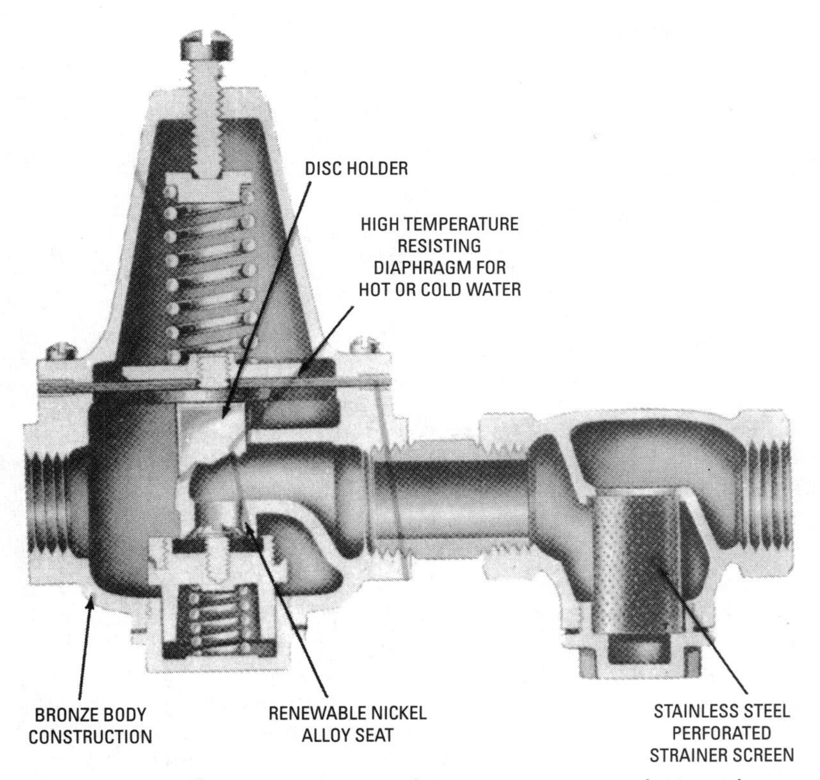

Figure 15-62 Basic components of a water-pressure-reducing valve with separate strainer unit. *(Courtesy Watts Regulator Co.)*

Combination Valves

Combination valves (also called *dual unit valves* or *dual control valves*) are designed for use in hot-water space heating systems. They combine a pressure-reducing/pressure-regulating valve and a positive relief valve in one body (Figures 15-64 and 15-65). Sometimes an integral bypass valve is also included in the same unit.

The purpose of a combination valve is to provide pressure regulation and safety control and to reduce boiler pressure and ensure automatic filling when conditions warrant. The valve is installed on the supply line of a boiler with the relief-valve section closest to the boiler.

The regulator side of the combination valve is designed to reduce the incoming water supply to the boiler to the required 14-psi operating pressure for a one-, two-, or three-story house. The pressure

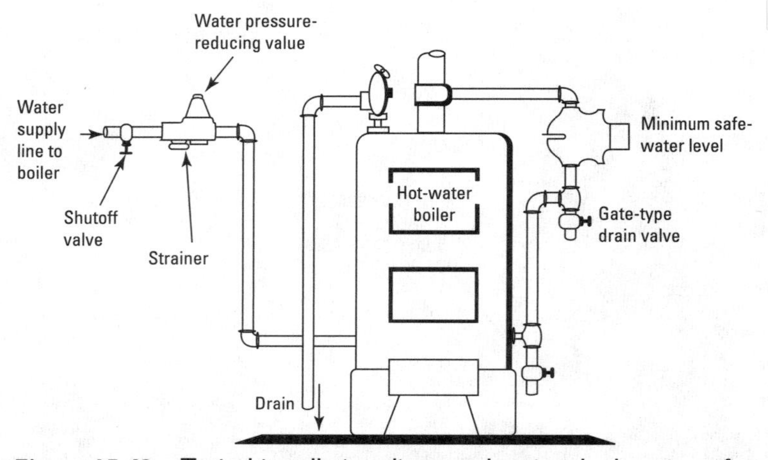

Figure 15-63 Typical installation diagram showing the location of a water-pressure-reducing valve. *(Courtesy Watts Regulator Co.)*

in the system builds up due to thermal expansion when the boiler is fired. Under normal operating conditions, the expansion tank will absorb the additional pressure. However, if the expansion tank is waterlogged or if the system has no expansion tank, the relief side of the valve will open at 23 psi and drop the pressure back to within safe limits. If the pressure should drop below 14 psi, the regulator will open again and automatically refill the system. A built-in check valve prevents the backflow of contaminated boiler water into the potable water supply.

Exploded views of combination valves are shown in Figures 15-66 and 15-67. The Cash-Acme Type CBL Valve (Figure 15-68) differs from the other two by having a built-in bypass valve. In all other respects, it is similar to the others except that it should be installed as close as possible to the top of the boiler (using close nipples). The principal advantages of having the bypass valve included are:

1. It allows rapid filling of the system.
2. It permits an easy high-pressure test for leaks and system purging.
3. It passes first filling dirty water around the valve seat, ensuring a clean, good seating surface.
4. It permits the use of a wide opening, small seated regulator that prevents wire drawing and rapid wearing of the seat.

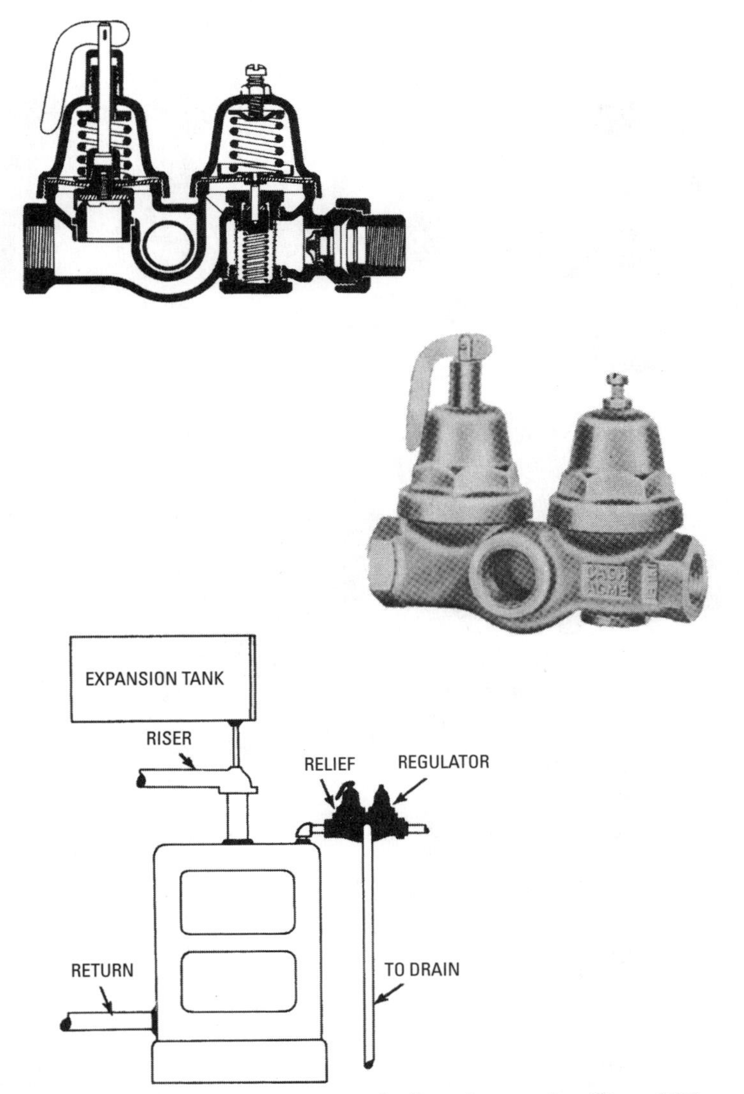

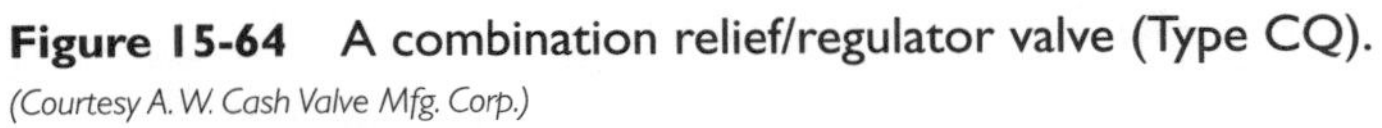

Figure 15-64 A combination relief/regulator valve (Type CQ).
(Courtesy A. W. Cash Valve Mfg. Corp.)

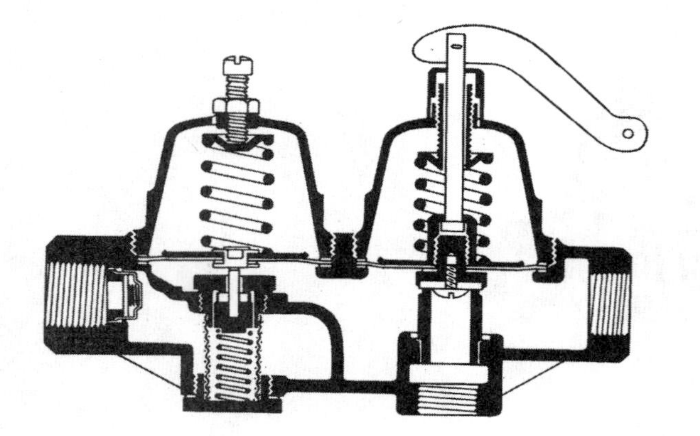

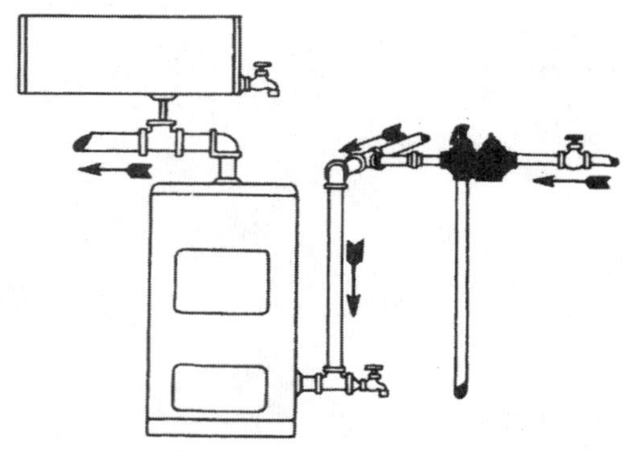

Figure 15-65 A combination relief/regulator valve (Type A-1).
(Courtesy A. W. Cash Valve Mfg. Corp.)

RELIEF SECTION

1. Lever pin
2. Lift lever
3. Pressure screw cap
4. Pressure screw
5. Locknut
6. Spring chamber
7. Spring button
8. Pressure spring, stainless steel
9. Pull rod nut
10. Pull rod
11. Lock washer
12. Pressure plate
13. Gasket, brass
14. Diaphragm
15. Composition seat shell
16. Composition seat
17. Composition seat screw
18. Body seat

REGULATOR SECTION

19. Closing cap
20. Pressure screw
21. Locknut
22. Spring chamber
23. Spring button
24. Pressure spring, black
25. Gasket, brass
26. Diaphragm assembly
27. Cylinder
28. Body
29. Piston assembly
30. Piston spring
31. Strainer
32. Gasket, red fiber
33. Bottom plug

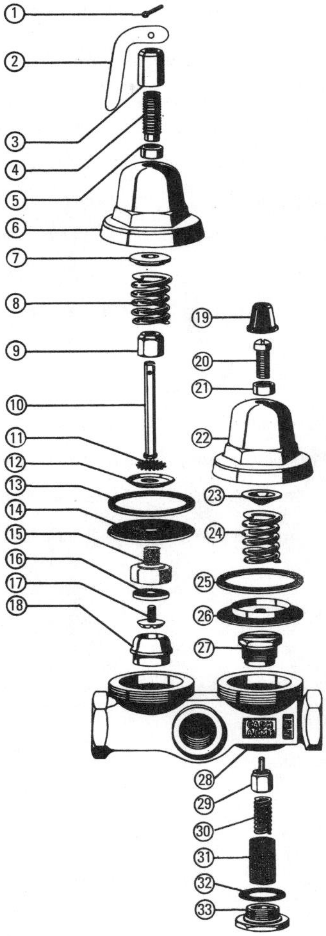

Figure 15-66 An exploded view of Type CQ combination valve.
(Courtesy A. W. Cash Valve Mfg. Corp.)

The most common types of problems encountered when using combination valves are the following:

1. Relief valve drips or cannot be shut off tightly.
2. Boiler pressure rises above the reducing valve setting.

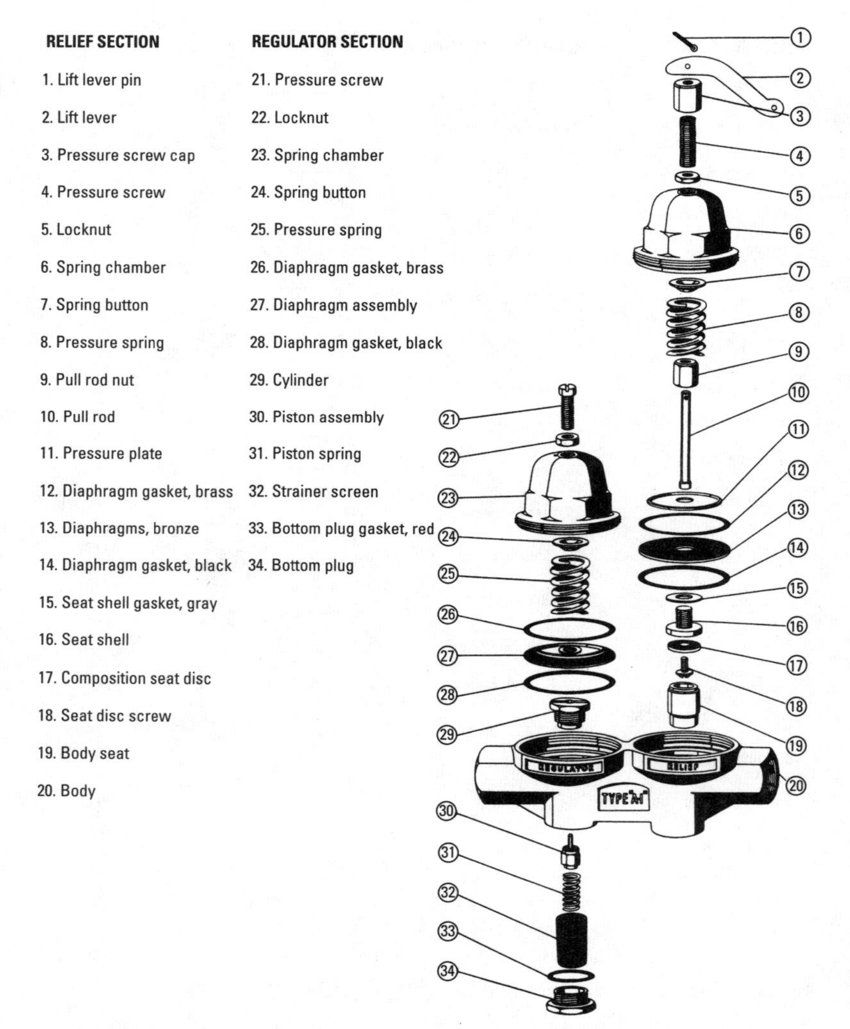

Figure 15-67 An exploded view of Type A-1 combination valve. *(Courtesy A. W. Cash Valve Mfg. Corp.)*

3. Water escapes from a weep hole on top of the regulator.
4. Water escapes from feeder side of the combination valve.
5. Foreign matter collects on the seat of the relief or regulator section of the valve.

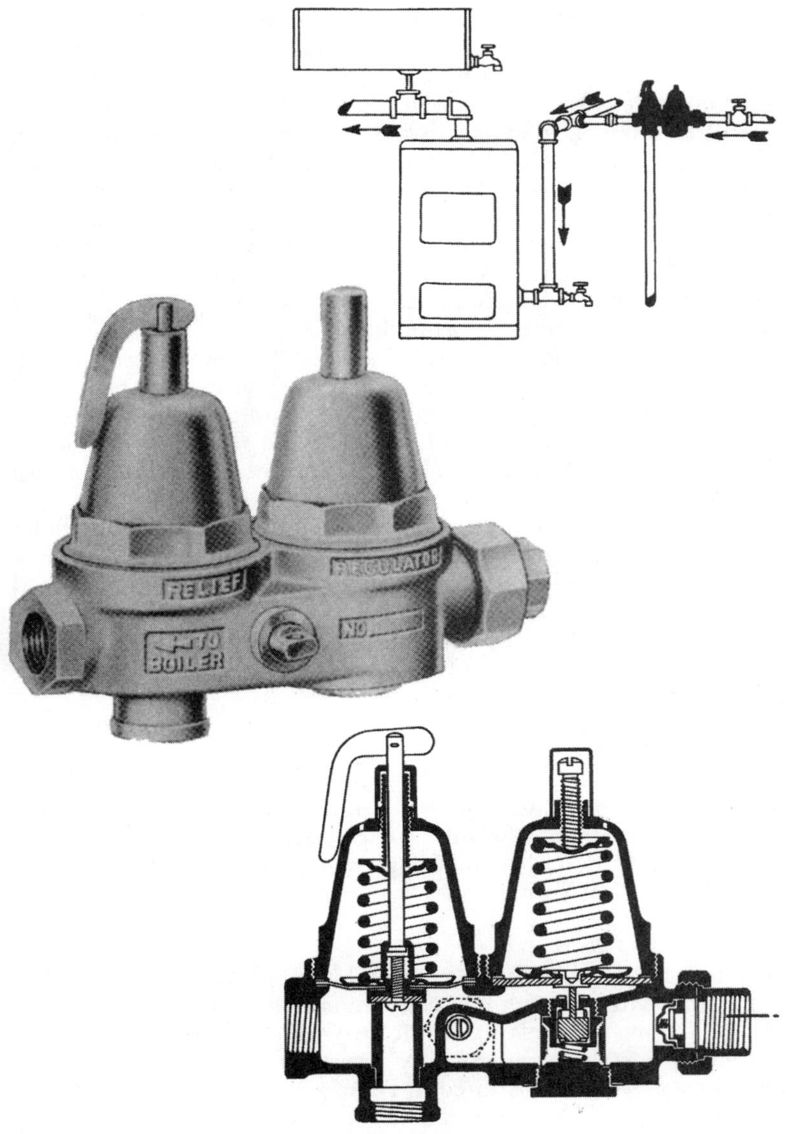

Figure 15-68 A combination relief/regulator valve (Type CBL).
(Courtesy A. W. Cash Valve Mfg. Corp.)

Some of the problems listed above (particularly the first two) may *not* be the fault of the valve itself. For example, a dripping relief valve can be caused by an expansion tank being completely filled up with water, an undersized expansion tank, or a leak in the coil of a tankless or indirect water heater installed in the boiler. These possibilities should be checked *before* attempting to service or repair the valve. The procedures are outlined in this chapter.

If, by process of elimination, the problem can be traced to the valve, try tapping the side of the valve with a wrench. Sometimes a piece of foreign matter becomes lodged and causes the regulator piston to stick. A sharp tap with a wrench may dislodge it and allow the valve to function properly.

Foreign matter such as dirt, pipe scale, or chips often cause a valve to malfunction by lodging on a seating surface, or nicking or chipping the surface. Valve manufacturers often provide replacements or instructions for field servicing and repairing. The latter should be attempted only by a skilled and experienced worker with the proper tools and gauges.

Water seeping from the regulator "weep hole" or from the feeder side of a combustion valve is usually an indication that there is a rupture or leak in the diaphragm. Follow the procedures described in the preceding paragraph for repairing the valve. An occasional flushing of the relief side of a combination valve will reduce the possibility of the type of lime or scale buildup that can cause the valve to fuse shut.

Balancing Valve

Balancing valves are used in a hot-water (hydronic) heating system to equalize the pressure drop in multiple piping circuits. The valves are installed on the return side of each circuit. Balancing valves are illustrated and described in Chapter 10, "Steam and Hot-Water Line Controls" in Volume 2.

Backflow Preventer

Sometimes boiler back-siphonage and back pressure will cause the boiler water to mix with and contaminate the domestic water supply. This unwanted mixing of the two water supplies can be prevented by installing a backflow preventer. The Bell & Gossett backflow preventer illustrated in Figures 15-69 and 15-70 consists of two independently operated check valves and an intermediate atmospheric air vent contained in the same housing.

When a back-siphonage condition occurs, the atmospheric vent opens to allow air to enter and break the siphon. Leakage is vented

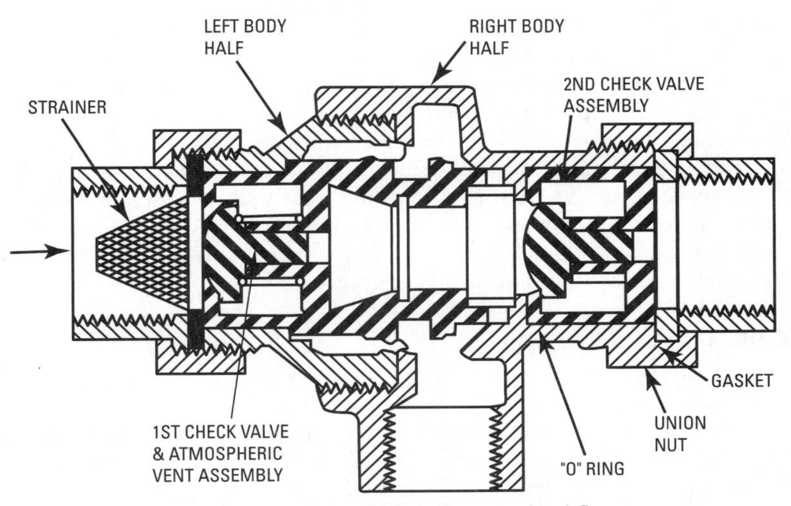

Figure 15-69 Schematic of the Bell & Gossett backflow preventer. *(Courtesy ITT Bell & Gossett)*

through the air vent if back-pressure occurs with a fouled second check valve.

Altitude Gauges

An altitude gauge mounted on the boiler is used to indicate the water level in *open* hot-water heating systems. The level (altitude) of the water is indicated on the gauge by the relative positions of two

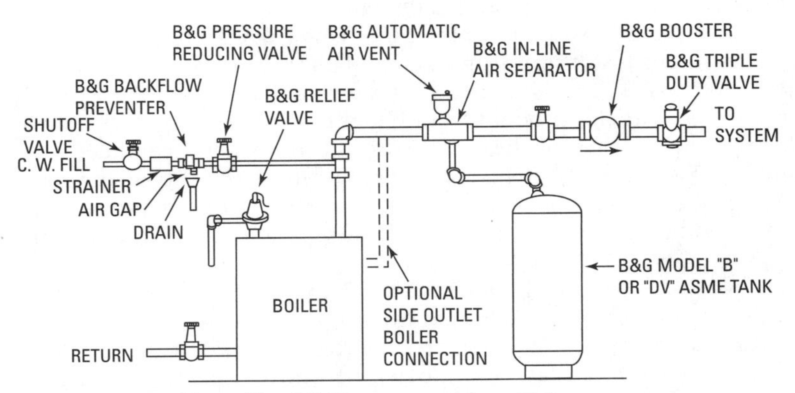

Figure 15-70 Typical backflow preventer installation. *(Courtesy ITT Bell & Gossett)*

hands or arrows. One hand (usually black) is stationary and is permanently set when the boiler is filled. The movable hand (usually red or white) is initially set in the same position as the stationary hand after the boiler has been filled with water. Its position will change as the water level in the boiler changes. This movable hand indicates the true level of the water. Efficient operation is being provided as long as the movable hand is directly above the stationary one. In *closed* hot-water heating systems, automatic valves are used to control boiler water level (see *Pressure Relief Valves* and *Pressure-Reducing Valves* in this chapter).

Circulator (Pump)

The circulator (circulating pump) in a hot-water space heating system operates in a sealed piping circuit (loop) and is always filled with water. Recommendations for their service, maintenance, troubleshooting, and repair are contained in Chapter 10, "Steam and Hot-Water Line Controls" in Volume 2.

Air Separator

When a hot-water heating system is filled with cold water, the water contains some air dissolved in solution. The air emerges from solution when the water is heated and moves rapidly through the pipes and radiators making noises in the pipes and radiators. In some cases, air pockets become trapped in the farthest radiators, which prevents them from heating. An *air separator* is used to trap, separate, and remove this trapped air from the water before it enters the system.

Some systems use inline separators installed in the piping close to the boiler. The inline separator shown in Figure 15-71 consists of two chambers slightly wider than the pipe. Water containing trapped air enters the separator and slows down slightly as it expands into the first chamber. The slowing down of the water causes the air trapped in it to separate, form bubbles, and float to the top of the first chamber. The water, free of the air, passes into the second chamber and then into the system piping. The air tapped at the top of the first chamber is vented out of the system through an automatic air vent (installed in the tapping on top of the air separator) or passed into an expansion tank.

The air separator shown in Figure 15-72 is screwed into the boiler supply tapping. The separator traps the air in the top of the boiler section, where the water is hottest and where it travels a long horizontal path at low velocity, permitting the air bubbles to separate. The trapped air escapes through a ¾-in tapping into an

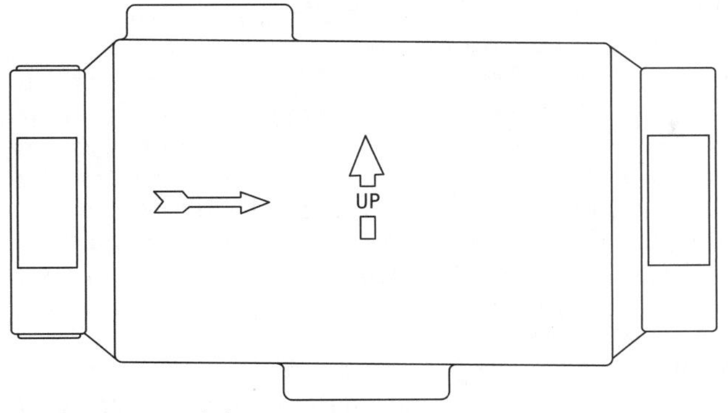

Figure 15-71 Inline air separator. *(Courtesy ITT Bell & Gossett)*

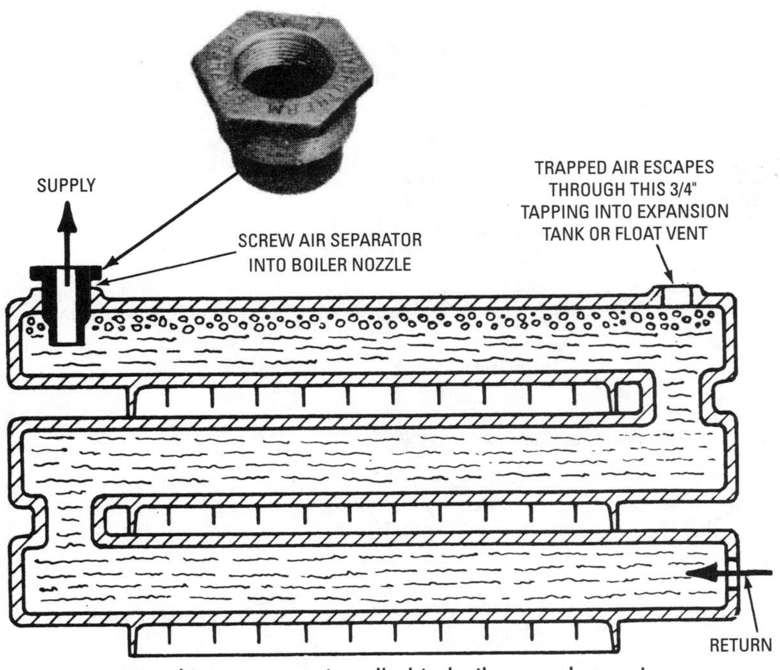

Figure 15-72 Air separator installed in boiler supply tapping.
(Courtesy Hydrotherm, Inc.)

air-cushion expansion tank or through an automatic float vent on systems using diaphragm expansion tanks.

An air separator can only expel air that reaches the boiler. However, sometimes air pockets remain trapped in the piping or radiation, thereby impeding the water flow and reducing the heating performance of the boiler. Such air pockets can be eliminated by installing several different purge fittings in the heating system. A manual air vent is installed on the highest point of the system (automatic air vents are not generally required); see the following section.

Some boilers are constructed with built-in air separators, which make the addition of an external unit unnecessary. Figure 15-73 illustrates the working principles of one of these integral units. In this design, the air is diverted to an automatic air vent. A variation of this design provides for the diversion of air to an expansion tank. The manufacturer refers to this device as an *air eliminator*. Its

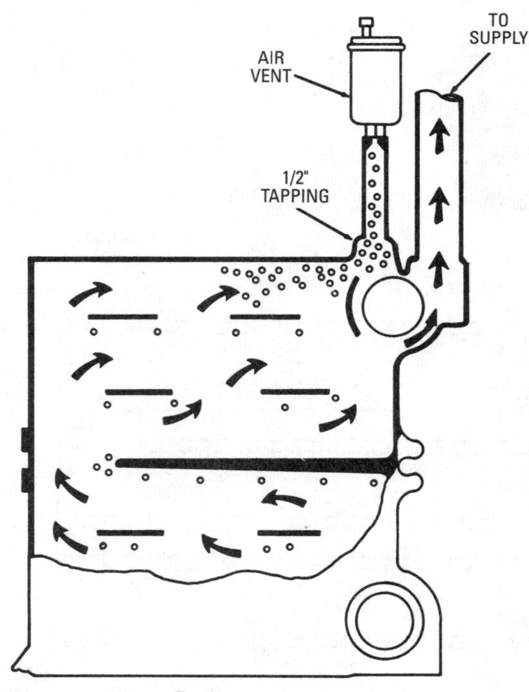

Figure 15-73 Built-in air separator.
(Courtesy Weil-McLain)

function is similar to the air eliminator used on pipelines, but it differs in design and construction.

Purging Air from the System

Built-in or externally installed air separators are designed to separate air from the water in an *operating* boiler and vent it either into an expansion tank or to the atmosphere through an automatic air vent.

Air can also be removed *from a heating system* by purging it from the piping connections to the boiler. This type of purging is standard when the system is being filled with water during setup. Purging instructions are usually included in the owner's manual or the manufacturer's installation guide.

Expansion Tanks

A hot-water heating system is completely filled with water. When the water is heated, it expands in volume by as much as 5 percent. The function of the expansion tank in a hot-water heating system is to provide a space to accommodate the increased volume of water. There are two types of expansion tanks: steel tanks and diaphragm tanks (Figure 15-74). Both are described in considerable detail in Chapter 10, "Steam and Hot-Water Line Controls" in Volume 2.

Air Supply and Venting

An adequate air supply must be provided for combustion to boilers that are fired by gas, oil, or coal, and the products of combustion must be vented to the outside atmosphere.

Combustion air is normally supplied through venting ducts or openings in the walls, if the appliance relies on natural ventilation for its air. The requirements for the air supply will depend upon the location and enclosure of the boiler.

Boilers installed in unconfined areas usually obtain adequate air for combustion by means of normal infiltration. However, normal air infiltration will be inadequate for this purpose if the construction of the structure is unusually tight. Under these circumstances, provision must be made for the entry of additional air from outside the building. Unobstructed openings with a total free area of not less than 1 in^2 per 5000 Btu/h of the total input of the boiler are necessary to provide an adequate air supply for combustion purposes.

Some boilers are installed in boiler rooms or enclosures supplied with combustion air from *inside* the structure. The air supply must enter and leave the boiler room through two openings in an *interior*

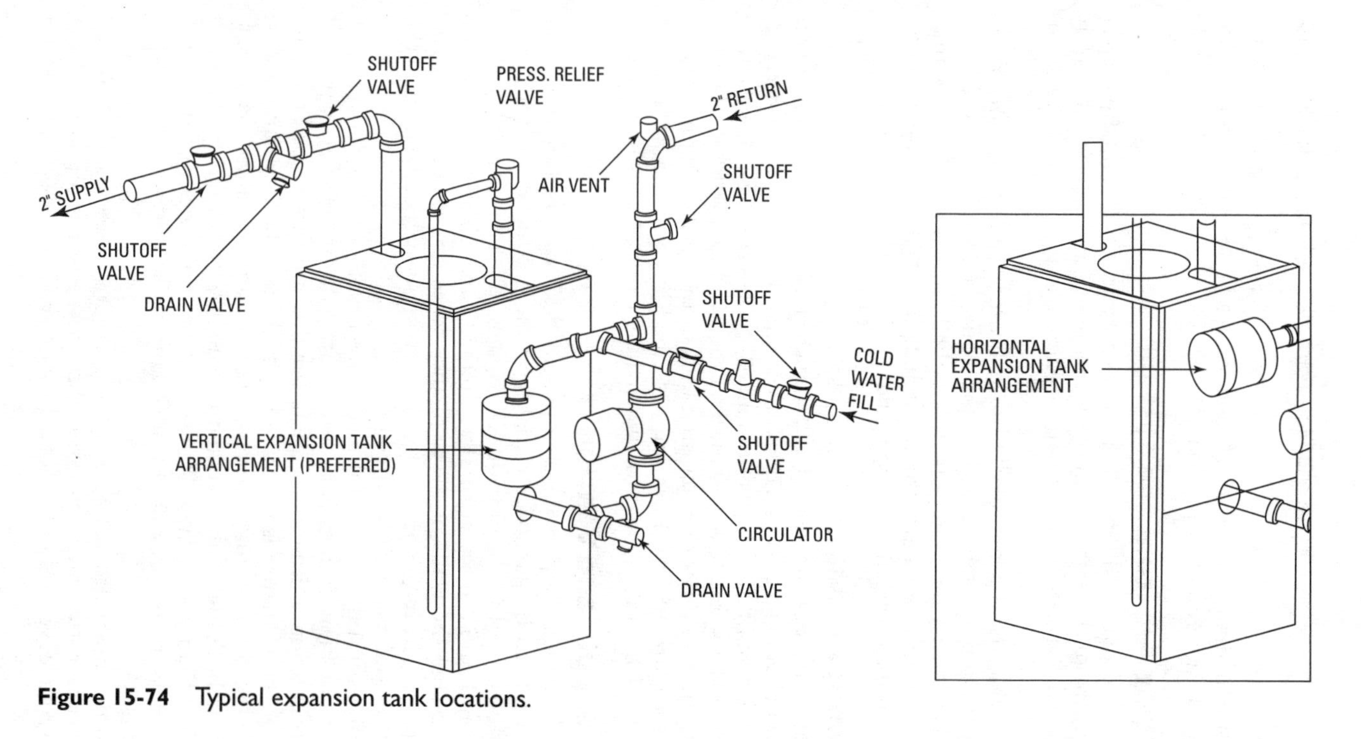

Figure 15-74 Typical expansion tank locations.

wall or door. One opening is located near the ceiling, the other opening near the floor. Each opening should have a free area of not less than 1 in^2 per 1000 Btu/h of the total input of the boiler.

If a boiler is located in a boiler room or similar enclosure that receives its air supply from *outside* the building, the two air openings are located on an *exterior wall*. Each opening must have a free area of not less than 1 in^2 per 4000 Btu/h of the total input of the boiler.

The products of combustion must be vented to the outside atmosphere. In order to accomplish this, the boiler must be connected to a suitable venting system, which should include a flue pipe and a chimney or stack of adequate size and capacity.

The chimney height is usually governed by the height of the structure. As a general rule-of-thumb, the chimney should extend beyond the high-pressure area caused by the passage of 10 ft of the flashing. If the chimney is too short, it will not extend beyond the high-pressure area caused by the passage of winds over the structure. As a result, a downdraft caused by the pressure difference in the high-pressure area will push air down into the chimney, which will block the escape of flue gases and interfere with the combustion process. Figures 15-75 and 15-76 illustrate these principles. The *correct* chimney height is shown by the dotted lines.

Figure 15-77 shows a type of chimney construction referred to as a *Type A vent*. This chimney can be either masonry or factory-built construction, but it must be designed and constructed in accordance with the standards set forth in national codes. Type A vents are suitable for venting of all gas appliances. These are the *only*

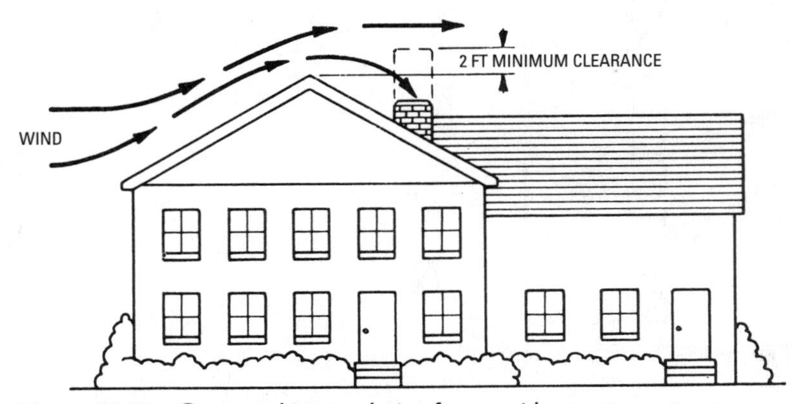

Figure 15-75 Correct chimney design for a residence. *(Courtesy Hydrotherm, Inc.)*

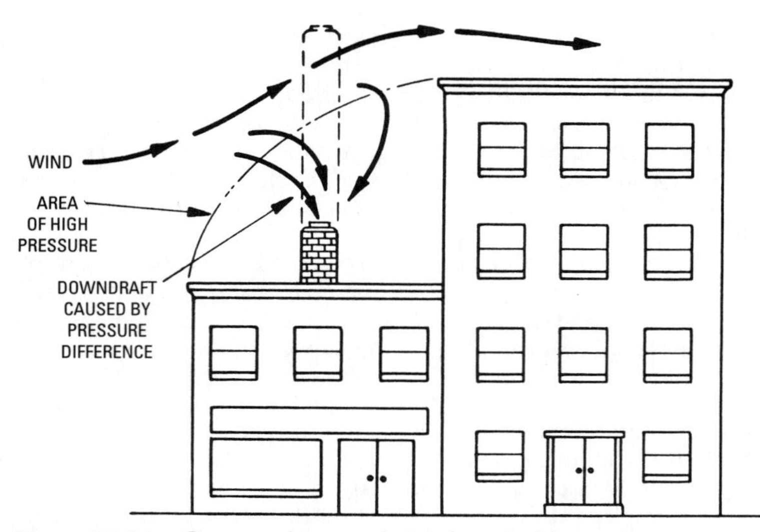

Figure 15-76 Correct chimney design for a building. *(Courtesy Hydrotherm, Inc.)*

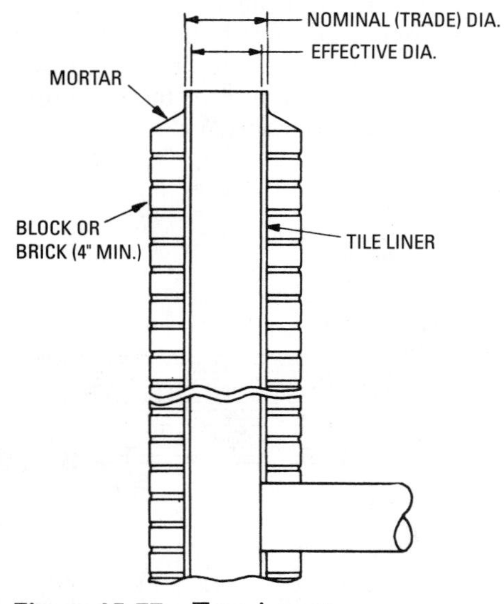

Figure 15-77 Type A vent.
(Courtesy Hydrotherm, Inc.)

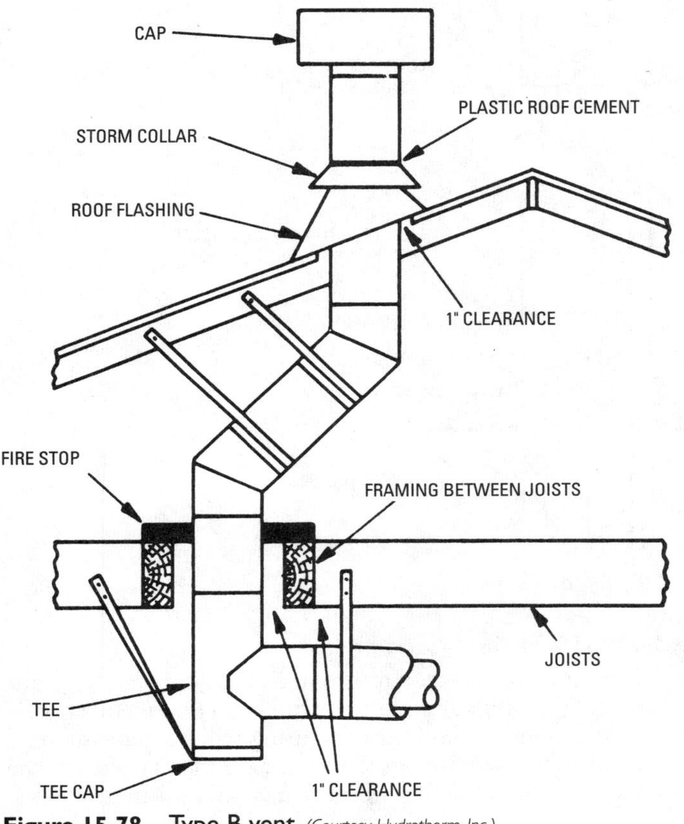

Figure 15-78 Type B vent. *(Courtesy Hydrotherm, Inc.)*

vents suitable for venting oil-burning equipment. Gas-fired boilers that produce temperatures at the draft diverter not in excess of 550°F may also use a *Type B vent* (Figure 15-78), which employs a double-wall metal vent pipe. A *Type C vent* (Figure 15-79) is used to vent gas appliances in attic installations.

The boiler should be located so that the length of flue pipe connecting the boiler (or draft diverter) to the chimney is as short as possible. Additional information about chimneys and flues is contained in Chapter 11, "Gas Furnaces."

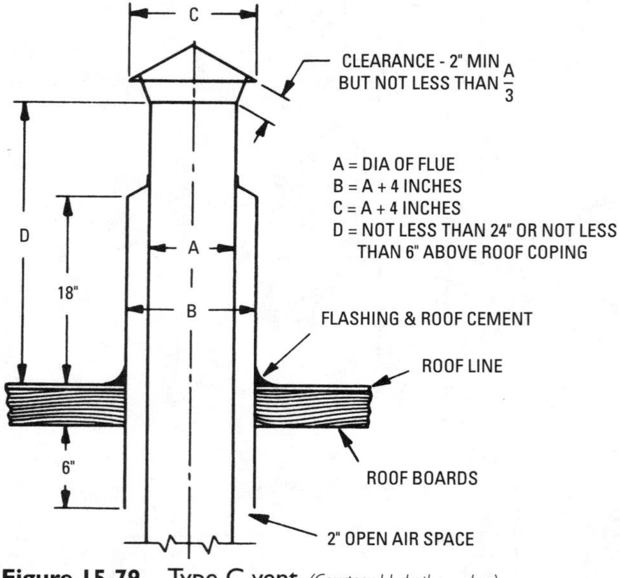

Figure 15-79 Type C vent. *(Courtesy Hydrotherm, Inc.)*

Induced-Draft Fans

Sometimes a chimney is too small to provide enough updraft to convey the products of combustion from the boiler to the outside atmosphere. When it is evident that natural venting will be inadequate, a mechanical draft inducer can be used to increase the capacity of the chimney. These devices are used with both gas- and oil-fired boilers.

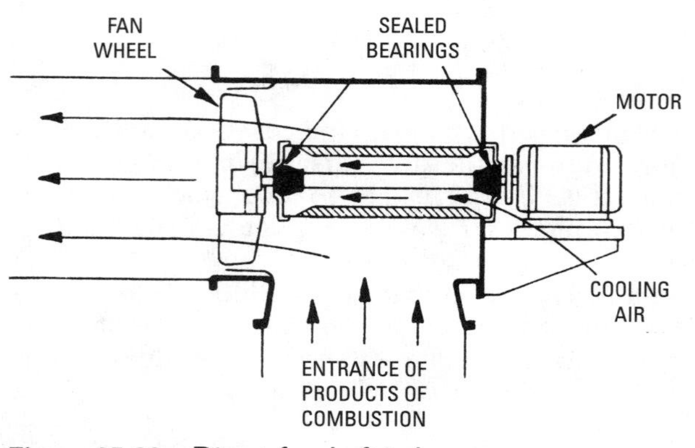

Figure 15-80 Direct-fan draft inducer. *(Courtesy Hydrotherm, Inc.)*

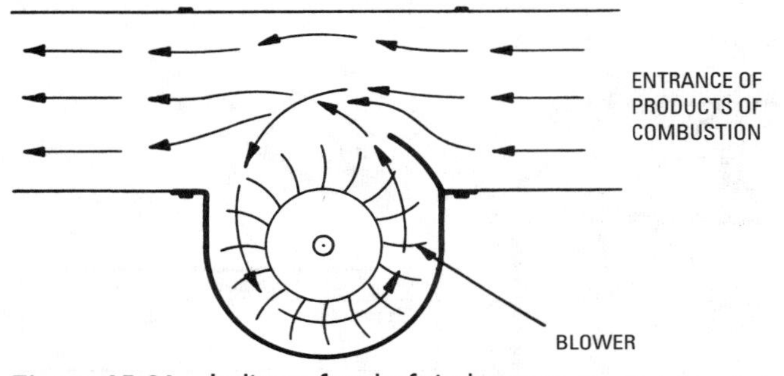

Figure 15-81 Indirect-fan draft inducer. *(Courtesy Hydrotherm, Inc.)*

The two principal types of induced-draft fans are (1) direct-fan draft inducer and (2) induced-flow draft inducer. When a direct-fan draft inducer is used, the gases are drawn through an inlet located on the side, top, or bottom of the unit and discharged as shown in Figure 15-80. The moving parts of an indirect fan draft inducer (Figure 15-81) are located outside the path of the hot flue gases. Cool outside air is introduced through the blower section of the unit. This creates suction at the throat of the inducer body, which, in turn, increases the flow rate of the flue gases.

The draft created by the induced-draft fan should closely match the demand. If the size of the draft inducer is correctly estimated, any dilution and excessive cooling of the flue gases (a condition that may result in condensation) will be greatly reduced. Many boiler manufacturers recommend the size draft inducer to be used with specific boiler models and provide data for making the appropriate calculations. Manufacturers of draft inducers are similarly helpful.

Controlling Excessive Draft (Gas-Fired Boilers)

Sometimes a boiler installation will have excessive draft conditions as a result of oversized chimneys or other factors. Draft conditions in excess of the draft design limits of a gas-fired boiler will reduce its combustion efficiency, resulting in higher fuel costs and possible pilot problems.

A suggested method of controlling the draft on such installations is to install a sheet-metal baffle, or restrictor, in the flue connection at the boiler (Figure 15-82). This baffle may be inserted in a tee, as shown, and secured with stove bolts after the proper draft is obtained as determined by measurements with a draft gauge. If the barometric control is installed in the side of the breeching by means of a draft control collar, the same method of baffling may be used

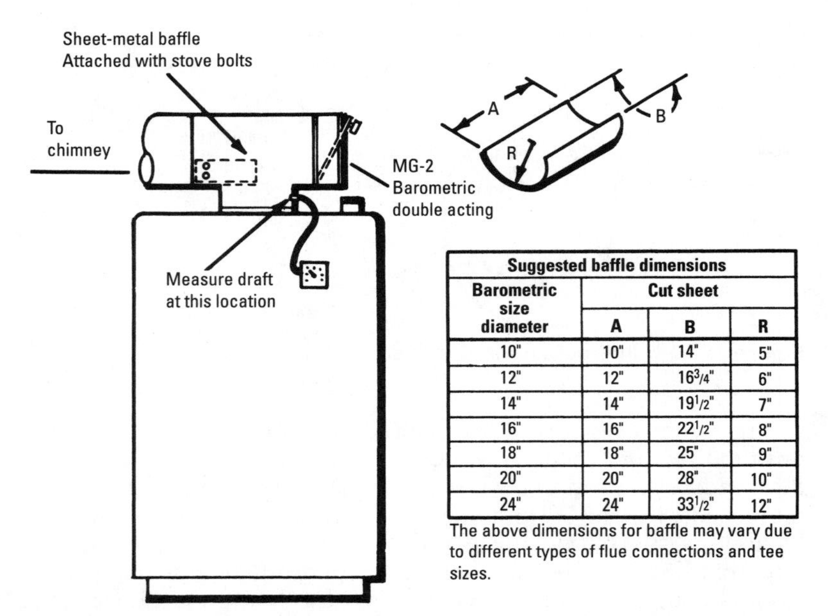

Suggested baffle dimensions			
Barometric size diameter	Cut sheet		
	A	B	R
10"	10"	14"	5"
12"	12"	16¾"	6"
14"	14"	19½"	7"
16"	16"	22½"	8"
18"	18"	25"	9"
20"	20"	28"	10"
24"	24"	33½"	12"

The above dimensions for baffle may vary due to different types of flue connections and tee sizes.

Figure 15-82 Control of excessive draft of gas-fired boilers.
(Courtesy Bryan Steam Corp.)

with the baffle restricting the breeching between the boiler and the barometric control. In this method of baffling, the breeching between the barometric control and breeching should not be restricted.

The baffle installation is correct if the gate of the barometric control is approximately half open while the gas burner is operating and the draft measurement is at or below the operating design of the boiler. The draft limit for a boiler can be obtained from the manufacturer's specifications.

Always check with the local gas company regarding the use of baffles. Caution should be exercised to ensure against overrestricting the flue. Always be certain that there is a very slight draft after the baffle is secured in position.

Tankless Water Heaters

Some boilers are manufactured with the option of using a *tankless water heater.* This device consists of an immersion coil inserted in a steam or hot-water space heating boiler to provide domestic and commercial service hot water. It is called a tankless water heater because no storage tank is used to store the heated water during periods of low demand.

The immersion coil is made of small-diameter copper tubes that are either straight with U-bends at the end or formed in the shape of a spiral (see Figure 15-10). The tankless heater is installed in the nipple port on cast-iron hot-water boilers. On steam boilers, it is installed in the left-hand side of a special back section at a point well below the water line. A uniform water temperature is maintained by a thermostatic three-way mixing valve installed in the supply line leading from the immersion coil.

Because these immersion coils function as a heat exchanger-type device and operate on an indirect heating principle, they are also variously referred to as an *indirect water heater, indirect heat exchanger,* or *heat exchanger coil.* In other applications, immersion coils provide hot-water radiant heat from a steam boiler, hot water for snow melting, heated water for pools, or industrial process water.

Additional information about water heaters, including suggestions for estimating the hot-water allowance, can be found in Chapter 4, "Water Heaters" of Volume 3.

Leaking Coils

On occasion, the coil of a tankless or indirect water heater installed in a boiler will leak, allowing high-pressure water into the boiler water and resulting in a rise in the boiler pressure. This, in turn, can cause the relief valve to open and fail to close again tightly, or it can cause the boiler pressure to rise above the setting of the pressure-reducing valve. These symptoms are often wrongly attributed to a valve malfunction when the actual cause may be a leaking coil.

You can determine whether the problem is a leaking coil by taking the following steps:

1. Shut off the feed valve and/or break the connection to the coil.
2. Check the pressure reading in the boiler.
3. Wait about 8 hours, and check the boiler pressure reading again.

If the pressure reading has remained approximately the same over the 8-hour period, it is a strong indication that the coil is leaking. If the boiler pressure continues to rise, the problem may lie elsewhere in the system.

Blowing Down a Boiler

Foam sometimes forms on the surface of the boiler water and is usually indicated by drops of water appearing with the steam. This condition is caused by the presence of oil, dissolved salts, or similar organic matter in the water. One method of eliminating this problem is by draining off part of the water in the boiler and adding an equal amount of fresh, clean water. Another method consists of blowing the foam from the water with a specially connected pipe or hose. The boiler should have a blowdown tapping for this purpose. The first method is the easiest. The second one (i.e., blowing down the boiler) requires considerable experience.

Boiler Operation, Service, and Maintenance

Always follow the manufacturer's instructions for boiler operation, service, and maintenance. These instructions are commonly contained in the owner's manual left with the boiler by the installer. If you do not have these instructions, you should contact the local dealer for advice or write to the manufacturer.

The sections that follow contain recommendations for operating and maintaining boilers. Because many of these recommendations specifically pertain to either steam or hot-water space heating boilers, they were divided and listed accordingly.

Note

All service and maintenance work must be performed by those trained in the proper application, installation, and maintenance of plumbing, steam, and electrical equipment and/or systems in accordance with all applicable codes and ordinances. If available, the boiler manufacturer's service and maintenance instructions should be followed

Safe and Reliable Boiler Performance

Poor boiler performance and unsafe operating conditions are caused by failure to follow proper maintenance procedures, the existence of low water conditions, the use of outdated or damaged safety devices, or operator error. The following recommendations will go a long way toward providing trouble-free boiler operation:

- Always follow the ASME code maintenance standards.
- Maintain an inspection and maintenance log for the boiler. The log should include the date and specific form of maintenance performed.
- Always follow the boiler manufacturer's operating instructions for initial setup and startup.
- Never use outdated controls. Many will have a date code stamped on them. Replace the control if its age exceeds the stamped date.

- Always operate the boiler with code safety relief valves.
- Always use safety relief valves rated for the boiler.
- Never bypass the existing boiler controls.

Caution

To prevent serious burns, always allow the boiler sufficient time for its temperature to cool down to at least 80°F (27°C) and for its pressure to drop to 0 psi (0 bar) before servicing. Drain the boiler to a level below that of the lower gauge glass tapping on the low-water cutoff control.

Caution

To prevent electrical shock, always shut off the electrical power to the boiler before making any electrical connections.

Steam Boilers

Operating and maintaining a steam boiler differs in certain respects from the procedures followed for hot-water space heating boilers. Many of these differences are evident in the recommendations that follow.

1. Keep a steam boiler filled to the water level recommended by the manufacturer when not in use.
2. Check the water level in the boiler before starting it. The heating surface can be damaged if the water level is too low.
3. Keep the water level at the center of the water gauge glass during operation. If the water level is too low, use the manual feed valves to add more water. These valves are found in most systems.
4. Always add water to the boiler gradually. If at all possible, avoid adding water to a hot boiler. *Never* add water to an operating cast-iron boiler. Cold water fed rapidly into such boilers may come in contact with the hot surface of the cast-iron heat exchanger, causing it to crack. Shut off the boiler and wait until it has cooled down before adding water. After the boiler has had sufficient time to cool, slowly add water through the cold-water feed line.
5. Check the low-water cutoff at regular intervals. Sediment or rust accumulating under the float of the low-water cutoff can cause it to malfunction. As a result, a drop in the boiler water level would not register properly.

6. Check all boiler accessories to make certain they are functioning properly. Movable parts should be inspected and oiled regularly.
7. Never allow a low-pressure steam boiler to exceed the upper safe pressure limits recommended by the manufacturer (usually 3- to 4-lb of pressure).

Hot-Water (Hydronic) Boilers

Recommendations for operating and maintaining hot-water (hydronic) boilers are as follows:

1. Keep the boiler *and the pipes* in the heating system filled with water when not in use. Keeping the pipes filled with water reduces the possibility of rust and corrosion.
2. Check the water level in the boiler before starting it. The heating surface can be damaged if the water level is too low.
3. Always add water to a boiler gradually. Never add water to a hot boiler. Shut the boiler down and allow it to cool first.
4. Check all boiler accessories to make certain they are functioning properly. Movable parts should be inspected and oiled regularly. Such maintenance should also include the pump in a forced-hot-water heating system.
5. When operating a hot-water space heating boiler, make certain all flow valves are open.
6. Never allow a boiler to exceed the upper safe temperature limit recommended by the manufacturer (usually about 200°F).

Boiler Water

Boiler water should be clean and kept clean for efficient operation. This is true for all boilers. *Never* add dirty or rusty water to a boiler. Even hard water may eventually interfere with the efficient operation of a boiler and should therefore be chemically treated before being added.

The Steel Boiler Institute adopted a chemical conditioning compound for treating water used in low-pressure steam and hot-water space heating boilers. Many manufacturers provide this chemical compound with their boilers. *Always* treat the water immediately after the boiler and the heating system have been cleaned.

Note

Boiler water contains suspended solids that are held in suspension during boiler operation by the circulating water and the action of treatment chemicals. Unless care is taken when draining the boiler to remove these suspended solids along with the water, they remain in the boiler, dry and stick to the heating surfaces, and require chemical cleaning to remove.

A system compatible antifreeze, such as propylene or ethylene glycol, can be used in the water of a hot-water (hydronic) boiler, but *only* when absolutely necessary. Additional information about the use of antifreeze in boilers is contained in The Hydronics Institute's *Technical Topics Number 2A* publication.

Some boiler heat exchangers are made of aluminum instead of steel. Only antifreeze solutions certified for use in aluminum boilers should be used.

Warning

Never use an RV-type antifreeze protection solution or an automotive-type antifreeze. Both types can damage the boiler and other system components.

Cleaning Boilers

A boiler should be inspected *at least once* every year for the accumulation of soot and other deposits that could impair its operation. This inspection should take place *before* the start of the heating season.

The accumulation of soot will result in improper combustion. The soot can be removed with a chemical cleaner or a flue brush. Many boiler manufacturers provide access to the heating surfaces through a removable top jacket panel and cover plates. The flue brush is used to push the soot down between the sections or fins to collect in the combustion chamber below. It can then be removed from the combustion chamber without much difficulty.

Oil burners, gas burners, and coal stokers should also be inspected for dust accumulations and cleaned. Always follow the manufacturer's recommendations for cleaning the boiler and automatic firing equipment. If you do not have an owner's manual, contact a field representative or write to the manufacturer.

Troubleshooting Boilers

Boilers are subject to numerous problems, many of which are due to improper type and design or poor servicing. The following are some of the problems usually encountered:

1. *Boiler does not deliver enough heat.* This is a very common complaint, but the boiler may not be causing the problem. The problem may be caused by a problem with the burner or the automatic controls. Even a pipeline with improper pitch will

cause the heat from the boiler to be blocked or trapped. If the boiler is found to be the cause of insufficient heat, the problem can be traced to any one (or more) of the following causes:

- **a.** Boiler too small for the heating system
- **b.** Improper arrangement of boiler sections in cast-iron boilers
- **c.** Poor draft
- **d.** Poor fuel
- **e.** Heating surfaces covered with soot

2. *Boiler delivers no heat.* The automatic controls should be checked first. Sometimes a low-water cutoff on steam boilers will shut off the burner or stoker before enough steam has formed. Another possible cause is that too much water is being fed into a steam boiler. As a result, not enough space is provided for the steam to form at the top of the boiler. Too little water in a hot-water space heating boiler can be caused by the limit control moving down to a lower setting.

3. *Too much time is required to get up steam in a steam boiler.* This common problem can be traced to the following possible causes:

- **a.** Too little or badly arranged heating surface
- **b.** Heating surfaces covered with soot
- **c.** Heating passages too short
- **d.** Poor fuel or fuel firing
- **e.** Poor draft
- **f.** Boiler too small
- **g.** Boiler defective

4. *Boiler is slow to respond to the operation of the dampers.* Slow response to damper operation can be caused by any of the following:

- **a.** Air leakage into the chimney or stack
- **b.** Poor fuel or fuel firing
- **c.** Boiler too small
- **d.** Clinkers on grate or ashpit full of ashes (coal-fired boilers)

5. *Water line is unsteady.* The problem of an unsteady water line may simply be due to connecting the water column to an extremely active section of the boiler. Therefore, it is extremely unlikely that the actual water level in the boiler can be read from the water column. Other possible causes of this problem are:

a. Dirt or grease in the water

b. Varying pressure differences on the system

c. Excessive boiler output

6. *Water disappears from the gauge glass.* The complete loss of water from the gauge glass could be due to priming (i.e., water globules being carried over into the steam). Other causes include:

a. Foaming

b. Pressure drop too great in return line

c. Improper water gauge connection

d. Valve closed in the return line

7. *Water is carried over into the steam main.* This problem is usually caused by one of the following:

a. Priming or foaming

b. Water line is too high

c. Outlet connections from boiler too small

d. Steam-liberating surface too small

e. Boiler output excessive

8. *Flues require cleaning too frequently.* A frequent buildup of soot and dirt in flues can be caused by any of the following:

a. Combustion rate too slow

b. Poor draft

c. Smoky combustion

d. Excess air in firebox

9. *Low carbon dioxide.* This condition can generally be traced to one of the following causes:

a. Air leakage between cast-iron sections

b. Improper conversion job

c. Problem with burners

10. *Smoke from boiler fire door.* The following conditions may be the cause of this problem:

a. Dirty or clogged flues

b. Incorrect setting of dampers

c. Poor or defective draft in the chimney

d. Incorrect reduction in the breeching size

Boiler Repairs

Repairs to the boiler itself should be done by an experienced and skilled worker. They should never be attempted while the boiler is under pressure. Always shut down the boiler first and allow it time to cool before beginning any repairs.

Installing Boilers

All new boilers are shipped with a complete set of installation instructions. Usually these instructions will also contain an inventory list of the contents in the crates and cartons. *Always* check the contents off against this list before you do anything else. Missing or damaged parts should be immediately reported to the manufacturer or its local representative.

A boiler should always be located so that the connecting flue pipe between the boiler (or draft diverter) and the chimney is as short as possible. Another important consideration is the recommended minimum clearances between the boiler and combustible materials. These factors plus the design considerations of the system will dictate where the boiler is located.

The manufacturer's installation guide will also contain instructions on how to light and operate the boiler. The instructions will differ in accordance with the automatic fuel-burning equipment used (e.g., oil burner and gas burner).

After completing the installation of the boiler, inspect all the controls to make certain they are operating properly. Start and stop the burner or stoker several times by moving the room thermostat setting.

All local codes and regulations take precedence over the installation instructions provided by the manufacturer. In the absence of local codes, the installation must conform with the boiler manufacturer's installation instructions plus regulations of the National Fire Protection Association, and the provisions in the latest editions of the *National Electrical Code* (ANSI/NFPA70) and the *National Gas Code* (ANSI Z223.1).

Chapter 16

Boiler and Furnace Conversion

Sometimes it becomes necessary to convert an existing boiler or furnace to another fuel. This is usually the case when the fuel for which it was originally designed has become too expensive relative to others or does not meet the desired level of efficiency. During the early 1950s, when coal was becoming more and more expensive as a domestic heating fuel and before oil and gas burners were offered as integral parts of heating units by manufacturers, converting boilers and furnaces was a much more widespread practice than it is today.

A number of manufacturers have made available gas-fired and oil-fired burners specifically designed for boiler and furnace conversions. It would be impossible within the space limitations of this chapter to cover all the different models of conversion burners offered by these manufacturers. Consequently, this chapter will emphasize all the *ramifications* involved in converting a boiler or furnace from one fuel to another instead of examining the specifications and operating characteristics of one or more conversion burners.

Initially, most coal burning furnaces and boilers were converted to oil or gas. This was done not only for convenience (the coal-burning appliances required more hands-on maintenance) but also for heating efficiency and lower energy costs. Tighter environmental regulations covering oil storage tanks (the tanks would often develop leaks as they became older) and the sudden rise in heating oil costs as the result of the 1970s oil crisis motivated a widespread conversion from oil to gas.

Before deciding to convert from one fuel to another, consult with a technician from a reputable heating dealer about the ramifications, practicality, and safety factors involved in converting. You may be able to lower energy costs by increasing the insulation levels in the home and by replacing the furnace or boiler with a high-efficiency appliance burning the same fuel. That is, better results might be obtainable by upgrading the system instead of converting.

If there is no alternative to converting from one fuel to another, then operating *safety* must be a primary consideration, followed by heating efficiency and energy costs. There are some potential safety hazards involved in converting from an oil-fired furnace or boiler to a gas-fired one. If the technician making the conversion is knowledgeable about these hazards, then proper precautionary measures can be taken to ensure safe operation. When using gas burners in a converted oil heating system, consider the following:

- Gas burners require greater amounts of excess combustion air than an oil burner does to ensure safe operation. Lower amounts can lead to the production of odorless, colorless, and deadly carbon monoxide gas. ALWAYS install a carbon monoxide gas detector near a gas combustion burner.
- Gas burners produce much more water vapor than oil burners. These water vapors will condense on the walls of the chimney and flue as they cool. This condensed water is acidic and can damage the chimney mortar, flue surfaces, and even the heating appliance itself. Consequently, some state and local codes have mandated the upgrading or replacement of the chimney and flue along with the installation of a gas-conversion burner. Upgrading the chimney involves lining it with an acid-resistant material. It is often necessary to install a new vent system meeting the requirements of gas-burner combustion.
- The gas burner flame pattern must match the size and shape of the combustion chamber, which was originally designed for an oil burner flame. The two kinds of flame have different sizes and characteristics. If this is not taken into consideration when the gas conversion burner is installed, the gas flame may contact the internal surfaces of the combustion chamber, resulting in the production of carbon monoxide gas.

Note

ALWAYS measure the carbon monoxide emission of a gas conversion burner immediately after it is installed. According to the ANSI Standard, CO emissions must be below 0.03 of the exhaust gases.

Preparation for Conversion

Before constructing a combustion chamber and installing a burner, the heating system should be carefully checked for defects and cleanliness. A boiler or furnace in need of repairs will not give satisfactory results after the burner is installed. Be certain that *all* flue passages are cleaned so that the maximum amount of heat generated is absorbed by the boiler. Soot and ash are good insulators, and both are always undesirable.

All doors should fit tightly, and all other openings or cracks should be tightly cemented shut. The stack from the furnace to the chimney should have tight joints. Dampers should not close off more that 80 percent of the cross-sectional area of the stack. Stacks should be inspected for leaks and obstructions of any kind.

Remove the old fuel-handling parts (e.g., the coal-handling parts in a coal-fired furnace or boiler) and thoroughly clean the interior surfaces, removing all soot, scale, tars, and dirt.

Any cracks in the heat exchanger should be repaired or replaced. On boilers, any water leaks should be sealed, broken gauges replaced, and loose boiler doors repaired or replaced. Air leaks along the floor should be grouted.

No positive catches should be used on firing doors. File off any catches so that the firing door will open easily to relieve pressure. A spring-loaded door holder is recommended.

If a conversion burner with a direct spark ignition system is installed, no provision need be made for top venting or horizontal or downdraft furnace. Check the local codes for further information.

Figure 16-1 illustrates some of the typical problems encountered when converting a coal-fired *boiler* to oil. It is highly doubtful that

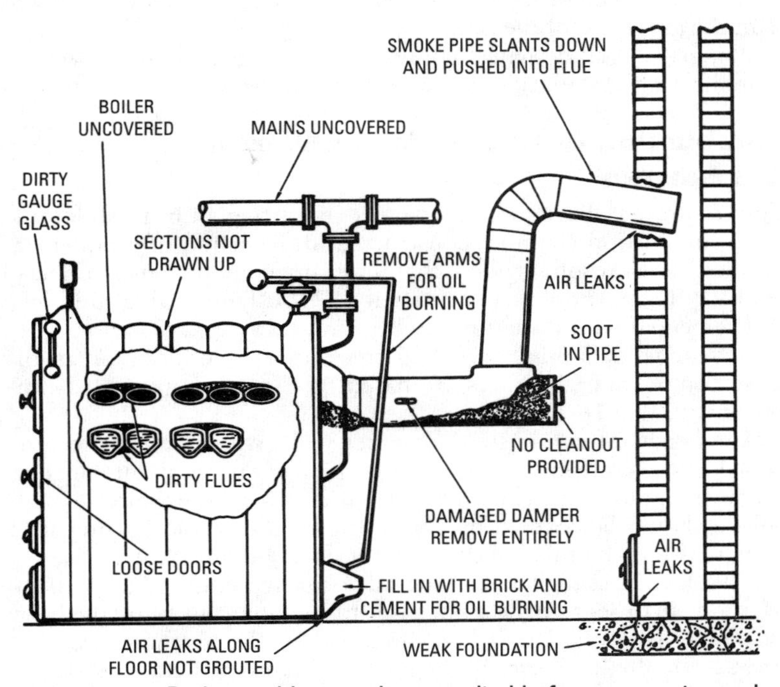

Figure 16-1 Boiler problems to be remedied before converting and installing an oil burner.

a boiler would have all these problems at the same time, but for your convenience they are included in the illustration to serve as a reference.

Basic Combustion Chamber Requirements

An essential requirement for combustion chambers is that the flame be in the presence of refractory materials so that it will not come into contact with any of the relatively cold heating surfaces of the boiler.

When the flame is burned in suspension in a combustion chamber of refractory material, the refractory walls will reflect the heat back into the flame and thereby increase its temperature. This increase in flame temperature due to reflected heat greatly increases the rapidity of combustion and thereby makes possible the burning of every particle of fuel with the zone combustion.

The combustion chamber walls should be high enough so that the flame will not come in contact with the relatively cool walls of the boiler. At the burner end, the refractory wall need not be higher than the grate line of the boiler.

Finally, a well-designed combustion chamber should provide rapid heating at the beginning of each firing cycle.

Combustion Chambers for Conversion Gas Burners

Conversion gas burners are designed for firing into a refractory lined combustion chamber constructed in the ashpit of a boiler or furnace. It is recommended that built-up combustion chambers used for this purpose be constructed of 2300°F insulating firebrick and cemented with insulating firebrick mortar.

Recommended *minimum* wall thicknesses and floor areas of these combustion chambers are based on the maximum rated Btu capacity (Table 16-1).

The height of combustion chamber walls will usually be determined by the grate line. Build the side and front walls about 2 inches above the grate line. The grate lugs and base of the water legs of boilers should be covered by about 3 or 4 in to avoid heating sections that may be filled with sediment. Carry the back wall one or two brick courses higher than the side or front walls and allow it to overhang in order to deflect hot gases from impingement on the heat exchanger surfaces. Use a *hard* firebrick for the overhang section, because the high-velocity gases moving against it tend to erode a softer brick.

Table 16-1 Recommended Minimum Wall Thicknesses of Combustion Chambers Using Conversion Gas Burners

Input Btu/h	*Floor Area Sq. In*	*Preferred Width and Length*	*Recommended Minimum Wall Thickness*	*Recommended Minimum Floor Construction*
100,000	180	12 × 15	2½ insulating firebrick plus backup of 1½ or more loose insulation	2½ insulating firebrick plus ½ insulating material recommended by firebrick manufacturer.
150,000	200	12 × 16		
200,000	220	13 × 17		
250,000	235	13 × 18		
300,000	260	13 × 20		
350,000	270	14 × 21		
400,000	330	15 × 22		
500,000	400	15 × 27		
600,000	460	15 × 31		
700,000	520	15 × 35		
800,000	600	18 × 33	4½ insulating firebrick plus backup of 1½ or more loose insulation	4½ insulating firebrick plus ½ insulating material recommended by the firebrick manufacturer.
900,000	650	18 × 36		
1,000,000	700	22 × 32		
1,100,000	750	22 × 34		
1,200,000	800	22 × 36		
1,600,000	1152	26 × 44		
2,000,000	1440	29 × 50		
2,500,000	1760	32 × 55		
3,000,000	2100	35 × 60		
3,500,000	2470	38 × 65		
4,000,000	3120	40 × 68		

Courtesy Magic Servant Products Co.

If the combustion chamber is to be placed directly on the floor of the ashpit, a layer of suitable insulating material or refractory brick recommended by the firebrick manufacturers (½-in minimum thickness) should underlay the combustion chamber. Ordinary brick may be used in the floor of the combustion chamber. Ordinary brick may be used in the floor of the ashpit. The remaining open space should be filled with vermiculite or some other suitable loose insulation.

Combustion Chambers for Conversion Oil Burners

A combustion chamber for a conversion oil burner should be constructed of lightweight 2000°F insulating firebrick. With lightweight refractory materials, there is no long smoky delay waiting for the firebox to reach the temperature necessary for complete combustion.

Different types of combustion chamber construction are shown in Figures 16-2 and 16-3. The shape of the combustion chamber will be determined by the shape of the boiler or furnace.

The combustion chamber area (space) is equal to the inside width of the boiler times the length times the distance from the combustion chamber floor to the crown sheet. It is recommended that 1 ft^3 be provided for each boiler horsepower, or 3 ft^3 per gph oil-fired.

The floor area of the combustion chamber should be based on 100 in^2 per gph of oil (see Table 16-2 and Figures 16-4 and 16-5). The minimum side wall height should conform to the recommended measurements listed in Table 16-2.

An overhang in the form of stepped corbels will deflect hot gases from impingement on the heat exchanger surfaces, increase combustion chamber temperatures, and thereby promote good combustion.

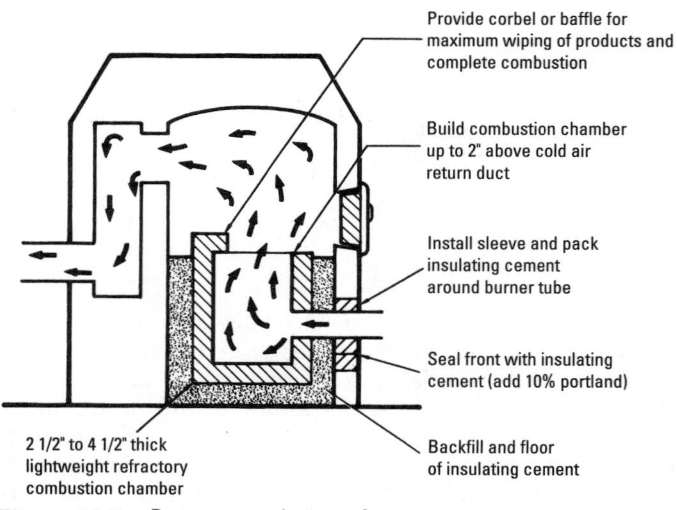

Figure 16-2 Recommendations for constructing a combustion chamber in a furnace. *(Courtesy Stewart-Warner.)*

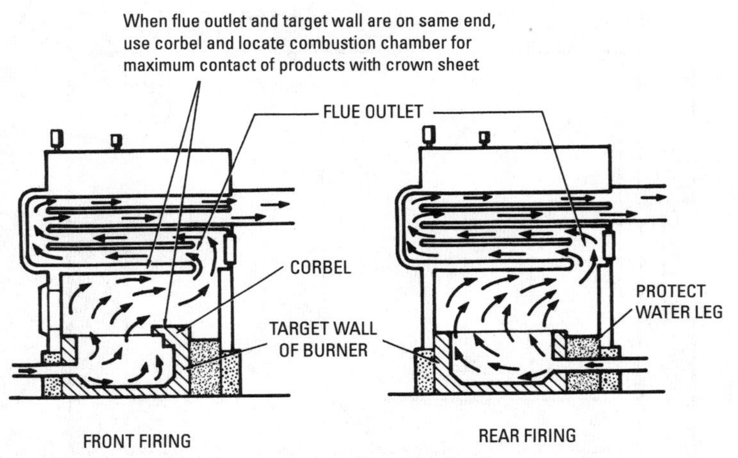

Figure 16-3 Combustion chambers for boilers. *(Courtesy Stewart-Warner.)*

High-flame-retention burners provide an intense flame that eliminates the need for refractory combustion chambers above 3.0 gph and has greatly reduced the need for such chambers in the 1.0- to 3.0-gph range.

Table 16-2 Combustion Chamber Floor Area Dimensions (100 in² per gph)

Gph	*Round I.D.*	*Rectangular W × L*	*Nozzle Height N*	*Min. Side Height H*
0.50	8"	7" × 8"	5"	12"
0.75	9½"	8" × 9"	5"	12"
0.85	10½"	9" × 9½"	5"	12"
1.00	11½"	9" × 11"	5½"	13"
1.10	12"	10" × 11"	5½"	13"
1.25	12"	10½" × 12"	6"	14"
1.35	13"	11" × 12"	6"	14"
1.50	14"	12" × 12½"	6½"	15"
1.75	15"	13" × 13½"	6½"	15"
2.00	16"	14" × 14"	6½"	16"
2.25	17"	14" × 16"	7"	16"
2.50	18"	14" × 18"	7"	17"

Courtesy Stewart-Warner

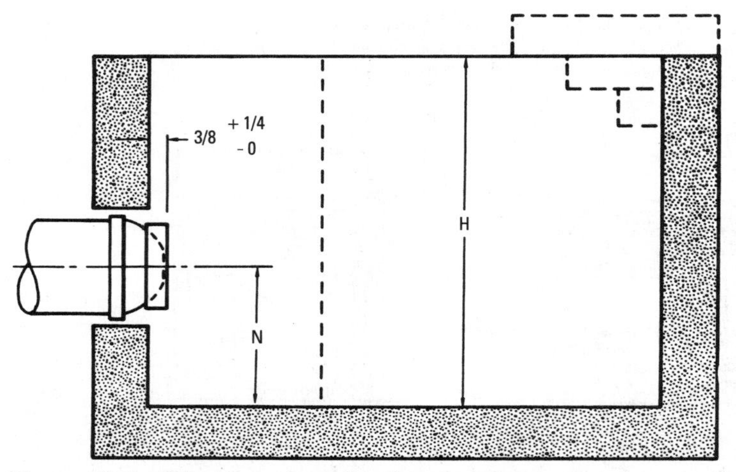

Figure 16-4 Side view of combustion chamber. *(Courtesy Stewart-Warner.)*

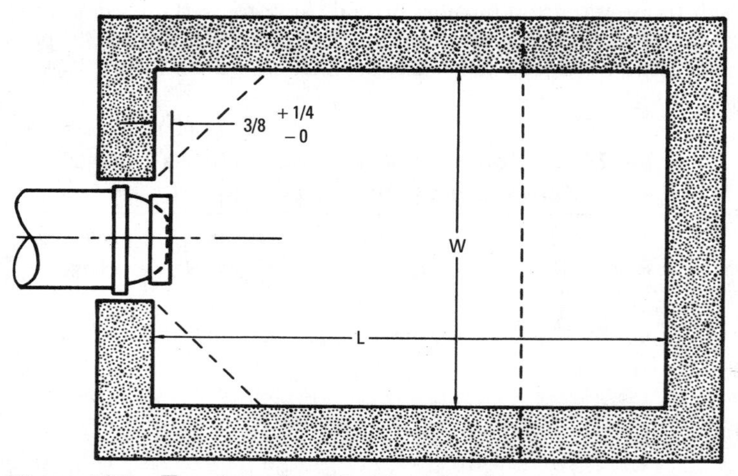

Figure 16-5 Top view of combustion chamber. *(Courtesy Stewart-Warner.)*

Construction Materials

The construction materials used in building combustion chambers for conversion burners should consist of the best grade of insulating firebrick and a good, high-temperature refractory mortar.

Insulating firebricks are available in different standard shapes and sizes for residential, commercial, and industrial applications. Their primary function is to provide a lining capable of withstanding the high temperatures in furnace and boiler combustion chambers, flues, chimneys, stacks, and fireplaces. To function in these high temperature environments, firebricks have upper-limit melting points that range from 2800°F (1540°C) for a brick made of fireclay to 4000°F (2200°C) for one made of silicon carbide.

Firebrick materials

The materials used in firebricks should make them capable of withstanding not only high temperatures, but also slag chemicals and spalling (i.e., flaking) under temperature changes. Firebricks are made from fire clay or kaolin, silica (silicon carbide), alumina, and magnesite (magnesia).

- **Clay firebricks.** Firebricks made from fire clay or kaolin are the most common type, but they are expensive and not good insulators. The principal advantage of the clay firebrick is that it creates a thin, lightweight combustion chamber wall. This results in a quick rise of temperature that produces fuel savings. On the down side, clay firebricks have a relatively low melting point of 1600°F (871°C).
- **Silica firebricks.** Silica firebricks retain their strength at high temperatures and do not react with ash, which makes them resistant to deposits. On the other hand, they are subject to spalling (flaking). They are also very expensive.
- **High alumina firebricks.** High alumina firebricks are capable of withstanding both high load deformation levels and high temperatures.
- **Magnesite (Magnesia) firebricks.** Magnesite firebricks resist the effects of alkalis.

Note:

Modern firebricks are asbestos free. If you have stocks of older firebricks containing asbestos do NOT use them, Contact a local EPA office for instructions about their safe disposal.

Firebricks can be machined or drilled at the factory in custom shapes and sizes to meet the requirements of specific applications. To mention only a few possibilities, custom firebricks are available with machined grooves, tapers, radii, tongue-and-groove edges, and drilled holes.

Insulating firebricks are produced in different grades to meet the requirements of specific applications. The temperature use limits are an important factor in making a selection. Another important factor to consider when selecting a suitable firebrick is its mean temperature; that is to say, the temperature at the midpoint of the firebrick. These and other specifications for each grade of firebrick are provided by the refractory material manufacturer.

Always use the refractory mortar furnished or recommended by the brick manufacturer for cementing the insulating firebrick. The mortar should be thinned to the consistency of a very thick cream before it is used. The usual method for applying the mortar is to dip each brick into it and set the brick in place as you lay each course.

Note:

Additional support is possible by using metal anchors projecting from the metal casing.

The shape of the bricks is also important, with arch or circled brick where 4½-inch wall thickness is preferred. The mortar-filled joints between the bricks should be not more than ⅛ inch.

The insulating firebricks are usually backed by a second, separate insulating layer, which may consist of one of the following materials:

1. Common brick
2. Magnesia block insulation
3. Hard firebrick
4. Expanded mica products (e.g., vermiculite or zonolite)
5. Dry sand
6. Dry sand mixed with an expanded mica product

The spaces between the outer edges of the firebrick should be filled with high-temperature refractory mortar and small pieces of firebrick to obtain firm construction and prevent infiltration of vapors through the wall.

Building a Combustion Chamber

Figure 16-6 illustrates the construction of a typical custom-built combustion chamber. The steps involved in constructing this chamber are as follows:

1. Place split brick *A* on the ashpit floor to form the floor of the combustion chamber.
2. Mix the refractory binder in a shallow pan to the consistency of a heavy cream.

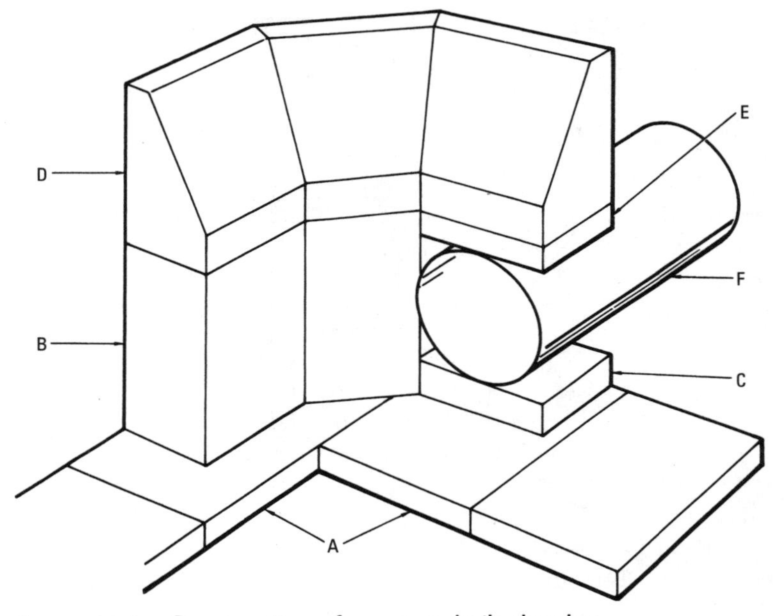

Figure 16-6 Construction of a custom-built chamber.

3. Dip the bottom and sides of brick *B* in the binder and place it in a circular position on the combustion chamber floor. You should start at the back and work toward the front of the chamber.
4. Follow the same procedure with the rest of the bricks in the tier (i.e., those on the same level as brick *B*).
5. When the tier is finished, tighten it in position with bands suitable for this purpose.
6. Fill behind the first tier with rock wool.
7. Install top tier *D* in the same manner described in steps 1 through 5; however, use only one band to tighten it.
8. Cap the chamber with a mixture of 3 parts refractory mortar and 1 part fire clay, and trowel smooth.
9. Place metal sleeve *F* in the opening, and cement it tightly in place.
10. Insert the draft tube in metal sleeve, and pack with loose rock wool or a suitable insulating material.

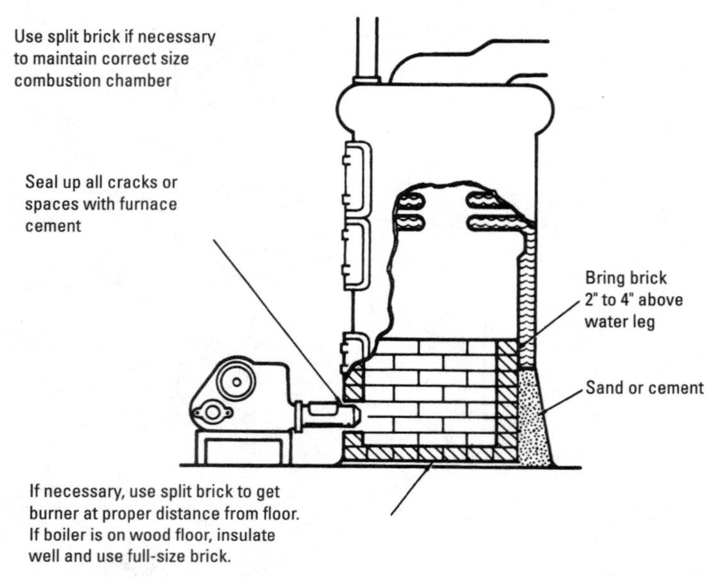

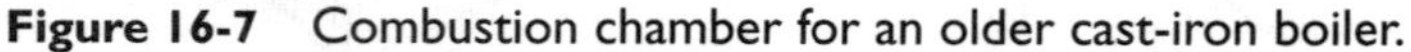

Figure 16-7 Combustion chamber for an older cast-iron boiler.

Figure 16-7 shows an old, round, cast-iron boiler that has been converted to oil. Note the height of the combustion chamber wall. The procedure for setting up the oil burner to the combustion chamber is illustrated in Figure 16-8.

Ventilation Requirements

If the existing boiler or furnace is located in an open area (basement or utility area) and the ventilation is relatively unrestricted, there should be sufficient supply of air for combustion and draft-hood dilution. On the other hand, if the heating unit is located in an enclosed furnace room or normal air infiltration is effectively reduced by storm windows or doors, then certain provisions must be made to correct this situation. Figure 16-9 illustrates the type of modification that should be made in a furnace room to provide an adequate supply of air. As shown, two permanent grilles are installed in the walls of the furnace room, each of a size equal to 1 in^2 of free area per 1000 Btu/h of burner output. One grille should be located near the ceiling, and the other near the floor.

If the boiler or furnace is located in an area of particularly tight construction, the heating unit should be *directly* connected to an

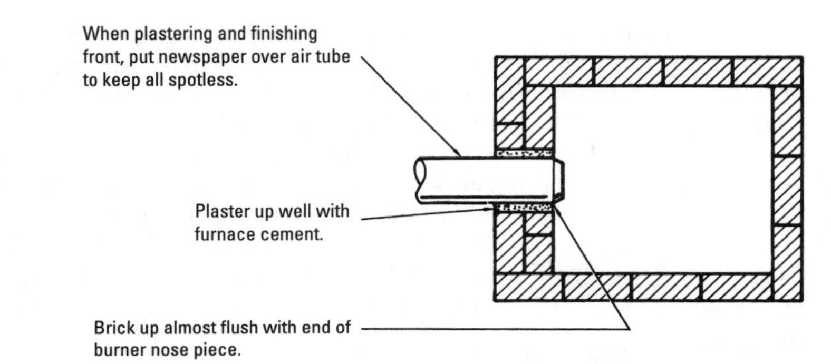

Figure 16-8 Procedure for setting up burner to combustion chamber.

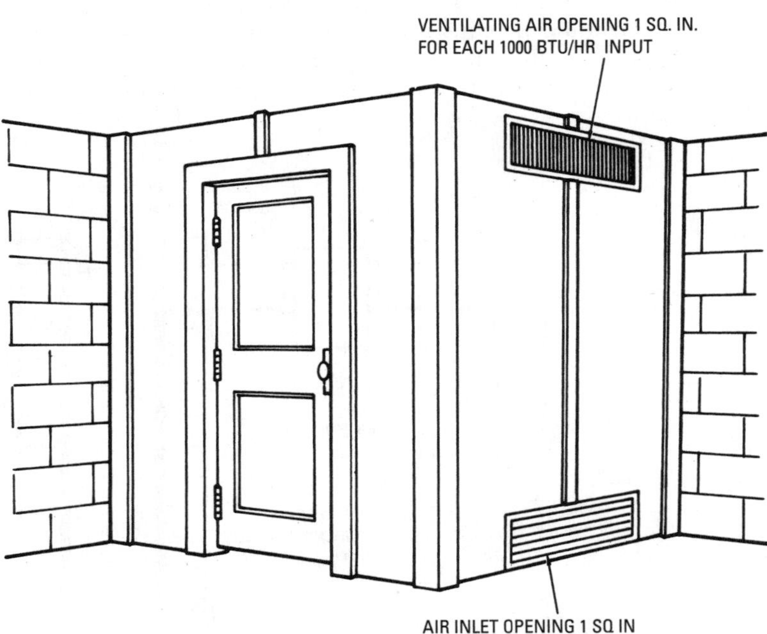

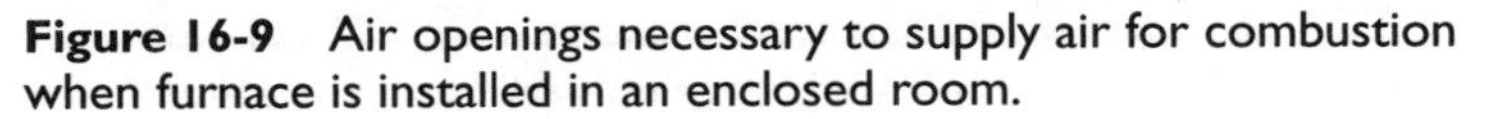

Figure 16-9 Air openings necessary to supply air for combustion when furnace is installed in an enclosed room.

outdoor source of air. A permanently open grille sized for *at least* 1 in^2 of free area per 5000 Btu/h of burner output should also be provided. Connection to an outside source of air is also recommended if the building contains a large exhaust fan.

A conversion burner equipped with a spark ignition system is not suited for use in furnaces or boilers located in areas subjected to sustained reverse draft or where large ventilating fans operate in combination with insufficient makeup air. There is always the danger of drawing flue gases into the structure. (See *Installing a Conversion Gas Burner* in this chapter.)

Flue Pipe and Chimney

The *flue pipe* is a pipe connecting the smoke outlet of the furnace or boiler with the flue of a chimney (Figures 16-10 and 16-11). It should *not* extend beyond the inner liner of the chimney and should *never* be connected with the flue of an open fireplace. Furthermore, flue connections from two or more sources should never enter the chimney at the same level from opposing sides.

Flue pipe should be constructed from corrosive resistant metal and be designed with as few sharp turns as practical. The straightest and shortest possible passage should be provided for the existing flue gases.

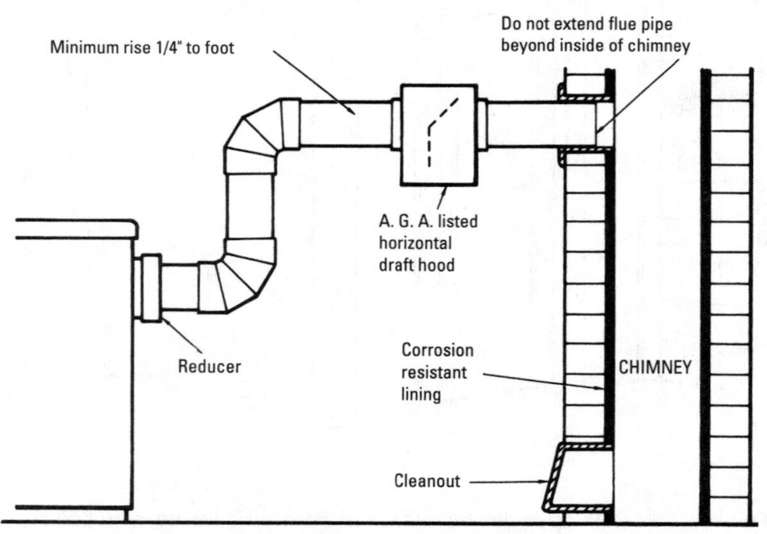

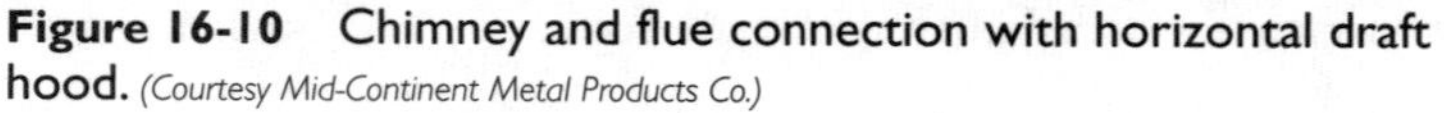

Figure 16-10 Chimney and flue connection with horizontal draft hood. *(Courtesy Mid-Continent Metal Products Co.)*

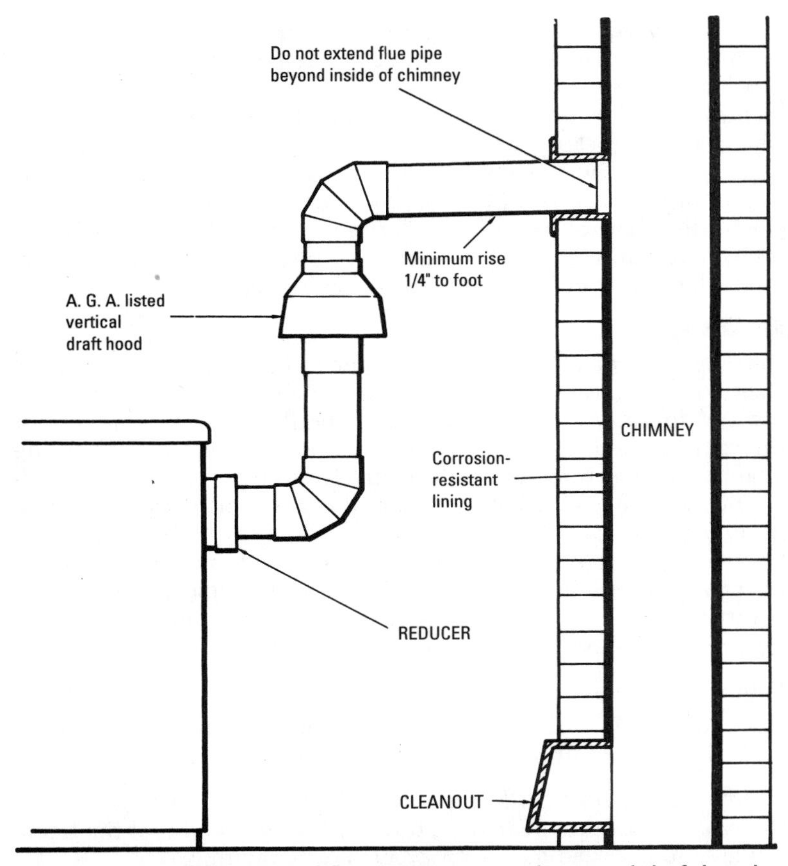

Figure 16-11 Chimney and flue connection with vertical draft hood.
(Courtesy Mid-Continent Metal Products Co.)

Pitch the flue pipe with a rise toward the chimney of at least ¼ in per foot. At approximate intervals, fasten the flue pipe securely with sheet-metal screws to prevent sagging.

Masonry is recommended for the construction of chimneys (prefab chimneys are also found suitable). Outside metal stacks are generally unsuitable for oil-fired burners.

The ordinary chimney *must* be at least 3 ft higher than the roof or 2 ft higher than any ridge within 10 ft of the building in order to avoid downdrafts.

Aside from adapting the furnace or boiler for a conversion burner, changes must also be made in the passages formed by the heating surface.

Extra-large flue passages are *not* suited to the high-temperature flue gas encountered with oil. In boilers having these large flue passages, baffling must be used to slow down the high velocities of the extra-hot gases; otherwise unburned particles of oil may lodge on the heating surface, resulting in carbon deposits. Because carbon is an excellent insulator, the efficiency of the heating surface is lowered whenever it collects. (See the following section.)

Baffling

Baffling is a type of obstruction designed to deflect and regulate the speed of flue gases. Baffles are especially necessary for round boilers and furnaces constructed so that the flue passages are almost direct from the firebox. Furthermore, boilers designed for burning coal are usually provided with relatively large flue passages, which are not normally suited to the higher flue gas velocities encountered in oil firing.

Baffling will, in most cases, help to overcome the inefficient operation resulting from the usual excessively high stack-gas temperatures. In such instances it is advisable to experiment with various methods of baffling as shown in Figure 16-12.

Note

Any baffling of the flue passages that prevents efficient operation of the burner should not be done.

Another form of baffling is a *corbel,* or stepout, arrangement of the brickwork of the rear wall, forming a target wall that the flame strikes and is curled back to prevent short circuiting.

Gas Piping and Piping Connections

Figures 16-13 and 16-14 illustrate the piping connections necessary for two different models of gas-fired conversion burners.

Note:

All local codes and ordinances take precedence over the instructions in the burner manufacturer's installation manual. If there is no manual available, the installation of the burner must conform to the manufacturer's instructions, the regulations of the National Fire Protection Association, and the provisions in the latest editions of the *National Electrical Code* (ANSI/NFPA70) and the *National Gas Code* (ANSI Z223.1).

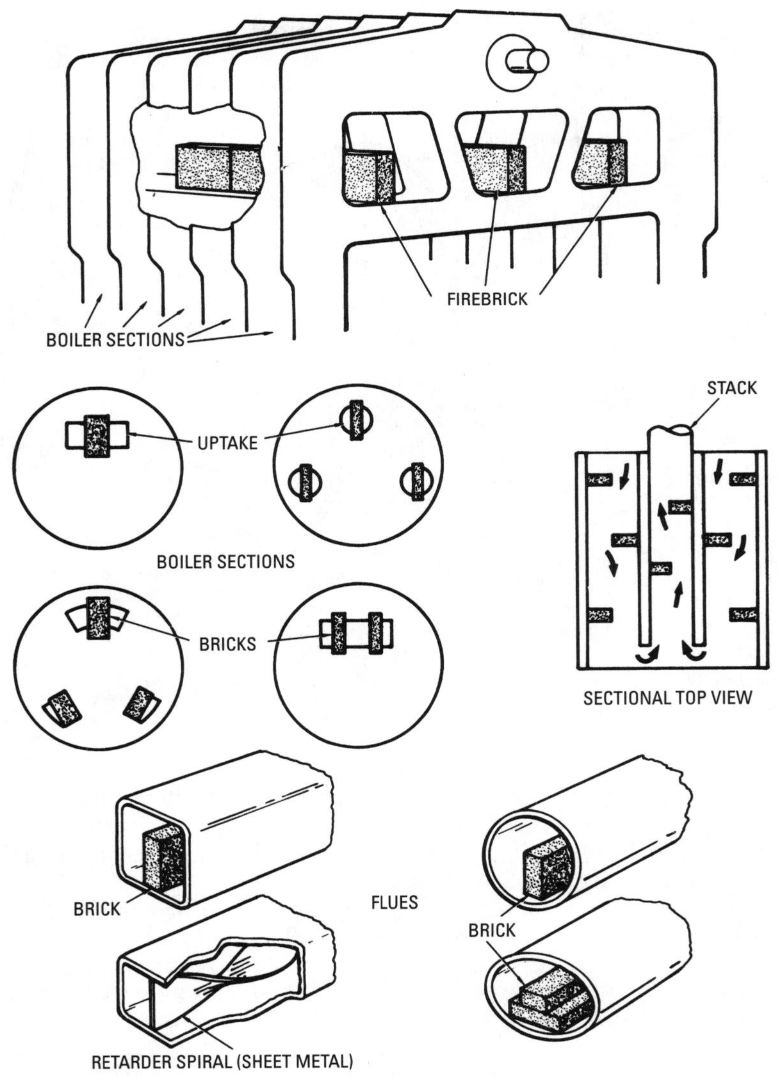

Figure 16-12 Various methods of baffling for sectional cast-iron boilers.

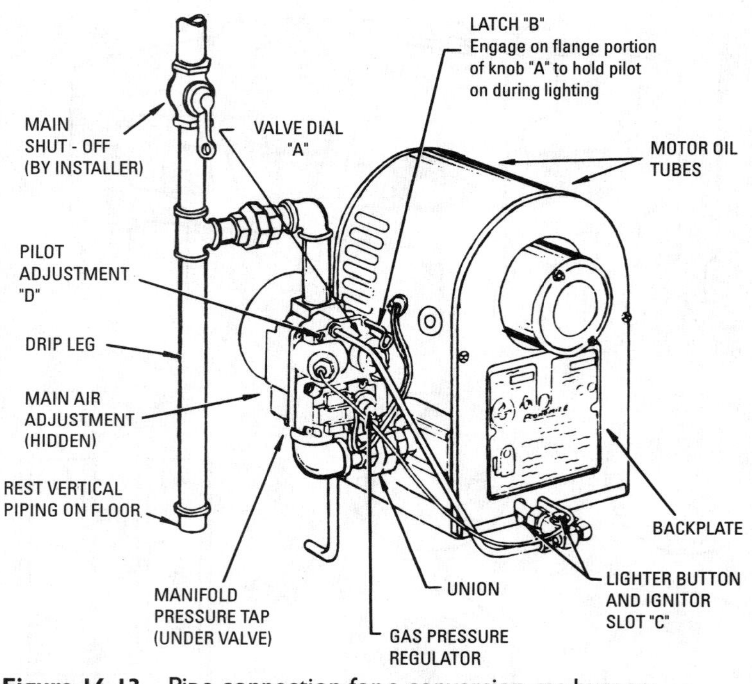

Figure 16-13 Pipe connection for a conversion gas burner (with pilot light). *(Courtesy Mid-Continent Metal Products Co.)*

No matter what type of gas-fired conversion burner you decide to use, it *must* be allowed to develop its rated capacity. This can be accomplished by making certain the burner is connected to a gas supply containing sufficient pressure.

In addition to providing for a gas supply with sufficient pressure, the following recommendations are also offered:

1. Provide for a separate gas-supply line of 1-in pipe size direct from the meter to the burner (ample for runs up to 60 ft long).
2. Install an intermediate regulator if the line pressure exceeds 13.5-in W.C (water column).
3. Connect the burner to the piping as shown in Figures 16-13 and 16-14.

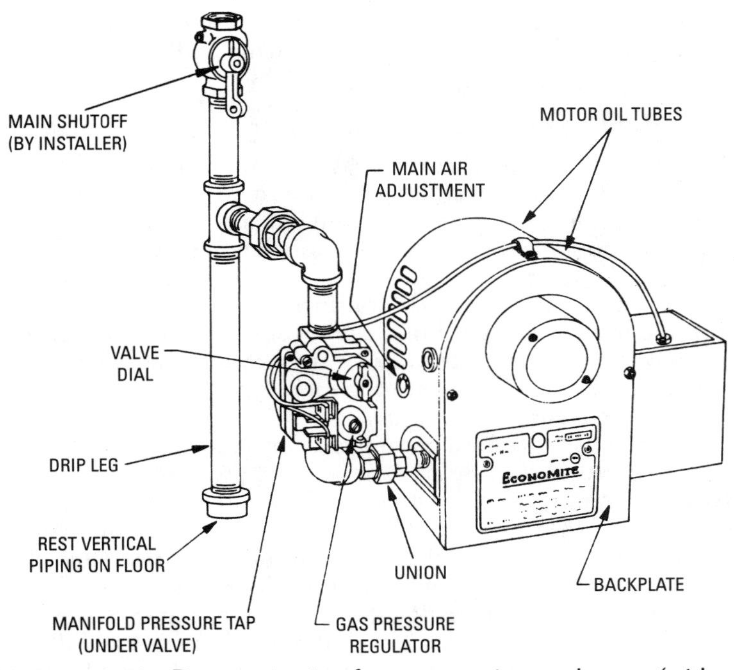

Figure 16-14 Pipe connection for a conversion gas burner (without pilot light). *(Courtesy Mid-Continent Metal Products Co.)*

4. Install a manually operated main shutoff valve 4 to 5 ft above the floor on the vertical pipe.
5. Make certain the pipe is clean and free of any scale.
6. Use malleable iron fittings.
7. Remove all burrs and scales from the pipe and clean it before installing it.
8. Do not tap off from the bottom of horizontal runs when branching from a pipeline.

Gas Input Setting

The gas conversion burner illustrated in Figure 16-13 is manufactured to operate on natural gas, but kits are available for converting to propane. A very important factor to be considered when installing these burners is a correct determination of the gas input setting, which should be based on the following two factors:

1. The heating rate of the structure
2. The rated maximum input of the burner

The *heating rate* of a structure is essentially the amount of warm air that the heating unit must deliver to heat it adequately. Methods for calculating the amount of heat required (i.e., the heating rate) are described in Chapter 4, "Sizing Residential Heating and Air Conditioning Systems."

Another important factor to consider is the *rated maximum input* of the burner. The gas input *cannot exceed* this rating. If the burner maximum input of the burner is below the required minimum of heat indicated by the heating rate, then the burner is inadequate for your needs.

The gas input setting for a natural gas conversion burner is regulated by the spud (orifice fitting) size and the pressure regulator. A burner is commonly shipped with three different spud sizes, one for each of three capacity ranges (high, intermediate, and low; Table 16-3). The capacity range is changed by using a different spud sizing. After the appropriate spud has been installed, the final gas input setting is made by adjusting the pressure regulator. Be sure to read the burner manufacturer's installation instructions carefully.

A propane gas conversion burner operates directly on 11-in wg gas supply pressure as determined by the propane-tank regulator. The portion of the combination control on the standard natural gas burner is blanked off with a plate provided in the propane conversion kit. Two spuds are provided: a minimum-size main spud and a pilot spud. The main spud must be redrilled for the required capacity (see Table 16-4).

Table 16-3 Spud Sizing for Natural Gas

Natural			
Capacity Btu/h		*Spud Size*	
Max. (3.5" W.C. Manifold Pressure)	*Min.*	*Drill No.*	*Dia.*
75,000	50,000	#20	0.161
125,000	75,000	# 4	0.209
200,000	125,000	17/64	0.266

Courtesy Mid-Continent Metal Products Co.

Table 16-4 Spud Sizing for Propane

Propane		
Capacity Btu/h	*Spud Size*	
11" W.C. Manifold Pressure	*Drill No.*	*Dia.*
50,000	#46	0.081
75,000	#40	0.098
100,000	#32	0.116
125,000	⅛"	0.125
150,000	#29	0.136
175,000	#26	0.147
200,000	#19	0.166

Courtesy Mid-Continent Metal Products Co.

Installing a Conversion Gas Burner

The conversion gas burners illustrated in Figures 16-13 and 16-14 are used to convert coal-fired or oil-fired central heating plants to gas. They are adaptable for use in forced-warm-air furnaces, gravity (updraft) furnaces, or boilers (Figures 16-15 through 16-18).

In addition to the basic combustion chamber requirements (see the appropriate section in this chapter), the following recommendations *must* be followed when installing a conversion gas burner.

1. The diameter of the burner opening into the combustion chamber must *not* be oversize. If the opening is oversize, slip the stainless steel sleeve (included with the burner) over the nozzle and fill in the remaining space with refractory cement.
2. Position the end of the burner nozzle so that it is at least 1 inch short of the inside surface of the combustion chamber. *Never* allow the nozzle to extend into the combustion chamber.
3. If the burner nozzle is too short to reach the combustion chamber, it may be lengthened with an extension tube.
4. Check the burner ports and pilot before permanently setting the burner in place, and remove any foreign matter blocking the openings.

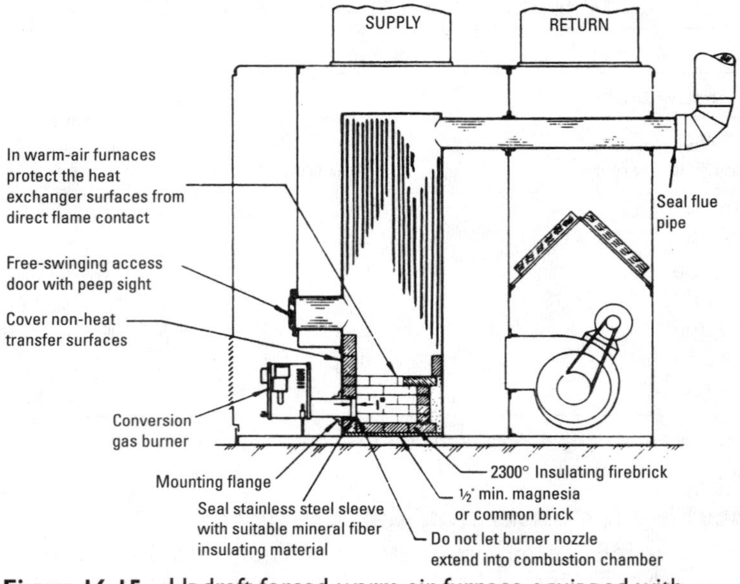

Figure 16-15 Updraft forced-warm-air furnace equipped with conversion gas burners. *(Courtesy Mid-Continent Metal Products Co.)*

Starting a Conversion Gas Burner (with Pilot Light)

This section covers the starting of conversion gas burners equipped with pilot lights. Before attempting to start this type of burner, the piping should be checked for leaks. The basic procedure for doing this is as follows:

1. Shut off all other gas appliances (gas water heaters, etc.) that tie into the same system.
2. Turn on the main shutoff valve (Figure 16-13).
3. Turn valve dial A to OFF (Figure 16-13).
4. Turn on gas pressure to the gas supply line.
5. Watch the meter test dial.

If the meter test dial shows no movement for *at least* 5 minutes, it is safe to say that there are no leaks in the piping. Movement of the meter test dial, however, means that you must locate the leak (or leaks) before starting the burner. The soapsuds test is a simple and effective method for doing this.

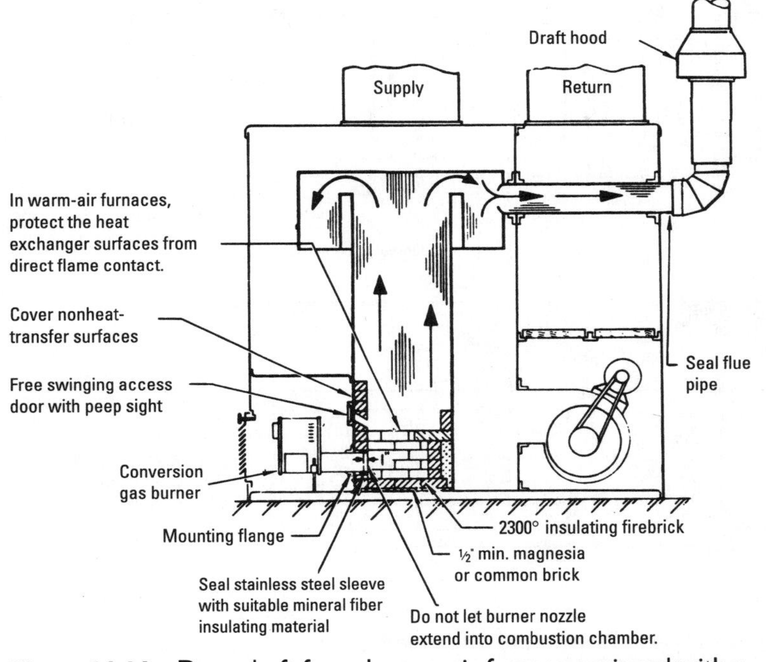

Figure 16-16 Downdraft forced-warm-air furnace equipped with a conversion gas burner. *(Courtesy Mid-Continent Metal Products Co.)*

The gas line must be purged before starting the burner. To do this, the following steps are recommended:

1. Disconnect the pilot tubing.
2. Depress valve dial A as far as it will go, and turn it *counter-clockwise* to PILOT (Figure 16-13). Do *not* release the dial at this point.
3. With valve dial A still fully depressed; lock the dial in position by engaging the lighting latch B (Figure 16-13).
4. Allow the air to escape until the gas line is completely purged (which will be indicated by gas replacing the air issuing from the pilot connection). Caution: *Never* purge the gas line into the combustion chamber.
5. Turn off valve dial A (depress and turn clockwise).
6. Reconnect the pilot tubing.

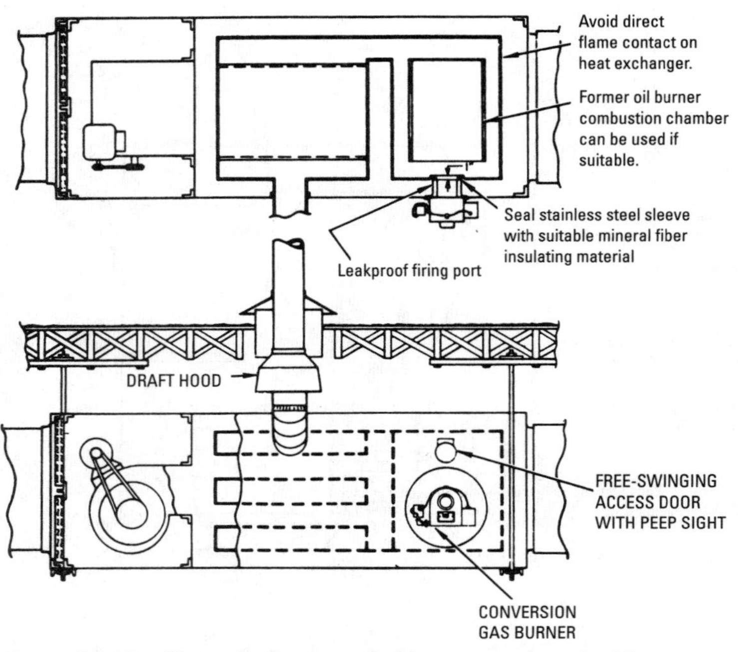

Figure 16-17 Downdraft suspended heater equipped with a conversion gas burner. *(Courtesy Mid-Continent Metal Products Co.)*

After you have checked the piping for leaks and have completely purged the gas lines, you are now ready to start the burner. First add a few drops of oil (#20 SAE oil is recommended) to the burner motor bearings. Then, check the setting of valve dial A. At this point, it should be turned to OFF (see step 5 in the purging procedure just described).

You are now ready for the initial startup procedure, which is accomplished as follows:

1. Set the main air shutter at approximately one-quarter open.
2. Turn on the main line switch.
3. Set the room thermostat *above* room temperature (this will cause the burner motor to start).
4. The burner motor should be allowed to run for about 5 minutes to ensure complete purging of the combustion chamber.

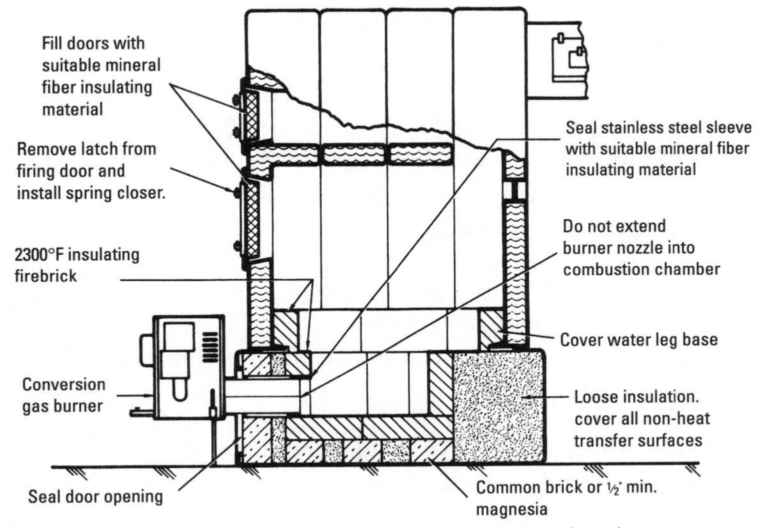

Figure 16-18 Hot water or steam boiler equipped with a conversion gas burner. *(Courtesy Mid-Continent Metal Products Co.)*

5. Shut off the burner motor by adjusting the thermostat setting *below* room temperature.
6. Fully depress valve dial A and turn it *counterclockwise* to PILOT. Keep it depressed.
7. Lock valve dial A in its depressed position by engaging the lighting latch B.
8. Light a match and hold the flame under the igniter valve and at igniter slot C while at the same time depressing button C for at least 10 seconds.
9. Release the lighter button. The pilot should now ignite. If it does not, then repeat Step 8 above.
10. After the pilot has been burning for at least 1 minute, release valve dial A and turn it *counterclockwise* to the ON setting.

Starting a Conversion Gas Burner (Pilotless)

Most of the instructions given in the previous section for starting a conversion gas burner equipped with a pilot light will also apply here. The major difference is the use of direct spark ignition rather than a pilot flame.

The basic procedure for starting a pilotless conversion gas burner is as follows (Figure 16-14):

1. Check the piping for leaks.
2. Bleed (purge) the gas line.
3. Set the air shutter about three-quarters open.
4. Depress the valve dial and turn it on the ON position.
5. Turn on the main line switch.
6. Set the room thermostat *above* the room temperature. This should cause the burner motor to start. If it does, burner ignition should occur after 30 seconds.
7. If the burner fails to light, turn off the main line switch and then turn it on again.
8. Readjust the air shutter for a quiet, soft flame. The flame should be blue at the burner, changing to orange at the tips.
9. Start and stop the burner several times with the thermostat to check its operation.

Servicing a Conversion Gas Burner

Always be sure that the manual gas valve and burner switch are turned off before attempting to service a gas burner. *Never* attempt to remove any parts for service before taking this precaution.

In direct spark ignition burners, the nozzle and electrode assembly can usually be removed as a unit. For example, the nozzle-and-electrode assembly of the Economite Model DS20A conversion gas burner can be completely removed in the following manner:

1. Remove the burner backplate.
2. Disconnect the pipe union.
3. Remove the curved orifice pipe.
4. Withdraw the nozzle assembly.
5. Disconnect the electrode leads.
6. Remove the unit.

The nozzle-and-electrode assembly can be reinstalled by reversing steps 1 through 6. When reassembling, make certain the orifice pipe enters the nozzle casting.

With the nozzle-and-electrode assembly completely removed, you are now in a position to inspect and clean it. Check the electrode for insulator cracks and evidence of serious burning.

Removal of the burner backplate also provides access to the motor and blower, motor relay, low-voltage transformer, and terminal board. These can be removed as a unit for servicing, but you will have to disconnect the terminal board and unplug the electrical pass-through fitting to do this. The following service may be required for this unit:

1. Cleaning of the blower wheel.
2. Cleaning the motor air vents.

Does the motor run without further burner sequencing taking place? This may indicate trouble in the centrifugal interlock switch. Some burners are provided with removable motor end caps to provide ready access to the centrifugal switch. If this should be the case, remove the cap and clean the contacts by burnishing. The contacts must be open when the motor is not operating. If necessary, the burner motor can be removed and replaced as follows:

1. Remove the blower wheel.
2. Remove the retainer clips at the motor grommets.
3. Pull the motor out of the brackets.
4. Repair or substitute a new motor.
5. Place motor in brackets.
6. Make certain grommets fit well in the bracket and replace retainer clips.
7. Replace blower wheel.

The sealed motor relay and low-voltage transformer normally do not require servicing. They should be replaced when defective.

Check gas lines and valves for leaks with the soapsuds test. *Never* use a flame to locate a gas leak. If the leakage occurs around the valve, the valve seat may need cleaning. Disassemble the valve, clean the seat, and reassemble it. If the valve malfunction is due to causes other than the seat, replace the entire valve.

Replace the gas pressure regulator if it fails to maintain a constant pressure. On the conversion gas burner illustrated in Figure 16-14, the regulator is part of a combination valve (in this case, a Robertshaw 24-volt combination gas valve), but it is designed so that it can be replaced without replacing the entire assembly.

Pressure adjustment required for setting intermediate capacities can be accomplished as follows:

1. Remove the adjustment screw cap.
2. Take a screwdriver and turn the adjustment screw *counter-clockwise* to reduce pressure.
3. Measure the pressure through the manometer connection adjacent to the outlet tapping.

Conversion gas burners equipped with a pilot light require inspection and servicing of the following:

1. The pilot orifice size will depend upon the type of gas used. For the conversion gas burner illustrated in Figure 16-13, the pilot orifice will have a diameter of 0.018 in for natural gas and 0.012 in for propane.
2. Set the end of the burner nozzle at least 1 inch short of the inside of the combustion chamber (see Figures 16-15 through 16-18). It must *not* be set flush with the chamber or be allowed to extend into the chamber (the pilot is often snuffed out in such cases by recirculated flue products).
3. Check for a high or low gas pressure. If over 7 inches W.C., reset the pilot adjustment (see Figure 16-13) to reduce the size of the pilot flame and to increase the heat on the thermocouple.

The thermocouple and pilot-safety section of the main valve should be tested to determine whether they are operating efficiently. The testing procedure is as follows:

1. Turn the burner switch to OFF.
2. Allow the combustion chamber to cool and the pilot light to operate for *at least* 5 minutes.
3. Turn off the pilot and listen for a "click" from the main valve.
4. If the "click" of the main valve is heard *more* than 30 seconds after you have shut off the pilot, then both the thermocouple and the pilot-safety section of the main valve are operating efficiently.
5. If the "click" occurs *less* than 30 seconds after you have shut off the pilot, then either the thermocouple or pilot-safety section of the main valve is faulty.

6. Test the thermocouple by disconnecting it from the main valve and checking it with a millivoltmeter. Under normal conditions (i.e., when heated by a standby pilot), it should develop at least 15 millivolts.
7. If the thermocouple develops at least 15 millivolts but the main valve "click" still occurs less than 30 seconds after you have shut off the pilot, the pilot-safety section is probably malfunctioning.
8. The thermocouple may also be tested while still connected to the valve (a closed-circuit connection) by using an adapter to connect the millivoltmeter. Under normal operating conditions, the thermocouple should develop at least 7 millivolts and the valve 4 millivolts.

The wiring diagram for a conversion gas burner equipped with a pilot light is shown in Figure 16-19. The installation and service instructions from most manufacturers will include a wiring diagram for the burner. Study it carefully before servicing or making repairs.

Oil Tanks and Oil Piping

Local codes and the National Board of Fire Underwriters will provide information for the installation of oil tanks and piping.

In oil piping, copper tubing is commonly used for suction and return lines. A ⅜-in OD minimum is recommended for lines under 30 ft long. If the suction and return lines exceed 30 ft, then ½-in OD copper tubing is recommended.

Other suggestions for the installation of the oil storage tank and piping include:

1. Use a good-quality pipe joint compound and make your pipe thread connections as tight as possible.
2. Keep the lines between the oil tank and burner as straight as possible (avoid kinks and traps).
3. Keep the connections in the return line to a minimum.
4. Check tubing flares and connections for air leaks (an indication of a poor seal resulting from improper meeting).

The two basic types of piping arrangements are: (1) the single-line system and (2) the two-line system. These two piping arrangements are distinguished from one another primarily by the location of the oil storage tank with respect to the oil-fired conversion burner.

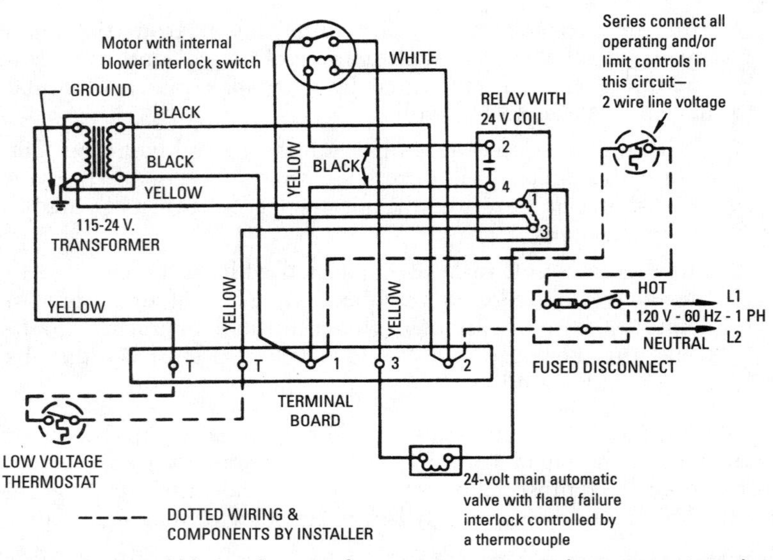

Figure 16-19 Wiring diagram for a conversion gas burner equipped with a pilot light. *(Courtesy Mid-Continent Metal Products Co.)*

A *single-line piping* arrangement is used only where the oil storage tank is located *above* the level of the oil burner. No return line is necessary. The oil is taken from an outlet in the bottom of the storage tank and fed by gravity to the burner. The slope of the pipe should be gradually downward (approximately ½ in per foot) to a point directly below the burner connection. This downward slope in the line is designed to prevent the formation of air pockets and bubbles. The installation of a shutoff valve in the line is recommended.

A *two-line piping* arrangement is required where the oil storage tank is buried at a level *below* that of the burner.

Installing a Conversion Oil Burner

Conversion oil burners should be installed in accordance with the provisions of the National Fire Protection Association and local codes and regulations. Local codes and regulations will always take precedence. Read these regulations and the manufacturer's installation instructions before making any attempt to install the burner.

If the furnace or boiler was originally designed to burn solid fuel and the ashpit is *not* used as a part of the combustion chamber, the ash door should be removed to prevent the accumulation of vapors in the ashpit. If removal of the ash door is not feasible, then some other means of bottom ventilation must be provided.

Before installing the burner, check the condition of the boiler or furnace. It must be in good condition or repair. The flue gas passages and combustion chamber must be tight against leaks. Reseal or reset the sections of a cast-iron boiler or furnace. Replace any damaged parts. The boiler or furnace should also be as clean as possible before installing the burner.

A combustion chamber must be provided that is in accordance with the specifications of the burner manufacturer. Read the appropriate sections in this chapter (e.g., *Basic Combustion Chamber Requirements*) for more information.

If you are satisfied that the boiler or furnace with which you are working satisfactorily meets the conditions mentioned in the previous paragraphs, then you are ready to install the conversion oil burner. The basic steps recommended for installation are as follows:

1. Position the burner on its mounting flange or pedestal so that the burner air tube is flush with the inside surface of the combustion chamber front wall. Do *not* allow the burner air tube to extend into the combustion chamber.
2. Adjust the burner tube on pedestal-mounted burners so that it pitches downward about 1½ in. Any fuel in the burner tube will then drain into the combustion chamber.
3. On flange-mounted burners, the distance from the flange to the end of the air cone will depend upon the requirements of each installation. Remove the air cone before installing the flange.
4. The distance between the center of the nozzle and the floor of the combustion chamber must be correct. If it is too close, it will result in flame impingement and carbonization. On the other hand, a nozzle placed too high will result in excessive flame noise and poor combustion.
5. The fuel pump should be adjusted for the oil pressure recommended by the manufacturer.
6. Adjust the air supply to the burner with the air inlet band. Rotate the air inlet band to the required position (the one

which will deliver the smallest amount of air and still maintain clean combustion), and secure it in position.

7. Make certain the size and spray angle of the nozzle is correct for the installation (check the UL rating plate).
8. Install the nozzle, and check the electrode tip position in accordance with the manufacturer's instructions.
9. Prime the fuel pump (see the manufacturer's instructions).

Starting a Conversion Oil Burner

Before attempting to start the oil burner, you should first carefully read the manufacturer's instructions. Unfortunately, these are not always available on older heating installations. If this is the case and you do not possess the necessary expertise to operate the burner without an instruction manual, it is strongly recommended that you call a professional for services.

Always test the oil lines for possible leaks and make the necessary repairs before starting the oil burner. Consult the manufacturer's instructions for bleeding the fuel pump. The bleed valve on the fuel pump is designed to facilitate air purging, cleaning, and priming. See Chapter 1, "Oil Burners" of Volume 2 for additional information about fuel pumps.

Check the electrical wiring diagram (if one exists) to make certain that the burner is correctly wired. Problems with an oil burner can often be traced to incorrect wiring. All wiring must be in accordance with the *National Electrical Code* as well as local codes, standards, and regulations.

It is important to make certain that the primary and limit controls are operating properly. Test the primary control by first removing the motor lead from the burner and then energizing the ignition circuit by throwing the switch. If the primary control is operating properly, the ignition will shut off *within 2 minutes* after the switch is thrown. If the ignition fails to shut off after 2 minutes, the primary control should be replaced.

Check the fuel level in the oil storage tank. There must be enough oil to operate the burner. Open both oil valves (at the oil storage tank and at the boiler or furnace).

You *must* have an adequate air supply for efficient combustion. The basic ventilation requirements (see *Ventilation Requirements* in this chapter) must be provided for when installing the oil burner. It is also important to determine whether the secondary air setting is correct. If smoke or soot is present when the oil burner is operating, the air setting must be corrected.

If all the suggestions in the previous paragraphs have been followed and the necessary adjustments or repairs made, the oil burner can now be started. The procedure for doing this is basically as follows:

1. Make certain that all controls have been set in normal starting position.
2. Make certain that the oil valve in the oil supply line is open.
3. Set the room thermostat 10° above the room temperature.
4. Check the reset button on the primary control to make certain it has been reset.
5. Open the air control band on the burner to about half open.
6. Throw the main electrical switch to ON. The burner should now start, ignite, and burn. If it does not, recheck steps 1 through 5. If you still experience difficulty in starting the burner, consult *Troubleshooting Oil Burners* in Chapter 1, "Oil Burners" of Volume 2.
7. Allow the oil burner to operate for about 15 minutes, then shut it down and give it time to cool off. After it has cooled off, restart it to be sure it operates properly on a cool start.

While the oil burner is operating (step 7), you should take the opportunity to make a number of tests. Check the fuel-pump pressure with an oil pressure gauge. These should be factory set. Adjust the air control band on the burner until 0+ smoke is obtained with a smoke tester. Adjust the draft control (if necessary) to obtain at least 0.02 in W.C. of draft over the fire.

Servicing a Conversion Oil Burner

Protecting the oil burner motor from unnecessary wear is extremely important. *At least* once a year (but not more than three times a year), preferably at the beginning of each heating season, place about 10 drops of #20 SAE oil in each oil port of the burner motor. After you have oiled the burner motor, clean the fuel strainer and change the oil-filter cartridge. Clogged fuel strainers are a common source of trouble. In rotary- and gun-type oil burners, these components are usually located either at the point at which the oil line connects to the tank or at the oil line connection to the burner. Follow the manufacturer's instructions for removing the fuel strainer, and clean it with a stiff wire brush or a blast of compressed air.

Be sure that you are using the correct weight of oil. If the oil is too heavy, the burner motor will start, but it may fail to establish a flame. Continued use of the wrong oil may cause damage to burner components.

Soot should be periodically removed from the burner and combustion chamber. Commercial soot removers are available for this purpose. When burning out a layer of soot, take care not to ignite a soot fire in the chimney. This could be very dangerous because it may lead to the igniting of combustible materials near the chimney. The hard layer of carbon that frequently forms on the bottom of the firepot in burners should also be removed.

Other servicing recommendations for conversion oil burners are as follows:

1. Check the furnace or boiler for possible air leaks. Any that are found should be sealed (a refractory mortar is recommended) because they interfere with combustion efficiency.
2. Check the burner nozzle in gun-type burners. If it is dirty, remove and clean it according to the manufacturer's instructions.
3. Check the position of the electrodes in rotary- and gun-type oil burners. They should be positioned near, but out of, the direct spray of the oil.

Troubleshooting Oil and Gas Conversion Burners

Operating problems, their symptoms and causes, and the suggested remedies for oil and gas conversion burners are covered in Chapter 1, "Oil Burners" and Chapter 2, "Gas Burners" in Volume 2.

Appendix A

Professional and Trade Associations

Many professional and trade associations have been formed to develop and provide research materials, services, and support for those working in the heating, ventilating, and air conditioning trades. The materials, services, and support include:

1. Formulating and establishing specifications and professional standards.
2. Certifying that equipment and materials meet or exceed minimum standards.
3. Certifying that technicians have met education and training standards.
4. Conducting product research.
5. Promoting interest in the product.
6. Providing education and training.
7. Publishing books, newsletters, articles, and technical papers.
8. Conducting seminars and workshops.

A great deal of useful information can be obtained by contacting these associations. With that in mind the names and addresses of the principal organizations have been included in this Appendix. They are listed in alphabetical order.

Air-Conditioning and Refrigeration Wholesalers International (ARWI)

(*See* Heating, Air Conditioning & Refrigeration Distributors International)

Air-Conditioning and Refrigeration Institute (ARI)

4100 North Fairfax Drive, Suite 200

Arlington, Virginia

Phone: (703) 524-8800

Fax: (703) 528-3816

Email: ari@ari.org

Web site: www.ari.org

A national trade association of manufacturers of central air conditioning, warm-air heating, and commercial and industrial refrigeration equipment, ARI publishes ARI standards and guidelines, which can be downloaded free from its web site. ARI is an approved certifying organization for the EPA Technician Certification Exam. ARI also provides a study manual for those taking the EPA Technician Certification Exam. ARI developed the Curriculum Guide in collaboration with HVACR instructors, manufacturing training experts, and other industry professionals for use in all school programs that educate and train students to become competent, entry-level HVACR technicians.

Air Conditioning Contractors of America (ACCA)

2800 Shirlington Road

Suite 300

Arlington, Virginia 22206

Phone: (703) 575-4477

Fax: (703) 575-4449

Email: info@acca.org

Web site: www.acca.org

A national trade association of heating, air conditioning, and refrigeration systems contractors, ACCA (until 1978, the National Environmental Systems Contractors Association) publishes a variety of different manuals useful for those working in the HVAC trades, including residential and commercial equipment load calculations, residential duct system design, and system installation. ACCA also publishes training and certification manuals. The ACCA publications can be purchased by both members and nonmembers. Check the web site for a list of the ACCA publications, because it is very extensive.

Air Diffusion Council (ADC)

1000 E. Woodfield Road

Suite 102

Schaumburg, Illinois 60173

Phone: (847) 706-6750

Fax: (847) 706-6751

Email: info@flexibleduct.org

Web site: www.flexibleduct.org

The Air Diffusion Council (ADC) was formed to promote the interests of the manufacturers of flexible air ducts and related air distribution equipment. The ADC supports the maintenance and development of uniform industry standards for the installation, use, and performance of flexible duct products. It encourages the use of those standards by various code writing groups, government agencies, architects, engineers, and heating and air conditioning contractors.

Air Filter Institute

(*See* Air-Conditioning and Refrigeration Institute)

Air Movement and Control Association International, Inc. (AMCA)

30 West University Drive

Arlington Heights, Illinois 60004

Phone: (847) 394-0150

Fax: (847) 253-0088

Email: amca@amca.org

Web site: www.amca.org

The Air Movement and Control Association International, Inc. (AMCA) is a trade association of the manufacturers, wholesalers, and retailers of air movement and control equipment (fans, louvers, dampers, and related air systems equipment). The AMCA Certified Ratings Programs are an important function of the association. Their purpose is to give the buyer, specification writer, and user of air movement and control equipment assurance that published ratings are reliable and accurate. The AMCA publishes current test standards for fans, louvers, dampers, and shutters. It also issues a variety of AMCA certified rating seals for different types of air movement and control equipment. It publishes a newsletter various technical specifications for members and those who work with air movement and control systems.

American Boiler Manufacturers Association (ABMA)

4001 North 9th Street

Suite 226

Arlington, Virginia 22203

Phone: (703) 522-7350

Fax: (703) 522-2665

Email: randy@abma.com

Web Site: www.abma.com

The American Boiler Manufacturers Association (ABMA) is a national association representing the manufacturers of commercial, industrial, and utility steam generating and fuel burning equipment, as well as suppliers to the industry. The primary goal of ABMA is topromote the common business interests of the boiler manufacturing industry and to promote the safe, environmentally friendly use of the products and services of its members. Publishes technical guides and manuals.

American Gas Association (AGA)

151400 North Capitol Street, N.W.

Washington, DC 20001

Phone: (202) 824-7000

Fax: (202) 824-7115

Email: Fax: krogers@aga.org

Web site: www.aga.org

The AGA develops residential gas operating and performance standards for distributors and transporters of natural, manufactured, and mixed gas.

American Society of Heating, Refrigeration, and Air-Conditioning Engineers (ASHRAE)

1791 Tullie Circle NE

Atlanta, GA 30329

Phone: (800) 527-4723 (toll free)

Phone: (404) 636-8400

Fax: (404) 321-5478

Email: ashrae@ashrae.org

Web site: www.ashrae.org

The American Society of Heating, Refrigeration, and Air-Conditioning Engineers (ASHRAE) is an international professional association concerned with the advancement of the science and technology of heating, ventilation, air conditioning, and refrigeration through research, standards writing, continuing education, and publications. Membership in ASHRAE is open to any person associated with heating, ventilation, air conditioning, or refrigeration. There are several different types of membership depending on the individual's background and experience in the different HVAC and refrigeration fields. An important benefit of belonging to ASHRAE is access to numerous technical publications.

American Society of Mechanical Engineers (ASME)

Three Park Avenue

New York, New York 10016

Phone: (800) 843-2763 or (973) 882-1167

Fax: (973) 882-1717

Email: infocentral@asme.org

Web site: www.asme.org

A nonprofit technical and education association, ASME develops safety codes and standards and has an extensive list of technical publications covering pressure vessels, piping, and boilers. ASME offers educational and training services and conducts technology seminars and on-site training programs.

Better Heating-Cooling Council

(*See* Hydronics Institute)

Fireplace Institute

(Merged with Wood Energy Institute in 1980 to form the Wood Heating Alliance. *See* Wood Heating Alliance) Gas Appliance Manufacturers Association (GAMA)

2701 Wilson Boulevard, Suite 600

Arlington, Virginia 22201

Phone: (703) 525-7060

Fax: (703) 525-6790

Email: information@gamanet.org.

Web Site: www.gamanet.org

GAMA is a national trade association of manufacturers of gas-fired appliances, and certain types of oil-fired and electrical appliances, used in residential, commercial, and industrial applications. An important service provided by GAMA to its members is a testing and certification program. GAMA will test the rated efficiency and capacity of a manufacturer's product and, if it passes the testing criteria, certify it. The manufacturer can then market the product with the appropriate certification label. Program participants and their products are listed in the ratings directories.

Heating and Piping Contractors National Association

(*See* Mechanical Contractors Association of America)

Heating, Airconditioning & Refrigeration Distributors International (HARDI)

1389 Dublin Road
Columbus, Ohio 43215
Phone: (888) 253-2128 (toll free)
Phone: (614) 488-1835
Fax: (614) 488-0482
Email: HARDImail@HARDInet.org
Web site: www.hardinet.org

Heating, Airconditioning & Refrigeration Distributors International (HARDI) is national trade association of wholesalers and distributors of air conditioning and refrigeration equipment. It was formed by merging the Northamerican Heating, Refrigeration & Airconditioning Wholesalers (NHRAW) and the Air-conditioning & Refrigeration Wholesalers International (ARWI). Among products and services provided to its members are self-study training materials, statistical studies, as well as training and reference materials. See Appendix B (Education, Training, and Certification) for a description of the HARDI Home Study Institute. The HARDI publications are available to both members and nonmembers.

Home Ventilating Institute (HVI)

30 West University Drive
Arlington Heights, Illinois 60004
Phone: (847) 394-0150
Fax: (847) 253-0088
Email: hvi@hvi.org
Web site: www.hvi.org

The Home Ventilating Institute (HVI) is a nonprofit trade association representing national and international manufacturers of residential ventilation products. HVI is primarily concerned with developing performance standards for residential ventilating equipment. It has created a number of certified ratings programs that provide a fair and credible method of comparing ventilation performance of similar products. HVI publishes a number of interesting and informative articles on ventilation that can be downloaded from its Web site.

The Hydronics Foundation, Inc. (THFI)

The Hydronics Foundation, Inc. (THFI) was chartered in 1997 as a nonprofit organization to disseminate knowledge about hydronic equipment and technology. Manufacturers of HVAC equipment also contribute material from installation manuals, specification sheets, and product reviews.

119 East King Street
P.O. Box 1671
Johnson City, TN 37606
Phone: (800) 929-8548
Fax: (800) 929-9506
Email: jdhowell@jdhowell.com
Web site: www.hydronics.com or www.hydronics.org

Hydronic Heating Association (HHA)

P.O. Box 388
Dedham, MA 02026
Phone: (781) 320-9910
Fax: (781) 320-9906
Email: info@comfortableheat.net
Web site: www.comfortableheat.net

The Hydronic Heating Association is an organization of independent contractors, wholesalers, and manufacturers established to promote the latest hydronic technology, set uniform industrial standards, educate HVAC contractors, and inform the public of the benefits of having a quality hot-water system installed. Their Web site contains useful articles and essays on hydronic equipment and systems. It also offers many useful links to manufacturers of hydronic system products.

The Hydronics Institute Division of GAMA

P.O. Box 218

Berkley Heights, New Jersey 07922

Phone: (908) 464-8200

Fax: (908) 464-7818

Email: information@gamanet.org

Web site: www.gamanet.org

The Hydronics Institute was originally formed by a merger of the Better Heating-Cooling Council and the Institute of Boiler and Radiator Manufacturers. It represents manufacturers, suppliers, and installers of hot-water and steam heating and cooling equipment. It is now a division of the Gas Appliance Manufacturers Association. The Hydronics Institute represents and promotes the interests of the manufacturers of hydronic heating equipment. It also provides training materials for hydronic heating courses in schools and technical publications for technicians in the field.

Institute of Boiler and Radiator Manufacturers

(*See* Hydronics Institute)

Mechanical Contractors Association of America, Inc. (MCAA)

1385 Piccard Drive

Rockville, MD 20850

Phone: 301-869-5800

Fax: 301-990-9690

Web site: www.mcaa.org

The Mechanical Contractors Association of America, Inc. (MCAA) is a national trade association for contractors of piping and related equipment used in heating, cooling, refrigeration, ventilating, and air conditioning.

National Association of Plumbing Heating Cooling Contractors (PHCC)

180 S. Washington Street

P.O. Box 6808

Falls Church, VA 22040

Phone: (800) 533-7694 (toll free) or (703) 237-8100

Fax: (703) 237-7442

Email: naphcc@naphcc.org

Web site: www.phccweb.org

The National Association of Plumbing Heating Cooling Contractors (PHCC) is a trade association of local plumbing, heating, and cooling contractors. There are 12 regional chapters.

National Environmental Systems Contractors Association

(*See* Air Conditioning Contractors of America)

National Warm Air Heating and Air Conditioning Association

(*See* Air Conditioning Contractors of America)

North American Heating Refrigerating Air conditioning Wholesalers (NHRAW)

(*See* Heating, Airconditioning & Refrigeration Distributors International)

Radiant Panel Association (RPA)

P.O. Box 717

1399 South Garfield Avenue

Loveland, CO 80537

Phone: (800) 660-7187 (toll free) or (970) 613-0100

Fax: (970) 613-0098

Email: info@rpa-info.com

Web site: www.radiantpanelassociation.org

The Radiant Panel Association provides downloadable technical papers and notes on a variety of different topics concerning radiant panel heating and cooling systems. Links to several manufacturers of radiant heating equipment are also available at their Web site.

Refrigeration and Air Conditioning Contractors Association

(*See* Air Conditioning Contractors of America)

Refrigeration Service Engineers Society (RSES)

1666 Rand Road

Des Plaines, Illinois 60016

Phone: (800) 297-5660 (toll free) or (847) 297-6464

Email: general@rses.org.

Web site: www.rses.org

The Refrigeration Services Engineers Society (RSES) is an international association of refrigeration, air conditioning, and heating equipment installers, service persons, and sales persons. The Society conducts educational meetings, seminars, workshops, technical qualification and examination programs, instructor-led and self-study training courses. It offers a variety of certification program for technicians.

Sheet Metal and Air Conditioning Contractors National Association (SMACNA)

4201 Lafayette Center Dr.

Chantilly, Virginia 20151

Phone: (703) 803-2980

Fax: (703) 803-3732

Email: info@smacna.org

Web site: www.smacna.org

The SMACNA is an international trade association of union contractors who install ventilating, warm-air heating, and air-handling equipment and systems. SMACNA publishes technical papers, answers technical question, and provides distance learning courses for its members. American National Standards Institute has accredited SMACNA as a standards-setting organization matter.

Steam Heating Equipment Manufacturers Association

(*defunct*)

Steel Boiler Institute

(*defunct*)

Underwriters Laboratories, Inc. (UL)

Northbrook Division

Corporate Headquarters

333 Pfingsten Road

Northbrook, Illinois 60062
Phone: (847) 272-8800
Fax: (847) 272-8129
Email: northbrook@us.ul.com
Web site: www.ul.com

The Underwriters Laboratories is an independent, nonprofit product safety testing and certification organization. It promotes safety standards for equipment through independent testing.

Wood Energy Institute

(Merged with Fireplace Institute in 1980 to form Wood Heating Alliance.)

Other National and International Professional and Trade Associations

The following associations also provide support, services, technical publications, and/or training to its members who are involved in the manufacture, sale, or installation and repair of heating, ventilating, and air conditioning systems and equipment. Because space is limited, only their names are listed. Contact information can be obtained by accessing the Internet and entering the association name or by visiting the reference room of your local library and using the *Encyclopedia of Associations*.

Air Distribution Institute (ADI)

American National Standards Institute (ANSI)

American Society for Testing and Materials (ASTM)

Australian Home Heating Association (AHHA)

Australian Institute of Refrigeration, Air Conditioning and Heating (AIRAH)

Heating Alternatives, Inc

Heating, Refrigeration and Air Conditioning Contractors of Canada (HRAC)

Heating, Refrigeration and Air Conditioning Institute of Canada (HRAE)

Institute of Heating & Air Conditioning, Inc (IHACI).

Insulation Contractors of America (ICA)

International Energy Association (IEA)—Solar Heating and Cooling Programme

National LP-Gas Association (NLPGA)
National Oil Fuel Institute
National Oilheat Research Alliance
Plumbing-Heating-Cooling Contractors—National Association (PHCC)
Plumbing-Heating-Cooling Information Bureau (PHCIB)
Wood Heating Alliance (WHA)

Appendix B

Manufacturers

Adams Manufacturing Company

9790 Midwest Avenue
Cleveland, OH 44125
(216) 587-6801
(216) 587-6807 (Fax)
www.gamanet.org

Amana Refrigeration, Inc.

1810 Wilson Parkway
Fayetteville, TN 37334
(800) 843-0304
(931) 433-6101
(931) 433-1312
www.amana.com

American Standard Companies Inc.

One Centennial Ave.
Piscataway, NJ 08855
(732) 980-6000
(732) 980-3340 (Fax)
www.americanstandard.com

A.O. Smith Water Products Company

600 E. John Carpenter Freeway #200
Irving, TX 75062-3990
(972) 719-5900
(972) 719-5960 (Fax)
www.hotwater.com

Bacharach Inc.

621 Hunt Valley
New Kensington, PA 15068
(724) 334-5000

(724) 334-5001 (Fax)
www.bacharach-inc.com

Bard Manufacturing Co
P.O. Box 607
Bryan, OH 43506
(419) 636-1194
(419) 636-2640 (Fax)
www.bardhvac.com

R.W. Beckett Corporation
P.O. Box 1289
Elyria, OH 44036
(800) 645-2876
(440) 327-1060
(440) 327-1064 (Fax)
www. beckett.com

Bell & Gossett
(*See* ITT Bell & Gossett)

Bryan Boilers/Bryan Steam Corporation
P.O. Box 27
783 N. Chili Avenue
Peru, IN 46970
(765) 473-6651
(765) 473-3074 (Fax)
www.bryanboilers.com

Burnham Hydronics
U.S. Boiler Co., Inc.
P.O. Box 3079
Lancaster, PA 17604
(717) 397-4701
(717) 293-5827 (Fax)
www.burnham.com

Carrier Corporation
World Headquarters
One Carrier Place
Farmington, CT 06034
(860) 674-3000
www.carrier.com

Cash Acme
2400 7th Avenue S.W.
Cullman, Alabama 35055
(256) 775-8200
(256) 775-8238 (Fax)
www.cashacme.com

Coleman Corporation
Unitary Products Group
5005 York Dr.
Norman, OK 73069
(405) 364-4040
www.colemanac.com

Columbia Boiler Company
P.O. Box 1070
Pottstown, PA 19464
(610) 323-2700
(610) 323-7292 (Fax)
www.columbiaboiler.com

Danfoss A/S
DK-6430 Nordborg
Denmark
+45 7488 2222
+45 7449 0949 (Fax)
www.danfoss.com

Domestic Pump
(*See* ITT Domestic Pump)

Dornback Furnace

9545 Granger Road
Garfield Heights, OH 44125
(216) 662-1600
(216) 587-6807
www.gamanet.org

Ernst Gage Co.

250 S. Livingston Ave.
Livingston, NJ 07039 4089
973-992-1400
888-229-4243
973-992-0036 (Fax)

General Filters Inc.

43800 Grand River Ave.
Novi, MI
(248) 476-5100
(248) 349-2366 (Fax)
www.generalfilters.com

Goodman Manufacturing Corp

2550 North Loop West #400
Houston, TX 77092
(713) 861-2500
(888) 593-9988
www.goodmanmfg.com

Heat Controller, Inc.

1900 Wellworth Avenue
Jackson, MI 49203
(517) 787-2100
(517) 787-9341
www.heatcontroller.com

Hoffman Specialty

(*See* ITT Hoffman Specialty)

Honeywell, Inc.
101 Columbia Road
Morristown, NJ 07962
(973) 455-2000
(800) 328-5111
(983) 455-4807 (Fax)
www.honeywell.com

Hydro Therm, A Division of Mastek, Inc.
260 North Elm Street
Westfield, MA 01085
(413) 564-5515
www.hydrotherm.com

Invensys Building Systems, Inc.
1354 Clifford Ave.
P.O. Box 2940
Loves Park, IL 61132-2940
(815) 637-3000
(815) 637-5350 (Fax)
www.invensys.com

ITT Bell & Gossett
8200 North Austin Avenue
Morton Grove, IL 60053
(847) 966-3700
(847) 966-9052
www.bellgossett.com

ITT Domestic Pump
8200 N. Austin Ave.
Morton Grove, IL 60053
(847) 966-3700
(847) 966-9052
www.domesticpump.com

ITT Hoffman Specialty
3500 N. Spaulding Avenue
Chicago, IL 60618
(773) 267-1600
(773) 267-0991
www.hoffmanspecialty.com

ITT McDonnell & Miller
3500 N. Spaulding Avenue
Chicago, IL 60618
(723) 267-1600
(773) 267-0991
www.mcdonnellmiller.com

Janitrol Air Conditioning and Heating
www.janitrol.com

(*See* Goodman Manufacturing Company)

S.T. Johnson Company
Innovative Combustion Technologies, Inc.
925 Stanford Avenue
Oakland, CA 94608
(510) 652-6000
(510) 652-4302 (Fax)
www.johnsonburners.com

Johnson Controls, Inc.
5757 North Green Bay Avenue
Milwaukee, WI 53209
(262) 524-3285
www.johnsoncontrols.com
www.jci.com

Lennox Industries Inc.
2100 Lake Park Boulevard
Richardson, TX 75080
(972) 497-5000

(972) 497-5392 (Fax)
www.davelennox.com

Marathon Electric, Inc.
P.O. Box 8003
Wausau, WI 54402
(715) 675-3359
(715) 675-8050 (Fax)

McDonnell & Miller
(*See* ITT McDonnell & Miller)

Midco International Inc.
4140 West Victoria Street
Chicago, IL 60646-6790
(773) 604-8700
(773) 604-4070 (Fax)
www.midco-intl.com

Nordyne
P.O. Box 8809
O'Fallon, MO 63366
(636) 561-7300
(800) 222-4328
(636) 561-7365 (Fax)
www.nordyne.com

Raypak
2151 Eastman Ave.
Oxnard, CA 93030
(805) 278-5300
(805) 278-5468 (Fax)
www.raypak.com

RBI, Mestek Canada, Inc.
1300 Midway Blvd.
Mississauga Ontario L5T 2G8
(905) 670-5888
www.rbimestek.com

Rheem Manufacturing
5600 Old Greenwood Road
Fort Smith, AR 72908
(479) 646-4311
(479) 648-4812 (Fax)
www.rheemac.com

Robertshaw
(See Invensys)

Smith Cast Iron Boilers
260 North Elm Street
Westfield, MA 01085
(413) 562-9631
www.smithboiler.com

SpacePak
125 North Elm Street
Westfield, MA 01085
(413) 564-5530
www.spacepak.com

Spirax Sarco Inc.
Northpoint Business Park
1150 Northpoint Blvd
Blythewood, SC 29016
(803) 714-2000
(803) 714-2222
www.spiraxsarco.com

Sterling Hydronics
260 North Elm Street
Westfield, MA 01085
(413) 564-5535
www.sterlingheat.com

Sterling HVAC
125 North Elm Street
Westfield, MA 01085

(413) 564-5540
www.sterlinghvac.com

Suntec Industries Incorporated
2210 Harrison Ave
P.O. Box 7010
Rockford IL 61125-7010
(815) 226-3700
(815) 226-3848 (Fax)
www.suntecpumps.com

Thermo Pride
P.O. Box 217
North Judson, IN 46366
(574) 896-2133
(574) 896-5301
www.thermopride.com

Trane
P.O. Box 9010
Tyler, TX 75711-9010
(903) 581-3200
www.trane.com

Triangle Tube/Phase III Company, Inc.
Blackwood, NJ
(856) 228-1881
(856) 228-3584 (Fax)
www.triangletube.com

Vulcan Radiator (Mastec)
515 John Fitch Blvd
South Windsor, CT 06074
(413) 568-9571
www.mestec.com

Water Heater Innovations, Inc.
3107 Sibley Memorial Highway
Eagan, MN 55121

(800) 321-6718
www.marathonheaters.com

Watts Regulator Company
815 Chestnut Street
North Andover, MA 01845
(976) 688-1811
(978) 794-848 (Fax)
www.wattsreg.com

Wayne Combustion Systems
801 Glasgow Ave.
Fort Wayne, IN 46803
(800) 443-4625
www.waynecombustion.com

Weil-McLain
500 Blaine St.
Michigan City, IN 46360
(219) 879-6561
(219) 879-4025
www.weil-mclain.com

White-Rodgers, Div. of Emerson Electric Co.
9797 Reavis Rd.
St. Louis, MO 63123
(314) 577-1300
(314) 577-1517
www.white-rodgers.com

Wm. Powell Company
2503 Spring Grove Avenue
Cincinnati, OH 45214
(513) 852-2000
(513) 852-2997 (Fax)
www.powellvalves.com

York International Corporation
P.O. Box 1592-232F
York, PA 17405

(717) 771-7890
(717) 771-7381 (Fax)
www.york.com

Yukon-Eagle Wood & Multifuel Furnaces
10 Industrial Blvd.
P.O. Box 20
Palisade, MN 56469
(800) 358-0060
(800) 440-1994 (Fax)
www.yukon-eagle.com

John Zink Company
Gordon-Piatt Group
11920 East Apache
Tulsa, OK 74116
(800) 638-6940
www.johnzink.com

Appendix C

HVAC/R Education, Training, Certification, and Licensing

A career in the heating, ventilating, air-conditioning, and refrigeration (HVAC/R) trades requires special education and training. Formal education and training in the HVAC/R trades is available in many local public colleges and proprietary schools. Certification requires passing a standardized test indicating a thorough knowledge of the subject matter. The states also require HVAC/R technicians and contractors to take and pass licensing examinations.

HVAC/R Education and Training Programs

HVAC/R education and training programs are offered by four-year colleges, community colleges, proprietary schools, professional and trade associations, and manufacturers of HVAC/R appliances and system components.

One way to find a local school offering courses in HVAC/R education and training is to go online to the Cool Careers Web site at www.coolcareers.org. On their "Schools with HVACR Programs" page, you will find a list of all fifty states. Each state has the names of all the schools in that state offering programs in HVAC/R training. According to Cool Careers, their database contains the names of over 1300 training schools. You will have to contact the schools to enquire about entrance requirements, course content, class schedules, and financial aid.

Note

> If you don't have a computer, or know how to use one, go to the reference section in your local public library and ask the reference librarian to download the information from the internet and provide a printout. They will be willing to do this for you. It's part of the many services offered by the local public library.

Some of these schools offer only courses; others offer both courses and degrees. The least expensive courses are found at community colleges. The level of instruction will vary, depending on the school and the instructors. Your best source of information in this regard is local word of mouth. If you are already working as an entrance level trainee with a local HVAC firm, they should be able to help you find the best school and courses. After all, they often hire the graduates.

Cool Careers-Hot Jobs

The Cool Careers-Hot Jobs web site was created in 2000 by a coalition of organizations representing the heating, air conditioning, refrigeration, and plumbing industry. Its purpose is to provide information about education, training, jobs, and careers in the HVAC/R trades.

HVAC/R Certification

Certification means that the individual has taken and passed a standardized examination that certifies the individual's knowledge level. After the basic certification has been obtained, the technician can then study for and take certification exams at more advanced levels.

The following four organizations provide guidance and/or testing for the certification of HVAC/R technicians. Their addresses, telephone numbers, and web site addresses are listed in Appendix A (Professional and Trade Associations).

1. **Air-Conditioning and Refrigeration Institute (ARI).** ARI is a national trade association whose members represent most of the manufacturers of central air conditioning and refrigeration equipment. ARI administers the Industry Competencies Exam (ICE), which is given primarily to students from vocational school HVAC/R programs. The ARI also provides textbooks and training materials for preparing for both the ICE and EPA certification exams.
2. **North American Technician Excellence, Inc. (NATE).** NATE is a nonprofit organization established in 1997 by members of the HVAC/RE industry to test and certify technicians working in the heating ventilation, air-conditioning, and refrigeration trades. The tests are intended for experienced technicians.
3. **Refrigeration Service Engineers Society (RSES).** The RSES Educational Foundation was established in 1983 as a separate nonprofit organization to develop a comprehensive *voluntary* technician certification program (NTC). The program guides, tests, and certifies members through each of five levels of HVAC/R technician competency ranging from Level I (Technician) to Level V (Mastertech specialist).

4. **Air Conditioning Contractors of America (ACCA).** ACCA works in conjunction with RSES and NATE to provide a national certification program for HVAC/R technicians.

HVAC/R State Licensing

HVAC/R work is regulated at the state level by law. The law requires that a licensing exam must be taken and passed before working in an HVAC/R trade. It is the responsibility of the individual to contact the appropriate state office and obtain the necessary information about the state licensing examination. An easy way to locate the state office charged with licensing HVAC/R technicians and contractors is to ask the reference librarian at your local public library. You could also phone the state government and ask the operator to connect you to the office.

Appendix D

Data Tables

Table D-1 Equivalent Length of New Straight Pipe for Valves and Fittings for Turbulent Flow

Fittings			1/4	3/8	1/2	3/4	1	1 1/4	1 1/2	2	2 1/2	3	4	5	6	8	10	12	14	16	18	20	24
Regular 90° Ell	Screwed	Steel	2.3	3.1	3.6	4.4	5.2	6.6	7.4	8.5	9.3	11	13	—	—	—	—	—	—	—	—	—	—
		C. I.	—	—	—	—	—	—	—	—	—	9	11	—	—	—	—	—	—	—	—	—	—
	Flanged	Steel	—	—	0.92	1.2	1.6	2.1	2.4	3.1	3.6	4.4	5.9	7.3	8.9	12	14	17	18	21	23	25	30
		C. I.	—	—	—	—	—	—	—	—	—	3.6	4.8	—	7.2	9.8	12	15	17	19	22	24	28
Long Radius 90° Ell	Screwed	Steel	1.5	2	2.2	2.3	2.7	3.2	3.4	3.6	3.6	4	4.6	—	—	—	—	—	—	—	—	—	—
		C. I.	—	—	—	—	—	—	—	—	—	3.3	3.7	—	—	—	—	—	—	—	—	—	—
	Flanged	Steel	—	—	1.1	1.3	1.6	2	2.3	2.7	2.9	3.4	4.2	5	5.7	7	8	9	9.4	10	11	12	14
		C. I.	—	—	—	—	—	—	—	—	—	2.8	3.4	—	4.7	5.7	6.8	7.8	8.6	9.6	11	11	13
Regular 45° Ell	Screwed	Steel	0.34	0.52	0.71	0.92	1.3	1.7	2.1	2.7	3.2	4	5.5	—	—	—	—	—	—	—	—	—	—
		C. I.	—	—	—	—	—	—	—	—	—	3.3	4.5	—	—	—	—	—	—	—	—	—	—
	Flanged	Steel	—	—	0.45	0.59	0.81	1.1	1.3	1.7	2	2.6	3.5	4.5	5.6	7.7	9	11	13	15	16	18	22
		C. I.	—	—	—	—	—	—	—	—	—	2.1	2.9	—	4.5	6.3	8.1	9.7	12	13	15	17	20
Tee-Line Flow	Screwed	Steel	0.79	1.2	1.7	2.4	3.2	4.6	5.6	7.7	9.3	12	17	—	—	—	—	—	—	—	—	—	—
		C. I.	—	—	—	—	—	—	—	—	—	9.9	14	—	—	—	—	—	—	—	—	—	—
	Flanged	Steel	—	—	0.69	0.82	1	1.3	1.5	1.8	1.9	2.2	2.8	3.3	3.8	4.7	5.2	6	6.4	7.2	7.6	8.2	9.6
		C. I.	—	—	—	—	—	—	—	—	—	1.9	2.2	—	3.1	3.9	4.6	5.2	5.9	6.5	7.2	7.7	8.8
Tee-Branch Flow	Screwed	Steel	2.4	3.5	4.2	5.3	6.6	8.7	9.9	12	13	17	21	—	—	—	—	—	—	—	—	—	—
		C. I.	—	—	—	—	—	—	—	—	—	14	17	—	—	—	—	—	—	—	—	—	—
	Flanged	Steel	—	—	2	2.6	3.3	4.4	5.2	6.6	7.5	9.4	12	15	18	24	30	34	37	43	47	52	62
		C. I.	—	—	—	—	—	—	—	—	—	7.7	10	—	15	20	25	30	35	39	44	49	57

Pipe Size header spans columns 1/4 through 24.

180° Return Bend	Screwed	Steel	2.3	3.1	3.6	4.4	5.2	6.6	7.4	8.5	9.3	11	13	—	—	—	—	—	—	—	—	—	—
		C. I.	—	—	—	—	—	—	—	—	—	9	11	—	—	—	—	—	—	—	—	—	—
	Reg. Flanged	Steel	—	—	0.92	1.2	1.6	2.1	2.4	3.1	3.6	4.4	5.9	7.3	8.9	12	14	17	18	21	23	25	30
		C. I.	—	—	—	—	—	—	—	—	—	3.6	4.8	—	7.2	9.8	12	15	17	19	22	24	28
	Long Rad. Flanged	Steel	—	—	1.1	1.3	1.6	2	2.3	2.7	2.9	3.4	4.2	5	5.7	7	8	9	9.4	10	11	12	14
		C. I.	—	—	—	—	—	—	—	—	—	2.8	3.4	—	4.7	5.7	6.8	7.8	8.6	9.6	11	11	13
Globe Valve	Screwed	Steel	21	22	22	24	29	37	42	54	62	79	110	—	—	—	—	—	—	—	—	—	—
		C. I.	—	—	—	—	—	—	—	—	—	65	86	—	—	—	—	—	—	—	—	—	—
	Flanged	Steel	—	—	38	40	45	54	59	70	77	94	120	150	190	260	310	390	—	—	—	—	—
		C. I.	—	—	—	—	—	—	—	—	—	77	99	—	150	210	270	330	—	—	—	—	—
Gate Valve	Screwed	Steel	0.32	0.45	0.56	0.67	0.84	1.1	1.2	1.5	1.7	1.9	2.5	—	—	—	—	—	—	—	—	—	—
		C. I.	—	—	—	—	—	—	—	—	—	1.6	2	—	—	—	—	—	—	—	—	—	—
	Flanged	Steel	—	—	—	—	—	—	—	2.6	2.7	2.8	2.9	3.1	3.2	3.2	3.2	3.2	3.2	3.2	3.2	3.2	3.2
		C. I.	—	—	—	—	—	—	—	—	—	2.3	2.4	—	2.6	2.7	2.8	2.9	3	3	3	3	3
Angle Valve	Screwed	Steel	12.8	15	15	15	17	18	18	18	18	18	18	—	—	—	—	—	—	—	—	—	—
		C. I.	—	—	—	—	—	—	—	—	—	15	15	—	—	—	—	—	—	—	—	—	—
	Flanged	Steel	—	—	15	15	17	18	18	21	22	28	38	50	63	90	120	140	160	190	210	240	300
		C. I.	—	—	—	—	—	—	—	—	—	23	31	—	52	74	98	120	150	170	200	230	280
Swing Check Valve	Screwed	Steel	7.2	7.3	8	8.8	11	13	15	19	22	27	38	—	—	—	—	—	—	—	—	—	—
		C. I.	—	—	—	—	—	—	—	—	—	22	31	—	—	—	—	—	—	—	—	—	—
			—	—	3.8	5.3	7.2	10	12	17	21	27	38	50	63	90	120	140	—	—	—	—	—
		Steel	—	—	—	—	—	—	—	—	—	22	31	—	52	74	98	120	—	—	—	—	—
	Flanged	C. I.	0.14	0.18	0.21	0.24	0.29	0.36	0.39	0.45	0.47	0.53	0.65	—	—	—	—	—	—	—	—	—	—

(continued)

Table D-1 (continued)

Fittings			Pipe Size																				
			1/4	3/8	1/2	3/4	1	1 1/4	1 1/2	2	2 1/2	3	4	5	6	8	10	12	14	16	18	20	24
Coupling or Union	Screwed	Steel																					
		C. I.																					
Bell Mouth Inlet		Steel	—	—	—	—	—	—	—	—	—	0.44	0.62	—	—	—	—	—	—	—	—	—	—
		C. I.	0.04	0.07	0.1	0.13	0.18	0.26	0.31	0.43	0.52	0.67	0.95	1.3	1.6	2.3	2.9	3.5	4	4.7	5.3	6.1	7.6
Square Mouth Inlet		Steel	—	—	—	—	—	—	—	—	—	0.55	0.77	—	1.3	1.9	2.4	3	3.6	4.3	5	5.7	7
		C. I.	0.44	0.68	0.96	1.3	1.8	2.6	3.1	4.3	5.2	6.7	9.5	13	16	23	29	35	40	47	53	61	76
			—	—	—	—	—	—	—	—	—	5.5	7.7	—	13	19	24	30	36	43	50	57	70
Reentrant Pipe		Steel	0.88	1.4	1.9	2.6	3.6	5.1	6.2	8.5	10	13	19	25	32	45	58	70	80	95	110	120	150
		C. I.	—	—	—	—	—	—	—	—	—	11	15	—	26	37	49	61	73	86	100	110	140
Y-Strainer			—	4.6	5	6.6	7.7	18	20	27	29	34	42	53	61								

V1 V2 Sudden Enlargement

$$h = \frac{(V_1 - V_2)^2}{(2g)} \text{ Feet of Liquid; If } V_2 = 0 \; h = \frac{V^2}{(2g)} \text{ Feet of Liquid}$$

Courtesy The Hydraulic Institute (reprinted from the Standards of the Hydraulic Institute, Eleventh Edition, Copyright 1965)

Table D-2 Schedule 80 Pipe Dimensions

	Diameters		Nominal	Transverse Areas			Length of Pipe Per Square Foot of				
Size in	External in	Internal in	Thickness in	External in^2	Internal in^2	Metal in^2	External Surface ft	Internal Surface ft	Cubic Feet per ft of Pipe	Weight per ft Pounds	Number Threads per in of Screw
1/8	0.405	0.215	0.095	0.129	0.036	0.093	9.431	17.75	0.00025	0.314	27
1/4	0.54	0.302	0.119	0.229	0.072	0.157	7.073	12.65	0.0005	0.535	18
3/8	0.675	0.423	0.126	0.358	0.141	0.217	5.658	9.03	0.00098	0.738	18
1/2	0.84	0.546	0.147	0.554	0.234	0.32	4.547	7	0.00163	1	14
3/4	1.05	0.742	0.154	0.866	0.433	0.433	3.637	5.15	0.003	1.47	14
1	1.315	0.957	0.179	1.358	0.719	0.639	2.904	3.995	0.005	2.17	11 1/2
1 1/4	1.66	1.278	0.191	2.164	1.283	0.881	2.301	2.99	0.00891	3	11 1/2
1 1/2	1.9	1.5	0.2	2.835	1.767	1.068	2.01	2.542	0.01227	3.65	11 1/2
2	2.375	1.939	0.218	4.43	2.953	1.477	1.608	1.97	0.02051	5.02	11 1/2
2 1/2	2.875	2.323	0.276	6.492	4.238	2.254	1.328	1.645	0.02943	7.66	8
3	3.5	2.9	0.3	9.621	6.605	3.016	1.091	1.317	0.04587	10.3	8
3 1/2	4	3.364	0.318	12.56	8.888	3.678	0.954	1.135	0.06172	12.5	8
4	4.5	3.826	0.337	15.9	11.497	4.407	0.848	0.995	0.0798	14.9	8
5	5.563	4.813	0.375	24.3	18.194	6.112	0.686	0.792	0.1263	20.8	8
6	6.625	5.761	0.432	34.47	26.067	8.3	0.576	0.673	0.181	28.6	8
8	8.625	7.625	0.5	58.42	46.663	12.76	0.442	0.501	0.3171	43.4	8

(continued)

Table D-2 (continued)

Size	Diameters		Nominal	Transverse Areas			Length of Pipe Per Square Foot of		Cubic Feet	Weight	Number Threads
Size in	External in	Internal in	Thickness in	External in^2	Internal in^2	Metal in^2	External Surface ft	Internal Surface ft	per ft of Pipe	per ft Pounds	per in of Screw
10	10.75	9.564	0.593	90.76	71.84	18.92	0.355	0.4	0.4989	64.4	8
12	12.75	11.376	0.687	127.64	101.64	26	0.299	0.336	0.7058	88.6	
14	14	12.5	0.75	153.94	122.72	31.22	0.272	0.306	0.8522	107	
16	16	14.314	0.843	201.05	160.92	40.13	0.238	0.263	1.112	137	
18	18	16.126	0.937	254.85	204.24	50.61	0.212	0.237	1.418	171	
20	20	17.938	1.031	314.15	252.72	61.43	0.191	0.208	1.755	209	
24	24	21.564	1.218	452.4	365.22	87.18	0.159	0.177	2.536	297	

Courtesy Sarco Company, Inc.

Table D-3 Schedule 40 Pipe Dimensions

	Diameters			Transverse Areas			Length of Pipe Per ft^2 of				
Size in	External in	Internal in	Nominal Thickness in	External in^2	Internal in^2	Metal in^2	External Surface ft	Internal Surface ft	Cubic Feet per ft of Pipe	Weight per ft Pounds	Number Threads per in of Screw
⅛	0.405	0.269	0.068	0.129	0.057	0.072	9.431	14.199	0.00039	0.244	27
¼	0.54	0.364	0.088	0.229	0.104	0.125	7.073	10.493	0.00072	0.424	18
⅜	0.675	0.493	0.091	0.358	0.191	0.167	5.658	7.747	0.00133	0.567	18
½	0.84	0.622	0.109	0.554	0.304	0.25	4.547	6.141	0.00211	0.85	14
¾	1.05	0.824	0.113	0.866	0.533	0.333	3.637	4.635	0.0037	1.13	14
1	1.315	1.049	0.133	1.358	0.864	0.494	2.904	3.641	0.006	1.678	11½
1¼	1.66	1.38	0.14	2.164	1.495	0.669	2.301	2.767	0.01039	2.272	11½
1½	1.9	1.61	0.145	2.835	2.036	0.799	2.01	2.372	0.01414	2.717	11½
2	2.375	2.067	0.154	4.43	3.355	1.075	1.608	1.847	0.0233	3.652	11½
2½	2.875	2.469	0.203	6.492	4.788	1.704	1.328	1.547	0.03325	5.793	8
3	3.5	3.068	0.216	9.621	7.393	2.228	1.091	1.245	0.05134	7.575	8
3½	4	3.548	0.226	12.56	9.886	2.68	0.954	1.076	0.06866	9.109	8
4	4.5	4.026	0.237	15.9	12.73	3.174	0.848	0.948	0.0884	10.79	8
5	5.563	5.047	0.258	24.3	20	4.3	0.686	0.756	0.1389	14.61	8
6	6.625	6.065	0.28	34.47	28.9	5.581	0.576	0.629	0.2006	18.97	8
8	8.625	7.981	0.322	58.42	50.02	8.399	0.442	0.478	0.3552	28.55	8

(continued)

Table D-3 (continued)

	Diameters			Transverse Areas			Length of Pipe Per ft^2 of				
Size in	External in	Internal in	Nominal Thickness in	External in^2	Internal in^2	Metal in^2	External Surface ft	Internal Surface ft	Cubic Feet per ft of Pipe	Weight per ft Pounds	Number Threads per in of Screw
10	10.75	10.02	0.365	90.76	78.85	11.9	0.355	0.381	0.5476	40.48	8
12	12.75	11.938	0.406	127.64	111.9	15.74	0.299	0.318	0.7763	53.6	
14	14	13.125	0.437	153.94	135.3	18.64	0.272	0.28	0.9354	63	
16	16	15	0.5	201.05	176.7	24.35	0.238	0.254	1.223	78	
18	18	16.874	0.563	254.85	224	30.85	0.212	0.226	1.555	105	
20	20	18.814	0.593	314.15	278	36.15	0.191	0.203	1.926	123	
24	24	22.626	0.687	452.4	402.1	50.3	0.159	0.169	2.793	171	

Courtesy Sarco Company, Inc.

Table D-4 Properties of Saturated Steam

	Gauge Pressure psig	Temperature °F	Heat in Btu/lb			Specific Volume ft³/lb	Gauge Pressure psig	Temperature °F	Heat in Btu/lb			Specific Volume ft³ per lb
			Sensible	Latent	Total				Sensible	Latent	Total	
In Vac.	25	134	102	1017	1119	142	150	366	339	857	1196	2.74
In Vac.	20	162	129	1001	1130	73.9	155	368	341	885	1196	2.68
In Vac.	15	179	147	990	1137	51.3	160	371	344	853	1197	2.6
In Vac.	10	192	160	982	1142	39.4	165	373	346	851	1197	2.54
In Vac.	5	203	171	976	1147	31.8	170	375	348	849	1197	2.47
	0	**212**	**180**	**970**	**1150**	**26.8**	**175**	**377**	**351**	**847**	**1198**	**2.41**
	1	215	183	968	1151	25.2	180	380	353	845	1198	2.34
	2	219	187	966	1153	23.5	185	382	355	843	1198	2.29
	3	222	190	964	1154	22.3	190	384	358	841	1199	2.24
	4	224	192	962	1154	21.4	195	386	360	839	1199	2.19
	5	**227**	**195**	**960**	**1155**	**20.1**	**200**	**388**	**362**	**837**	**1199**	**2.14**
	6	230	198	959	1157	19.4	205	390	364	836	1200	2.09
	7	232	200	957	1157	18.7	210	392	366	834	1200	2.05
	8	233	201	956	1157	18.4	215	394	368	832	1200	2
	9	237	205	954	1159	17.1	220	396	370	830	1200	1.96
	10	**239**	**207**	**953**	**1160**	**16.5**	**225**	**397**	**372**	**828**	**1200**	**1.92**
	12	244	212	949	1161	15.3	230	399	374	827	1201	1.89
	14	248	216	947	1163	14.3	235	401	376	825	1201	1.85

(continued)

Table D-4 (continued)

Gauge Pressure psig	Temper-ature °F	Heat in Btu/lb			Specific Volume ft³/lb	Gauge Pressure psig	Temper-ature °F	Heat in Btu/lb			Specific Volume ft³ per lb
		Sensible	Latent	Total				Sensible	Latent	Total	
16	252	220	944	1164	13.4	240	403	378	823	1201	1.81
18	256	224	941	1165	12.6	245	404	380	822	1202	1.78
20	**259**	**227**	**939**	**1166**	**11.9**	**250**	**406**	**382**	**820**	**1202**	**1.75**
22	262	230	937	1167	11.3	255	408	383	819	1202	1.72
24	265	233	934	1167	10.8	260	409	385	817	1202	1.69
26	268	236	933	1169	10.3	265	411	387	815	1202	1.66
28	271	239	930	1169	9.85	270	413	389	814	1203	1.63
30	**274**	**243**	**929**	**1172**	**9.46**	**275**	**414**	**391**	**812**	**1203**	**1.6**
32	277	246	927	1173	9.1	280	416	392	811	1203	1.57
34	279	248	925	1173	8.75	285	417	394	809	1203	1.55
36	282	251	923	1174	8.42	290	418	395	808	1203	1.53
38	284	253	922	1175	8.08	295	420	397	806	1203	1.49
40	**286**	**256**	**920**	**1176**	**7.82**	**300**	**421**	**398**	**805**	**1203**	**1.47**
42	289	258	918	1176	7.57	305	423	400	803	1203	1.45
44	291	260	917	1177	7.31	310	425	402	802	1204	1.43
46	293	262	915	1177	7.14	315	426	404	800	1204	1.41
48	295	264	914	1178	6.94	320	427	405	799	1204	1.38
50	**298**	**267**	**912**	**1179**	**6.68**	**325**	**429**	**407**	**797**	**1204**	**1.36**
55	300	271	909	1180	6.27	330	430	408	796	1204	1.34

Table D-4 (continued)

Gauge Pressure psig	Temperature °F	Heat in Btu/lb			Specific Volume ft^3/lb	Gauge Pressure psig	Temperature °F	Heat in Btu/lb			Specific Volume ft^3 per lb
		Sensible	Latent	Total				Sensible	Latent	Total	
60	307	277	906	1183	5.84	335	432	410	794	1204	1.33
65	312	282	901	1183	5.49	340	433	411	793	1204	1.31
70	316	286	898	1184	5.18	345	434	413	791	1204	1.29
75	**320**	**290**	**895**	**1185**	**4.91**	**350**	**435**	**414**	**790**	**1204**	**1.28**
80	324	294	891	1185	4.67	355	437	416	789	1205	1.26
85	328	298	889	1187	4.44	360	438	417	788	1205	1.24
90	331	302	886	1188	4.24	365	440	419	786	1205	1.22
95	335	305	883	1188	4.05	370	441	420	785	1205	1.2
100	**338**	**309**	**880**	**1189**	**3.89**	**375**	**442**	**421**	**784**	**1205**	**1.19**
105	341	312	878	1190	3.74	380	443	422	783	1205	1.18
110	344	316	875	1191	3.59	385	445	424	781	1205	1.16
115	347	319	873	1192	3.46	390	446	425	780	1205	1.14
120	350	322	871	1193	3.34	395	447	427	778	1205	1.13
125	**353**	**325**	**868**	**1193**	**3.23**	**400**	**448**	**428**	**777**	**1205**	**1.12**
130	356	328	866	1194	3.12	450	460	439	766	1205	1
140	361	333	861	1194	2.92	500	470	453	751	1204	0.89
145	363	336	859	1195	2.84	550	479	464	740	1204	0.82
						600	489	475	728	1203	0.74

Courtesy Sarco Company, Inc.

Table D-5 Friction Loss for Water in Feet per 100 Feet of Schedule 40 Steel Pipe

U.S. gal/min	Velocity ft/sc	hf Friction	U.S. gal/min	Vel. ft/sec	hf Friction
	3/8" Pipe			1/2" Pipe	
1.4	2.25	9.03	2	2.11	5.5
1.6	2.68	11.6	2.5	2.64	8.24
1.8	3.02	14.3	3	3.17	11.5
2	3.36	17.3	3.5	3.7	15.3
2.5	4.2	26	4	4.22	19.7
3	5.04	36.6	5	5.28	29.7
3.5	5.88	49	6	6.34	42
4	6.72	63.2	7	7.39	56
5	8.4	96.1	8	8.45	72.1
6	10.08	136	9	9.5	90.1
7	11.8	182	10	10.56	110.6
8	13.4	236	12	12.7	156
9	15.1	297	14	14.8	211
10	16.8	364	16	16.9	270
	3/4" Pipe			1" Pipe	
4	2.41	4.85	6	2.23	3.16
5	3.01	7.27	8	2.97	5.2
6	3.61	10.2	10	3.71	7.9
7	4.21	13.6	12	4.45	11.1
8	4.81	17.3	14	5.2	14.7
9	5.42	21.6	16	5.94	19
10	6.02	26.5	18	6.68	23.7
12	7.22	37.5	20	7.42	28.9
14	8.42	50	22	8.17	34.8
16	9.63	64.8	24	8.91	41
18	10.8	80.9	26	9.65	47.8
20	12	99	28	10.39	55.1
22	13.2	120	30	11.1	62.9
24	14.4	141	35	13	84.4
26	15.6	165	40	14.8	109
28	16.8	189	45	16.7	137
			50	18.6	168

(continued)

Table D-5 (continued)

U.S. gal/min	Velocity ft/sc	hf Friction	U.S. gal/min	Vel. ft/sec	hf Friction
	1¼" Pipe			1½" Pipe	
12	2.57	2.85	16	2.52	2.26
14	3	3.77	18	2.84	2.79
16	3.43	4.83	20	3.15	3.38
18	3.86	6	22	3.47	4.05
20	4.29	7.3	24	3.78	4.76
22	4.72	8.72	26	4.1	5.54
24	5.15	10.27	28	4.41	6.34
26	5.58	11.94	30	4.73	7.2
28	6.01	13.7	35	5.51	9.63
30	6.44	15.6	40	6.3	12.41
35	7.51	21.9	45	7.04	15.49
40	8.58	27.1	50	7.88	18.9
45	9.65	33.8	55	8.67	22.7
50	10.7	41.4	60	9.46	26.7
55	11.8	49.7	65	10.24	31.2
60	12.9	58.6	70	11.03	36
65	13.9	68.6	75	11.8	41.2
70	15	79.2	80	12.6	46.6
75	16.1	90.6	85	13.4	52.4
			90	14.2	58.7
			95	15	65
			100	15.8	71.6
	2" Pipe			2½" Pipe	
25	2.39	1.48	35	2.35	1.15
30	2.87	2.1	40	2.68	1.47
35	3.35	2.79	45	3.02	1.84
40	3.82	3.57	50	3.35	2.23
45	4.3	4.4	60	4.02	3.13
50	4.78	5.37	70	4.69	4.18
60	5.74	7.58	80	5.36	5.36
70	6.69	10.2	90	6.03	6.69
80	7.65	13.1	100	6.7	8.18
90	8.6	16.3	120	8.04	11.5

(continued)

Table D-5 (continued)

U.S. gal/min	Velocity ft/sc	hf Friction	U.S. gal/min	Vel. ft/sec	hf Friction
	2" Pipe			**2½" Pipe**	
100	4.34	2.72	200	5.04	12.61
120	11.5	28.5	160	10.7	20
140	13.4	38.2	180	12.1	25.2
160	15.3	49.5	200	13.4	30.7
			220	14.7	37.1
			240	16.1	43.8
	3" Pipe			**4" Pipe**	
50	2.17	0.762	100	2.52	0.718
60	2.6	1.06	120	3.02	1.01
70	3.04	1.4	140	3.53	1.35
80	3.47	1.81	160	4.03	1.71
90	3.91	2.26	180	4.54	2.14
100	3.34	2.75	200	5.04	2.61
120	5.21	3.88	220	5.54	3.13
140	6.08	5.19	240	6.05	3.7
160	6.94	6.68	260	6.55	4.3
180	7.81	8.38	280	7.06	4.95
200	8.68	10.2	300	7.56	5.63
220	9.55	12.3	350	8.82	7.54
240	10.4	14.5	400	10.1	9.75
260	11.3	16.9	450	11.4	12.3
280	12.2	19.5	500	12.6	14.4
300	13	22.1	550	13.9	18.1
350	15.2	30	600	15.1	21.4
	5" Pipe			**6" Pipe**	
160	2.57	0.557	220	2.44	0.411
180	2.89	0.698	240	2.66	0.482
200	3.21	0.847	260	2.89	0.56
220	3.53	1.01	300	3.33	0.733
240	3.85	1.19	350	3.89	0.98
260	4.17	1.38	400	4.44	1.25
300	4.81	1.82	450	5	1.56

(continued)

Table D-5 *(continued)*

U.S. gal/min	*Velocity ft/sc*	*hf Friction*	*U.S. gal/min*	*Vel. ft/sec*	*hf Friction*
350	5.61	2.43	500	5.55	1.91
400	6.41	3.13	600	6.66	2.69
450	7.22	3.92	700	7.77	3.6
500	8.02	4.79	800	8.88	4.64
600	9.62	6.77	900	9.99	5.81
700	11.2	9.13	1000	11.1	7.1
800	12.8	11.8	1100	12.2	8.52
900	14.4	14.8	1200	13.3	10.1
1000	16	18.2	1300	14.4	11.7
			1400	15.5	13.6

Courtesy Sarco Company, Inc.

Table D-6 Flow of Water through Schedule 40 Steel Pipe

Pressure Drop 1000 Feet of Schedule 40 Steel Pipe, in Pounds per Square Inch

Discharge gal/min	Velocity ft/sec	Pressure Drop	Velocity ft/sec	Pressure Drop	Velocity ft/sec	Pressure Drop	Velocity ft/sec	Pressure Drop	Velocity ft/sec	Pressure Drop	Velocity ft/sec	Pressure Drop	Velocity ft/sec	Pressure Drop	Velocity ft/sec	Pressure Drop	Velocity ft/sec	Pressure Drop
	1"																	
1	0.37	0.49	1¼"															
2	0.74	1.7	0.43	0.45	1½"													
3	1.12	3.53	0.64	0.94	0.47	0.44												
4	1.49	5.94	0.86	1.55	0.63	0.74	2"											
5	1.86	9.02	1.07	2.36	0.79	1.12												
6	2.24	12.25	1.28	3.3	0.95	1.53	0.57	0.46										
8	2.98	21.1	1.72	5.52	1.26	2.63	0.76	0.75	2½"									
10	3.72	30.8	2.14	8.34	1.57	3.86	0.96	1.14	0.67	0.48								
15	5.6	64.6	3.21	17.6	2.36	8.13	1.43	2.33	1	0.99	3"							
20	7.44	110.5	4.29	29.1	3.15	13.5	1.91	3.86	1.34	1.64	0.87	0.59	3½"					
25			5.36	43.7	3.94	20.2	2.39	5.81	1.68	2.48	1.08	0.67	0.81	0.42				
30			6.43	62.9	4.72	29.1	2.87	8.04	2.01	3.43	1.3	1.21	0.97	0.6	4"			
35			7.51	82.5	5.51	38.2	3.35	10.95	2.35	4.49	1.52	1.58	1.14	0.79	0.88	0.42		
40					6.3	47.8	3.82	13.7	2.68	5.88	1.74	2.06	1.3	1	1.01	0.53		
45					7.08	60.6	4.3	17.4	3	7.14	1.95	2.51	1.46	1.21	1.13	0.67		
50					7.87	74.7	4.78	20.6	3.35	8.82	2.17	3.1	1.62	1.44	1.26	0.8		
60							5.74	29.6	4.02	12.2	2.6	4.29	1.95	2.07	1.51	1.1	5"	
70							6.69	38.6	4.69	15.3	3.04	5.84	2.27	2.71	1.76	1.5	1.12	0.48

Table D-6 (continued)

Pressure Drop 1000 Feet of Schedule 40 Steel Pipe, in Pounds per Square Inch																		
Dis-charge gal/min	Veloc-ity ft/sec	Pres-sure Drop	Veloc-ity ft/sec	Pres-sure Drop	Veloc-ity ft/sec	Pres-sure Drop	Veloc-ity ft/sec	Pres-sure Drop	Veloc-ity ft/sec	Pres-sure Drop	Veloc-ity ft/sec	Pres-sure Drop	Veloc-ity ft/sec	Pres-sure Drop	Veloc-ity ft/sec	Pres-sure Drop	Veloc-ity ft/sec	Pres-sure Drop
80		6"					7.65	50.3	5.37	21.7	3.48	7.62	2.59	3.53	2.01	1.87	1.28	0.63
90							8.6	63.6	6.04	26.1	3.91	9.22	2.92	4.46	2.26	2.37	1.44	0.8
100	1.11	0.39					9.56	75.1	6.71	32.3	4.34	11.4	3.24	5.27	2.52	2.81	1.6	0.95
125	1.39	0.56							8.38	48.2	5.45	17.1	4.05	7.86	3.15	4.38	2	1.48
150	1.67	0.78							10.06	60.4	6.51	23.3	4.86	11.3	3.78	6.02	2.41	2.04
175	1.94	1.06		8"					11.73	90	7.59	32	5.67	14.7	4.41	8.2	2.81	2.78
200	2.22	1.32									8.68	39.7	6.48	19.2	5.04	10.2	3.21	3.46
225	2.5	1.66	1.44	0.44							9.77	50.2	7.29	23.1	5.67	12.9	3.61	4.37
250	2.78	2.05	1.6	0.55							10.85	61.9	8.1	28.5	6.3	15.9	4.01	5.14
275	3.06	2.36	1.76	0.63							11.94	75	8.91	34.4	6.93	18.3	4.41	6.22
300	3.33	2.8	1.92	0.75							13.02	84.7	9.72	40.9	7.56	21.8	4.81	7.41
325	3.61	3.29	2.08	0.88									10.53	45.5	8.18	25.5	5.21	8.25
350	3.89	3.62	2.24	0.97									11.35	52.7	8.82	29.7	5.61	9.57
375	4.16	4.16	2.4	1.11									12.17	60.7	9.45	32.3	6.01	11
400	4.44	4.72	2.56	1.27									13.78	77.8	10.7	41.5	6.82	14.1
425	4.72	5.34	2.27	1.43									12.97	68.9	10.08	39.7	6.41	12.9
450	5	5.96	2.88	1.6		10"							14.59	87.3	11.33	46.5	7.22	15
475	5.27	6.66	3.04	1.69	1.93	0.3									11.96	51.7	7.62	16.7

(continued)

Table D-6 (continued)

Pressure Drop 1000 Feet of Schedule 40 Steel Pipe, in Pounds per Square Inch

Discharge gal/min	Velocity ft/sec	Pressure Drop	Velocity ft/sec	Pressure Drop	Velocity ft/sec	Pressure Drop	Velocity ft/sec	Pressure Drop	Velocity ft/sec	Pressure Drop	Velocity ft/sec	Pressure Drop	Velocity ft/sec	Pressure Drop	Velocity ft/sec	Pressure Drop	Velocity ft/sec	Pressure Drop
500	5.55	7.39	3.2	1.87	2.04	0.63									12.59	57.3	8.02	18.5
550	6.11	8.94	3.53	2.26	2.24	0.7									13.84	69.3	8.82	22.4
600	6.66	10.6	3.85	2.7	2.44	0.8612"									15.1	82.5	9.62	26.7
650	7.21	11.8	4.17	3.16	2.65	1.01											10.42	31.3
700	7.77	13.7	4.49	3.69	2.85	1.18	2.01	0.48									11.22	36.3
750	8.32	15.7	4.81	4.21	3.05	1.35	2.15	0.55									12.02	41.6
800	8.88	17.8	5.13	4.79	3.26	1.54	2.29	0.62	14"								12.82	44.7
850	9.44	20.2	5.45	5.11	3.46	1.74	2.44	0.7	2.02	0.43							13.62	50.5
900	10	22.6	5.77	5.73	3.66	1.94	2.58	0.79	2.14	0.48							14.42	56.6
950	10.55	23.7	6.09	6.38	3.87	2.23	2.72	0.88	2.25	0.53							15.22	63.1
1,000	11.1	26.3	6.41	7.08	4.07	2.4	2.87	0.98	2.38	0.59							16.02	70
1,100	12.22	31.8	7.05	8.56	4.48	2.74	3.16	1.18	2.61	0.68	16"						17.63	84.6
1,200	13.32	37.8	7.69	10.2	4.88	3.27	3.45	1.4	2.85	0.81	2.18	0.4						
1,300	14.43	44.4	8.33	11.3	5.29	3.86	3.73	1.56	3.09	0.95	2.36	0.47						
1,400	15.54	51.5	8.97	13	5.7	4.44	4.02	1.8	3.32	1.1	2.54	0.54						
1,500	16.65	55.5	9.62	15	6.1	5.11	4.3	2.07	3.55	1.19	2.73	0.62						
1,600	17.76	68.1	10.26	17	6.51	5.46	4.59	2.36	3.8	1.35	2.91	0.71	18"					
1,800	19.98	79.8	11.54	21.6	7.32	6.91	5.16	2.98	4.27	1.71	3.27	0.85	2.58	0.48				

Table D-6 (continued)

Pressure Drop 1000 Feet of Schedule 40 Steel Pipe, in Pounds per Square Inch																		
Dis-charge gal/min	Veloc-ity ft/sec	Pres-sure Drop	Veloc-ity ft/sec	Pres-sure Drop	Veloc-ity ft/sec	Pres-sure Drop	Veloc-ity ft/sec	Pres-sure Drop	Veloc-ity ft/sec	Pres-sure Drop	Veloc-ity ft/sec	Pres-sure Drop	Veloc-ity ft/sec	Pres-sure Drop	Veloc-ity ft/sec	Pres-sure Drop	Veloc-ity ft/sec	Pres-sure Drop
2,000	22.2	98.5	12.83	25	8.13	8.54	5.73	3.47	4.74	2.11	3.63	1.05	2.88	0.56				
2,500			16.03	39	10.18	12.5	7.17	5.41	5.92	3.09	4.54	1.63	3.59	0.88	20"			
3,000			19.24	52.4	12.21	18	8.6	7.31	7.12	4.45	5.45	2.21	4.31	1.27	3.45	0.73		
3,500			22.43	71.4	14.25	22.9	10.03	9.95	8.32	6.18	6.35	3	5.03	1.52	4.03	0.94	24"	
4,000			25.65	93.3	16.28	29.9	11.48	13	9.49	7.92	7.25	3.92	5.74	2.12	4.61	1.22	3.19	0.51
4,500					18.31	37.8	12.9	15.4	10.67	9.36	8.17	4.97	6.47	2.5	5.19	1.55	3.59	0.6
5,000					20.35	46.7	14.34	18.9	11.84	11.6	9.08	5.72	7.17	3.08	5.76	1.78	3.99	0.74
6,000					24.42	67.2	17.21	27.3	14.32	15.4	10.88	8.24	8.62	4.45	6.92	2.57	4.8	1
7,000					28.5	85.1	20.08	37.2	16.6	21	12.69	12.2	10.04	6.06	8.06	3.5	5.68	1.36
8,000							22.95	45.1	18.98	27.4	14.52	13.6	11.48	7.34	9.23	4.57	6.38	1.78
9,000							25.8	57	21.35	34.7	16.32	17.2	12.92	9.2	10.37	5.36	7.19	2.25
10,000							28.63	70.4	23.75	42.9	18.16	21.2	14.37	11.5	11.53	6.63	7.96	2.78
12,000							34.38	93.6	28.5	61.8	21.8	30.9	17.23	16.5	13.83	9.54	9.57	3.71
14,000									33.2	84	25.42	41.6	20.1	20.7	16.14	12	11.18	5.05
16,000											29.05	54.4	22.96	27.1	18.43	15.7	12.77	6.6

Courtesy Sarco Company, Inc.

Table D-7 Warmup Load in Pounds of Steam per 100 Feet of Steam Main (Ambient Temperature 70°F)[a]*

Steam Pressure	Main Size													0°F Correction
(psig)	2"	2½"	3"	4"	5"	6"	8"	10"	12"	14"	16"	18"	20"	24"	Factor[a]
0	6.2	9.7	12.8	18.2	24.6	31.9	48	68	90	107	140	176	207	208	1.5
5	6.9	11	14.4	20.4	27.7	35.9	48	77	101	120	157	198	233	324	1.44
10	7.5	11.8	15.5	22	29.9	38.8	58	83	109	130	169	213	251	350	1.41
20	8.4	13.4	17.5	24.9	33.8	43.9	66	93	124	146	191	241	284	396	1.37
40	3.9	15.8	20.6	29.3	39.7	51.6	78	110	145	172	225	284	334	465	1.32
60	11	17.5	22.9	32.6	44.2	57.3	86	122	162	192	250	316	372	518	1.29
80	12	19	24.9	35.3	47.9	62.1	93	132	175	208	271	342	403	561	1.27
100	12.8	20.3	26.6	37.8	51.2	66.5	100	142	188	222	290	366	431	600	1.26
125	13.7	21.7	28.4	40.4	54.8	71.1	107	152	200	238	310	391	461	642	1.25
150	14.5	23	30	42.8	58	75.2	113	160	212	251	328	414	487	679	1.24
175	15.3	24.2	31.7	45.1	61.2	79.4	119	169	224	265	347	437	514	716	1.23
200	16	25.3	33.1	47.1	63.8	82.8	125	177	234	277	362	456	537	748	1.22
250	17.2	27.3	35.8	50.8	68.9	89.4	134	191	252	299	390	492	579	807	1.21
300	25	38.3	51.3	74.8	104	142.7	217	322	443	531	682	854	1045	1182	1.2
400	27.8	42.6	57.1	83.2	115.7	158.7	241	358	493	590	759	971	1163	1650	1.18
500	30.2	46.3	62.1	90.5	125.7	172.6	262	389	535	642	825	1033	1263	1793	1.17
600	32.7	50.1	67.1	97.9	136	186.6	284	421	579	694	893	1118	1367	1939	1.16

**Loads based on Schedule 40 pipe for pressures up to and including 250 psig and on Schedule 80 pipe for pressures above 250 psig.*

[a]For outdoor temperature of 0°F, multiply load value in table for each main size by correction factor corresponding to steam pressure.

Courtesy Sarco Company, Inc.

Table D-8 Condensation Load in Pounds per Hour per 100 Feet of Insulated Steam Main (Ambient Temperature 70°F; Insulation 80% Efficient)[a]*

Steam Pressure (psig)	Main Size														0°F Correction Factor[a]
	2"	2½"	3"	4"	5"	6"	8"	10"	12"	14"	16"	18"	20"	24"	
10	6	7	9	11	13	16	20	24	29	32	36	39	44	53	1.58
30	8	9	11	14	17	20	26	32	38	42	48	51	57	68	1.5
60	10	12	14	18	24	27	33	41	49	54	62	67	74	89	1.45
100	12	15	18	22	28	33	41	51	61	67	77	83	93	111	1.41
125	13	16	20	24	30	36	45	56	66	73	84	90	101	121	1.39
175	16	19	23	26	33	38	53	66	78	86	98	107	119	142	1.38
250	18	22	27	34	42	50	62	77	92	101	116	126	140	168	1.36
300	20	25	30	37	46	54	68	85	101	111	126	138	154	184	1.35
400	23	28	34	43	53	63	80	99	118	130	148	162	180	216	1.33
500	27	33	39	49	61	73	91	114	135	148	170	185	206	246	1.32
600	30	37	44	55	68	82	103	128	152	167	191	208	232	277	1.31

**Chart loads represent losses due to radiation and convection for saturated steam.*

[a]For outdoor temperature of 0°F, multiply load value in table for each main size by correction factor corresponding to steam pressure.

Courtesy Sarco Company, Inc.

Table D-9 Flange Standards (All dimensions are in inches)

	125-lb Cast Iron		*ASA B16.1*												
Pipe Size	**½**	**¾**	**1**	**1¼**	**1½**	**2**	**2½**	**3**	**3½**	**4**	**5**	**6**	**8**	**10**	**12**
Diameter of Flange			4¼	4⅝	5	6	7	7½	8½	9	10	11	13½	16	19
Thickness of Flange (min)[1]			7/16	½	9/16	⅝	11/16	¾	13/16	15/16	15/16	1	1⅛	1 3/16	1¼
Diameter of Bolt Circle			3⅛	3½	3⅞	4¾	5½	6	7	7½	8½	9½	11¾	14¼	17
Number of Bolts			4	4	4	4	4	4	8	8	8	8	8	12	12
Diameter of Bolts			½	½	½	⅝	⅝	⅝	⅝	⅝	¾	¾	¾	⅞	⅞

	250-lb Cast Iron		ASA B16.2												
Pipe Size	**½**	**¾**	**1**	**1¼**	**1½**	**2**	**2½**	**3**	**3½**	**4**	**5**	**6**	**8**	**10**	**12**
Diameter of Flange			4⅞	5¼	6⅛	6½	7½	8¼	9	10	11	12½	15	17½	20½
Thickness of Flange (min)[2]			11/16	¾	13/16	⅞	1	1⅛	1 3/16	1¼	1⅜	1 7/16	1⅝	1⅞	2
Diameter of Raised Face			2 11/16	3 1/16	3 9/16	4 3/16	4 15/16	6 5/16	6 5/16	6 15/16	8 5/16	9 11/16	11 15/16	14 1/16	16 7/16
Diameter of Bolt Circle			3½	3⅞	4½	5	5⅞	6⅝	7¼	7⅞	9¼	10⅝	13	15¼	17¾
Number of Bolts			4	4	4	8	8	8	8	8	8	12	12	16	16
Diameter of Bolts			⅝	⅝	¾	⅝	¾	¾	¾	¾	¾	¾	⅞	1	1⅛

(continued)

Table D-9 (continued)

	150-lb Bronze		ASA B16.24												
Pipe Size	**½**	**¾**	**1**	**1¼**	**1½**	**2**	**2½**	**3**	**3½**	**4**	**5**	**6**	**8**	**10**	**12**
Diameter of Flange	3½	3⅞	4¼	4⅝	5	6	7	7½	8½	9	10	11	13½	16	19
Thickness of Flange (min)[3]	5/16	11/32	⅜	13/32	7/16	½	9/16	⅝	11/16	11/16	¾	13/16	15/16	1	1 1/16
Diameter of Bolt Circle	2⅜	2¾	3⅛	3½	3⅞	4¾	5½	6	7	7½	8½	9½	11¾	14¼	17
Number of Bolts	4	4	4	4	4	4	4	4	8	8	8	8	8	12	12
Diameter of Bolts	½	½	½	½	½	⅝	⅝	⅝	⅝	⅝	¾	¾	¾	⅞	⅞

	300-lb Bronze		ASA B16.24												
Pipe Size	**½**	**¾**	**1**	**1¼**	**1½**	**2**	**2½**	**3**	**3½**	**4**	**5**	**6**	**8**	**10**	**12**
Diameter of Flange	3¾	4⅝	4⅞	5¼	6⅛	6½	7½	8¼	9	10	11	12½	15		
Thickness of Flange (min)[4]	½	17/32	19/32	⅝	11/16	¾	13/16	29/32	31/32	1 1/16	1⅛	1 3/16	1⅜		
Diameter of Bolt Circle	2⅝	3¼	3½	3⅞	4½	5	5⅞	6⅝	7¼	7⅞	9¼	10⅝	13		
Number of Bolts	4	4	4	4	4	8	8	8	8	8	8	12	12		
Diameter of Bolts	½	⅝	⅝	⅝	¾	⅝	¾	¾	¾	¾	¾	¾	⅞		

(continued)

Table D-9 (continued)

Pipe Size	150-lb Steel		ASA B16.5												
	½	¾	1	1¼	1½	2	2½	3	3½	4	5	6	8	10	12
			4¼	4⅝	5	6	7	7½	8½	9	10	11	13½	16	19
Thickness of Flange (min)[5]			7⁄16	½	9⁄16	⅝	13⁄16	¾	13⁄16	15⁄16	15⁄16	1	1⅛	1 3⁄16	1¼
Diameter of Raised Face			2	2½	2⅞	3⅝	4⅛	5	5½	6 3⁄16	7 5⁄16	8½	10⅝	12¾	15
Diameter of Bolt Circle			3⅛	3½	3⅞	4¾	5½	6	7	7½	8½	9½	11¾	14¼	17
Number of Bolts			4	4	4	4	4	4	8	8	8	8	8	12	12
Diameter of Bolts			½	½	½	⅝	⅝	⅝	⅝	⅝	¾	¾	¾	⅞	⅞

Pipe Size	300-lb Steel		ASA B16.5												
	½	¾	1	1¼	1½	2	2½	3	3½	4	5	6	8	10	12
Diameter of Flange	4⅞	5¼	6⅛	6½	7½	8¼	9	10	11	12½	15	17½	20½		
Thickness of Flange (min)[6]	11/16	¾		13/16	⅞	1	1⅛	1 3/16	1¼	1⅜	1 7/16	1⅝	1⅞	2	
Diameter of Raised Face	2	2½	2⅞	3⅝	4⅛	5	5½	6 3/16	7 5/16	8½	10⅝	12¾	15		
Diameter of Bolt Circle	3½	3⅞	4½	5	5⅞	6⅝	7¼	7⅞	9¼	10⅝	13	15¼	17¾		
Number of Bolts	4	4	4	8	8	8	8	8	8	12	12	16	16		
Diameter of Bolts	⅝	⅝	¾	⅝	¾	¾	¾	¾	¾	¾	⅞	1	1⅛		

(continued)

Table D-9 (continued)

	400-lb Steel		ASA B16.5												
Pipe Size	½	¾	1	1¼	1½	2	2½	3	3½	4	5	6	8	10	12
Diameter of Flange	3¾	4⅝	4⅞	5¼	6⅛	6½	7½	8¼	9	10	11	12½	15	17½	20½
Thickness of Flange (min)[7]	9/16	⅝	11/16	13/16	⅞	1	1⅛	1¼	1⅜	1⅜	1½	1⅝	⅞	2⅛	2¼
Diameter of Raised	1⅜	1 11/16	2	2½	2⅞	3⅝	4⅛	5	5½	6 3/16	7 5/16	8½	10⅝	12¾	15
Diameter of Bolt Circle	2⅝	3¼	3½	3⅞	4½	5	5⅞	6⅝	7¼	7⅞	9¼	10⅝	13	15¼	17¾
Number of Bolts	4	4	4	4	4	8	8	8	8	8	8	12	12	16	16
Diameter of Bolts	½	⅝	⅝	⅝	¾	¾	⅞	⅞	⅞	⅞	1	1⅛	1¼		

	600-lb Steel		ASA B16.5												
Pipe Size	**½**	**¾**	**1**	**1¼**	**1½**	**2**	**2½**	**3**	**3½**	**4**	**5**	**6**	**8**	**10**	**12**
Diameter of Flange	3¾	4⅝	4⅞	5¼	6⅛	6½	7½	8¼	9	10¾	13	14	16½	10	12
Thickness of Flange (min)[8]	9/16	⅝	11/16	13/16	⅞	1	1⅛	1¼	1⅜	1½	1¾	1⅞	2 3/16	2½	2⅝
Diameter of Raised	1⅜	1 11/16	2	2½	2⅞	3⅝	4⅛	5	5½	6 3/16	7 5/16	8½	10⅝	12¾	15
Diameter of Bolt Circle	2⅝	3¼	3½	3⅞	4½	5	5⅞	6⅝	7¼	8½	10½	11½	13¼	17	19½
Number of Bolts	4	4	4	4	4	8	8	8	8	8	8	12	12	16	20
Diameter of Bolts	½	⅝	⅝	⅝	¾	⅝	¾	¾	⅞	⅞	1	1	1⅛	1¼	1¼

[1]*125-lb. flanges have plain faces.*

[2]*250-lb. flanges have a 1/16" raised face, which is included in the flange thickness dimensions.*

[3]*150-lb. bronze flanges have plain faces with two concentric gasket-retaining grooves between the port and the bolt holes.*

[4]*300-lb. bronze flanges have plain faces with two concentric gasket-retaining grooves between the port and the bolt holes.*

[5]*150-lb. steel flanges have a 1/16" raised face, which is included in the flange thickness dimensions.*

[6]*300-lb. steel flanges have a 1/16" raised face, which is included in the flange thickness dimensions.*

[7]*400-lb. steel flanges have a ¼" raised face, which is NOT included in the flange dimensions.*

[8]*600-lb. steel flanges have a ¼" raised face, which is NOT included in the flange dimensions.*

Table D-10 Pressure Drop in Schedule 40 Pipe

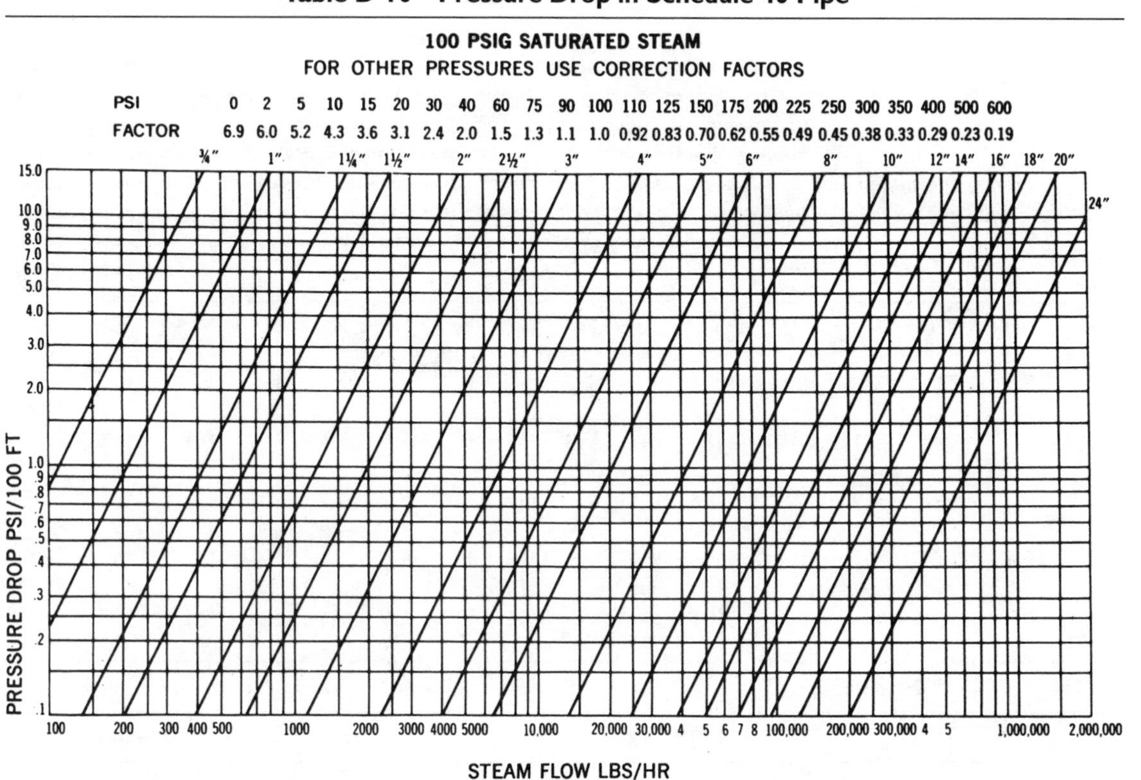

Table D-11 Steam Velocity Chart

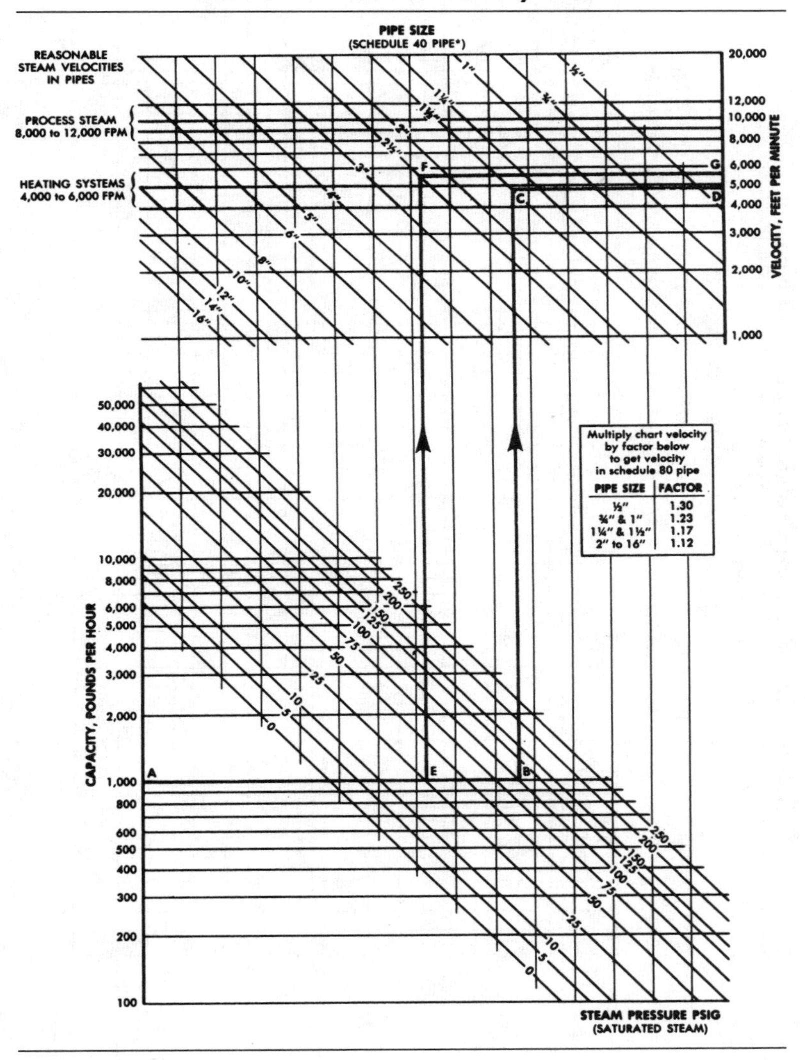

PIPE SIZE	FACTOR
½"	1.30
¾" & 1"	1.23
1¼" & 1½"	1.17
2" to 16"	1.12

Table D-12 Pressure Drop in Schedule 80 Pipe

100 PSIG SATURATED STEAM

FOR OTHER PRESSURES USE CORRECTION FACTORS

PSI	0	2	5	10	15	20	30	40	60	75	90	100	110	125	150	175	200	225	250	300	350	400	500	600
FACTOR	6.9	6.0	5.2	4.3	3.6	3.1	2.4	2.0	1.5	1.3	1.1	1.0	0.92	0.83	0.70	0.62	0.55	0.49	0.45	0.38	0.33	0.29	0.23	0.19

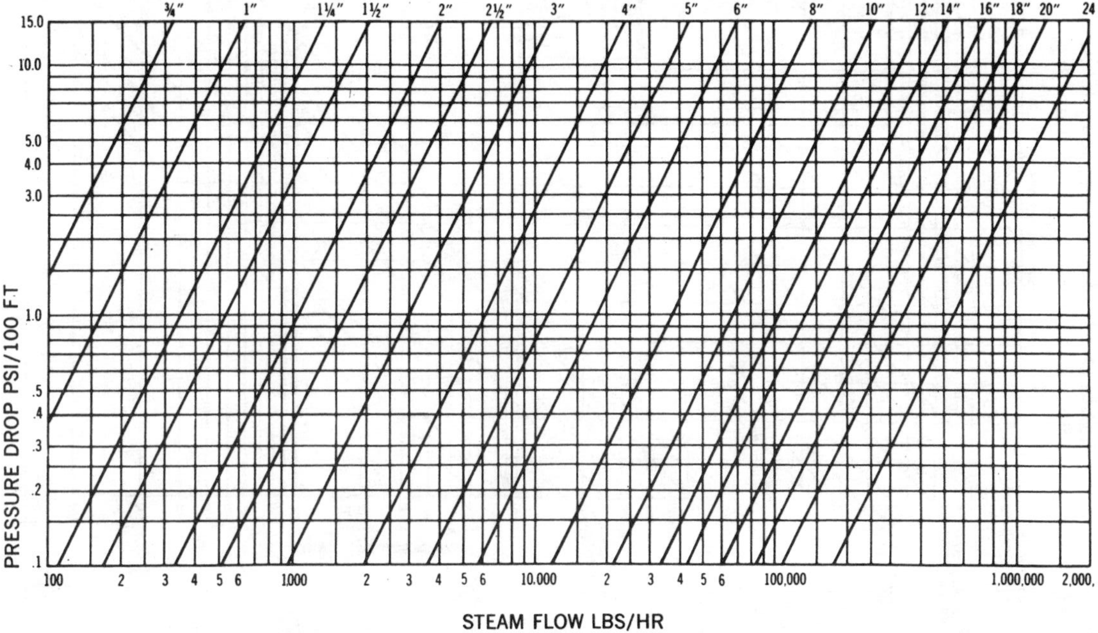

Table D-13 Chimney Connector and Vent Connector Clearance from Combustible Materials

Description of Appliance	*Minimum Clearance, Inches**
Residential Appliances	
Single-Wall, Metal Pipe Connector	
Electric, gas, and oil incinerators	18
Oil and solid-fuel appliances	18
Oil appliances listed as suitable for use with Type L venting system, but only when connected to chimneys	9
Type L Venting-System Piping Connectors	
Electric, gas, and oil incinerators	9
Oil and solid-fuel appliances	9
Oil appliances listed as suitable for use with Type L venting systems	[a]
Commercial and Industrial Appliances	
Low-Heat Appliances	
Single-Wall, Metal Pipe Connectors	
Gas, oil, and solid-fuel boilers, furnaces, and water heaters	18
Ranges, restaurant type	18
Oil unit heaters	18
Other low-heat industrial appliances	18
Medium-Heat Appliances	
Single-Wall, Metal Pipe Connectors	
All gas, oil, and solid-fuel appliances	36

**These clearances apply except if the listing of an appliance specifies different clearances, in which case the listed clearance takes precedence.*

[a]*If listed Type L venting-system piping is used, the clearance may be in accordance with the venting-system listing.*

If listed Type B or Type L venting-system piping is used, the clearance may be in accordance with the venting-system listing.

The clearances from connectors to combustible materials may be reduced if the combustible material is protected in accordance with Table 1C.

Courtesy National Oil Fuel Institute

Table D-14 Standard Clearances for Heat-Producing Appliances in Residential Installations

Residential-Type Appliances for Installation in Rooms That Are Large[*]		*Appliance*				
		Above Top of Casing or Appliance	*From Top and Sides of Warm-Air Bonnet or Plenum*	*From Front*[a]	*From Back*	*From Sides*
Boilers and Water Heaters						
Steam boilers—15 psi; water boilers—250°F; water heaters—200°F; all water walled or jacketed	Automatic oil or combination gas-oil	6	—	24	6	6
Furnaces—Central						
gravity, upflow, downflow, horizontal and duct. warm-air—250°F max.	Automatic oil or combination gas-oil	6[b]	6[c]	24	6	6

Table D-14 (continued)

Residential-Type Appliances for Installation in Rooms That Are Large*		Appliance: Above Top of Casing or Appliance	Appliance: From Top and Sides of Warm-Air Bonnet or Plenum	Appliance: From Front[a]	Appliance: From Back	Appliance: From Sides
Furnaces—Floor						
For mounting in combustible floors	Automatic oil or combination gas-oil	36	—	12	12	12

**Rooms that are large in comparison to the size of the appliance are those having a volume equal to at least 12 times the total volume of a furnace and at least 16 times the total volume of a boiler. If the actual ceiling height of a room is greater than 8 ft, the volume of a room shall be figured on the basis of a ceiling height of 8 ft.*

[a]*The minimum dimension should be that necessary for servicing the appliance including access for cleaning and normal care, tube removal, etc.*

[b]*For a listed oil, combination gas-oil, gas, or electric furnace, this dimension may be 2 in if the furnace limit control cannot be set higher than 250°F or 1 in if the limit control cannot be set higher than 200°F.*

Courtesy National Oil Fuel Institute

Table D-15 Standard Clearances for Heat-Producing Appliances in Commercial and Industrial Installations

Commercial-Industrial Type Low Heat Appliance (Any and All Physical Sizes Expect As Noted)		*Appliance*				
		*Above Top of Casing or Appliance**	*From Top and Sides of Warm-Air Bonnet or Plenum*	*From Front*	*From Back**	*From Sides**
Boiler and Water Heaters						
100 ft³ or less, any psi, steam	All fuels	18	—	48	18	18
50 psi or less, any size	All fuels	18	—	48	18	18
Unit Heaters						
Floor mounted or suspended—	Steam or hot water	1	—	—	1	1
Suspended—100 ft³ or less	Oil or combination gas-oil	6	—	24	18	18
Suspended —over 100 ft³	All fuels	18	—	48	18	18
Floor mounted any	All fuels	18	—	48	18	18

**If the appliance is encased in brick, the 18 in clearance above and at sides and rear may be reduced to not less than 12 in*

Courtesy National Oil Fuel Institute

Table D-16 Clearance (in Inches) with Specified Forms of Protection

Input heat units	Efficiency %	Usable Btu/h	gph 100°F rise	Tank size, gal	Available Hot-Water Storage Plus Recovery				
					100°F Rise				Continuous Draw, gph
					15 min	30 min	45 min	60 min	
Electricity, kW									
1.5	92.5	4,750	5.7*	20	21.4	22.8	24.3	25.7	5.7
2.5	92.5	7,900	9.5*	20	32.4	34.8	37.1	39.5	9.5
4.5	92.5	14,200	17.1*	30	44.3	48.6	52.9	57.1	17.1
4.5	92.5	14,200	17.1*	50	54.3	58.6	62.9	67.1	17.1
6	92.5	19,000	22.8*	66	71.6	77.2	82.8	88.8	22.8
7	92.5	22,100	26.5	80	86.6	93.2	99.8	106.5	26.5
Gas, Btu/h									
34,000	75	25,500	30.6	30	37.7	45.3	53	55.6[a]	25.6[a]
42,000	75	31,600	38	30	39.5	49	58.8	61.7[a]	31.7[a]
50,000	75	37,400	45	40	51.3	62.6	73.9	77.6[a]	37.6[a]
60,000	75	45,000	54	50	63.5	77	90.5	95.0[a]	45.0[a]
Oil, gph									
0.5	75	52,500	63	30	45.8	61.6	77.4	82.5[a]	52.5[a]
0.75	75	78,700	94.6	30	53.6	77.2	100.8	109.0[a]	79.0[a]
0.85	75	89,100	107	30	57.7	83.4	110.1	119.1[a]	89.0[a]
1	75	105,000	126	50	81.5	13	144.5	155.0[a]	105.0[a]
1.2	75	126,000	151.5	50	87.9	125.8	163.7	176.0[a]	126.0[a]
1.35	75	145,000	174	50	93.5	137	180.5	195.0[a]	145.0[a]
1.5	75	157,000	188.5	85	132.1	179.2	226.3	242.0[a]	157.0[a]
1.65	75	174,000	204.5	85	136.1	187.2	238.4	259.0[a]	174.0[a]

**Assumes simultaneous operation of upper and lower elements.*

[a]Based on 50 minute-per-hour operation.

Courtesy National Oil Fuel Institute

Appendix E

Conversion Tables

To Convert	*Into*	*Multiply By*
Acres	square feet	43,560.0
Atmospheres	centimeters of mercury	76.0
Atmospheres	feet of water (at 4°C)	33.90
Atmospheres	inches of mercury (at 0°C)	29.92
Atmospheres (atm)	kilograms/ square centimeter	1.0333
Atmospheres	pounds/square inch	14.70
Barrels (U.S., liquid)	gallons	31.5
Barrels (oil) (bbl)	gallons (oil)	42.0
Btu	foot-pounds	778.3
Btu	gram-calories	252.0
Btu	horsepower-hour	3.931×10^{-4}
Btu	kilowatt-hours	2.928×10^{-4}
Btu per hour (Btu/h)	horsepower	3.931×10^{-4}
Btu/h	watts	0.2931
Calories, gram (mean)	Btu (mean)	3.9685×10^{-3}
Degrees centigrade or Celsius (°C)	degrees Fahrenheit	$(°C + 40)\ 9/5 - 40$
Centimeters (cm)	feet	3.281×10^{-2}
Centimeters	inches	0.3937
Centimeters	mils	393.7
Centimeters of mercury (cm Hg)	atmospheres	0.01316
Centimeters of mercury	feet of water	0.4461
Centimeters of mercury	pounds/square inch	0.1934
Circumference	radians	6.283
Cubic centimeters (cm^3)	cubic feet	3.531×10^{-5}
Cubic centimeters	cubic inches	0.06102
Cubic centimeters	gallons (U.S. liquid)	2.642×10^{-4}
Cubic feet (ft^3)	cubic centimeters	28,320.0
Cubic feet	cubic inches	1,728.0

(continued)

To Convert	*Into*	*Multiply By*
Cubic feet	gallons (U.S. liquid)	7.481
Cubic feet	liters	28.32
Cubic feet	quarts (U.S. liquid)	29.92
Cubic feet per minute	gallons/second	0.1247
Cubic feet per minute	pounds of water/ minute	62.43
Cubic inches (in^3)	cubic centimeters	16.39
Cubic inches	gallons	4.329×10^{-3}
Cubic inches	quarts (U.S. liquid)	0.01732
Cubic meters	cubic feet	35.31
Cubic meters	gallons (U.S. liquid)	264.2
Cubic yards (yd^3)	cubic feet	27.0
Cubic yards	cubic meters	0.7646
Cubic yards	gallons (U.S. liquid)	202.0
Degrees (angle)	radians	0.01745
Drams (apothecaries' or troy)	ounces (avoirdupois)	0.13714
Drams (apothecaries' or troy)	ounces (troy)	0.125
Drams (U.S., fluid or apothecary)	cubic centimeter	3.6967
Drams	grams	1.772
Drams	grains	27.3437
Drams	ounces	0.0625
Degrees Fahrenheit (°F)	degrees centigrade	(°F + 40) 5/9 − 40
Feet (ft)	centimeters	30.48
Feet	kilometers	3.048×10^{-4}
Feet	meters	0.3048
Feet	miles (nautical)	1.645×10^{-4}
Feet	miles (statute)	1.894×10^{-4}
Feet of water (ft H_2O)	atmospheres	0.02950
Feet of water	inches of mercury	0.8826
Feet of water	kilograms/ square centimeter	0.03045
Feet of water	kilograms/ square meter	304.8

Feet of water	pounds/square foot	62.43
Feet of water	pounds/square inch	0.4335
Foot-pounds (ft-lb)	Btu	1.286×10^{-3}
Foot-pounds	gram-calories	0.3238
Foot-pounds	horsepower-hours	5.050×10^{-7}
Foot-pounds	kilowatt-hours	3.766×10^{-7}
Foot-pounds/minute (ft-lb/min)	Btu/ minute	1.286×10^{-3}
Foot-pounds/minute	horsepower	3.030×10^{-5}
Foot-pounds/second (ft-lb/sec)	Btu per hour	4.6263
Furlongs	miles (U.S.)	0.125
Furlongs	feet	660.0
Gallons (gal)	cubic centimeter	3,785.0
Gallons	cubic feet	0.1337
Gallons	cubic inches	231.0
Gallons	cubic meters	3.785×10^{-3}
Gallons	cubic yards	4.951×10^{-3}
Gallons	Liters	3.785
Gallons (British Imperial liquid)	gallons (U.S. liquid)	1.20095
Gallons (U.S.)	gallons (Imperial)	0.83267
Gallons of water	pounds of water	8.3453
Gallons per minute (gal/min, gpm)	cubic feet/ second	2.228×10^{-3}
Gallons per minute	liters/second	0.06308
Gallons per minute	cubic feet/hour	8.0208
Grains (troy) (gr)	grains (avdp.)	1.0
Grains (troy)	grams	0.06480
Grains (troy)	ounces (avdp.)	2.286×10^{-3}
Grains (troy)	pennyweight (troy)	0.04167
Grains per U.S. gallon	parts/million	17.118
Grains per U.S. gallon	pounds/ million gallons	142.86
Grains oer Imperial gallon	parts/million	14.286

(continued)

To Convert	*Into*	*Multiply By*
Grams (g)	grains	15.43
Grams	ounces (avdp.)	0.03527
Grams	ounces (troy)	.03215
Grams	poundals	0.07093
Grams	pounds	2.205×10^{-3}
Grams per liter (g/L)	parts/million	1,000.0
Gram-calories (g-cal)	Btu	3.9683×10^{-3}
Gram-calories	foot-pounds	3.0880
Gram-calories	kilowatt-hour	1.1630×10^{-6}
Gram-calories	watt-hour	1.1630×10^{-3}
Horsepower (hp)	Btu/minute	42.40
Horsepower	foot-pounds/minute	33,000.0
Horsepower	foot-pounds/second	550.0
Horsepower (metric) (542.5 ft-lb/sec)	horsepower (550 ft-lb/sec)	0.9863
Horsepower (550 ft-lb/sec)	horsepower (metric) (542.5 ft-lb/sec)	1.014
Horsepower	kilowatts	0.7457
Horsepower	watts	745.7
Horsepower (boiler)	Btu/hour	33.520
Horsepower (boiler)	kilowatts	9.803
Horsepower-hour (hp-hr)	Btu	2,547
Horsepower-hour	foot-pounds	1.98×10^{6}
Horsepower-hour	kilowatt-hours	0.7457
Inches (in)	centimeters	2.540
Inches	meters	25.40
Inches	millimeters	2.540×10^{-2}
Inches	yards	22.778×10^{-2}
Inches of mercury (in Hg)	atmospheres	0.03342
Inches of mercury	feet of water	1.133
Inches of mercury	kilograms/ square centimeters	0.03453

Inches of mercury	kilograms/ square meter	345.3
Inches of mercury	pounds/square foot	70.73
Inches of mercury	pounds/square inch	0.4912
Inches of water (at 4°C) (in H_2O, in wg, in W.C.)	atmospheres	2.458×10^{-3}
Inches of water (at 4°C)	inches of mercury	0.07355
Inches of water (at 4°C)	kilograms/ square centimeter	2.538×10^{-3}
Inches of water (at 4°C)	ounces/ square inch	0.5781
Inches of water (at 4°C)	pounds/ square foot	5.204
Inches of water (at 4°C)	pounds/ square inch	0.03613
Joules (J)	Btu	9.480×10^{-4}
Kilograms (kg)	grams	1,000.0
Kilograms	pounds	2.205
Kilograms per cubic meter (kg/m^3)	pounds/ cubic foot	0.06243
Kilograms per cubic meter	pounds/ cubic inch	3.613×10^{-5}
Kilograms per square centimeter (kg/cm^2)	atmospheres	0.9678
Kilograms per square centimeter	feet of water	32.84
Kilograms per square centimeter	inches of mercury	28.96
Kilograms per square centimeter	pounds/ square foot	2,048.0
Kilograms per square centimeter	pounds/ square inch	14.22
Kilograms per square meter (kg/m^2)	atmospheres	9.678×10^{-5}
Kilograms per square meter	feet of water	3.281×10^{-3}
Kilograms per square meter	inches of mercury	2.896×10^{-3}

(continued)

To Convert	*Into*	*Multiply By*
Kilograms per square meter	pounds/ square foot	0.2048
Kilograms per square meter	pounds/ square inch	1.422×10^{-3}
Kilograms per square millimeter	kilograms/ square meter	10^{6}
Kilogram-calories	Btu	3.968
Kilogram-calories	foot-pounds	3,088.0
Kilogram-calories	horsepower-hour	1.560×10^{-3}
Kilogram-calories	kilowatt-hour	1.163×10^{-3}
Kilogram meters	Btu	9.294×10^{-3}
Kilometers	centimeters	10^{5}
Kilometers	feet	3,281.0
Kilometers	miles	0.6214
Kilowatts	Btu/minute	56.87
Kilowatts	foot-pounds/minute	4.426×10^{4}
Kilowatts	foot-pounds/second	737.6
Kilowatts	horsepower	1,341.0
Kilowatts	watts	1,000.0
Kilowatt-hour	Btu	3,413.0
Kilowatt-hour	foot-pounds	2.655×10^{6}
Kilowatt-hour	horsepower-hour	1,341
Knots	statute miles/hour	1.151
Liters (L)	cubic centimeters	1,000.0
Liters	cubic feet	0.03531
Liters	cubic inches	61.02
Liters	gallons (U.S. liquid)	0.2642
Meters (m)	centimeters	100.0
Meters	feet	3.281
Meters	inches	39.37
Meters	millimeters	1,000.0
Meters	yards	1.094
Microns (μ, μm)	inches	39.37×10^{-6}
Microns	meters	1×10^{-6}
Miles (statute) (mi)	feet	5,280.0
Miles (statute)	kilometers	1.609
Miles per hour (mi/hr, mph)	centimeters/second	44.70

Miles per hour	feet/minute	88.0
Mils	inches	0.001
Mils	yards	2.778×10^{-5}
Nepers	decibels	8.686
Ohms (ω)	megohms	10^{-6}
Ohms	microhms	10^{6}
Ounces (avoirdupois) (oz)	drams	16.0
Ounces (avoirdupois)	grains	437.5
Ounces (avoirdupois)	grams	28.35
Ounces (avoirdupois)	pounds	0.0625
Ounces (avoirdupois)	ounces (troy)	0.9115
Ounces (troy)	grains	480.0
Ounces (troy)	grams	31.10
Ounces (troy)	ounces (avdp.)	1.09714
Ounces (troy)	pounds (troy)	0.08333
Parts per million (ppm)	grains/ U.S. gallon	0.0584
Parts per million	grains/ Imperial gallon	0.07016
Parts per million	pounds/ million gallons	8.33
Pounds (avoirdupois) (lb)	ounces (troy)	14.58
Pounds (avoirdupois)	drams	256.0
Pounds (avoirdupois)	grains	7,000.0
Pounds (avoirdupois)	grams	28.35
Pounds (avoirdupois)	kilograms	0.02835
Pounds (avoirdupois)	ounces	16.0
Pounds (avoirdupois)	tons (short)	0.0005
Pounds (troy)	ounces (avdp.)	13.1657
Pounds of water	cubic feet	0.01602
Pounds of water	cubic inches	27.68
Pounds of water	gallons	0.1198
Pounds of water per minute	cubic feet/ second	2.670×10^{-4}
Pounds per cubic foot (lb/ft^3)	kilogram/ cubic meter	0.01602
Pounds per cubic foot	grams/ cubic centimeter	16.02

(continued)

To Convert	*Into*	*Multiply By*
Pounds per cubic foot	pounds/ cubic inch	5.787×10^{-4}
Pounds per cubic inch	pounds/ cubic foot	1.728.0
Pounds per square foot (lb/ft^2)	atmospheres	4.725×10^{-4}
Pounds per square foot	feet of water	0.01602
Pounds per square foot	inches of mercury	0.01414
Pounds per square inch (lb/in^2, psi)	atmospheres	0.06804
Pounds per square inch	feet of water	2.307
Pounds per square inch	inches of mercury	2.036
Pounds per square inch	kilograms/ square meter	703.1
Pounds per square inch	pounds/ square foot	144.0
Radians	degrees	57.30
Revolutions per minute (rpm)	degrees/ second	6.0
Revolutions per minute	radians/ second	0.1047
Revolutions per minute	revolutions/ second	0.01667
Square centimeters (cm^2)	square feet	1.076×10^{-3}
Square centimeters	square inches	0.1550
Square centimeters	square meters	0.0001
Square centimeters	square millimeters	100.0
Square feet (ft^2)	acres	2.296×10^{-5}
Square feet	square centimeters	929.03
Square feet	square inches	144.0
Square feet	square miles	3.587×10^{-8}
Square inches (in^2)	square centimeters	6.452
Square inches	square feet	6.944×10^{-3}
Square inches	square yards	7.716×10^{-4}
Square meters	square feet	10.76

Square meters	square inches	1,550.0
Square meters	square millimeters	10^6
Square meters	square yards	1.196
Square millimeters (mm^2)	square inches	1.550×10^{-3}
Square yards (yd^2)	square feet	9.0
Square yards	square inches	1,296.0
Square yards	square meters	0.8361
Temperature (°C) + 273.15	absolute temperature (kelvins or °C)	1.0
Temperature (°C) + 17.78	temperature (°F)	1.8
Temperature (°F) + 459.67	absolute temperature (°Rankine)	1.0
Temperature (°F) − 32	temperature (°C)	5/9
Tons (long)	kilograms	1,016.0
Tons (long)	pounds	2,240.0
Tons (long)	tons (short)	1.120
Tons (metric)	kilograms	1,000.0
Tons (metric)	pounds	2,205.0
Tons (short)	kilograms	907.2
Tons (short)	pounds	2,000.0
Tons (short)	tons (long)	0.89287
Tons of water per 24 hours	pounds of water/hour	83.333
Tons of water per 24 hours	gallons/minute	0.16643
Tons of water per 24 hours	cubic feet/hour	1.3349
Watts (W)	Btu/h	3.4129
Watts	Btu/minute	0.05688
Watts	horsepower	1.341×10^{-3}
Watts	horsepower (metric)	1.360×10^{-3}
Watts	kilowatts	0.001
Watts (abs.)	Btu (mean)/minute	0.056884
Watt-hours	Btu	3.413
Watt-hours	horsepower-hours	1.341×10^{-3}
Yards (yd)	centimeters	91.44
Yards	kilometers	9.144×10^{-4}
Yards	meters	0.9144

Index

C

I

J

K

L

O

P